AF767691

Global Media Convergence and Cultural Transformation:

Emerging Social Patterns and Characteristics

Dal Yong Jin
Advanced Institute of Science and Technology, South Korea

Information Science REFERENCE

INFORMATION SCIENCE REFERENCE
Hershey · New York

Director of Editorial Content: Kristin Klinger
Director of Book Publications: Julia Mosemann
Acquisitions Editor: Lindsay Johnston
Development Editor: Michael Killian
Publishing Assistant: Casey Conapitski
Typesetter: Casey Conapitski
Production Editor: Jamie Snavely
Cover Design: Lisa Tosheff

Published in the United States of America by
Information Science Reference (an imprint of IGI Global)
701 E. Chocolate Avenue
Hershey PA 17033
Tel: 717-533-8845
Fax: 717-533-8661
E-mail: cust@igi-global.com
Web site: http://www.igi-global.com

Copyright © 2011 by IGI Global. All rights reserved. No part of this publication may be reproduced, stored or distributed in any form or by any means, electronic or mechanical, including photocopying, without written permission from the publisher. Product or company names used in this set are for identification purposes only. Inclusion of the names of the products or companies does not indicate a claim of ownership by IGI Global of the trademark or registered trademark.

Library of Congress Cataloging-in-Publication Data

Global media convergence and cultural transformation : emerging social patterns and characteristics / Dal Yong Jin, editor.
p. cm.
Includes bibliographical references and index.
Summary: "This book aims to engage the complex relationship between technology, culture, and socio-economic elements by exploring it in a transnational, yet contextually grounded, framework, exploring diverse perspectives and approaches, from political economy to cultural studies, and from policy studies to ethnography"--Provided by publisher.
ISBN 978-1-60960-037-2 (hardcover) -- ISBN 978-1-60960-039-6 (ebook) 1. Popular culture. 2. Popular culture and globalization. 3. Mass media and globalization. 4. Technological innovations--Economic aspects. 5. Technological innovations--Social aspects. I. Jin, Dal Yong, 1964-
HM621.G579 2011
306--dc22

2010024363

British Cataloguing in Publication Data
A Cataloguing in Publication record for this book is available from the British Library.

All work contributed to this book is new, previously-unpublished material. The views expressed in this book are those of the authors, but not necessarily of the publisher.

Editorial Advisory Board

Jack Bratich, *Rutgers University, USA*
Mahmoud Eid, *University of Ottawa, Canada*
Peter-Chow White, *Simon Fraser University, Canada*
Larissa Hjorth, *RMIT University, Australia*
Soochul Kim, *Soongsil University, Korea*

Table of Contents

Detailed Table of Contents

Section 2
Convergence of New Media

This chapter analyzes the contents of the Laughing Gor Facebook group, 100 blog articles discussing the phenomenon, and news coverage of the phenomenon. It also discussed the characteristics and significance of the Laughing Gor phenomenon within the specific context of television and society in Hong Kong, as well as the broad theoretical concern of television and organization of public attention in the Internet age.

This chapter argues that the aestheticization of suffering is a dominant and integral part of our culture and new media technologies illuminate and offer new ways to commodify and engage with suffering. It argues that the commodification of suffering on video streaming platforms on the Internet are spaces of both perverse obsession with pain and suffering but equally they are performative counter-sites which offer different ways to record and narrate human suffering.

This chapter explores the role of QQ through a case study in Shanghai. It points out that QQ has become a rite of passage for the migrating youth leaving home to study in another city or country, because as the first generation to grow up within the IT educational reforms that saw a growing importance placed on new media and Internet literacy, these students are passing on their knowledge to their family and friends back home, demonstrating that SNS is far from the prerogative of Western, middle-class youth.

This chapter argues that it is almost impossible to have a right to communicate with the presence of a modern form of Pontifice Maximum: the media conglomerates by drawing from the work of Harold Innis, as well as Jean d'Arcy—the father of the right to communicate. It highlights the various possibilities of enforcing a human right to communicate with the introduction of the new social networking sites.

This chapter argues that engaging citizens in developing regions in information creation and information sharing leverages peoples' existing social networks to facilitate transmission of critical information, exchange of ideas, and distributed problem solving. It describes Picture Talk, a mobile social computing application framework designed to facilitate local information sharing in regions with sparse Internet connectivity, low literacy rates and having users with little prior experience with information technology.

This chapter explores how the keitai Internet has come to dominate Japan, marginalizing the PC-based Internet. The discussion focuses on the country's cultural and ideological aspects that have worked as driving forces behind its mobile Internet boom. It examines how Japan has developed a qualitatively different Internet use pattern that cannot be revealed in statistics like "number of Internet subscribers" or "household.

This chapter investigates the influence of mobile phone diffusion among Zimbabwean migrants at the individual, community, and governmental levels, given that from the early 2000s, the population mobility from Zimbabwe has drastically increased due to the collapse of the national economy and political instability.

Preface

Convergence has been a cornucopia in the communication industries, and it has been touted as the biggest leap in the digital era in the 21st century. From content producers, such as Viacom and Disney to the Internet and telecommunication firms, including AOL and AT&T, media and telecommunication firms have pursued convergence due largely in part to synergy effects, economies of scale, and the production and distribution of diverse cultural products. Converging media technologies and culture have therefore become some of the most significant trends, and convergence has rapidly become a major corporate strategy in the communication sector.

Media convergence can be categorized in three major areas: the flow of content across multiple media platforms, the cooperation between multiple media industries, and the migratory behavior of media audience who will go almost anywhere in search of the kinds of entertainment experience they want (Jenkins, 2006, 2-3). Convergence can also be categorized as the integration of different technologies, cultures and industries. Whilst these forms of convergence are not isolated, cultural convergence—a mix of two different cultures—for example, between American culture and Chinese culture, is increasingly a unique form of convergence in the globalized world. In other words, as convergence is an elusive term that is used in multiple contexts and is often ambiguous in its definition, one may need to understand convergence as not a simple meaning but multiple meanings.

There are several significant drivers for the swift growth of convergence in the communication sector; however, the contemporary media convergence has been primarily fuelled by two major elements: one is increasingly neoliberal communication policies, and the other is pervasive digital technologies. Most of all, convergence has been expedited with the rapid employment of neoliberal communication policies began in the early 1980s and continued throughout the 1990s. Before the introduction of and its integration with neoliberal globalization, the majority of media companies had focused their own core business areas, partially because the government policies, including antitrust laws and cross-ownership restraints, sought to define them distinctly and kept them separate (Baldwin et el., 1996). For example, Time Inc. was known as a publication company, and Viacom was famous for its TV syndication and cable businesses, whereas News Corporation was considered a newspaper company based in Australia.

However, pro-business neoliberal communication policies have lifted the barriers of new investment opportunities, which has resulted in the concentration of ownership through media convergence to the hands of a few media giants (Jin, 2009). Privatization of public broadcasting and telecommunications and the liberalization of domestic markets have blurred the distinction between the old media and the new media, and media companies have become a multi-media company through mergers and acquisitions (M&As) in the midst of neoliberal reform (McPhail, 2006). Deregulation, in particular, the loosing of regulations in cross-ownership among communication industries has become one of the most significant factors in expanding their business areas beyond the traditional core business realms in many media companies (Thussu, 2006).

Meanwhile, the idea of convergence emphasizes the importance of the technological upheaval in communications and validly stresses that communications infrastructures are becoming multifunctional as they assimilated versatile digital electronics (Schiller, 2007). With the emergence of what is commonly referred to as Web 2.0 technology, convergence has been a must take business model in the communication industries. The Internet and mobile communications and their applications, including online gaming, have made a huge impact on political participation, business, education, culture, social and family relations. Several regions, including Asia and North America are marked by diverse penetration rates of high-speed internet and gaming (e.g., online, mobile, and console games), and mobile and broadband technologies, which are subject to local culture and socio-economic nuances. Young people are techno-savvy, and their use of new technology has been culturally remarkable. Media policy makers, software designers, mobile technicians, and computer corporations around the world are eager to learn, not only the result of the rapid growth of new technologies and their implications, but also the primary driver of the development of advancement of new technologies, as a reflection of the importance of convergence of technology and culture. Convergence is especially a challenge for many old media who attempt to venture into the Internet market, and it has fundamentally changed the structure and system of both small media firms and global media giants.

Finally, cultural convergence has become popular in the midst of globalization. The term cultural convergence can be used to describe mixed cultures or the process of mixing genres within a culture. Previously culture flowed mainly from Western countries to non-Western countries; however, with the rapid growth of local culture in Asia and Latin America, the direction of cultural flow has partially changed to between local (non-western) and local, as well as local and global (western) as in the case of the recent growth of Korean cultural penetration in East Asia. The emerging cultural exchanges in Asia and Latin America have proved that cultural convergence is becoming one of the major drivers for and a consequence of transculturla flow in several cultural genres, including film, music and television program, and gaming.

The overall mission of the book is to engage the complex relationship between technology and culture, as well as technology and socio-economic elements by exploring it in a transnational yet contextually grounded framework. This book employs diverse perspectives and approaches, from political economy to cultural studies, and from policy studies to ethnography in order to reflect several different focuses and areas in new media technologies and their convergence with culture. It also provides understanding of macro and micro level factors currently affecting convergence from both a developing and a developed nation's perspective, while offering cross-country analysis of media and technology convergence that can be applied in various geographical settings. We expect that the documentation of our experience of technological change in economy and culture that may shed light on the more general trends of the shifting global media, culture and technology. In order to cater to the information needs of a diverse spectrum of readers and at the same time effectively present this global topic, this book is structured into five sections, which each section including a number of chapters. A brief description of each chapter is provided below.

Chapter 1, *Super-sized Fandom: Globalization, Popular Culture, and the Peculiar Case of Miss Macdonald* by Andrew Kirk, looks at "A Day in the Life of Miss McDonald"—a photography exhibit produced in the Philippines and disseminated online—to discuss how globalization, coupled with advances in Internet technologies, has transformed the manner in which popular culture acquires meaning.

The case of Miss McDonald offers a useful point of entry for understanding popular culture's past as well as its global future.

Chapter 2, *Urban Screens and Transcultural Consumption between South Korea and Australia* by Audrey Yue and Sun Jung, examines urban screens as sites of media convergence and transcultural consumption. Using two case studies in Melbourne (Australia) and Songdo (Incheon, Korea), it considers how these screens have emerged through technological innovations led by cultural planning and urban regeneration. It critically examines the augmentation of these spaces as sites for cultural citizenship and transcultural consumption.

Chapter 3, *Sharing Bites on Global Screens: The Emergence of Snack Culture* by Hudson Moura, presents and analyses a variety of media snacks that form and circulate as snack culture, and it also elucidates some of those current changes that are shaping today's relationship between society and media. It articulates that snack culture is the new phenomenon that shrinks media cultural products and can be easily shared on social networks of the Internet; therefore, it can be consumed in a reduced amount of time circulating instantly all over the globe.

Chapter 4, *Global Media and Information Ethics: Challenges Re-Examined* by D. Ndirangu Wachanga, argues that the emerging global phenomenon problemitizes the Cartesian plane of oppositions – center vs. periphery, North vs. South, global vs. local, which has been the forte of globalization studies until recently. It pays attention to the antagonistic mechanics informing the domination and rejection of intangible ethical principles.

Chapter 5, *Public Discourses on the Korean Wave in Taiwan* by Lihyun Lin, analyzes how the interpretive communities have formed their interpretive frameworks, what they have conceived from the Korean Wave and what policy implications have been drawn. This chapter suggests that public discourses should address media problems in order to enhance communications at home and abroad.

Chapter 6, *The Emerging Media Exchange in the Cultural Regionalization of Asia* by Peichi Chung, focuses on the emerging media regionalization that takes place in Asia in 2000s. It selects three cases to present the detailed image of cultural standardization in Asia's media regionalization. Among these, the first centers on the circulation of media text in television drama, emphasizing on Korean wave and the particular TV series, Boys Over Flowers. The second case discusses Taiwanese popular music and its influence on Mandopop in the Chinese communities.

Chapter 7, *New Imagined Community by Cultural Participation: A Study on Taiwan's Digital Cultural Heritage* Pi-Chun Chang, analyzes the characteristics of National Digital Archives Program and how these characteristics promote a sense of participation. It discusses the convergence of new media theory and histories of Taiwan's public museums and concludes that as digitized, web-based display practices are invented, the task of transforming the basis of cultural identity from ethnicity to participation is also invested.

Chapter 8, *New-Media-Based Attention Implosion and Television Publicness in the Era of Internet-TV Convergence* by Francis Lee, analyzes the contents of the Laughing Gor Facebook group, 100 blog articles discussing the phenomenon, and news coverage of the phenomenon. It also discussed the characteristics and significance of the Laughing Gor phenomenon within the specific context of television and society in Hong Kong, as well as the broad theoretical concern of television and organization of public attention in the Internet age.

Chapter 9, *Global Media Convergence and Cultural Transformation Bearing Witness through Technology War and Visual Culture in a Converged World* by Yasmin Ibrahim, argues that the aestheticization of suffering is a dominant and integral part of our culture and new media technologies illuminate and offer new ways to commodify and engage with suffering. It argues that the commodification of suffering on video streaming platforms on the Internet are spaces of both perverse obsession with pain and suffering but equally they are performative counter-sites which offer different ways to record and narrate human suffering.

Chapter 10, *Mum's the Word: A Case Study of Intergenerational New Media Literacy in Shanghai By Students* Larissa Hjorth, explores the role of QQ through a case study in Shanghai. It points out that QQ has become a rite of passage for the migrating youth leaving home to study in another city or country, because as the first generation to grow up within the IT educational reforms that saw a growing importance placed on new media and Internet literacy, these students are passing on their knowledge to their family and friends back home, demonstrating that SNS is far from the prerogative of Western, middle-class youth.

Chapter 11, *The Electronic Pontifices Maximi: Social Networking and the Hope of a Right to Communicate* by Aliaa Dakroury, argues that it is almost impossible to have a right to communicate with the presence of a modern form of Pontifice Maximum: the media conglomerates by drawing from the work of Harold Innis, as well as Jean d'Arcy—the father of the right to communicate. It highlights the various possibilities of enforcing a human right to communicate with the introduction of the new social networking sites.

Chapter 12,*The Impact of Social Software Inside Organizations* by Robert Farrell et al., argues that engaging citizens in developing regions in information creation and information sharing leverages peoples' existing social networks to facilitate transmission of critical information, exchange of ideas, and distributed problem solving. It describes Picture Talk, a mobile social computing application framework designed to facilitate local information sharing in regions with sparse Internet connectivity, low literacy rates and having users with little prior experience with information technology.

Chapter 13, *Going Online in the PC Graveyard: The Sociocultural Evolution of Japan's Mobile Internet* by Inkyu Kang, explores how the keitai Internet has come to dominate Japan, marginalizing the PC-based Internet. The discussion focuses on the country's cultural and ideological aspects that have worked as driving forces behind its mobile Internet boom. It examines how Japan has developed a qualitatively different Internet use pattern that cannot be revealed in statistics like "number of Internet subscribers" or "household

Chapter 14, *Exploring the Diffusion of Mobile Phone among Zimbabwean Migrants in Botswana* by Do Kyun Kim, investigates the influence of mobile phone diffusion among Zimbabwean migrants at the individual, community, and governmental levels, given that from the early 2000s, the population mobility from Zimbabwe has drastically increased due to the collapse of the national economy and political instability.

Chapter 15, *The Triumph of Shanzhai: No Name Brand Mobile Phones and Youth Identity in China* by Mei Wu and Hongye Li, explores the connection between the popularity of shanzhai mobile phones among urban youth and their social characteristics in contemporary China. Through qualitative analyses of in-depth interviews both shanzhai and non-shanzhai users and selected online group discussions of shanzhai workers, this connection is examined in five realms: political attitudes, practical rationale, social values, interest-seeking and self identification.

Chapter 16, *Convergence of Mobile TV in Everyday Life* by Seung-Hyun Lee, explores the diffusion, use, and convergence of mobile TV with everyday life and culture and to characterize the use of mobile TV in a wide range of everyday practices by employing the approach of diffusion of innovation. It discovers what factors influencing the adoption and use of mobile TV-DMB, how mobile TV-DMB influences both personal and mass media consumption behavior by overcoming the spatial and temporal limitations of traditional media.

Chapter 17, *Call Centers, India, and a New Politics: Cultural Interpretations* by Maheswar Satpathy, interrogates the relative benefits and weigh them with the demerits that it has accumulated on the loyal workers. It examines the motives of western countries (or at least the cause behind such accusation) to sustain their monopolistic cultural imperialism through multifarious means, as evidenced in the case of Call centers in India.

Chapter 18, *Mobile Games in China: Formation, Ferment, and Future* by Zixue Tai and Haifang Zeng, presents a critical analysis of China's nascent but fast-evolving mobile game market. Through a penetrating examination of the historical evolution of the market from the late 1990s to the dawn of the 3G era, it offers insight on the changing dynamics of interplay as well as major trends of concentration and internationalization among key sectors of the Chinese mobile market structure.

Chapter 19, *Cinema of the Gun: Science and Technology Studies and the First Person Shooter* by Sam Hinton, argues for the use of STS in games research and uses an STS approach to explore the evolution of a 'genre' of computer games – the first person shooter, or FPS – and in particular the cinematic turn that has taken place within this genre over the past decade.

Chapter 20, *Playing as Producing: Convergence Culture and Localization of EA Digital Games in Taiwan* by Ying-Chia H. Lin, explores the dynamics and interactions between foreign producers, media technologies, and local consumers in the process of globalization through a discussion of the localization of digital games.

Chapter 21, *TV Box on the Internet: The Interplay between Politics and Market in China* by Anthony Fung and Luzhou Li, presents a special case in which political forces become a potential shaping force for media convergence and in the authoritarian environment of China, politics still reigns over technologies and the market. From a political economy perspective, the dominance of politics can be explained in terms of the process of regulatory spatialization.

Chapter 22, *Corporate Strategies in Media Convergence: A Comparative Study of Sony vs. Samsung as Transnational Cultural Industries* by Dal Yong Jin, investigates the cause of the changes and growth of Sony and Samsung, and it discusses the similarities and differences between them, with a focus on corporate policies and business strategies in convergence. It articulates the ways in which changing corporate policies have played a critical role in the growth of local-based transnational cultural corporations.

Chapter 23, *Fostering Innovation in Converging Audiovisual Media Services: Do Mergers and Acquisitions Really Help*? by Sergio Sparviero, examines whether the digitalization of media content and communication processes has been used as the main justification for an important turn in the design of regulations affecting various audiovisual media services and information industries, more in general. A brief case study presented in this chapter testifies that from the 1950s until the 1990s the United States' regulator facilitated the introduction of fundamental innovations by preventing vertical mergers between different audiovisual media services.

Chapter 24, *The Three Cs of Key Music Sector Trends Today: Commodification, Concentration and Convergence* by Paschal Preston and Jim Rogers, explores the evolving relationships between new digital media technologies, socio-economic factors and media cultures as we enter the second decade of the twenty-first century. It examines the implications of these trends with regard to three fundamental concepts in the analysis of culture, namely commodification, concentration and convergence.

Dal Yong Jin
Advanced Institute of Science and Technology, South Korea

REFERENCES

Baldwin, T. F., McVoy, D. S., & Steinfield, C. (1996). *Convergence: Integrating media, information, and communication*. Thousand Oaks, CA: Sage.

Jenkins, H. (2006). *Convergence Culture*. New York: University Press.

Jin, D. Y. (2009). De-convergence: a shifting business trend in the U.S. digital media industries. *Journal of Media Economics and Culture, 7*(1), 7-44.

McPhail, T. (2006). *Global Communication: theories, stakeholders, and trends*. 2nd ed. Oxford, UK: Blackwell.

Schiller, D. (2007). *How to think about Information*. Urbana, IL: University of Illinois Press.

Thussu, D. (2006). *International Communication: change and continuity*, 2nd ed. London: Arnold.

Section 1
Cultural Convergence and Globalization

Chapter 1
Super-Sized Fandom:
Globalization, Popular Culture, and the Peculiar Case of "Miss McDonald"

Andrew J. Kirk
Southern Illinois University - Carbondale, USA

ABSTRACT

This chapter looks at "A Day in the Life of Miss McDonald"—a photography exhibit produced in the Philippines and disseminated online—to discuss how globalization, coupled with advances in Internet technologies, has transformed the manner in which popular culture acquires meaning. In sum, these advances have allowed fans to wrangle control of the meanings bestowed to popular culture artifacts back from dominant producers, regardless of national origin. This chapter begins by examining traditional conceptualizations of popular culture before discussing how globalization and consumers like Miss McDonald have helped fans become original meaning makers rather than the meaning re-makers, always necessarily limited by the control of the producers. The case of Miss McDonald offers a useful point of entry for understanding popular culture's past as well as its global future.

INTRODUCTION

As a testament to Ronald McDonald's enduring appeal, a young woman in the Philippines known only as "Miss McDonald" decided to dress up as the beloved clown for Halloween in 2004, and that idea blossomed into a wildly popular, weekly, year-long photography project that she ultimately posted to the World Wide Web ("A Day in the Life", 2004). Given her demonstrated passion and commitment alone, it would certainly seem that Miss McDonald is an enormous fan of the McDonald's corporation. As illustrated by a number of scholars, fandom is characterized by a wildly diverse set of practices where "identification, both with the object of fandom (e.g., a celebrity) and the community of fans, is central" (Soukup, 2006, p. 322). However, while engaging in fan practices, consumers oftentimes "actively produce [their own] cultural meanings through creative appropriations and recontextualizations of mass cultural commodities" (Coombe & Her-

DOI: 10.4018/978-1-60960-037-2.ch001

Copyright © 2011, IGI Global. Copying or distributing in print or electronic forms without written permission of IGI Global is prohibited.

man, 2001, p. 926). This trend frequently results in oppositional interpretations of popular culture texts, thus highlighting the "the agency and inventiveness of subordinated cultures" (Rogers, 2006, p. 484) that have submitted to the pleasures of fandom.

In mapping out the primary contours of this trend, John Fiske (1989) suggests a model of popular culture that revolves around acts of excorporation and incorporation. First, consumers will "make their own culture out of the resources and commodities provided by the dominant system" (p. 15), thus establishing their own subversive meanings for those products, i.e., excorporation. However, in order to manage such reappropriations, thereby maintaining some semblance of control over their consumers, corporations often try to reintegrate those ideas, at least when possible. This re-integration is referred to as incorporation, and it allows a dominant producer to absorb "the signs of resistance...[back] into the dominant system" (p. 18), thus reclaiming any oppositional meanings from subordinated groups.

While audiences the world over are free to actively construct meanings for texts they consume based on their own experiences and frames of knowledge (Hirschman & Thompson, 1997, p. 45), their rejection and reappropriation of old meanings coupled with their ability to publicly disseminate new meanings on the Internet problematize Fiske's original conceptualizations about the incorporation and excorporation of popular culture. As illustrated through the case of Miss McDonald, the advent and availability of Internet technologies across the globe enables a truly international body of fans to become *new* meaning makers rather than meaning *re*-makers, with the latter serving as the common conclusion in traditional audience analyses. Ultimately, fans are able to use the Internet to sidestep the original producers entirely, bestowing contemporary consumers with the power to communicate directly with other consumers who can then re-appropriate and re-disseminate texts like Miss McDonald's as *they* see fit.

In this chapter, I ruminate on "A Day in the Life of Miss McDonald" to problematize Fiske's arguably dated (yet still foundational) model of popular culture production and consumption to illustrate the meaning-making that can occur when Western popular culture meets Eastern audiences (and vice versa). I begin by providing an overview of globalization and fandom, including the ways that Internet technologies have helped shape contemporary fan practices throughout the world, before reviewing Fiske's model of popular culture production and consumption. With dramatic increases in Internet usage and availability, as well as the onslaught of globalization, I ultimately use the peculiar case of Miss McDonald as a way to demonstrate the need for revisions to Fiske's model in order to reflect our current global reality.

BACKGROUND

Globalization and the Golden Arches

While the term is increasingly utilized in academic, economic, and social circles, discourses abound as to what exactly "globalization" entails (Koh, 2005; Pieterse, 2004). It has been addressed politically, economically, historically, and through myriad other foci, and this multidimensional accessibility has resulted in numerous "textbooks on globalization...[with] no textbook consensus" (Pieterse, 2004, p. 15) about what, if anything, globalization actually is. My goal with this section is not to arrive at a definitive characterization, but to explore how global consumerism has, in part, shaped contemporary popular culture.

Despite its complexities, a cursory overview of how scholars have conceptualized globalization is both necessary and useful. Belay (1996) argues that globalization is a process that reshapes the world "into a single interactional space," one that "does not eliminate the role of place or time...

[but] restructures their influence" (p. 342). From a business standpoint, Sturken and Cartwright (2001) suggest that globalization involves both "the increased conglomeration of corporate interests on an international scale" and "the movement of cultural products across national boundaries" (p. 318). More specifically, Carroll (2007) writes that globalization "is an advanced stage of capitalism" that makes the world "a much 'smaller place' in terms of the amount of time it takes to move information, people…goods, jobs, investment capital, fashions, corporations, services, and so forth around it" (p. 131). In other words, globalization represents a multi-faceted process resulting in greater connectivity between otherwise disparate cultures, brought on by such factors as greater media availability, more accessible lines of communication between cultures via technology, and the overseas expansion of transnational corporate interests.

Mike Featherstone (1995) identifies two views of culture that result from globalization processes. In one, multiple cultures are compressed, with various cultural elements "pil[ing] on top of each other in heaps without obvious organizing principles" (p. 6). This version of culture is characterized by diversity and fragmentation, with myriad cultures existing side-by-side. With the other view, "heterogeneous cultures become incorporated and integrated into a dominant culture which eventually covers the whole world" (p. 6). This version aligns more with cultural imperialism, often typified as Western culture increasingly using its power and resources to dominate, absorb, and ultimately erase Eastern culture. Here, scholars view globalization as "consum[ing] the core of non-Western and traditional Western cultures and spread[ing] mainstream Western consumer culture along its way" (Adu-Febiri, 2006, p. 36). One way this can occur is through the introduction of Western media texts to non-Western audiences, including television programs, films, and magazines. Such texts consequently transmit Western ideals about various matters, such as beauty, status, glamour, etc., as well as the Western "requirements" for obtaining such things, resulting in a sort of "cultural inferiority complex" (p. 39). For example, the influx of mediated images of Western women is believed to be at least partly responsible for the increasing number of plastic surgeries in Japan with female patients requesting facial features that appear more Western (Darling-Wolf, 2004; Rogers, 2006).

Two different but related subsets of globalization also warrant discussion: McDonaldization and Americanization. George Ritzer (2000) defines McDonaldization as "the process by which the principles of the fast-food restaurant are coming to dominate more and more sectors of American society as well as of the rest of the world" (p. 1). These principles are inextricably tied to systems that promote efficiency, calculability, control, and predictability, and his analysis decries the ease with which these decidedly Western principles have spread throughout the non-Western world (not to mention *the Western world*), re-shaping everything from health care to education. He argues that while "most globalization theorists contend that we are witnessing greater heterogenization," the adoption and increased popularity "of McDonaldized systems around the world points toward homogenization" (p. 177). Ritzer acknowledges that some scholars view McDonaldized systems as locally adaptable to the cultures adopting them, but he seems to view it more as a cultural imposition, as a "one-sided process of 'cultural imperialism'" (p. 172).

Yet, McDonaldized establishments do more than simply deliver, promote, and spread McDonaldized systems; they also represent the exportation, in a sense, of America itself. In other words, while McDonaldization owes its name and spirit to McDonald's restaurants, there is not necessarily any additional connection (i.e., the "golden arches" do not appear anywhere in streamlined hospital computer files). However, the presence of such restaurants on foreign soil does frequently connote the United States. As Ritzer elaborates, McDonald's, like other American brands, "is simply a product that retains its American identity even when it is sold and consumed in other countries" (p. 178). The imposition of such products internationally is known as Americanization, or the "processes in which economic, technological, political, social, cultural, and/or sociopsychological influences emanating from America or Americans impinge on values, norms, belief systems, mentalities, habits, rules, technologies, practices, institutions, and behaviors of non-Americans" (van Elteren, 2006, p. 345). Consequently, foreign consumers shopping or dining at American companies abroad, such as McDonald's, Pizza Hut, the Gap, etc., are given access to "the dominant ideology of American capitalism," made possible through "specific systems of representations built into consumption" (p. 347).

To illustrate, while Ronald McDonald's somewhat garish appearance—bright red hair and lips, white face, yellow shirt and slacks with red and white striped sleeves and socks—might simply trigger thoughts of cheeseburgers in the United States, scholars have found that foreign consumers often view him as "symbol[izing] status, modernity, Western civilization, Americana, and even gender equality" (Zhou & Hui, 2003, p. 43). In locations like China and the Philippines, the jovial feelings and emotions evoked by Ronald McDonald, especially among "his friends, the children" (Fishwick, 1995, p. 14), often transform him from a mere clown to an icon of the idealized *American* way of life. In this sense, Ronald McDonald is a *representative of* the United States of America, providing foreign consumers with all of the perks they have learned to associate with the American culture when they consume his products (Sturken & Cartwright, 2001; Zhou & Hui, 2003). How they manage those engagements with "Americana" after they leave those restaurants and shops, however, is the issue.

Fandom and Textual Engagement

Lawrence Grossberg (1997) writes that fans are individuals who "make an affective investment into the objects of their taste," with those investments allowing them to "construct...a consistent but necessarily temporary affective identity" (p. 247). Henry Jenkins (1988) focuses his conceptualization more on the individuals who *become* fans, arguing that fandom itself "is a vehicle for marginalized subcultural groups (women, the young, gays, etc.) to pry open space for their cultural concerns within dominant representations" (p. 87). Regardless, fandom encompasses a variety of people, all of whom demonstrate their fan appreciation in a number of ways. Moreover, fandom can develop around almost any aspect of a text, such as a novel, one of the characters within the novel, the actor who portrays the character in its filmed adaptation, and/or the author or director him-/herself (Greenfeld, 2005). The everyday pleasure an enamored consumer (as opposed to an *indifferent* viewer) feels when consuming the object of his/her fan affection "constitutes culture for the fan" (Markman, 2005, p. 7), with their pleasure rooted in engagement that moves "beyond the private act of viewing and reflects an enhanced emotional involvement" (Bielby, Harrington, & Bielby, 1999, p. 35).

As this groundwork might insinuate, this chapter accepts and relies on the notion that fans are not "cultural dupes, social misfits, and mindless consumers," but "active producers and manipulators of meaning" (Jenkins, 1992, p. 23). In other words, consumers who identify as fans do not passively consume texts and then set them aside once the credits roll, the final page is turned, or the McDonald's commercial ends. Rather, fans actively invest themselves in the characters, the storylines, the franchises, etc., and they may enact that investment by, for example, reading fan-centric magazines, sending letters to the stars and producers of their favorite programs, becoming active in fan clubs, traveling to fan conventions, and/or discussing their favorite characters, television series, and films in online forums (Bielby, Harrington, and Bielby, 1999). However, fans can also channel their pleasure through their own creative outlets, demonstrating a unique autonomy that allows them "to interact in an active way with existing cultural materials, to recreate and reshape them, and to express one's own voice through a dialogue with those of others" (Bracha, 2007, p. 1847). The results can take many forms, including fan fiction, erotic fan fiction (e.g., "slash" fiction), fan art, fan film, etc., suggesting that "this ability to transform personal reaction into social interaction, spectator culture into participatory culture, is one of the central characteristics of fandom" (Jenkins, 1988, p. 88).

In recent years, the Internet has helped make fandom even more dynamic and social. Yet, the Internet also enables fan-based practices to "undermine the traditional capacities of companies to control their images and manage their imagery"; it creates

conditions for consumers to challenge the very forms of commodity fetishism (erasures of both conditions of production and the conditions under which symbolic value is produced) on which the corporate persona as an asset has for so long relied. (Coombe & Herman, 2001, p. 919)

In other words, the advent and proliferation of Internet technologies in recent decades has greatly complicated the already tense relationship between producers and consumers. It provides consumers with opportunities to embrace "cyberdemocracy," giving producers and consumers alike the chance to initiate and participate in cultural discourse (Coleman, 2003).

Historically, consumers who have taken part in the creation and dissemination of textual reappropriations have faced costly repercussions, including lawsuits. Artwork by Barbara Kruger, Hans Haacke, and others who have re-fashioned advertisements and circulated them, or artists such as Paul Hanson, who created images of Mattel's popular Barbie Doll in satirical, often tasteless scenarios (e.g., "Drag Queen Barbie," "White Trash Barbie," etc.) and displayed them, all irked the corporations they poached their source material from. Mattel even sued Paul Hanson for $1.2 billion dollars in damages because of his reappropriated dolls. Likewise, the relatives of John Wayne sought the removal of his image from greeting cards sold at gay bookstores, all featuring the tagline, "It's such a bitch being butch" (Coombe & Herman, 2001; Striphas & McLeod, 2006).

Because companies can fight such appropriations in the courts, they continue to maintain an impressive amount of cultural power in their ability to control and limit the alternative forms of textual engagement taken up by subordinate consumers (Coombe & Herman, 2001). Yet, the tactics these corporations use to enact this control presents a paradox: they *want* consumers to invest emotionally in their products, texts, characters, etc., but they also want that investment to be strictly *consumptive* rather than *creative* (Murray, 2004). Regardless, these acts of appropriating and recoding continue, and the online tools for circulating these new-texts are increasingly accessible. Thus, these fan activities continue to represent "the most promising avenue for diverse, independent, and original creations in a world of concentrated mass media and powerful control of cultural meaning by corporate hierarchies" (Bracha, 2007, p. 1851).

In the sections that follow, I unpack Fiske's model and focus my discussion specifically on the photographs of Miss McDonald, theorizing the nature and nuances of her popular culture creation in light of our current technological and globalized condition. My goal here is not to challenge Fiske, per se, but to insert a footnote on his original model in order to better suit our contemporary needs.

Producing, Disseminating, and "Popping" Culture

Much like globalization, the concept of popular culture continues to invoke a number of different meanings (Irwin, 2007; Kidd, 2007; Storey, 2003). Generally, the texts that ultimately fit this classification are "for the people" and are often "*by* the people." They are the texts that walk away from the "battleground...between social groups and associated identities that fall on either side of the dividing line of domination" (Giulianotti, 2007, p. 257). As with the previously cited examples offered by Grossberg (1997), Jenkins (1992), etc., much of what we recognize as popular culture, then, is at least partially a consequence of consumer engagement, with subordinate consumers and fans taking up dominant texts and celebrating them as altogether special.

John Fiske (1989a, 1989b) provides one of the more thoroughly explicated discussions of what popular culture is and how it functions. In discussing culture itself, Fiske (1989a) writes that it "is a constant succession of social practices; it is therefore inherently political, [and] it is centrally involved in the distribution and possible redistribution of various forms of social power," and in the U.S., this is inexorably tied to "white, patriarchal capitalism" (p. 1). It produces meanings about self as well as meanings about social experience. Popular culture, on the other hand, represents a "shadow" economy (Fiske, 1993) where fans invest in alternative claims to the resources put forth by the dominant producers, with these resources including anything from television programs to fashion, language to music, etc. As Fiske (1989a) describes it, popular culture "is made by various formations of subordinated or disempowered people out of the resources, both discursive and material, that are provided by the social system that disempowers them" (Fiske, 1989a, p. 1). Furthermore,

> *the resources...carry the interests of the economically and ideologically dominant; they have lines of force within them that are hegemonic and that work in favor of the status quo. But hegemonic power is necessary, or even possible, only because of resistance, so these resources must also carry contradictory lines of force that are taken up and activated differently by people situated differently within the social system. (p. 2)*

The result of these efforts by the subordinated classes is popular culture, and it can take many different forms.

However, Fiske's ideas about popular culture also reflect the aforementioned struggles of corporations to control consumers, resulting in a model of popular culture that is dialectical in nature (Rogers, 2006). As Fiske (1989b) articulates it in his seminal book *Understanding Popular Culture*, consumers will "make their own culture out of the resources and commodities provided by the dominant system" (p. 15), allowing them to establish their own subversive meanings for those products, i.e., the process of excorporation. However, in order to manage such reappropriations, thereby maintaining control over consumers, corporations often reintegrate those ideas, at least when possible. This reintegration is referred to as incorporation, and it allows producers to absorb "signs of resistance...[back] into the dominant system and...rob [subordinated groups] of any oppositional meanings" (p. 18). To illustrate, Fiske uses the example of blue jeans, noting how consumers often purchase these denim products and then rip or otherwise deface them (i.e., excorporation). Corporations take note of this consumer behavior, and, thus, start marketing *pre*-ripped and *pre*-defaced jeans (i.e., incorporation). The cycle continues endlessly, resulting in a circular, back-and-forth model of production and consumption, meaning making and meaning re-making, in which subordinate consumers are always repurposing the products of the producers to create their own popular culture and the producers are always reappropriating those products back into their dominant systems, wrestling control back from unruly and creative consumers in the process.

Central to this model is the notion that consumers, and fans in particular, act as *meaning re-makers*. Given textual "gifts" from dominant producers, consumers refashion them into new ideas and products, thus altering the original meanings and connotations along the way. However, as Fiske (1989b) posits, the producers' constant reclamation prevents consumers from truly becoming *new* meaning makers; they are always already working with texts and meanings that are already in circulation. Moreover, as Condit (1989) illustrates with fans of the television program *Cagney and Lacey*, there are limits to the range of meanings that consumers can arrive at, suggesting that the producers maintain *some* semblance of control during and after consumption. Yet, both Fiske's and Condit's discussions predate the figurative explosion of the Internet as a ubiquitous facet of our collective daily routine. The World Wide Web offers unprecedented access to cultural texts as well as limitless opportunities to communicate within and amongst cultures about those texts. Various computer programs provide those same consumers with the tools necessary for altering these texts and/or creating altogether original works. As such, Fiske and Condit were not *in*correct in their assertions; they were incapable of predicting the development, evolution, and impact of the Internet on these processes.

The Peculiar Case of Miss McDonald

As previously stated, "Miss McDonald" is a self-identified fan of Ronald McDonald and the company he serves, and her photography project presents an interesting point of entry to—and point of contention for—Fiske's initial conception. Miss McDonald's project started innocently enough as a simple Halloween costume. However, upon finding greater inspiration in the actual act of embodying the iconic clown, she decided to embrace the attire and makeup fully for a yearlong photography project in which she would document her version of Ronald McDonald living an ordinary daily existence. Once Miss McDonald completed her project, she held public showings at various shops and galleries throughout the Philippines and eventually posted the images to her online journal (http://miss-mcdonald.livejournal.com/). Coverage within the "blogosphere"—the myriad websites that are collectively centered on "blogging," or online journal publication—and general word-of-mouth publicity resulted in an enormous influx of visitors to her website, ultimately causing her journal to temporarily "crash" due to depleted bandwidth. However, despite her obvious appropriation of a widely recognized corporate icon, the website continues to remain accessible, exposing countless new online visitors to one of the biggest Internet "folk heroes" of 2005 (Ebenkamp & Hein, 2005).

The images produced by Miss McDonald are paradoxically both ordinary and outrageously peculiar. The 28 photographs show Miss McDonald performing mundane, domestic tasks (e.g., paying bills, going to the dentist, raking leaves, napping, shopping, tending laundry, etc.), socializing (e.g., playing on swing sets and sliding boards, playing basketball, watching cartoons with children, etc.), or just sitting on benches near or inside various McDonald's restaurants, oftentimes next to a statue of her muse. While these photographs comprise the bulk of the series—indeed, each of the first 27 photographs falls into one of the three categories—the final picture produced shows Miss McDonald in a far more sinister light.

Unlike the previous images, the final photograph—produced for another Halloween, no less—shows Miss McDonald crouched in a dimly lit industrial area, surrounded by desolation and ruin. While the setting is not disturbing in and of itself, her actions more than make up for it, as she is shown to be eating a severed human hand with a bloody, dismembered foot lying nearby. Miss McDonald has blood on her chin and glove, and the look on her face—wide eyes, mouth agape—suggest an element of surprise, as if she did not anticipate the documentation of her nocturnal, cannibalistic feast. The bulk of the images show Miss McDonald to be a happy, functioning member of society, but the final photograph reveals a darker side, one that suggests that everything viewers knew and loved about Ronald McDonald was merely a façade that masked more primal and deadly motives. Still, despite the content and tone of any particular image, the collection as a whole subverts the character's original purpose.

The fact that Miss McDonald recognizes, adores, and has an affective response to Ronald McDonald and the company he serves arguably is not unusual, but it does speak volumes regarding popular culture and corporate globalization, especially in the age of the Internet. As posited by Fiske (1989b), Miss McDonald's images present an unambiguous example of excorporation, whereby she embraced a dominant producer's (McDonald's Corporation) trademarked image (Ronald McDonald) and reappropriated it to suit her own subordinate desires. She made her "own culture out of the resources and commodities provided by the dominant system" (Fiske, 1989b, p. 15) by "poaching" (Jenkins, 1992) Ronald McDonald and "rewriting" him through the act of embodiment, by depicting him beyond the traditional context of McDonald's restaurants, and by re-imagining how the character spends his free time.

On the other hand, Miss McDonald's photographs expose contemporary flaws in Fiske's dialectical model of popular culture, especially with regards to the act of incorporation. Certainly, companies often reappropriate the subversive products that result from consumer excorporation by embracing and redeploying these products as their own in order to maintain a dominant role in the relationship. In situations where incorporation is undesirable, such as cases where the re-imagined product is distasteful or inescapably critical of the company, lawsuits are filed and "cease and desist" letters are sent, allowing companies to erase public access to the consumer text in question and, therefore, maintain control. With Miss McDonald's images, the McDonald's Corporation (apparently) did nothing. The company has not adopted her images or her subversive approach to re-imagining the character in any of their promotions or marketing efforts. They have made no efforts to stop the dissemination of her project, or at least their efforts failed dramatically given the continued presence of her collection on the Internet and the innumerable links to her images still littered across the World Wide Web.

As the case of Miss McDonald suggests, incorporation, at least as scholars have traditionally conceived it, is not only thwarted, but its very nature with regards to *who* incorporates and *how* they do so is also transformed. Rather than cycling back to the producers, the availability of these pictures to literally anybody with Internet access harkens a "tree" of incorporation rather than a circular model, with "branches" producing new branches, and so forth. In other words, consumers are able to intercept her images, and they are able to develop their own affective responses free and clear of any interference by the original producers. Hypothetically, Russian fans could take Miss McDonald's inspired appropriation, reappropriate it in their own way, and re-mediate their creative efforts however they see fit, eventually leading to the process that cycles entirely amongst consumers rather than between producers and consumers. In fact, two years after Miss McDonald's project ended, another Philippine fast food restaurant, Jollibee, had to contend with a popular fan video featuring the Jollibee mascot fornicating with one of his co-mascots (Ebenkamp, 2007). Whether these fans found inspiration in "A Day in the Life of Miss McDonald" (i.e., "A day in the [*sex*] life of the Jollibee mascot") is unclear, but a "cyber-genealogy" appears to link these sorts of creative fan products, and like Miss McDonald's project, this video also remains accessible online. With popular culture configured in this manner, the producers, for all intents and purposes, become increasingly obsolete once they put forth the original image or text.

As companies take their messages abroad and consumers post appropriated texts online—a realm that virtually transcends the demarcated borders of nation-states, although various restrictions on usage are enforced in some countries—a more global audience of consumers encounters these messages, leading to a wider range of affective reactions and active responses. In other words, as globalization continues, and as more people understand and learn the specifics of online publishing, we might expect a range of responses—a surprisingly diverse range, in fact—not just for Miss McDonald's images, but for any text produced and "let loose" on the World Wide Web. Foreign consumers oftentimes relish the chance to have the McDonald's experience, yet just as many find them to be intrusive, as another way for the West to impinge upon the East. How might these viewers read Miss McDonald's collection? What might they do with those responses?

When I first encountered Miss McDonald's project, I instantly wanted to assume a critical perspective. The last image was the first image I encountered, and while her pictures are largely benign in tone, the final picture is anything but benevolent. As a result, it transformed the overall meaning of the collection into something far more depraved, at least for me. As previously described, the last photograph shows Miss McDonald ingesting human body parts. The surroundings appear war torn and decrepit. The entire atmosphere is grim and gritty. As a capstone to this visual collection—as the consumers literally become *the consumed*—the suggested innocence of Miss (or Ronald) McDonald's previous engagements with society appears rather dubious and deceptive. Hypothetically, the character could have been interacting with the populace as a way to gain their trust and acceptance, to infiltrate by way of their invitation and embrace. Mr./Miss McDonald could be an embodiment of the ultimate capitalistic colonizer, mimicking "normalcy" until all signs of the active and vibrant host society that embraced him/her have been chewed apart and swallowed—literally.

Furthermore, Miss McDonald's version of the iconic figure could be read to suggest that his active participation in society is not only an attempt to establish trust, but also a tactic designed to help him become ubiquitous. Indeed, he *wants* to be everywhere at once, and he *will* be waiting for you at the store, the doctor's office, the gymnasium, the playground, etc. He will even pay his bills in the meantime to ensure his continued acceptance. Not surprisingly, fast food is not inherently associated with such things as an active lifestyle or parental responsibility, especially with regards to feeding children healthy meals. Yet, Ronald McDonald's activities appear to make these correlations. Unfortunately, as the final photo might suggest, cultures that blindly accept his deception will inadvertently help him attain his ultimate goal—to dominate and destroy everything (and everyone) around him. He will outlive us all, even if our rampant consumerism helped ensure his omnipresence in the first place.

Of course, this is but one way to read "A Day in the Life of Miss McDonald." Interestingly, it was my affective response to a mediated text that was produced in the Philippines, but that was itself created in response to an American corporation—one that I have much familiarity with, one that I resent for distorting my own waistline, one that I see looming every day at the end of my neighborhood block. The fact that countless others have engaged with and celebrated these photographs—others literally scattered across the globe that are free to do with them as they choose—hints at the diverse range of responses that these images have the potential to elicit. Consequently, this range and this inability and/or refusal of dominant producers to corral the images and the products stemming from them illustrates that a "new" model of popular culture is at play. As Internet technologies continue advancing, and as more and more individuals gain access to and knowledge of those technologies, increasingly innovative forms of engagement will continue to prosper. Popular culture will cease being a predominantly Western phenomenon, and a truly globalized and dynamic popular culture will emerge.

CONCLUSION

As the peculiar case of Miss McDonald suggests, consumers are no longer simply meaning re-makers, or at least they do not have to be. Technology has enabled consumers across the globe to become *new* meaning makers, with each new consumer who encounters a fan-text capable of interpreting it, producing even newer meanings, and disseminating it to an even broader audience, with the process expanding and recycling within the consumer sphere alone. Approval is not needed, nor is validation, but the achievement of cultural popularity will likely bestow both.

Fiske's (1989b) basic model does not negate this potential, per se, but it does not seem to anticipate it fully, as should probably be expected seeing as his conceptualization appeared well before the Internet achieved its current prominence. Standard acts of incorporation and excorporation will continue, but as technology and consumer creativity become more innovative, we must also anticipate the novel ways in which these consumer responses will manifest, and how these new approaches can alter the very nature of popular culture production, consumption, and dissemination, at least as we have come to understand it. Our inability to do so steals some measure of consumer agency away from popular culture, and, when all is said and done, it *is* the consumer who makes that kind of culture so popular in the first place.

I feel it is important to reiterate that I do not mean to imply that the inimitable John Fiske, with his firm grasp on popular culture, was incorrect, or that his enviable work was without merit; conversely, I *rely* wholeheartedly and graciously on Fiske's foundation to support my own analytical extensions. Humbly, I see this project more as an *addendum* rather than a *correction*. Here, I argue that contemporary adaptations and transformations in fan behaviors, the changing face of technology (and its availability), and the unavoidable consequences of globalization necessitate a revision to Fiske's original conception. Consumers are not mindless, nor are they resourcefully destitute. They *can* engage with texts in ways that are consistently surprising, occasionally alarming, and shockingly inappropriate, and, invariably, some *do*. However, the incredible hodgepodge that results from such an ever-changing mish-mash begs for further inspection and discussion, and that is what I hope I have accomplished here.

"A Day in the Life of Miss McDonald" is likely not the first popular culture text to throw a wrench into the works of Fiske's model, but it does provide an accessible point of entry for discussing the myriad nuances that affect its contemporary form. Popular culture is (literally) what we make of it, but the most important facet of this new configuration is the "we." *We* are provided texts by dominant voices. And *we* are capable of making new meanings from them. We do not have to rely on the producers to provide the meanings we embrace; we can provide our own. We, after all, now have access to more active and effective means for doing so. Indeed, we can ingest and regurgitate dominant texts, and we can become popular culture "folk heroes" in the process.

REFERENCES

A day in the life of Miss McDonald. (2004). Retrieved July 22, 2007, from http://miss-mcdonald.livejournal.com/

Adu-Febiri, F. (2006). The destiny of cultural diversity in a globalized world. *Review of Human Factor Studies*, *12*(1), 30–64.

Belay, G. (1996). The (re)construction and negotiation of cultural identities in the age of globalization. In Mokros, H. (Ed.), *Interaction and identity: Information and behavior*. New Brunswick, NJ: Transaction Publishers.

Bielby, D., Harrington, C., & Bielby, W. (1999). Whose stories are they? Fans' engagement with soap opera narratives in three sites of fan activity. *Journal of Broadcasting & Electronic Media*, *43*(1), 35–51.

Bracha, O. (2007). Standing copyright law on its head? The Googlization of everything and the many faces of property. *Texas Law Review*, *85*, 1799–1869.

Carroll, N. (2007). Art and globalization: Then and now. *The Journal of Aesthetics and Art Criticism*, *65*(1), 131–143. doi:10.1111/j.1540-594X.2007.00244.x

Coleman, R. (2003). Elmo is black! Black popular communication and the marking and marketing of black identity. *Popular Communication*, *1*(1), 51–64. doi:10.1207/S15405710PC0101_7

Condit, C. (1989). The rhetorical limits of polysemy. *Critical Studies in Mass Communication*, *6*(2), 103–122. doi:10.1080/15295038909366739

Coombe, R., & Herman, A. (2001). Culture wars on the net: Intellectual property and corporate propriety in digital environments. *The South Atlantic Quarterly*, *100*(4), 919–947. doi:10.1215/00382876-100-4-919

Darling-Wolf, F. (2004). Sites of attractiveness: Japanese women and Westernized representations of feminine beauty. *Critical Studies in Media Communication*, *21*(4), 325–345. doi:10.1080/0739318042000245354

Ebenkamp, B. (2007, November 12). Jollibee is on fast (food) track in virgin markets. *Brandweek*, *48*(41), 14.

Ebenkamp, B., & Hein, K. (2005, November 21). A clown around town. *Brandweek*, *46*(42), 38.

Featherstone, M. (1995). *Undoing culture: Globalization, postmodernism, and identity*. London: Sage Publications.

Fishwick, M. (1995). Ray and Ronald girdle the globe. *Journal of American Culture*, *18*(1), 13–29. doi:10.1111/j.1542-734X.1995.1801_13.x

Fiske, J. (1989a). *Reading the popular*. Boston: Unwin Hyman.

Fiske, J. (1989b). *Understanding the popular*. Boston: Unwin Hyman.

Fiske, J. (1993). *Power plays, power works*. New York: Verso.

Giulianotti, R. (2007). Popular culture, social identities, and internal/external cultural politics: The case of Rangers supporters in Scottish football. *Identities (Yverdon)*, *14*(3), 257–284. doi:10.1080/10702890601162732

Greenfeld, M. (2005). *Serious play: How gossiping is accomplished in an online fandom forum*. Paper presented at the Annual Meeting of the International Communication Association, New York, NY.

Grossberg, L. (1997). *Dancing in spite of myself: Essays on popular culture*. Durham, NC: Duke University Press.

Hirschman, E., & Thompson, C. (1997). Why media matter: Toward a richer understanding of consumers' relationships with advertising and mass media. *Journal of Advertising*, *26*(1), 43–60.

Irwin, W. (2007, November/December). The pop culture manifesto. *Philosophy Now*, *64*, 11–15.

Jenkins, H. (1988). *Star Trek* rerun, reread, rewritten: Fan writing as textual poaching. *Critical Studies in Mass Communication*, *5*(2), 85–107. doi:10.1080/15295038809366691

Jenkins, H. (1992). *Textual poachers: Television fans and participatory culture*. New York: Routledge.

Kidd, D. (2007). Harry Potter and the functions of popular culture. *Journal of Popular Culture*, *40*(1), 69–89. doi:10.1111/j.1540-5931.2007.00354.x

Koh, A. (2005). "Heteroglossic" discourses on globalization: A view from the "East.". *Globalizations*, *2*(2), 228–239. doi:10.1080/14747730500202230

Markman, K. (2005). Star Trek, *fan film, and the Internet: Possibilities and constraints of fan-based vernacular cultures*. Paper presented at the Annual Meeting of the International Communication Association, New York, NY.

Murray, S. (2004). "Celebrating the story the way it is": Cultural studies, corporate media and the contested utility of fandom. *Continuum: Journal of Media and Cultural Studies*, *18*(1), 7–25. doi:10.1080/1030431032000180978

Pieterse, J. (2004). *Globalization and culture: Global mélange*. Lanham, MD: Rowman and Littlefield Publishers.

Ritzer, G. (2000). *The McDonaldization of Society*. Thousand Oaks, CA: Pine Forge Press.

Rogers, R. (2006). From cultural exchange to transculturation: A review and reconceptualization of cultural appropriation. *Communication Theory*, *16*(4), 474–503. doi:10.1111/j.1468-2885.2006.00277.x

Soukup, C. (2006). Hitching a ride on a star: Celebrity, fandom, and identification on the World Wide Web. *The Southern Communication Journal*, *71*(4), 319–337.

Storey, J. (2003). *Inventing popular culture: From folklore to globalization*. Malden, MA: Blackwell Publishing.

Striphas, T., & McLeod, K. (2006). Strategic improprieties: Cultural studies, the everyday, and the politics of intellectual properties. *Cultural Studies*, *20*(2/3), 119–144. doi:10.1080/09502380500495635

Sturken, M., & Cartwright, L. (2001). *Practices of looking: An introduction to visual culture*. Oxford: Oxford University Press.

van Elteren, M. (2006). Rethinking Americanization abroad: Toward a critical alternative to prevailing paradigms. *Journal of American Culture*, *29*(3), 345–367. doi:10.1111/j.1542-734X.2006.00376.x

Zhou, L., & Hui, M. (2003). Symbolic value of foreign products in the People's Republic of China. *Journal of International Marketing*, *11*(2), 36–58. doi:10.1509/jimk.11.2.36.20163

ADDITIONAL READING

Allen, M., & Sakamoto, R. (Eds.). (2008). *Popular culture, globalization, and Japan*. New York: Routledge.

Allison, A. (2006). *Millennial monsters: Japanese toys and the global imagination*. Berkeley: University of California Press.

Ang, I. (1985). *Watching Dallas*. London: Methuen.

Armbrust, W. (Ed.). (2000). *Mass mediations: New approaches to popular culture in the Middle East and beyond*. Berkeley: University of California Press.

Banks, J. (1997). MTV and the globalization of popular culture. *International Communication Gazette*, *59*(1), 43–60. doi:10.1177/0016549297059001004

Bennett, T., & Woollacott, J. (1987). *Bond and beyond: The political career of a popular culture hero*. New York: Methuen.

Bentley, J., Tinney, M., & Chia, B. (2005). Intercultural internet-based learning: Know your audience and what it values. *Educational Technology Research and Development*, *53*(2), 117–128. doi:10.1007/BF02504870

Bordwell, D. (1989). *Making meaning: Inference and rhetoric in the interpretation of cinema*. Cambridge, MA: Harvard University Press.

Crothers, L. (2007). *Globalization and American popular culture*. Lanham, MD: Rowman and Littlefield Publishers.

Davisson, A., & Booth, P. (2007). Reconceptualizing communication and agency in fan activity: A proposal for a projected interactivity model for fan studies. *Texas Speech Communication Journal*, *32*(1), 33–43.

Drushel, B. (2007). Pandora's box in cyberspace: The online alternative fan sites of *Hercules: The Legendary Journeys*. *Femspec*, *7*(2), 7–28.

Edelman, D. (2007). From the periphery to the core: As online strategy becomes overall strategy, marketing organizations and agencies will never be the same. *Journal of Advertising Research*, *47*(2), 130–134. doi:10.2501/S0021849907070146

Falzone, P. (2005). The final frontier is queer: Aberrancy, archetype and audience generated folklore in K/S slashfiction. *Western Folklore*, *64*(3/4), 243–261.

Gunn, J., & Brummett, B. (2004). Popular culture after globalization. *The Journal of Communication*, *54*, 705–721. doi:10.1111/j.1460-2466.2004.tb02651.x

Hall, S. (1981). Notes on deconstructing "the popular". In Samuel, R. (Ed.), *People's history and socialist theory*. London: Routledge.

Hall, S. (1995). New cultures for old. In Massey, D., & Jess, P. (Eds.), *A place in the world?: Places, cultures and globalization*. London: Oxford University Press.

Iwabuchi, K. (2002). *Recentering globalization: Popular culture and Japanese transnationalism*. Durham, NC: Duke University Press.

Katz, E., & Liebes, T. (1987). Decoding *Dallas*: Notes from a cross-cultural study. In Newcomb, H. (Ed.), *Television: The critical view* (4th ed., pp. 419–432). New York: Oxford University Press.

King, A. (Ed.). (1997). *Culture, globalization and the world-system: Contemporary conditions for the representation of identity*. Minneapolis: University of Minnesota Press.

Lasn, K. (1999). *Culture jam: How to reverse America's suicidal consumer binge—and why we must*. New York: HarperCollins.

Lee, M., & Cho, C. (1990). Women watching together: An ethnographic study of Korean soap opera fans in the U.S. *Cultural Studies*, *4*(1), 30–44. doi:10.1080/09502389000490031

Markman, K. (2005). Star Trek, *fan film, and the Internet: Possibilities and constraints of fan-based vernacular cultures*. Paper presented at the Annual Meeting of the International Communication Association, New York, NY.

Ohmae, K. (1990). *The borderless world: Power and strategy in the interlinked world economy*. New York: HarperBusiness.

Parameswaran, R. (2004). Global queens, national celebrities: Tales of feminine triumph in post-liberalization India. *Critical Studies in Media Communication*, *21*, 346–370. doi:10.1080/0739318042000245363

Radway, J. (1984). *Reading the romance: Women, patriarchy, and popular literature*. Chapel Hill, NC: University of North Carolina Press.

Schlosser, E. (2001). *Fast food nation: The dark side of the all-American meal*. Boston: Houghton-Mifflin.

Sholle, D. (1991). Reading the audience, reading resistance. *Journal of Film and Video, 43*(1/2), 80–89.

Steigerwald, D. (2004). *Culture's vanities: The paradox of cultural diversity in a globalized world*. Lanham, MD: Rowman and Littlefield Publishers.

Tomlinson, J. (1991). *Cultural imperialism*. Baltimore, MD: Johns Hopkins University Press.

Watson, J. (Ed.). (1997). *Golden arches east: McDonald's in east Asia*. Stanford, CA: Stanford University Press.

Woo, H., & Dominick, J. (2003). Acculturation, cultivation, and daytime TV talk shows. *Journalism & Mass Communication Quarterly, 80*(1), 109–128.

Chapter 2
Urban Screens and Transcultural Consumption between South Korea and Australia

Audrey Yue
University of Melbourne, Australia

Sun Jung
Victoria University, Australia

ABSTRACT

This chapter examines urban screens as sites of media convergence and transcultural consumption. Using two case studies in Melbourne (Australia) and Songdo (Incheon, South Korea), this chapter considers how these screens have emerged through technological innovations led by cultural planning and urban regeneration. Furthermore, using audience reception and cultural participation studies, this chapter critically examines the augmentation of these spaces as sites for cultural citizenship and transcultural consumption. Urban screens, this chapter argues, are new contact zones of mediascapes, social belonging and transcultural identities.

INTRODUCTION

On 7 August 2009, two large urban screens in Songdo (Incheon, South Korea) and Melbourne (Australia) simultaneously connected and communicated through digital art and public SMS texting. Audiences in Melbourne, at Federation Square, a large civic plaza where the screen is displayed, interacted with audiences in Songdo's Tomorrow City, also a large civic plaza housing a new urban screen. This event, organized and funded in partnership with an Australian Research Council Linkage grant and Seoul-based Art Center Nabi, materializes the "doing" of cultural citizenship and the potentials of a transnational public sphere.

This chapter uses this live telematic media event as a case study to critically examine how large urban screens are new media technologies that function as new social sites of transcultural consumption. Unlike most urban screens that focus predominantly on outdoor advertising, this

DOI: 10.4018/978-1-60960-037-2.ch002

Copyright © 2011, IGI Global. Copying or distributing in print or electronic forms without written permission of IGI Global is prohibited.

chapter will focus on screens that display public art and are situated in civic plazas for public use. Unlike the personalized nature of mobile media or the apparent passive reception of export cultural industries like film and television, this chapter suggests urban screens are highly localized and public, and yet also simultaneously transcultural.

This chapter first maps the development of urban screens in Australia and Korea through the frameworks of urban regeneration, cultural policy and media convergence. It shows how urban screens decenter the phenomenology of the cinematic screen to provide an embodied experience of interactivity that potentially enhances local cultural citizenship and transnational cultural consumption. Using the telematic event introduced above, it further considers the reception of digital art and public short message service (SMS) texting by Korean and Australian audiences to create a transnational public sphere and engender local cultural citizenship. In particular, this chapter discusses how such participatory interactions in South Korea reinforce the construction of a transcultural media consumption lifestyle. Urban screens, this chapter argues, are new contact zones of mediascapes, social belonging and transcultural identities.

URBAN SCREENS: MEDIA CONVERGENCE AND CULTURAL REVITALISATION

From Las Vegas, Berlin to Sao Paulo and Beijing, urban screens are new infrastructures in contemporary cityscapes. Defined as "yard- and perch-scale displays... deployed in public spaces in urban contexts" (MacColl and Richardson, 2008, p. 100), they are both public amenities and cinematic screens. As such, they engage multiple stakeholders, from designers, digital art producers, planners, architects, to the cultural histories of places and people, and are usually embedded in the larger project of urban regeneration and economic development. As Auerbach (2006) states, urban screens are made possible when the interests of those who control the exhibition space, technology, potential content streams and potential revenue streams converge.

Urban screens range in size, resolution, use and location. Central to the innovation in public screens, they make use of technological development in Liquid Crystal Display (LCD), Plasma Display (PDP), Digital Light Processing (DLP), Light Emitting Diode (LED) and Spectacolour High Definition. From projection broads, information terminals to intelligent buildings, they are used for a variety of purposes. The most common is entertainment, exemplified by the world's largest canopy screen and light show on Fremont Street in Las Vegas. For companies seeking cool branding, the use of such outdoor display advertising is increasingly commonplace, with Prada, Nike and Apple having launched high profile advertising campaigns through these platforms. In the finance sector, they are ubiquitous in central business districts, such as the NASDAQ ticker that wraps around the cylindrical cone of its building in Manhattan. Art and entertainment have also featured on the facades of commercial buildings. In Rotterdam and Seoul, the headquarters of their country's respective telecommunications company—the KPN building in the Kop van Zuid docklands and the SK telecom building in the art precinct around Insadong—also feature media walls and giant screens that can be seen kilometers afar. In the Millennium Park in Chicago and Federation Square in Melbourne, civic and recreational spaces are built around large screens to further engage the public user.

From live television broadcast to marketing and digital art, urban screens have become a universal visual culture in contemporary urban life (Lester, 2006). At sports stadiums, gas stations, shopping malls, supermarkets, schools, gyms, escalators, elevators and toilets, these screens, whether small mobile surfaces or big public monuments, are designed to reach the increasingly mobile population, people on the move. Such a changed lifestyle has attracted the attention of advertisers eager to tap into the pedestrian marketplace. In 2008, CBS reported that its revenue for outdoor operations rose 8% to US$1.1 billion. The same period saw advertisement spending on traditional television fell by 1.7% (Steinberg, 2008). In post-socialist consumer driven China, the use of these screens in commercial culture has recently risen exponentially. Focus Media, China's leading multi-platform digital media company, operates the largest network of outdoor advertising in the country with 190 000 screens installed across ninety cities (Balfour, 2008). It sells fifteen second commercials which air sixty times a day on 8,000 screens for US$41,000. Listed on the NASDAQ exchange, it is the country's second largest media advertising company, after the state-run Chinese Central Television Network (CCTV). In 2008, its net earnings soared to US$790 million (Focus Media, 2008), a jump of 179% in fiscal sales from 2007. With clients such as Armani Cosmetics, Motorola and Lenovo, these screens not only reach out to the purchasing potential of the urban elites; they also possess the potential to by-pass state regulation. No media license is needed for the re-edited television broadcast content, and content is quickly updated, cheaply and informally by a fleet of cyclists swapping memory cards.

Urban screens are new sites of media convergence. Slaatta (2006) uses the term, media infrastructure, to refer to the combination of media and infrastructure. He suggests buildings are now informational and transitional objects, with changing decorative and media capacities. Manovich (2006) describes this dynamic between architecture and information as the space of augmentation. In augmented space, data is overlaid onto physical space. In such electronic landscapes, the mobility of media is utilised to its full capacity. Whether it is digital media art specially commissioned for the large screen, the screening of a movie with sound relayed through a mobile portable music player, live television broadcasting, re-edited television sequences, or the invitation to display private SMS messages and images on the interactive public screen, this type of mobile media convergence has changed urban morphology, social relations and communications.

Buhlmann (2006) discusses how architecture has become virtualised through large sized media skins, and argues they are not simply surplus devices, but new infrastructures that constitute new social habits. The media in architectural media facades literally and physically introduce a separation between inside and outside to create a spatialization of media where communication takes place. New social relationships are structured by such a performative mediality in its topography. This type of embodiment can be considered a form of "cultural somatics," a phrase used to refer to a combined approach "which accounts for the imbrications of cultural specificity and context, corporeality, and our somatic involvement with technologies-in-use" (MacColl and Richardson, 2008, p. 100). Cultural somatics provides a more dynamic framework to rethink the socio-cultural boundaries between technology, body, community and place.

Augmented space also changes the experience of architecture by placing the viewer inside the object rather than the traditional cinema that creates an object where the viewer can look at (Manovich, 2006). To understand the potential of urban screens, Auerbach (2006) proposes acknowledging the dialectics of screens and their relations of representation: that is, their double and contradictory functions; their ability to display, conceal and transmit. He illustrates this using the role of video in the conditioning of urban subjects. In city regeneration projects where large screens are usually introduced, video installations and surveillance technology work hand in hand. Urban subjects are conditioned into being observed while also looking at themselves on the screens. The screen not only hides its panoptic potential; it also displays its narcissistic potential. It becomes the audience for the user to reform relationships, such as the one to many communications enabled by the shared spectacle, as well as through its interactive capacity to make public private photos and SMS messages. Through these capacities, national cultural institutions have the potential to redefine their roles to the audience, market and public.

Urban screens are attractive to large media corporations in their building projects. The screen in Melbourne houses a suite of other cultural institutions, such as a multicultural broadcaster, a national art gallery and state film theatres. Similarly, the screens in Rotterdam and Seoul are also commissioned as important elements of their respective telco headquarters. In these sites, new electronic landscapes mediate capital and labour socially and spatially (Slaata, 2006). As media is no longer by bounded by national space, these new screen technologies allow media institutions to reshape their roles. Audiences are re-territorialised in new spatial relationships between the market and the public. As mediators between the natural and electronic, the screen also enhances the symbiotic relationship to creativity. Media work can be produced inside the building and seamlessly exhibited outside the building. While it holds the potential for new creative clusters, it also potentially conceals its own logics and structures of production.

Media convergence has clearly changed the aesthetics and morphology of urbanism and its new relationships to sociality and communication. The screens in Australia and South Korea, two of the ten in the world permanently and continuously transmitting content in their public and civic roles, have emerged as a result of robust and regionally-specific cultural policy developments.

The large screen in downtown Melbourne, Australia, at Federation Square opened to the public in 2002. The Square sits at the intersection between Flinders and Swanston Streets, two of the major arterials in the southern grid of the city. It was built over disused rail yards and abandoned concrete towers formerly occupied by the Gas and Fuel Corporation and Victorian Employees Federation. Located opposite the city's central railway station and a historic church, the architecture's postmodern fractal façade of sandstone, zinc and glass was controversial. While some lamented its anachronistic design, others lauded the potential of its new urbanism to engage the user in a new experience of critical spatiality (Macarthur, Crist, Hartoonian & Stanhope, 2003; Rice, 2005; Rundle, 2002). The complex's meandering walkways followed the style of the city's iconic laneways but functioned as an antidote to the square grid of the major streets. The 38.5 square meters screen, the first of its kind in Australia, sits in the southern front of the complex overlooking the Yarra River and the iconic Princess Bridge. In the complex are other mixed-sized LED and LCD screens, in the foyers and along its corridors and escalators.

When the Square was conceived in the mid 1990s, Australia was in the throes of cultural revitalization. In 1994, Australia released its first cultural policy statement, *Creative Nation* (Department of Communication and the Arts, 1994). Amongst its innovations is the stress on multiculturalism and indigeneity as central to the nation's cultural heritage, as well as the focus on new managerial entrepreneurship and a shift of policy from supply to demand (Radbourne, 1997). Key to these is the emphasis on the creativity of new media technologies. Melbourne followed in this agenda with large-scale building projects, such as the interactive museum and an urban renewal of the riverside precinct. Amongst these developments is the installation of the large screen at Federation Square. Its state policy, *Arts 21*, captured by slogan, Victoria "On the Move," matched the zeitgeist that celebrated the new mobile experiences of the global city. Its strategic aims were to provide world-class facilities with a diversity of programming that would lead the state into the information age and deliver Australia to the world through customer-focused marketing (Stevenson, 2000). In 2003, these aims were further harnessed in an updated policy, *Creative Capacity +*, that aimed to build creative communities out of these amenities (Arts Victoria, 2003). In the seven years since opening, Federation Square has lived up to its potentials; it is now the most visited tourist destination in the state, and a central meeting hub for residents and tourists alike (Federation Square, 2009).

Like Tokyo and Taipei, Seoul is a city of large screens. Since the 1960s, South Korea has adopted an open door policy that has led to a high export-driven economic growth in the 1990s. In particular, it has quickly developed digital technology since the late 1990s and this has enhanced the urban screen infrastructure. South Korea is currently one of the most digitally well-networked countries. For the past decade, the country's broadband penetration rate matched the top five countries along with Netherlands, Denmark, Norway and Switzerland (OECD, 2009). The country now has 94.7% of households with broadband access, the highest among the OECD countries (OECD, 2008). These changes were developed to reinvigorate its economy after the 1997-8 financial crisis. Seoul became the incubator for government cultural and industrial policies on broadband, fashion, culture and consumerism. Building on "pervasive computing and communications" (Townsend, 2007, p. 396), this infrastructure develops a network between telecommunications and urbanization by harnessing its capacity for spatial decentralization and agglomeration. Its aim is to build cities as technological hubs for information flows. South Korean cyber culture has infused all aspects of everyday life, evident in the visible emergence of *PC bangs*—small cybercafés, which are also known as *game bangs* due to their significant function as LAN gaming centers—that provided cheap, fast and unabated access. Housing precincts such as the Digital Media City, with its wired apartments and superfast broadband, are also specially developed as media and living hubs. Large screens emerged against this ecology of urban cyber culture.

The mediascape in Seoul is largely dotted with corporate towers and outdoor advertising screens. Art Centre Nabi's networked urban screen, COMO, provides a stark contrast. Launched in 2004 and located in the city's SKT Tower and Daejeon SKT Building in the district of Ulgiro, COMO uses public programming to display artistic content. COMO is an extended media platform, consisting of three large LED screens: two located at SKT Tower, and the other at SKT Building. The first screen, Channel 1, measures 53 x 1 meters, is located as part of the exterior of the building. The second screen, Channel, 2, is located at the lobby, and features a large screen. The third screen, Channel 3, is located at the inner ceiling and column of the building. Programming changes on a monthly basis, and content is aimed at creating an interactive communication between the building and the public. Although Nabi is a privately funded art gallery, it uses the potential of its urban screens as a "live window (to present) interactive installations and networked art projects, which intervene in the commercial density of the Seoul ad-scape" (Papastergiadis, Scott & Martin, 2009, p. 49).

In August 2009, as part of the Global Fair and Festival, Nabi opened another screen in the new Songdo district of Incheon, a port city located about fifty minutes drive from Seoul. The screen is similar to Melbourne's Federation Square, located inside an open theatre plaza (*Keunullim* [Echo] Plaza) in a six-storey complex called Tomorrow City. Situated in the centre of the Songdo International Business District, Tomorrow City is an augmented building themed around the vision of future cities. Housing a public transport interchange and retail shops, the complex also includes a fun floor, an emotional bench and an i-wall, with travel master, port finder and bus maps plans at the touch of the fingertip. During the launch of the plaza, the rest of the complex was still in the final stage of construction. A temporary noodle exhibition, however, was staged in a row of shops next to the plaza. Stalls, accompanied by robot messengers as wait staff, not only sold noodles; some also had artwork showing the history and the globalization of noodles. Situated alongside the large screen, this installation further augmented the senses. Its visuality, tactility and orality replicate the new sensations of cultural somatics created by the large screen.

Songdo has been created as a global city, part of the Incheon Free Trade New Economic Zone (NEZ), where various international corporations will be located and English is the planned *lingua franca*. The development of Incheon mirrors Seoul's rapid high-tech modernization. It is hailed as the country's first NEZ, and aims to secure its position as the leading North East Asian hub into Russia and China. According to mayor Sang-Soo Ahn, Incheon has poised itself as "a leading international logistics city, an international business city, an education and information city, and a leisure and tourism city" (cited in *Global Fair and Festival*, 2009, np). Three precincts, housing advanced knowledge-based industry, international business, aviation logistics, maritime tourism and leisure, international finance and entertainment, will be completed by 2020. Companies including IBM, Morgan Stanley, Gudel, DHL, TNT, Schenker, and universities such as Macquarie, Yonsei and Seoul National, have already secured footholds there. Alongside these will be new residential developments with ubiquitous technology in the homes, on the streets, and in the shops. Planned as an ubiquitous city, it shows how a city is embedded with information technology, where all information systems are linked through technologies such as wireless networking and RFID (Radio-frequency identification) tags. Other u-amenities include "an aquarium, golf course, American-managed hospital and preparatory schools, a central park (like New York's), a system of canals (like Venice's) and pocket parks (like Savannah's), a self-described patchwork of elements gleaned from other cities" (O'connell, 2005).

The convergence of culture, business and technology has underpinned the development of creative industries spear-headed by both developed and developing countries eager to transform their weakening global economies. In 2009, at least thirty-five countries are known to have introduced these development plans (Evans, 2009). Incheon is part of South Korea's culture-led urban regeneration plan that includes other nominated cities of culture such as Busan, Kyongju, Jeonju and Gwangju. Examining how Gwangju has been stripped of its cultural heritage by massive large-scale developments, Kwang-Suk Lee (2007) argues these policies are driven by neoliberalism that creates incentives to reward foreign entrepreneurs and entice corporate investments which in turn ignore local urban ecology, bring economic exploitation and add to the rise of urban ghettos. Such "economic reductionism of culture," he suggests, commodifies cultural heritage and prevents the underprivileged from cultural participation (Lee, 2007, p. 336). His perspective differs from the study of cultural participation by Seo-Yong Kim and Hye-Sun Kim (2009) who show how arts participation and cultural experiences enhance the quality of life and subjective wellbeing. Compared to 2006, there was 1.5% increase in South Korean cultural participation in 2007-8 where at least 67.3% of the population visited art exhibits and performances (Ministry of Culture, Sports and Tourism & Korea Culture and Tourism Institute, 2008). This is a 12.5% increase compared to 2000. This increase was attributed to the quality and quantity of cultural infrastructure. In fact, the new South Korean government, setting and realizing the policy goal of becoming "one of the world top five cultural innovation industries," has recently provided full support for establishing new cultural infrastructure (Ministry of Culture, Sports and Tourism, 2009). While participation is contingent upon income and education, Kim and Kim show how cultural infrastructures are central to the frequency of public participation and quality of life. These results and debates bear implications for policy developments. In planning for culture-led urban regeneration, the quality and diversity of cultural content and cultural infrastructure are central to generating public participation and maintaining local cultural specificity. Urban screens fulfill both these demands, evident in the recent policy emphasis by the Korea Creative Content Agency (KOCCA).

Riding on the global popularity of the Korean wave, South Korea recently established KOCCA on 7 May 2009 under Article 31 of the Framework Act on Cultural Industry Promotion to promote and develop the country's cultural industry (KOCCA, 2009). This stage of creative economic development suits South Korea as it is, like advanced countries, currently in the third stage of its three-phase technology development trajectory. The three-phase model--acquisition, assimilation and improvement—was developed by Utterback and applied to South Korea by Linsu Kim (1980). More recently, Kim reformulates the third stage as "creative imitation," a transfer that generates "facsimile products with new performance features" (Kim, 2004, p. 345). The formation of KOCCA testifies to both the economic potential and cultural indigenization of the Korean wave as a set of cultural products that emerged from creative imitation.

KOCCA is South Korea's super body that covers broadcasting, cultural content, game development and digital contents business. Its aim is to increase domestic market development, job creation and export promotion. As content is national competitiveness, KOCCA's vision is to assist South Korea to become one of the top five content providers in the world. Unique to KOCCA and the Korean creative economy is the emphasis on the development of Culture Technology (CT). CT includes the planning, development and commercialization of content, as well as the promotion of digital broadcasting and visual industries. By using technologies to add value to cultural content, CT is integral to the development of urban screens. While CT emphasizes the use of public cultural resources as intellectual property, urban screens stress the embodied interaction of public participation. As Incheon's creative developments attest, the aim of CT is to add value to quality of life in the city (*Global Fair and Festival*, 2009, np). While CT promotes and exports what is purported to be the national cultural Korean "DNA" (e.g. in the international successes of *Winter Sonata* and *Daejanggeum*), urban screens potentially use the national cultural Korean "DNA" to foster civic inclusion and collective belonging.

The maintenance of cultural identity through the development of cultural policy has always underpinned cultural planning in South Korea (Yim, 2002). With a history of Japanese colonialism, national division, the Korean War and rapid modernization that have seen the influx of Western culture, South Korean cultural identity has always been seen as precarious. Whether it was an open or closed-door policy to Japanese culture in the 1940s, anti-communism and reunification in the 1950s, cultural welfarism in the 1960s and 1970s, or screen quotas in the 1980s and 1990s, regulation was always implemented to protect cultural identity. In the current era of media globalization and convergence, it is not surprising that cultural identity is also harnessed as products for cultural export as well as tools for sustainable development. Urban screens' new cultures of convergence, as well as their capacities for creative nationalism, can be seen in the programming and reception of Tomorrow City's live telematic event introduced at the start of this chapter.

THE LIVE TELEMATC EVENT: NEW MEDIA ART AND CULTURAL CITIZENSHIP

Australia and Korea have a history of established bicultural relations dating back to the 1950s when Australian soldiers fought to defeat communism during the Korean War. With shared commitments to democracy and regional security, this bilateral relationship has since flourished through trade and cultural ties. In 2008-09, the two countries exchanged over US$23 billion in trade (Austrade, 2009). South Korea is currently Australia's third largest export market, accounting for AU$19.2 billion in 2008-09, as well as the fourth biggest bilateral trading partner (Australian Department of Innovation, Industry, Science and Research of Australia, 2010). In general, education, tourism, cultural and media exchanges have increased over past decades, bolstered by strong support from the two governments, the Australian Korean community, Korean citizens in Australia and the growing Australian presence in South Korea (Australian Department of Foreign Affairs and Trade, 2010). The partnership between Melbourne and Seoul adds to the increasing levels of cultural exchange and knowledge transfer at the level of higher education and cultural institutions, and it is against this backdrop that the live telematic exchange between Nabi and Federation Square was conceived, as the first in an annual series of five.

The live telematic event consisted of several media art productions and performances. First was the screening of short films and media art by artists from both countries: Joon Kim, Joo-Myoung Song, Kye-Ryoon Han, Han-Sol Huh, Daniel Crooks and Sam Smith. Smith's short film features the iconic Melbourne tram as a motif for anchoring the relationship between transformation and transportation. Huh's work animates Korean characters (Hangeul). In these two examples, "place" is produced through interconnections between mobility and technology. Both embody local cultural specificity yet show how technology can also transcend the local and bridge the bicultural.

In Nicolas Cante and Gilles Toutevoix's piano performance, high culture is revisioned as high-tech culture through cybernetics. Teaming as Mekanik Kantatik, their performance is a "living installation" that revolves around a piano. Cante, performing as a post-human pianist, combines all the components of his instrument (keyboard, cords, wood and iron) with modern technology (computers, sensors) to create an electronic and profoundly organic music. It is unknown whether the artist controls the machine or the machine controls the artist. The song and the instruments are acoustic and robotic while the sound is mechanical and organic. In the poetry segment of the program, Korean traditional opera singer Sook-Seon Ahn mixes her classical art with video art by Min-Ha Yang. Here, convergence is evident through postmodern embodiment and the collapse of the distinctions between classical and media art. These artworks engage technology and transform place, body and culture.

Two participatory digital art performances featuring the use of SMS further highlight media convergence and demonstrate cultural citizenship. The first is "Value@ Tomorrow City," by Korea's Seung-Joon Choi. Audiences are asked to text their responses to the question: "As a member of the future city, what do you think is the most important value?" As audiences send their SMS, their responses appear, in real time, on the screen. The size of the word (or its equivalent graphic symbol) that appears on the screen changes accordingly, depending on how many similar words are sent. In other words, the more of the same values, the bigger the graphic text on the screen. This arrangement is open-ended and self-organizing. The rationale behind this is that if future city dwellers (or stake holders) increase their participation and become active in inculcating civic values, potential positive change is possible. This change will also be incremental and inclusive as more constituents are engaged in the same process, with the same values. During this event, words, or values, such as "women," "consistency," "happiness," "dignity," "relations," "party," "communication," and "lucky" are texted and dominated the screen. These words first appear as star-like sparkles, and are later recomposed, through graphics, as the roots of a tree.

The metaphor for this artwork speaks directly to cultural citizenship. Cultural citizenship refers to the process of how different groups make claims to their right to culture (Stevenson, 2003, p. 23). Cultural understandings of citizenship are concerned with how culture is made available to different population groups of the population. These discussions stress the politics of recognition and redistribution, as well as the capacity of difference and diversity to shape common culture and autonomy (Rosaldo 1999; Kymlicka 1995). In cultural policy studies, these theorizations focus on cultural participation; here cultural citizenship refers to "the capacity to participate effectively, creatively and successfully within a national culture" (Turner, 2002, p. 12). Cultural participation is considered a basic building block of cultural citizenship; it is "a proxy for citizenship in the nation-state—or; increasingly global agency. It is a way of 'measuring' the realization of cultural citizenship" (Murray, 2005, p. 40).

In this artwork, the text displays a range of common and diverse values desired of the city, from gender equality, harmonious co-existence, respect for each other, open networks of communication and relationships, to a fun place to live, work and play. They encapsulate the desire for recognition and resource allocation. In the cultural somatics of the texting, they also highlight the centrality of cultural participation. These practices—going to the event, actively engaging and transforming the art work on the screen, as well as individually and collectively voicing public and common values about the future city—are exemplary of cultural participation and cultural citizenship. They also highlight the new media culture of convergence: the use of mobile technology with situated technology; private messages as public voices; actual and virtual modes of information and communication. Together, they engender local belonging to place through embodiment and technology. Increased sociality is also evident through the incremental sizing of the text: as people see certain words transmitted on the screen, they also texted similar words, thereby increasing the potential of common values and culture, through the screen as an audience, and the space as a public sphere.

The second participatory SMS artwork builds on the first, to extend cultural citizenship as transnational. Entitled "SMS-Origins" by Australia's Leon Cmielewski, Josephine Starrs, and Adam Hinshaw, this work instructs the audience to "sms [their] family origins." Participants send text messages of their own and their parents' places of birth. Linked curved vectors are added to a public map on the screen that updates in real-time as it receives texts. At the end of this artwork, multiple curved vectors linked Korea and Australia to all parts of the world, such as China, Italy, Greece, Singapore, the USA and UK. Here, the similar practices of cultural citizenship experienced in the earlier event are further encoded with the potential for globalizing the local and transforming citizenship. Where citizenship refers to the formal processes of rights and representation, this networked globality shows how citizenship is transformed by literal migration and mobile technology. Undoubtedly, migration and technology have impacted nation-states and the territoriality enshrined to citizenship. This artwork makes explicit the changing cultural dimensions of citizenship to notions such as migration, belonging and birthright. Here, not only is citizenship transcultural, attachment and solidarity are also global.

These works show how urban screens are sites of transnationality and transculturality enabled by the media convergence. The concluding act, the media performance of AntiVJ, brings earlier discussions of place, body, culture and technology together as a media wall and an augmented space that literally inscribes the future city and its visions for cultural citizenship. AntiVJ is a visual label initiated by media performance group of European artists fascinated by the use of light and its influence on our perception of space. AntiVJ uses video projection to create pieces that are at the crossroads of installation work, live performance, graphic design and architecture, bringing two-dimensional images to tridimensional world. In this act, the plaza's lights are switched off, and a light show, showing architectural visions of the future city, is projected onto the outer walls of the six-storey building, enveloping the audience sitting at the plaza below. Here, time and space are abstracted to convey the illusion of a fully built future city populated by audience-citizens. Surrounded by the tridimensional images, the audience-citizens are placed inside the object and literally experienced the immersion of technology, media and space, and the hypermedality of sociality. Not only is Tomorrow City a media infrastructure, it is also momentarily, a live cityscape of moving images and future potentialities. In the following section, this chapter considers how audiences engage with the event, and evaluates their transcultural practices of consumption.

THE LIVE TELEMATIC EVENT: AUDIENCES AND TRANSCULTURAL CONSUMPTION

This section is based on survey analysis of collected questionnaires among audiences at the event venues. 57 questionnaires and 14 questionnaires were collected in Songdo and Melbourne respectively (see Appendix 1). The participants are randomly selected at the event venue, most of whom are major urban citizens in the 20-40 age group. Due to the considerably smaller number of the participants from Melbourne, comparative research is not employed. Instead, this section evaluates the Songdo audience group, focusing particularly on the ways the urban screen experience constitutes transcultural citizenship. The following analysis is limited by the broad context of the event, which focuses particularly on the 'digital media experiences in a new urban city'. This means that our respondents are mostly urban media consumers, especially young people who are attracted by the 'cool factor' of the activity. Non-urban and older audiences did not attend nor participate. These limitations highlight the myth of digital democraticisation and the inequitable access to the types of new cultural participation promulgated by the digital revolution and the new creative city. Transcultural citizenship refers to the way cultural citizenship has been transformed by the new consumption lifestyle practices of cross-border flows. In current contemporary culture, lifestyles, identities and citizenships are increasingly constructed by choice rather than tradition (Stevenson, 2003, p. 14). As Jan Pakulski suggests, cultural citizenship possesses the capacity to reorganize lifestyle and consumption practices (Pakulski, 1997, p. 80). Similarly, social practices are "constantly examined and reformed in the light of incoming information about those very practices, thus constitutively altering their character" (Giddens, 1990, p. 38). As such, it is important to look at how, through media art consumption practices, Songdo audiences express new lifestyles and, to some extent, cultivate transcultural citizenship. As Toby Miller posits, the discourse of cultural citizenship is developed "in response to the great waves of cross-class migration of the past fifty years and an increasingly mobile middle-class culture industry workforce..." (Miller, 2001, p. 2). Citizenship is no longer based on soil or blood; rather, it is founded on some variant of those qualities in connection with culture and the capitalist labour market (Miller, 2001, p. 4). This kind of cultural citizenship suggests "there is no longer any stability in the points of origin, no finality in the points of destination, and no necessary coincidence between social and national identities" (Cohen, 2008, p. 174). Stevenson's approach to cosmopolitanism in his study of cultural citizenship reflects this:

Cosmopolitanism is a way of viewing the world that among other things dispenses with national exclusivity, dichotomous forms of gendered and racial thinking and rigid separations between culture and nature. Such a sensibility would be open to the new spaces of political and ethical engagement that seeks to appreciate the ways in which humanity is mixed into intercultural ways of life (2003, p. 5).

Transnational media reinforces the cosmopolitan ways of life that enable the transmisson of cultural content across borders, opening up the possibility of what J. B. Thompson (1995) has described as 'intimacy at a distance'. As Nick Stevenson explains, "[f]or many, the media's capacity to enhance *connexity* holds out the possibility of enhancing surface forms of cosmopolitanism" (2003, p. 103). Here, *connexity* refers to the capacity of media technologies to create a cosmopolitan mentality (Mulgan, 1997). The development of media, digital media in particular, has diversified cosmopolitan ways of living that has enabled the building of links, solidarities and connections with others. As such, in this new media empowered global era, people with different cultural backgrounds share a cosmopolitan mentality as a result of transnationally flowing cultures, information, ideas and images. Such cosmopolitan media consumption practices have significantly accelerated the construction of transcultural citizenship. The analysis from the research conducted in Songdo has identified that audiences potentially cultivate transcultural citizenship through media art experiences. The three aspects of such a process are reflected in the questionnaire responses: (1) the transcultural space of Songdo, Tomorrow City and the urban screen; (2) the transcultural consumption of media art, and; (3) transcultural participatory interaction.

Transcultural Space

The first aspect observed from the questionnaires is the way audiences engage the transculturalism of the space. A transcultural space is produced when different cultures, shaped by overlapping socio-politico-cultural relationships, meet and are transformed through collective and individual cultural practices. While Songdo audiences are acutely aware of the top-down urban regeneration of Incheon and its capacity to engage transculturalism and globality, their responses to the outdoor space of Tomorrow City's open theatre show how a transcultural space is also created through their own specific and individual experiences of media consumption.

It was wonderful as [various genres of art works] got out of the galleries and carried out global communication.

Global.

Different scale and many people are participating here.

Digital art exhibitions in museums were monodirectional within a limited space. But tonite's event was held in an outdoor space where we could communicate with other people through art.

The previous participation is more like communication with limited people in an limited space, today's one is about exchanging experiences with many people in an open space.

Clearly, the newness of the planned global city, as a North East Asian hub, has transformed their experiences of the artwork and capacity to feel connected globally:

As it's an outdoor event, the scale is much bigger [than indoor ones] and I was impressed by the overwhelming screen images and sound ... I felt connected [to Melbourne].

It was wonderful as, unlike gallery exhibitions, the event was held in outdoor space which, I believe, stimulates active global communication.

It is interesting and fun to experience new media artworks in a public space. Especially, it is mean-

ingful since such digital media, without employing any extra mechanical device, bring us new global experiences within an existing space which we've already had.

Indeed, many participants expressed their enchantment towards the media artworks on the big outdoor screen:

Was overwhelmed by the freshness and the sublime which I haven't experienced before.

Amazing. Can't stop admiration.

[O]ur emotions had become maximized because the auditory stimuli were transformed into the visual.

The first impression was new and exciting. The visual performances which I've never experienced before stimulated all my senses.

I saw so many new and wonderful things.

Epochal and wonderful.

In the responses above, the experiences of enchantment and shock—as exemplary new experiences of modernity—are produced through the size and content of the screen. The size not only overwhelms the senses; these senses are experienced through the contact zone of the content—its capacity to bridge distance, and its collapse between analogue and digital. These mediating characteristics are features of the high modernity envisioned of Songdo as a planned global city, and have clearly impacted the audiences' experience of the screen.

The open-air atmosphere also amplifies the audiences' experience of exchange with people outside the country. Many audiences were able to cultivate a transnational connection with Melbourne:

It seems like a process to become one.

A sense of contact....

I felt difference, but I felt a sense of connection....

The urban screen as a transcultural space is reinforced through the audiences' direct participation in the artworks featuring SMS. As introduced in section 2, "Value @ Tomorrow City" requires audiences from the both cities to send values they wish to have, via SMS texting, for a future city. Of significance here is how the words that appeared on the screen are intermingled with each other, and un-identifiable which city they were sent from. The values sent from Songdo and Melbourne have combined to become a universal set of values for a future city. A transcultural space is formed when a mixture of values from two different cities is projected on the screen.

The second media art project, "SMS-Origins," more explicitly represents how a big screen can be a transcultural arena. At the beginning, it seemed the project was not suitable to be performed in the South Korean context as there would not have been many South Koreans born outside Korea, compared to Australia where one in four is born outside the country. For example, in the case of many ordinary Koreans, "my birthplace-my mom's-my dad's" will most probably be rendered through SMS as something like "Kor-Kor-Kor." This means there will not be many curved vectors across the world map, and is likely that the Korea represented on the screen will reflect a less culturally diverse society. However, a few minutes after the performance began, several curved vectors from South Korea to the rest of the world slowly began to appear on the screen. Songdo audiences, like Melbourne's, also claim places such as USA and France as places of birth and origin. In this segment, the screen brings the curved vectors from Australia and South Korea together, becoming a transcultural space of intersection for Songdo and Melbourne audiences.

Transcultural Consumption of Media Art

Another notable aspect from the questionnaire responses is the willing and appreciative consumption of foreign art. Mike Featherstone states that the "dual focus on a life of aesthetic consumption and the need to form life into an aesthetically pleasing whole on the part of artistic and intellectual countercultures should be related to the development of mass consumption in general and the pursuit of new tastes and sensations and the construction of distinctive lifestyles which has become central to consumer culture" (Featherstone, 2007, p. 66). The consumption of media art by Songdo audiences replicate today's consumer culture where consumers are continuously searching for new tastes and sensations that blur the barriers between art and everyday life. Such postmodern lifestyling resonates strongly with contemporary youth culture, evident from the dominant group of respondents who are in the twenties. Most of the respondents, for example, are unanimous in their appreciation of European artists such as the AntiVJ collective and Mekanik Kantatik.

It's so great to have such a fresh shock.

I like the both [AntiVJ and Mekanik Kantatik]. Especially I really liked the piano performance.

AntiVJ rocks!

They are also active consumers who are keen to travel to Songdo to explore foreign artworks to satiate their desire for new tastes and sensations. It is notable that most of the audiences were from outside Songdo as the city was still under the development at the time of the event. Responding to the question, "Why did you come here?," 61% of the respondents expressed their active intention to consume "the media art (event)." These respondents stressed they had particularly come "to experience the event" and "to see the performances." Many especially came to see "global" or "foreign" media art works:

[I came] to witness the development of global media art.

I wanted to see the performances of globally well-known artists. I was curious what kind of feelings, emotions and expressions they have.

These respondents have clearly come to see the event because they wanted to experience feelings, emotions and expressions of people from outside Korea, i.e. Australia and Europe. Such active consumption of foreign or global media artworks shows a willingness to cross the cultural borders through which transcultural citizenship might be engendered:

I came because I was told that Korea and Melbourne will be connected with each other through digital art. Also I was expecting to see the current flows [trends] of world art.

I was curious how Melbourne and Korea would be introduced to each other through art as a medium. I wanted to share various art [forms] and cultures with other people [outside Korea].

In particular, these two responses above demonstrate how Songdo audiences voluntarily come to this event expecting to engage with and share in the digital art experience with audiences in Melbourne's Federation Square. Another participant further commented that "I felt like I was with [the rest of] the world"; another stated, "I think it's a process of we are all becoming connected with each other." This awareness and/or a feeling of transcultural engagement and connectedness demonstrate how active media art consumption practices function as a potential site for engaging global and transcultural citizenship.

Transcultural citizenship is also cultivated from the way audiences cross cultural borders by embracing the foreign media artworks. According to them, these artworks are "new and fresh" and "strange and marvellous." Such opinions reflect cosmopolitan ways of thinking and living that are concerned with the transgression of boundaries and markers between the familiar "I" and strange "Others" (Stevenson, 2003, p. 5). Responding to the question, "What did you think of the creative works on the large screen tonight?," 75% of the respondents extolled: "overwhelming, amazing, shockingly fresh, innovative, original, new and unique." Songdo audiences are clearly keen to consume "new foreign artworks" that will potentially enable them to transcend pre-defined cultural boundaries. Cultural migrations occur not only between Songdo audiences and artworks transmitted from Melbourne, but also between Songdo audiences and AntiVJ's tridimensional installation. The digital mapping techniques and holographic illusions offered by AntiVJ are encountered in these ways:

Very impressive artwork which led the audiences to experience the space which seemed totally different from the one we just had had [before the performance began].

Mainly, I was deeply moved by AntiVJ's performance. I could empathize with the messages the artists wanted to deliver [to the audiences] in this digital era.

AntiVJ has created an alternative transcultural space by crossing different art forms as well as the boundary between Korean audiences and European art(ists). Many respondents had especially come to see AntiVJ's project, praising it as "amazing," "overwhelming," "fantastic," "the best," "my favourite," and "impressive." As one respondent succinctly professed:

I've got shocked by this new astonishing piece of artwork. This is my first time [to experience such an art performance] and it was amazing as the building was incorporated with the performance

By positively embracing the foreign art of AntiVJ, audiences potentially reinforce the ties of cultural experiences to the global through which they can deterritorialize themselves. As John Tomlinson suggests, modern deterritorialized localities has become increasingly penetrated by globalizing forces (2002, p. 152-53). Through actively immersing in AntiVJ's works, audiences have in effect become empowered by the increasingly mobile cosmopolitan ways of living and thinking. The willingness to embrace AntiVJ's media project reflects the audiences' ability to enjoy difference and a healthy respect for otherness, a form of cultural literacy central to transcultural citizenship.

Transcultural Participatory Interactions

Participatory interactions have also enabled Songdo audiences to potentially acquire transcultural citizenship. Responding to the question, "Did you engage with the works via SMS?," 79% answered "yes." 84% revealed that their participation engagement can be attributed as ranging from medium to high level. As suggested earlier, the urban screen has become a transcultural space when SMS text messages from the both cities are intermingled with each other. Underpinning this media practice is active transcultural participation. Active transcultural participation is evident in: mass participation and direct participation.

Many respondents commented how participating within a mass public amplified the experience:

The previous participation is more likely communicating with a limited number of people in a closed space, today's one is about exchanging experiences with many people in an open space – more fun.

The interactive artworks were pretty much like concerts where singers and mass audiences communicate with each other through singing and cheering.

The above suggests respondents have built some sense of connection with other people in a public space. As suggested in section 2, these participants, by sending their messages about the values they wish to have in Tomorrow City, have created a public sphere of critical communication (Habermas, 1998). Here, the public sphere is not only national, it is also transnational and transcultural.

Direct participation is also evident during the SMS texting performances:

It was very new that we could directly participate [in art performance] through mobile phones. My previous experience with media art was one-dimensional, where the screen images were changed responding to my movements.

It was fascinating to see that [I] could directly take part in the artworks through my texts.

It was fun. People in Melbourne and I could share each other's words [values].

Significant here is how participants "share their words [values]" with people in Melbourne which represents a way of sharing critical thoughts and ideas (cultural literacy) within a transnational public arena. Direct participation enhances international and cross-cultural dialogues, and constructed solidarity at a distance. Responding to the question, "Did you feel a sense of connection with people in Australia through the event tonight?," 39 of out 57 respondents said "yes" while 16 replied "no." One respondent remarked:

I felt very close to them as if I couldn't feel the physical distance. It's hard to say that a sense of connection has been created all of a sudden. However, I feel that we [Koreans] are a little bit connected to Australia's art and media through texting and visual screening.

These comments suggest media practices of participatory interactions reinforce a sense of community between people from different cultural backgrounds, which minimize their cultural distance while maximizing communicative exchanges. In other words, participatory interactions have created a transcultural public sphere where the binary oppositions between the East and the West, and the local and the global have been broken. It shows how participating in such a media event offers an opportunity to construct a global public sphere through which connections with others who are at a distance can be built.

CONCLUSION

This chapter began by suggesting that urban screens are contact zones of mediascapes, social belonging and identities. Contact zones are traditionally conceived as "social spaces where cultures meet, clash, and grapple with each other" (Pratt 1991, p. 33). In contemporary visual culture, the urban screen has become the exemplary site for the contact zone. As this chapter has shown, urban screens in South Korea and Australia are interfaces for mediating new creative industries and communities, both nationally and internationally. They are contact points between industries, governments and users; they mediate new visualities among viewers and audiences, and; by connecting disparate data including virtual, actual and cross-cultural, they create transcultural communities.

Using the logics of media convergence and creative cultural planning, this chapter has shown how the urban screen has emerged as a result of digital revolution and culture-led urban regeneration. As media architecture, urban screens combine information and infrastructure to produce a new space of augmentation. Such convergence has been facilitated by mediascapes, with global media companies (and governments) building new platforms, and through the circulation of global images. Augmented space has clearly redefined urban morphology, social relations and communication.

In Australia, Melbourne's urban screen at Federation Square has emerged against the policy shift in the 1990s that saw the country transforming itself to becoming a creative nation. What was initially controversial media architecture has been quickly embraced as a popular tourist destination and local meeting hub. In South Korea, a broadbanded cityscape in the early 2000s has also allowed urban screens to become public amenities with the potential to build engaged communities. Central to these sites is the increased capacity for new sociality and the quality of mediated communication. The interactivity of the screen, and its attendant performative mediality, has created a cultural somatics of embodied participation. The convergence of place, body and technology has enabled new cultures of citizenship that have redefined social identities and belonging.

Further using the live telematic event between Songdo and Melbourne, this chapter has shown how transculturalization is the key feature of the contact zone. The exchange between the two screens in Melbourne and Songdo has facilitated transcultural consumption. Transcultural consumption is evident in the ways users and audiences react to the space and respond to the content. The enchantment created by the high modernity of the large screen has transformed locality into a new transcultural space of interactive and global communication. The embrace of foreign media art reflects the new cosmopolitan lifestyles desired by the predominant Gen X and Gen Y audiences. In their direct and mass participatory interactions with the artworks, a transcultural and transnational public sphere is also potentially created.

In this study however, the emancipating qualities of transcultural experience are largely limited to the young—Gen X and Gen Y—urban media consumers. The specificity of the technology, the range of its screen content, and the particularity of this form of telematics, together with the geographical locality of the event, have inevitably excluded participants from non-urban and older backgrounds. Even among the questionnaire respondents, who are young urban citizens, some express how they feel excluded and alienated by the 'complex' modern or postmodern digital artworks. A few even complain that they do not see any reason for exhibiting such "strange and abstract stuff". Thus, other occasions and events must be further investigated to find a balanced theoretical framework to explain this realised potential of creating contact zones through transcultural urban screens.

While live telematics is transient and can be considered temporarily utopic, this event makes explicit the contact zone of the urban screen as a site of contestation. Its claims to urban revitalisation have masked the neoliberal agenda, and the politics of bicultural collaboration has also exposed the uneven geographies of cultural capital and resource allocation. In Melbourne, this event was scheduled as part of the calendrical programming of Federation Square, as one in the line-up of many, and competing against other activities taking place around the square. In Songdo, this event was widely promoted as a prestigious occasion that opened a new landmark in a new city, and was attended by international guests and politicians. The untranslated English-language poems performed by Melbourne artists were equally as alienating to Songdo audiences as the spectacle on the Melbourne screen, located in a different time zone three hours ahead late into the wintry night, without its community of audience and hive of festivities. The urban screen, ultimately, calls to task the architecture of media futures, as a site for mediating conflict and cohesion.

REFERENCES

Auerbach, A. (2006, February). Interpreting urban screens. *First Monday, 4*. Retrieved October 20, 2009, from http://www.uic.edu/htbin/cgiwrap/bin/ojs /index.php/fm/rt/printerFriendly/1546/1461

Austrade (2009). *Republic of Korea fact sheet*. Retrieved March 11, 2010, from http://www.dfat.gov.au/geo/fs/rkor.pdf

Australian Department of Foreign Affairs and Trade. (2010). *Republic of Korea country brief.* Retrieved March 11, 2010, from http://www.dfat.gov.au/geo/rok/brief_index.html

Australian Department of Innovation. Industry, Science and Research (2010). *Australia's exports fact sheet*. Retrieved March 11, 2010, from http://www.innovation.gov.au/Section/ AboutDIISR/FactSheets/Pages/Australia%27sExportsFactSheet.aspx

Balfour, F. (2008, February 11). Catching the eye of China's elite. *Business Week, 470*(4070), 55-56.

Buhlmann, V. (2006, February). Intelligent skin: Real virtual. *First Monday, 4*. Retrieved October 20, 2009, from http://www.uic.edu/htbin/cgiwrap/ bin/ojs/index.php/fm/rt/printerFriendly/1554/1469

Cohen, R. (2008). *Global diasporas: An introduction* (2nd ed.). London: Routledge.

Department of Communication and the Arts. (1994). *Creative nation: commonwealth cultural policy, October 1994*. Retrieved February 26, 2005, from http://www.nla.gov.au/creative. nation/contents.html

Evans, G. (2009). Creative Cities, Creative Spaces and Urban Policy. *Urban Studies (Edinburgh, Scotland), 46*(5&6), 1003–1040. doi:10.1177/0042098009103853

Featherstone, M. (2007). *Consumer culture and postmodernism*. London: SAGE Publications.

Giddens, A. (1990). *The consequences of modernity*. Cambridge, UK: Polity Press.

Global Fair and Festival. (2009). *Incheon Metropolitan City*. South Korea: Incheon.

Habermas, J. (1989). *The structural transformation of the public sphere*. Cambridge, UK: Polity Press.

Kim, L. (1980). Stages of Development of Industrial technology in a developing country: a model. *Research Policy, 9*, 254–277. doi:10.1016/0048-7333(80)90003-7

Kim, L. (2004). The Multifaceted Evolution of Korean Technological Capabilities and its Implications for Contemporary. *Oxford Development Studies, 32*(3), 341–363. doi:10.1080/1360081042000260566

Kim, S., & Kim, H. (2009). Does Cultural Capital Matter? Cultural Divide and Quality of Life. *Social Indicators Research, 93*, 293–313. doi:10.1007/s11205-008-9318-4

Korea Creative Content Agency (KOCCA). (2009). *Introduction*. Retrieved September 1, 2009, from http://www.koreacontent.org/ weben/etc/kocca.jsp

Kymlicka, W. (1995). *Multicultural citizenship: A liberal theory of minority rights*. Oxford, UK: Oxford University Press.

Lee, K. (2007). Questioning A Neoliberal Urban Regeneration Policy: The rhetoric of "cities of culture" and the city of Gwangju, Korea. *International Journal of Cultural Policy, 13*(4), 335–347. doi:10.1080/10286630701683201

Lester, P. M. (2006, February). Urban screens: The beginning of a universal visual culture. *First Monday, 4*. Retrieved October 20, 2009, from http://www.uic.edu/htbin/cgiwrap/bin/ ojs/index.php/fm/rt/printerFriendly/1543/1458

MacCarthur, J., Crist, G., Hartoonian, G., & Stanhope, Z. (2003). Federation square: Melbourne's most significant set of new public buildings and spaces is now open. *Architecture Australia*, *92*(2), 46.

MacColl, I., & Richardson, I. (2008). A cultural somatics of mobile media and urban screens: Wiffiti and the IWALL prototype. *Journal of Urban Technology*, *15*(3), 99–116. doi:10.1080/10630730802678002

Manovich, L. (2006, February). The poetics of urban media surfaces. *First Monday*, *4*. Retrieved October 20, 2009, from http://www.uic.edu/htbin/cgiwrap/bin/ ojs/index.php/fm/rt/printerFriendly/1545/

Mary Louise Pratt, M. L. (1991). Arts of the Contact Zone. *Profession*, *91*, 33–40.

Media, F. (2009, March 23). *Focus media reports fourth quarter and full year 2008 results*. Retrieved October 20, 2009, from http://ir.focusmedia.cn/phoenix.zhtml?c= 190067&p=irol-news Article_print&ID=1268983&highlight

Miller, T. (2001). Introducing… cultural citizenship. *Social Text*, *19*(4), 1–5. doi:10.1215/01642472-19-4_69-1

Miller, T. (2006). *Cultural citizenship: Cosmopolitanism, consumerism, and television in a neoliberal age*. Philadelphia: Temple University Press.

Ministry of Culture. Sports and Tourism & Korean Culture and Tourism Institute. (2008). *Survey report on cultural enjoyment*. Retrieved March 15, 2010, from http://www.mcst.go.kr/web/dataCourt /statisticsData/statisticsView.jsp

Ministry of Culture. Sports and Tourism, South Korea. (2009). *Cultural Industry Report*. Retrieved March 15, 2010. From http://www.mcst.go.kr/web/data Court/research/researchView.jsp

Mulgan, G. (1997). *Connexity: How to live in a connected world*. London: Chatto and Windus.

Murray, C. (2005). Cultural participation: A fuzzy cultural policy paradigm. In Andrew, C., Gattinger, M., Jeannotte, M. S., & Straw, W. (Eds.), *Accounting for Culture: Thinking through Cultural Citizenship* (pp. 32–56). Ottawa: University of Ottawa Press.

O'connell, P. L. (2005, October 5). Korea's high-tech utopia, where everything is observed. *New York Times*. Retrieved November 15, 2009, from http://www.nytimes.com/2005/10/05/t echnology/techspecial/05oconnell.html

OECD. (2008). *Households with broadband access, 2004-2008. Percentage of all households*. Retrieved March 6, 2010, from http://www.oecd.org/sti/ict/broadband

OECD. (2009). Broadband penetration rates [from http://www.oecd.org/sti/ict/broadband]. *Top (Madrid)*, *5*, Retrieved March 6, 2010.

Pakulski, J. (1997). Cultural citizenship. *Citizenship Studies*, *1*(1), 73–86. doi:10.1080/13621029708420648

Papastergiadis, N., McQuire, S., & Martin, M. (2009). Large Screens, Cultural Creativity and the Global City. In Lim, W. S. (Ed.), *Collection of Essays on Asian Design Cultures* (pp. 42–53). Singapore: Select Books.

Radbourne, J. (1997). Creative nation – A policy for leaders or followers? An evaluation of Australia's 1994 cultural policy statement. *Journal of Arts Management. Law and Society*, *26*(4), 271–284.

Rice, C. (2005). Experience and criticality: Returning to Federation Square. *The Journal of Architecture*, *10*(3), 323–333. doi:10.1080/13602360500162253

Rosaldo, R. (1999). Cultural citizenship, inequality, and multiculturalism. In Torres, R. D., Miron, L. F., & Inda, J. X. (Eds.), *Race, identity and citizenship: A reader* (pp. 79–100). Cambridge, MA: Blackwell.

Rundle, G. (2002, August 27). Building the past. *The Bulletin with Newsweek, 120*(6339), 13.

Schieck, A. F. g. (2006, February). Towards an integrated architectural media space. *First Monday, 4*. Retrieved October 20, 2009, from http://www.uic.edu/htbin/cgiwrap/bin/o js/index.php/fm/rt/printerFriendly/1550/1465

Slaatta, T. (2006, February). Urban screens: Towards the convergence of architecture and audiovisual media. *First Monday, 4*. Retrieved October 20, 2009, from http://www.uic.edu/htbin/cgiwrap/bin/ojs/index.php/fm/rt/printerFriendly/1549/1464

Square, F. (2009). *Fed square annual report – June 2009*. Melbourne, Australia: Fed Square Pty Ltd.

Steinberg, B. (2008, August 11). Big 3 TV nets follow consumers out of the living room; ABC, NBC and CBS set up screens at grocery stores, gas stations, malls. *Advertising Age*, *79*(41), 4.

Stevenson, D. (2000). *Arts and organisation: making Australian cultural policy*. St Lucia, Australia: University of Queensland Press.

Stevenson, N. (2003). *Cultural citizenship: cosmopolitan questions*. Maidenhead, UK: Open University Press.

Thompson, J. B. (1995). *The media and modernity: A social theory of the media*. Cambridge, UK: Polity Press.

Tomlinson, J. (2002). Interests and identities in cosmopolitan politics. In Vertovec, S., & Cohen, R. (Eds.), *Conceiving cosmopolitanism: Theory, context, and practice* (pp. 240–253). Oxford, UK: Oxford University Press.

Townsend, A. (2007). Seoul: birth of a broadband metropolis. *Environment and Planning. B, Planning & Design*, *34*, 396–413. doi:10.1068/b32036t

Turner, B. S. (2002). Outline of a general theory of cultural citizenship. In Stevenson, N. (Ed.), *Culture and citizenship* (pp. 11–32). London: Sage.

Victoria, A. (2003). *Creative Capacity +*. Retrieved June 20, 2008, from http://www.arts.vic.gov.au/content/Public/About_Us/Who_We_Are/Policy_and_Planning/Creative_Capacity_.aspx

Yim, H. (2002). Cultural Identity and Cultural Policy in South Korea. *International Journal of Cultural Policy*, *8*(1), 37–48. doi:10.1080/10286630290032422

ADDITIONAL READING

Alasuutari (Ed.), *Rethinking the media audience: The new agenda* (pp. 1-21). London, Thousand Oaks and New Delhi: Sage Publications.

Anheier, H., & Isar, Y. R. (Eds.). (2008). *The cultural economy*. London: SAGE.

Brennan, T. (1997). *At home in the world: Cosmopolitanism now*. Cambridge: Harvard University Press.

Briggs, L. (2008). Transnationalism: A Category of analysis. *American Quarterly*, *60*(3), 625–648. doi:10.1353/aq.0.0038

Donald, S. H. (2008). *Branding cities: cosmopolitanism, parochialism, and social change*. Hoboken: Taylor & Francis.

Dyson, F. (2009). *Sounding new media: immersion and embodiment in the arts and culture*. Berkeley: University of California Press.

Flew, T. (2008). *New media: an introduction* (3rd ed. ed.). South Melbourne: Oxford University Press.

Florida, R. L. (2006). *Who's your city?: how the creative economy is making where to live the most important decision of your life*. New York: Basic Books.

Hannerz, U. (1996). *Transnational connections: Culture, people, places*. London, New York: Routledge.

Hartley, J. (2009). *The uses of digital literacy. St Lucia, Qld.* UQP.

Jenkins, H. (2006). *Convergence culture: Where old and new media collide.* New York, London: New York University Press.

Morley, D., & Robins, K. (1995). *Spaces of identity: Global media, electronic landscapes, and cultural boundaries.* London, New York: Routledge. doi:10.4324/9780203422977

Nightingale, V. A. (2007). *New media worlds: challenges for convergence.* South Melbourne, Vic.: Oxford University Press.

Nowicka, M., & Rovisco, M. (2009). *Cosmopolitanism in practice. Farnham and Burlington.* Ashgate.

Nye, J. S. (2004). *Soft power: The means to success in world politics.* New York: Public Affairs.

Pardue, D. (2007). Hip hop as pedagogy: A look into "Heaven" and "Soul" in Sao Paulo, Brazil. *Anthropological Quarterly, 80*(3), 673–709. doi:10.1353/anq.2007.0044

Pieterse, J. N. (2004). *Globalization and culture: Global melange. Lanham.* Boulder, New York, Toronto, Oxford: Rowman & Littlefield Publishers, Inc.

Poster, M. (2008). Global media and culture. *New Literary History, 39*(3), 685–703. doi:10.1353/nlh.0.0039

Smart, A., & Smart, J. (2003). Urbanization and the global perspective. *Annual Review of Anthropology, 32*, 263–285. doi:10.1146/annurev.anthro.32.061002.093445

Welsch, W. (1999). Transculturality: The puzzling form of cultures today. In Featherstone, M., & Lash, S. (Eds.), *Spaces of culture: city, nation, world* (pp. 194–213). London, Thousand Oaks, New Delhi: Sage Publications.

KEY TERMS AND DEFINITIONS

Audience Reception Studies: Audience reception studies emphasize the varied receptions of a text—which can be a book, a film, or other creative works—among audiences. This approach focuses on the scope of "negotiation" of the audiences and the "particularity" of each socio-cultural context. This means that audiences do not passively accept a text; instead, they interpret the meanings of the text based on their individual and socio-cultural backgrounds and life experiences. In short, the meaning of a text is not inherent within the text itself, but is created within the unique and dynamic relationships between the text and each individual audience.

Cultural Citizenship: Cultural citizenship generally refers to the process of how different groups make claims to their right to culture. It is a site for exposing the cultural politics of recognition and resource allocation. It is also a site for the making of common culture. Increasingly, it has reorganized lifestyles and also constructed alternative identities.

Cultural Policy: Cultural policy refers to the policies that shaped the governance of culture. Following the Foucaldian thesis, the term policy is understood is a more general sense, rather than the narrow definitions of statements and plans on culture by the government. In the more general sense, policies are understood as the discourses, practices and institutions for the social management of the population.

Transcultural Consumption: Transcultural consumption refers to a practice of global consumption which transcends the borders of traditional, national, racial, gender, and professional cultures. It highlights: breaking or minimizing the limitations and cultural barriers; and broadening and varying the cross-cultural consumption experiences of global consumers, which are hugely mobilized by new media technologies.

Transculturalism: Transculturalism refers to how cultures are shaped by continual interactions, by overlapping socio-politico-cultural relationships, and transformed through specific and individual cultural practices. The term describes how any culture or any society do not remain "pure" and never were "pure" in the first place, but is destined to change through the flows and intermingling of cultures over time. It emphasizes the multifaceted contextual and individual interactions to cultural experiences, rather than homogenization approach to culture.

Urban Regeneration: Urban regeneration covers a wide range of urban transformations, from parks, public housing, private conservation and large-scale government developments. It includes heritage preservation, ecological design, environmental renewal and the creation of new districts of residence and commerce. Culture-led urban regeneration has risen prominently in recent years as a result of government cultural policies that promote the use of culture as a driver for economic growth. Culture-led urban revival reflects the increased culturalisation of the contemporary urban economy and its multiple uses in leveraging social, economic and political inclusion.

Urban Screen: Urban screens range in size, resolution, use and location. They are both public amenities and cinematic screens deployed in public spaces in urban contexts. They make use of innovative LED and LCD technologies, and can be found on projection broads, information terminals and intelligent buildings. They are used for entertainment, live television broadcasting, marketing and exhibiting new media digital art.

APPENDIX 1

Table 1.

	Songdo, Korea	Melbourne, Australia
Age	- 20's: 36 - 30's: 12 - 40's: 5	- 18 – 24: 2 - 25 – 34: 4 - 35 – 50: 5 - 50 – 65: 2
Gender	- Male: 24 - Female: 30	- Male: 1 - Female: 8
Ethnic background	- Korea: 50 - Other Countries China: 2 India: 2 Algeria: 1 USA: 1	- Korea: 0 - Other Countries: India: 1 France: 1 Anglo-Australian: 3 Caucasian: 1 Greek-Australian: 1 African: 1 Sydney: 1 Australian: 2 Italian: 1
Visitor or resident	- Visitor: 2 - Resident: 49	- Visitor: 1, - Resident: 9 - Student: 1

Chapter 3

Sharing Bites on Global Screens: The Emergence of Snack Culture

Hudson Moura
Simon Fraser University, Canada

ABSTRACT

Snack culture is the new phenomenon that shrinks media cultural products and can be easily shared on social networks of the Internet. Thus, it can be consumed in a reduced amount of time circulating instantly all over the globe. These tiny and snappy materials are changing people's habits, transforming passive viewers into active users, and promoting equal access to all, and requiring no professional skills. Viewers now can also produce cultural and social content in widespread virtual communities (based on the Web 2.0) that are increasingly interactive. This chapter presents and analyses a variety of media snacks that form and circulate as snack culture; it also elucidates some of those current changes that are shaping today's relationship between society and media.

INTRODUCTION

In April 2007 I was interviewed[1] by *Folha de São Paulo*, one of the most influential Brazilian newspapers, regarding a new phenomenon on the Internet called *snack culture*. The term appeared the month before in a dossier entitled *Snack Attack!* in the celebrated magazine *Wired*, an American publication specializing in new technologies and electronic games. Pop culture—movies, TV shows, songs, games—according to their *minifesto* was now coming packaged like cookies or chips, in bite-sized bits for high-speed consumption. Snack culture or *snack media*, as I prefer to call it, involves making small "portions" of information and entertainment available in a great variety of forms and for multiple uses. The formation and consumption of these bites has become a habit, as much as is the time that we individually spend on blogs and on viewing streaming videos.

Sharing content over the Internet is a new and fast way to distribute and to consume culture and media via *snacks*. Those tiny and snappy media products are easier to create, to send, to receive, to

DOI: 10.4018/978-1-60960-037-2.ch003

Copyright © 2011, IGI Global. Copying or distributing in print or electronic forms without written permission of IGI Global is prohibited.

collect, or to consume instantly at any computer in any part of the world. Microblogs, such as Twitter and Jaiku, and social networks, such as Facebook and Orkut, are taking place of traditional media and imposing a new form of networking in global society. In the last decade, the superabundance of cultural and media productions (e.g., film, video, music, texts) has proliferated in the digital global environment. We could ask, has the globalization of culture, media, and art brought us any closer to understanding the particular and varied experiences of people in diverse parts of the world? At the very least, viewers have created a new form of communication where they comment and respond to each other via screens, where, in many cases, translation or professional audiovisual skills are not required. Thus, viewers are forming global, multiethnic cultural, media, and artistic production scenery on the World Wide Web.

Snacks can be original productions or a recycling and scattering (part, piece, or portion) of any media such as films, songs, or books. They don't require the viewer to spend as much time in front of just one single product on screen; instead viewers can enjoy a greater amount of entertainment and a greater number of diversified sources of information. Thus, like consuming any snack, the viewer can become addicted. In accumulating an impressive record of sharing and consuming snacks, the viewer ultimately expends more time than ever in front of a screen. The extent to which people are mediating their daily experience through screens is almost impossible to know, but it is considerable. What is relevant, however, is to identify and to understand what kinds of images and stories are emerging in this new "virtual" reality of experience.

In this chapter, I aim to present and analyze some of these contemporary forms of media production, sharing, and consumerism, and discuss their role in today's society. I then aim predict and, ultimately, to foresee what the media representations of specific cultural discourses tell us about the contingent nature of notions of global culture. Global media, as seen by snack culture, interprets, and corresponds to, an intermedial approach, a multifaceted intertwining of forms finally interfering and altering their discourses.

SHRINKING MEDIA

Small, Short, and in Motion

Complex intermedial relationships of digital media and the transformation of distinct cultural forms and practices are important issues in understanding the changing fabric of our society. New technologies, digital screens, and computer networks comprise important tools of communication, art and cultural experience within screens. Culture and media are being consumed as well as services are being provided via all kinds of screens and in diverse formats at the same moment all over the world.

In the mid-1950s screen culture was restricted to cinema and television, but now it has become an integral part of modern society on a global scale. We are living in a world saturated by mass media linked by images and texts through all sorts of screens. Individuals are being converted into screen-viewers—books, films, official documents, as well as banking, government, and many other services are being mediated through screens. Screen culture is transforming cultural habits and values and also influences our ways of telling stories and of creating cultural identity. We are living a new conception of the social community—one that is technologically mediated.

Clearly screens have been changing and transforming our sensory experience of ourselves and of social space. In part, thanks to the Internet, instantaneous entertainment comes from the proliferation of small doses of amusement and information called *snacks*. Media snacking is a new and fast mode of cultural and media consumption in the form of "bites." Originally from Middle Dutch, the term snack, that means to snap or bite, has been used related to food since the 17th century. Today, snack or fast food, is also called junk food for its low (or non-existent) levels of nutrients. However, fast food is now one of the most common sources of nourishment for an ordinary person in US, especially for youth, despite medical doctors and scientists arguing that it is one of the main causes of obesity[2]. The expression "snack culture" or "*snack-o-tainment*" was spread in 2007 by the magazine *Wired*, which specializes in technology. Snack culture can include mini-comic books; ten pieces of music of 30-seconds each composed especially to play by cell phone; short videos, of less than two minutes, made by a cell phone[3] or to be played on an portable media player (iPod, Zen, Archos, and Zune); or reduced segments from classics of cinema, such as a dream sequence from Christopher Nolan's *Inception* (2010). Even the business of news is changing, as mainstream media such as TV, radio, newspapers, and magazines are all creating, diversifying, and offering their services in small doses in blogs, podcasts, cell phones, and online aggregators (RSS, Atom, RDF). The term *snack culture* came into existence at the moment that the microblogs and smartphones (iPhone, BlackBerry, Windows, Symbian, and Android) with their personalized ring tunes were proliferating. In March 2007, according to specialized blog *Thws.cn* from China, 111 microblog sevices (such as Twitter, Jaiku, Tumblr, Quaiku, Plurk, Emote.in, identi.ca, among others) were offered in many parts of the world.

Today, in part due to the recession, newspapers all over the world are downsizing and some are closing their print shops and transferring their services and content onto the Internet. Radio listeners and television viewers are now able to follow online their favorite news broadcast one story at a time. Paraphrasing Brian Winston's question (1995, p. 55), what affect, if any, does snack culture have on the content, the output, of media? Certainly, the act of changing format, size, and platform of the news is affecting society. In North America, several television and radio stations are financially struggling so they are cutting back and exchanging their physical spaces for virtual ones.

Today, media and social spaces are not just becoming virtual, but mobile. Handheld devices, such as portable media players and smartphones, have been creating new virtual spaces in urban environments. These small and efficient gadgets have come to enlarge a series of transaction and advertising screens that have been ensconced in traditional public spaces such as subways, clubs, schools, museums, and streets. All such screens are transforming, implementing, and increasing, in an important way, the dynamics and imagery in one's experience of social space.

Urban people are changing their way of life by inhabiting the city virtually with new social networking communities, such as Facebook, Xing, LinkedIn, Orkut, MSN, Skype, MySpace among others. Coffee shops, for instance, which used to be an urban social gathering space, are now an example of *virtualization* within the city. In these spaces people are not communicating with each other in a traditional way, rather they are connected with others elsewhere through their screen devices while sharing media products more than ever before. This social interaction and consumption of cultural products allows users to send along a scene from a movie or a photo with comments instead of just discussing or describing them in words. Users are also able to manipulate and edit media content. Ultimately they determine what they want to see or listen to repeatedly. Technological social networking encourages users to complement their communication through media products such as videos, songs, photos, books, maps, and other digital documents.

The increase in cultural diversity and easy access to information are considered some of the positive aspects of snack culture. These products can be consumed in a much more dynamic way and have a lasting and extended life due to their repetitiveness and the ease of their being shared. However, the superficiality and the instantaneity of these products don't provide time for reflection on the content of what is being consumed.

Music, television, games, movies, fashion: We now devour our pop culture the same way we enjoy candy and chips – in conveniently packaged bite-size nuggets made to be munched easily with increased frequency and maximum speed. This is snack culture – and boy, is it tasty (not to mention addictive). (Miller at al., 2007)

Currently even a two-hour feature film can seem long. So why not transform something that is uninteresting into something that is dynamic and more personal? Could that be the minute that the protagonists kiss each other in the film; or when Justin Timberlake takes off Janet Jackson's bra; or, even a review of the spectacular scenes of 007? If the survival of the natural world depends on recycling, pop culture thrives in the moment of the replay, just as viewers crave to prolong the sensation of the moment of a goal in a TV hockey transmission or a child wants to play the same game repeatedly. Repetition is one of the main qualities of snack culture.

While this formatting by bites can be considered superficial or incomplete, it makes a series of exchanges possible and it allows for dynamism that in traditional ways is impossible to achieve. Thus snack culture must be understood in terms of its interactivity, exchange, accessibility, and discussions that spread freely and globally. The emergence of snack culture can be viewed as a way to undermine the complex monopoly of information and entertainment that, according to Robert W. McChesney (2001, p. 356) are controlled by a few giant conglomerates. The 'global commercial media system' in a neo-liberal world, as McChesney affirms, is part of a political and economic power structure maintained by a wealthy few.

SNACKING SOCIALLY

The Impact of Snack Culture in Daily Life

Globally on the Internet hundreds of millions of people are routinely creating and sharing cultural and social content through photos, writing, videos, and works of art. At the same time, the rapid growth of the number of "culture professionals" (artists, filmmakers, and publishers) who participate in global cultural productions and discussions has dramatically increased. Hundreds of thousands of artists and designers now have access to the same ideas, information, and tools. It is no longer possible to talk about centers and peripheries of art and culture.

Sharing information through the Internet is an essential component of today's culture, as computer engineer J.C.R. Licklider predicted 40 years ago. He is considered one of the original visionaries of the Internet. He has been promoting computer networking as being an undeniable fact for modern society rather than a futuristic prophecy. In the introduction of his 1968's paper *The Computer as a Communication Device*, written in collaboration with Robert W. Taylor, Licklider asserts "In a few years, men will be able to communicate more effectively through a machine than face to face" (1990, p. 21). Indeed, it did not take too long for us to be intimately accustomed to and to communicate through technology. And, the basis of this communication is the natural human desire to share information and experience.

As invoked by French philosophers Maurice Blanchot (*The Unavowable Community*) and Jean-Luc Nancy (*The Inoperative Community*), language is what forms and sustains any given community,

...as all community rests on language, and as the language in question is not the everyday language of the exchange of information, but rather the language that gives rise to community. Such

language articulates the relation between the members of the community, rather than merely designating things. (Haase & Large, 2001, p. 8)

Language in this sense inscribes and establishes the collective and the "social *duration*" of a community in the moment of communication, which is the sharing. Thus, the communication or the act of sharing with others precedes the community's own existence. In other words, without sharing either knowledge, information, cultural content, or social goods, communication doesn't exist as in the sense of community. As the act of sharing is an essential part of snack culture, it is possible to affirm that snacking involves a sense of community.

Snack culture is a trend of today's generation, and this philosophy and practice of snacking is present in all kinds of media—e.g., journalism, TV, etc.—, reflecting a contemporary social consumerist behavior. Reducing time is the most important challenge in contemporary society, people try to spend less time in doing ordinary daily activities, from eating to reading. Snack culture also involves reproducing the same effect of *zapping*[4] initiated a while ago. The principle is the same, to obtain maximum information in a shortest possible timeframe; both movements demand a lively engagement and an integrative, associative capacity. These capabilities in the viewers are stimulated by the intermedial and interactive quality of snack culture that features numerous media and formats. Thus, the integration of several small media products may create a unique understanding and discourse.

Snack culture promotes a revolution in understanding the cultural product while reducing everything to *immediate*. However, this *immediate* is not exclusively related to the immediacy of the world but to its "mediation." Thus, the proliferation of media challenges viewers' understanding as well as their behavior. Viewers have a great capacity to manage and to understand the virtual world (media) but have little capacity to comprehend the relationships that media establish with and mediate from the real world (immediacy). In this case it is important to make a distinction between media and culture.

Do viewers mediate culture through media or is media itself culture? Media is a way through which users can transmit and express their culture and it helps them to interact with their social group. However, viewers are learning to acknowledge media faster and interact with it more completely while integrating it into their culture. Microblogging or *tumblelogging* is one of those recent phenomena of social networking that is changing cultural and social habits. This Internet medium is a downsized version of blogs that restricts posts to a maximum of 140 characters.

Microblogging allows users to instantly publish texts stating what they are doing or thinking at that exact moment, replacing blogs that were considered personal diaries. Blogs have become more serious and professional and are often attached to mainstream media, whereas microblogging, such as Twitter, Jaiku, or Quaiku, work as a text-message service via the Internet, SMS (Short Message Service), or external applications. However, Twitter's developers claim it is not a social network, but rather it is an information network.

Information networks conveying social practices and creative projects reassemble artists all over the world. Independent cinema has achieved a rebirth because of the Internet, which today has become a new space used by filmmakers around the world to experiment, innovate, express, exhibit, share, and distribute audiovisual content. Some audiovisuals posted on the Internet may actually be viewed by a greater number than they could be through movie theatres or television. Also they will be accessible throughout the years. These information networks have created great opportunities for independent filmmakers interested in artistic expression and experimentation. Crowdsourcing is one of those projects that innovates by drawing on public participation and creativity to produce a unique piece, like the experimental project entitled *1 Second Film*.

The goal of *1 Second Film*, a collaborative art project, is to produce (with the aid of thousands of people around the world) an experimental animated movie of 1 second (12 frames-paintings) plus the final credits of 90-minutes in length, which show a *making-of* documentary that features the production process and acknowledges collaborators. However, other projects are claiming to be the beginning of crowdsourcing in more creative and innovative ways. Such projects are undertaken with the aid of thousands of people who not only produce, but also develop the screenplay and may even take part in the filming, as in the project *Swarm of Angels*. The copyrights of the movie will be gratuitous, being freely distributed under the license *creative commons*. Projects like these ones are invading all fields of knowledge, entertainment, and the arts including writing, photography, software, and games, among others, where the collective participation is more than welcome. On the Internet, one of the most successful examples of Web 2.0 is Wikipedia, a free encyclopedia, where anyone can 'anonymously' create, add, or correct the information in an open database-dictionary.

On the one hand, these open-projects are democratic in that they call upon participants to provide a product or service that represents diverse points of view and responds to everyone's expectations; but on the other hand there is no guarantee of the accuracy or the legitimacy of the source or the information itself. In a certain way, open-projects in the Internet are repeatedly recycling and reproducing almost the same material. This method of production involves double mediation (mediating a mediation) and losing track of reality. According to Heinrichs and Spielmann (2002, p. 6), the only verifiable information on the Internet is the source code of the programming, whereas the content of the information cannot be confirmed.

Because of its basic features, namely omni-directional, multi-dimensional and open-ended, the digital emerges as an essentially hybrid form that blurs distinctions between representation and presentation—between fact and fiction, between fake and simulation. [...] Hybrid forms of the digital are characterized as flexible, unstable, free-flowing, paradoxical and open-ended.

However, those open-projects are breaking two great barriers in the mass communication field: free access to information with the full participation of users and the creation of an alternative source of information. The advent of the Internet has been one of the biggest revolutions in the field of gratuitous mass communication. Since the launch of the Internet, global corporations and mass communication conglomerates have been trying to create obstacles to free access and to reduce popular participation.

The issue about democracy in global media, especially regarding the Internet, involves the question as to which definition of democracy is applicable in the World Wide Web? Democratization cannot be related only to the equality of access and sense of community, but must also include such notions as individual freedom and protection of the individual's copyright. As Will Straw states, global media is "to be imagined as an open-ended system of interconnection and passage, in which the distinct status of individual audiovisual texts is dissolved" (2000, 119). Even though snack culture promotes interaction and accessibility to information, it can be considered a threat to intellectual property rights, given that everything available in digital form is susceptible to being copied, recycled, manipulated, and multiplied. Nevertheless, snack culture does not completely guarantee one's freedom because ultimately a few huge conglomerates, such as Google, Bing and Yahoo!, control the Internet search system.

At the same time, social networking communities and interactive information sharing based on Web 2.0 (hosted services, web applications, social networking sites, video sharing sites, wikis, blogs, microblogging, mashups, and folksonomies) are the ways of rebellion.

The criminal conviction of Conrad Black, one of the most influential controllers of the global commercial media system, confirms that people should not place all their trust in the mainstream media. It also shows the necessity to create alternative sources of information and space for expressing diverse individual opinions. In doing so, society has access to and expands the avenues within interactive social networking and communication.

RE-REPRODUCTION

Images on a Variety of Sizes of Screens

The re-representation of one medium by another (cinema to computer, TV to streaming player, video to portable media player, videogame to emulation, etc.) is constituting a new cultural performance of the image. Even if we are not creating new images from or for these new mobile and technological devices, these screens are shaping the cultural existence of any given image. More and more audiovisual works, such as films, have been produced to perform in any screen size and shape. As Haydee Wasson argues, "Screens persistently and actively shape the images they yield and the experience of those who watch and listen to them. [...] Their shape, size, control buttons, and positioning reflect the logics of the systems and structures that produce and sustain them." (2007, p. 76) This means that screens affect the images' meaning. Although images are everywhere, they are not everywhere the same. Since the production and reproduction of the images are in a certain way controlled by some corporations, these images also reflect their expectations. However, the circulation of these images through a range of screens means they are fragmented, so it is almost impossible to predict how they will be perceived by the viewer when they are transposed from one screen to another. "Screens are nodes in complex networks (celluloid, electronic and digital images; institutions: urban, domestic and corporate; permanent and impermanent—body and shop windows) and cultural engagement ... more fluid; it is crucial not to lose sight of the persistent forms of materiality that undergird the meaning and experience of visual forms." (Wasson, 2007, p. 91) The complexity of the affected images resides in a dynamic created by intermediality—the convergence and interaction of media.

Gone are the boundaries between media when creating and presenting images. Images circulate from one medium to another, occupy a myriad of spaces, and are expected to perform all kind of tasks in our society. Will Straw points out this emerging task of the screens in society's daily life: "New screens partake of this persistence of tangibility, affixing audiovisual imagery to a variety of things scattered throughout our everyday environments" (2000, p. 119). Society mediates and has been mediated by screens via all kinds of platforms, places, and by acquiring various functionalities.

At the beginning of the twentieth century, Walter Benjamin emphasized how much the phenomenon of reproduction has affected the perception of the original object; multi-reproductions detach the object from the domain of tradition. Multi-reproductions are replacing a "unique existence":

> *And in permitting the reproduction to meet the beholder or listener in his own particular situation, it reactivates the object reproduced. These two processes lead to a tremendous shattering of tradition, which is the obverse of the contemporary crisis and renewal of mankind. Both processes are intimately connected with the contemporary mass movements. (1936, p. 21, italics added)*

If the phenomenon of reproduction was seen as a 'shattering of tradition', what can society acknowledge and expect today when the reproduction of the reproduction has become the main practice in mass media? Imagery within images comprises one of the most important tools of today's screen culture. Technology, according to Sobchack (2007:137), "is historically informed not only by its materiality but also by its political, economic, and social context, and thus it both co-constitutes and expresses not merely technological value but always also cultural values." Images on screens have become adjustable, mobile, functional, and interactive. "Technology is not merely used, never simply instrumental." They shape the image that they create and reproduce as well as viewers' understanding of that image, this also being affected by screens' sizes or formats.

Piracy and hacking on the Internet have created various sorts of images and screen formats. One of the most famous and interesting formats today, despite its poor quality and precarious image and inferior sound definitions is the *cam*. This new popular format on the Internet is shaping millions of viewers' film experience and it is worth a further research on how is it impacts and affects audience's perception. The *cam* is one "rip" made in the movie theatre—a video recording, normally with a digital camera, of a blockbuster movie. The *pirate* steals the movie and makes it available for download or online showing on the Internet regardless of all sorts of copyrights. The Internet community is well articulated and interactive—hackers and pirates from different places in the world help each other to provide on the Internet a new movie as soon as it has been released in theatres. For instance, a world premiere of an American movie can be recorded in a Russian movie theatre. It can be subtitled in any foreign language or have the original soundtrack in English added to replace the Russian dubbing dialogue, thus reestablishing the original soundtrack. Those formats are called R5 or TS (telesync).

Sometimes theses video recordings are made using a tripod, but most of the time this is not possible due to the theatre's security that the *pirate* must avoid, resulting in a shaking, handheld recording. Another physical difficulty that results in a specific aesthetic feature of the film copy is the angle of the camera in relation to the screen. Since available seating in the theatre is not always in the center, recordings may be filmed from disadvantaged angles. If the copy is adequately *cropped* and the film has no subtitles, the copy is difficult to differentiate from the original. However, many times a film in a *cam* format is presented as widescreen, with a width to height aspect ratio greater than the original screen, in order to hide some failures of the recording process, especially to eliminate silhouetted people's heads of viewers or tops of seats in front of the camera. Usually the sound is recorded with the camera's internal microphone, so the noise made by the audience is recorded, thus overlapping the film's soundtrack. This is true especially in popular films such as comedies and horror flicks, where laughs or screams can be heard in the *ripped*[5] film and give the rhythm of the film's pace. Actually, any sound coming from the audience close to the camera, such as conversations, the munching of popcorn, the unwrapping of candies, babies crying, or other peculiar noises, may be included in the final version of the movie. Due to these factors, the quality of sound and image are usually appalling, but sometimes, with luck, the cinema is almost empty and there is a relatively low level of noise in the *cam* copy.

MODERN PEOPLE

Inhabiting Mobile Spaces through Screens

What happens when our expressive technologies also become perceptive technologies transforming our comprehension of the world and our apprehension of ourselves, asks Vivian Sobchack (2007:135). Often in films that attempt to capture the idea of characters' immersion/concentration on a media gadget (computer, mobile, or portable media player), those films depict a representation/simulation penetration of the character inside the machine. The conveying of the experience of people interacting with this multitude of screens in daily life affects our perception of the world and ourselves; it could be integrating several scenarios.

In a video project entitled *Screen People*[6] that we have in development, we are working with two characters: a mailman and a secretary. We are putting them in situations related to their lives in order to create their personas based on their relationship with technology. The mailman, for instance, photographs his daily life—ordinary objects and spaces, such as an old book, a political brochure, a tree, or an empty alley as well as his own body. With this character we want to emphasize his embodiment through the screens—the media is in fact an extension of his body and gaze—and also the integration of his body inside these screens. As he doesn't recognize anything fully outside the screen, he feels the need to incorporate the world inside the media. As Elena del Río (1996, p. 97) argues, "Technology springs from the very human condition of embodiment and … the human imagination is of necessity a technologically drawn and grounded structure." He experiences the world by mediating his daily life. At same time, he photographs his own image in a mirror, as a form of dematerialization, recognition or virtualization of his own body. All this is immediately published on his blog and spread throughout his social network communities. Thus, these images assure himself about his own persona while create his conception of reality.

The character of the secretary is completely devoted to the new technology—cell phones, emails, MP3 player, and text messaging—and she uses it to interact with people. While her physical contact with others—outside her work—is rare, her contact with the world, through both verbal and visual virtual connections, is almost non-stop. She is completely absorbed in the media. When she wakes up, she checks her email and surfs the Internet; on her way to work she talks on the phone, writes text messages and hears songs; at work or in the city she sees a great number of screens as part of her landscape. These include advertising screens at the ferry and bus stop and in buildings as well as automatic machines in the subway and banks. With this we want to emphasize her ability to feel and connect with her technological environment and her friends through screens. In other words, she is fully integrated with her world via screens, rather than being isolated from the world. Others can no longer consider her inaccessible because of her relationship with the technology. She uses these devices to integrate herself and expand her circle of friends in a new conception of community.

A contemporary world that brings together technology and alienation is one of Chinese director Jia Zhangke's most cherished themes. In a remarkable short fiction film *Our Ten Years* (2007), Jia shows people's interaction with technology and its fast development (from radio to *Walkman* and cell phone to portable media player) in the last decade in China. Through a long sequence shot inside a train, Jia depicts the change in people's clothing and the evolution of electronic gadgets. In this way, he reveals the modernization of society, which involves the migration of people and advancement of technology.

Technology has taken on an important role in Jia's characters relationships, especially the cell phones with text-messages. It also is transforming the city as a huge illusory and animated screen intermediating Chinese people's thoughts and compensating for the lack of movement and experience in their lives. It also brings freedom and subjectivity to the impersonal world represented by the park as depicted in the film *The World* (2004).

The World is punctuated by animated daydreams sequences, created by animation artist Wang Bo (also known as Pisan), that materialize on the screen when the characters send each others text messages by cell phones. His flash animations,

Figure 1. Screen People (2011)

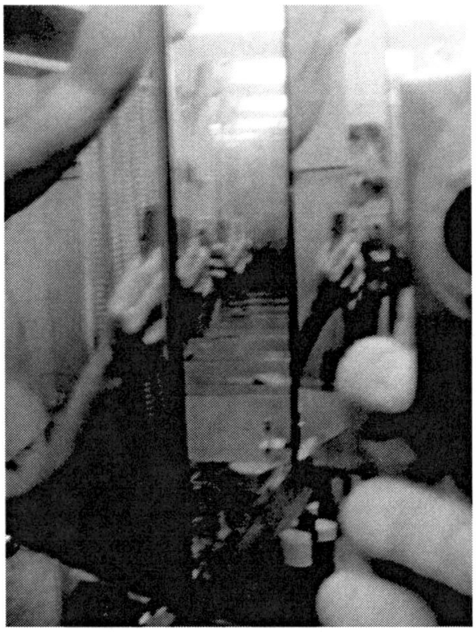

typically produced on the web are characterized by unpolished and simplistic images. He mixes painting and computer graphics that transpose the film's digital video images to animation and vice versa.

The cell phone screen becomes an animated image transforming the film's frame as an extension of the real world like a mobile screen replacing their experience of space and time through technology. Thus, new technologies are part of the characters' lives and world comprising important tools of communication. This transformation into flash animation becomes less evident in a digital environment. It seems digital video documents this moment in the development of modern society more accurately than film would, for it too is an intimate, immediate, and less-regulated medium.

When the characters send each other text messages by cell phone they create a virtual version of the world, an animated one. The enclosed park's world comprised of the miniature scale models is suddenly over-scaled and open. The flash animation transposes the real world to an imaginary and fantasized version of reality. However, this transposition is not directly from the characters' mind, nor is it a representation of their thoughts and dreams. This transposition comes from the real images because the world has been changed by the imposition of and the increase in the technology in the city's space. Digital CGI (Computer-Generated Imagery) effects and flash animation create an illusion and replace the actual images with the virtual. The resemblance or materialization of real images to animated versions of the same images has been achieved through computer programs more easily by using digital video. Digital video accomplishes this transformation from actuality (world's reality) to virtuality (media interaction). This morphing effect in the film results from direct reality and is not produced only by the characters' imagination.

The animated images are symbolic and fantasized representations while the park model simulacra images suggest social criticism regarding urban space and characters' social conditions. This symbolic idea of the world in the theme park is promoted by the integration of the technology into it—such as the illusion provoked by the video chromakey effect when the visitors create their own video of effect flying over the park or when they re-mediate the city's landmarks posing for photos in front of them.

The park is a re-creation from the real world, as well as being an image based on a media representation. The visitors want to experience the world as they imagined the world is, completely based on the media simulacra. Because of their saturation through continuous presence in the media, the Twin Towers have a greater importance among the park's sites than other cities' landmarks. The world outside maybe has been changed but is completely preserved inside of the park's space for the visitors' pleasure.

The theme park and the animated reveries of the world is far beyond the characters' inner world but the park is "representation," a composite image of the world, appearing as a screen. The fakeness is not, however, the main characteristic of the film; the actual modern world is intermediated by technology and is represented by images reproducing all spheres of our daily life. Many people turn to the web or virtual worlds created by video games and incorporate imagery from the web to their actual environment doubling their own experience of the space and time as an extension of the real world. Creating a digital approach to their reality by combining the animation sequences with SMS text messaging, electronic music and video souvenirs, the film has been able to show this intermediality between reality and fantasy by way of media subjectivity on the screen.

Jia has stated: the characters "don't really understand what the modern world is like. They are not yet thinking like modern people" (Havis, 2005, p. 59). Despite using the new technology, which is a symbol of the modern world, and which is part of their lives and allows them to "link" with each other, they are unable to fully understand the associated social and political implications.

CONCLUSION

Intermediality and the Process of Digitization

Accelerated digitization, the Internet, and the use of mobile technological devices have affected and shaped professional practices and forms of consumption in the audiovisual field. Media convergence has become an important feature of snack culture, the so-called intermediality. As Heinrichs and Spielmann explain, intermediality "represents a concept of dynamic change and transformation that alters existing media forms by inserting new elements" (2002, p. 6). Thus the increasing re-reproduction and assemblage of a variety of media forms describe a process during which previously distinct media merge with each other, "resulting in the creation of a new (art) form and shaping the form of a new medium." Having being characterized by the crossing, integration, contamination and interaction of media, intermediality consequently produces cultural texts that create and shape new conceptions of the world and ourselves.

Mariniello (2003, p. 48) suggests that "inter" in intermediality, as well as in *intertextuality*, refers one media practice to another, as well as suspending the space-time interval. The medium is where an event takes place, whereas the mediation indicates the way in which a meeting is possible between a subject and the world. On the other hand, the mediation could be between two subjects in a movement which, each time, the relationship with each other constitutes them. Intermediality also means the apparatus, which reveals the way in which the technique informs the meeting, as well as the becoming, which reveals the inseparable dynamics of any mediation.

Intermediality addresses issues relating to the rapid transformation of knowledge in an environment where media, users, and geography tend to merge. Snack culture—a ready-to-go media form—is a feature that privileges instant and rapid consumerism. However, because of its ephemeral quality, snack culture demands from the viewer a great effort to follow the flux and sequence of the information and to know the context at the moment. It's a dynamic way for viewers to communicate with each other and to produce new products by combining and recycling existing media forms. This merging occurs in an environment that, in spite of appearances, does not reveal "the multiple of being to and with regard to the meaning." Despite the lack of historical context and depth of content, snack culture tends to capture the moment and to integrate it in a comprehensible and self-sufficient package. The phenomenon of snacks privileges the dynamic, the movement, and the integration of the senses and the flow of information and goods over fractures, intervals, fragmentation, and conciseness, thus creating new forms of communication and consumerism.

REFERENCES

Acland, C. (2009). Curtains, cart, and mobile screens. *Screen, 50*(1), 148–166. doi:10.1093/screen/hjn071

Benjamin, W. (1981). The work of art in the age of mechanical reproduction. In *Selected writings* (pp. 319–334). Cambridge, MA: Harvard University Press.

Blanchot, M. (1988). *The unavowable community*. New York: Station Hill Press.

Del Rio, E. (1996). The body as foundation of the screen: Allegories of technology in Atom Egoyan's Speaking Parts. *Camera Obscura*, (Summer): 37–38.

Haase, U., & Large, W. (2001). *Maurice Blanchot*. New York: Routledge.

Havis, R. J. (2005). Illusory worlds: An interview with Jia Zhangke. *Cineaste*, (Fall), 58-59.

Heinrichs, J., & Spielmann, Y. (Eds.). (2002). Special issue: What is Intermedia? *Convergence*, *8*(4), 5–10.

Licklider, J. C. R., & Taylor, R. W. (1990). The computer as a communication device. In *Memoriam: J. C. R. Licklider 1915-1990*. Palo Alto, CA: Systems Research Center.

Mariniello, S. (2003). Commencements. *Intermédialités*, *1*, 47–62.

McChesney, R. (2001). Global Media and Democracy. In Weibel, P., & Druckrey, T. (Eds.), *Net Condition: Art and Global Media* (pp. 356–363). Cambridge, MA: The MIT Press.

Miller, N. at all. (2007). Snack Attack! *Wired*, 15 (3), March. Retrieved April 1, 2008, from http://www.wired.com/wired/ archive/15.03/snack.html

Nancy, J.-L. (1991). *The inoperative community*. Minneapolis, MN: University of Minnesota.

Sobchack, V. C. (2004). *Carnal thoughts: Embodiment and moving image culture*. Berkeley, CA: University of California Press.

Straw, W. (2000). Proliferating screens. *Screen, 41*(1), 115–119. doi:10.1093/screen/41.1.115

Wasson, H. (2007). The networked screen: moving images, materiality, and the aesthetics of size. In Marchessault, J., & Lord, S. (Eds.), *Fluid screens, expanded cinema* (pp. 74–95). Toronto, Canada: University of Toronto Press.

Winston, B. (1995). How are media born and developed? In Downing, J., Mohammadi, A., & Sreberny-Mohammadi, A. (Eds.), *Questioning the media. A critical introduction* (pp. 54–74). London: Sage.

ENDNOTES

1. D. Arrais & E. Bertoni. (2007) Cultura em Pedaços é febre na internet. *Folha de São Paulo*, Ilustrada, p. 8, June 17.
2. JOHNSTON, L. D. & O'Malley, P. M. "Obesity Among American Adolescents: Tracking the Problem and Searching for Causes," *Youth, Education, & Society*. Institute for Social Research. The University of Michigan. Ann Arbor, Michigan, May 2003. http://www.rwjf.org/files/research/YESOccPaper3.pdf
3. In France, a famous film festival called *Pocket Films Festival* is entirely dedicated to cell phone movies.
4. The act of moving abruptly and quickly from one television channel to another (channel surfing), or forwarding and rewinding a videotape by use of a remote control. Figuratively, it means rapidly changing an activity, idea, or opinion. See Daney, S. (1988). *Le salaire du Zappeur*, Paris: Ramsey; Bellamy, R. V., & Walker Jr., J. R. (1996). *Television and the remote control: Grazing on a vast*

wasteland. New York, NY: Guilford Press; and Jenkins, H. (2006). *Convergence Culture*. New York, NY: New York University Press.

5 CAM, TS, R5, TVrip, DVDrip, DVD-R, DVDscr, B-Urip, WP, among others.

6 Direction and Production: Hudson Moura and Nancy Shaw. Actors: Riley Maruyama and Valérie Lecomte.

Chapter 4
Global Media and Information Ethics:
Challenges Re-Examined

D. Ndirangu Wachanga
University of Wisconsin-Whitewater, USA

ABSTRACT

Any meaningful debate on global media and information ethics is burdened with the complexity of dissecting various disjunctive dynamics that characterize the complexity of emerging global relationships. The authors argue that the emerging global phenomenon problematizes the Cartesian plane of oppositions – center vs. periphery, North vs. South, global vs. local, which has been the forte of globalization studies until recently. It is against this background that the authors seek to examine challenges of having a global information and media ethics. The authors will pay attention to the antagonistic mechanics informing the domination and rejection of intangible ethical principles. In this discussion, they will be guided, partly, by Alleyne's (2009, p. 384) postulation on the need to pay attention to "changes in state power, the relationship between the market and the state, and modifications in the ideological assumptions about the optimum form of world order."

INTRODUCTION

In his article "*Disjuncture and difference in the global cultural economy*," Arjun Appadurai (1990), argues against the Cartesian plane of oppositions – center vs. periphery, North vs. South, global vs. local, which has been the forte of globalization studies until recently. He suggests the need to focus on global politics from a viewpoint that acknowledges the ceaseless interaction of global relations, which mutate constantly. If this mutation has led to a global culture, it is one that is "marked by an organization of diversity rather than a replication of uniformity" and "no total homogenization of systems of meaning and expression has occurred, nor does it appear likely that there will be one any time soon" (Hannerz, 1990, p. 237).

Any meaningful debate on global media and information ethics is burdened with the complex-

DOI: 10.4018/978-1-60960-037-2.ch004

Copyright © 2011, IGI Global. Copying or distributing in print or electronic forms without written permission of IGI Global is prohibited.

ity of dissecting the aforementioned dynamics of global relationships. There are scholars (See Alleyne, 2009 for example) who point at the futility of pursuing universal ethics, arguing that it is unachievable. Pursuing Alleyne's line of thoughts are primordialists who support cultural relativism, but unlike Alleyne, they fail to address both the susceptibility and potential degeneration of relativism to ethnocentrism. Thomas Cooper's (1996) ethnographic studies among the Shuswap, Polynesians and the Navajo have demonstrated not only a dedicated endeavor to understanding universal humanity, but also, from a socio-anthropological standpoint, the problematic and indefensibility of cultural relativism.

Other scholars (See Christians, 2007, 2008, Etzioni, 1996) have suggested communitarianism as a universal model of ethics. Universality in Christians and Etzioni's sense presuppose the existence of universal standards, especially because universalism has its anchorage on a broad postulation of commonalities of human experience. This presupposed common human experience does not obstruct espoused universal standards.

At the core of this search for universal media and information ethics, is the seemingly less pivotal role of non-Western ethical guidelines in challenging the Western models, especially when the canonicity of Western guidelines is presented as worth of emulation at the expense of unique traits of post-colonial societies. (Although we do not pose the question "*is there a political-theoretical horizon beyond the existing discourse of postcoloniality*?" with Gayatri Spivak (2009), we share her concern on the limitations of post-colonial discourse)

Attempts by scholars such as Rao & Wassermann (2007), Land (2006), and Christians (2004), to incorporate ethical concepts from non-Western contexts in the global media and information ethics debate have altered neither the debate nor the fundamentals informing Western theoretical paradigms. There is, however, noticeable scholarly effort in challenging not only the dominance of Western-based paradigms, but also a rising rejection of "abstract ethical principles conceived in the Western mind but presumed true for all cultures" (Christians 2005, p. 3).

It is against this background that we seek to examine challenges of having a global information and media ethics. We will pay attention to the antagonistic dynamics characterizing the domination and rejection of the said "abstract ethical principles." In this discussion, we will be guided, partly, by Alleyne's (2009, p. 384) postulation on the need to pay attention to "changes in state power, the relationship between the market and the state, and modifications in the ideological assumptions about the optimum form of world order."

MEDIA ETHICS AND GLOBALIZATION

The foundation of abstract principles, at least in media and information ethics, emphasizes the centrality of individual autonomy. It is laced with freedom as its dominant motif, a prevailance of "the cult of human personality in all its freedom" (Christians, Ferrè, & Fackler, 1993, p. 21). It is a foundation whose building blocks are mostly borrowed from the Enlightenment vocabulary, part of it emanating from the Renaissance, consisting of terms such as human and individual rights, freedom, justice, democracy, sovereignty, unrestrained speech, unbridled "marketplace of ideas." Upon this lexicon is a predicate for a Euro-American grand narrative, which was globally assumed to exhibit internal coherence in the way it tied representation and the public sphere.

However, this perspicuity was altered, if not entirely disrupted, by global changes that visited the nineteenth century, producing a loosely structured landscape whose nation-states were systematized around a redefined lexicon and consequent varying political narratives. At once, there emerged tension in global relations as nation-states sought to adjust to influences from without while maintaining what they perceived as "indigenous," even as they engaged in a series of activities seeking to "indigenize" what they borrowed.

The resultant- and increasing changing- global landscape, consequently, experienced fundamental constitutional fluidity, whose architects, though not entirely responsible for, consist of the Foucauldian binary typology of power and knowledge - where "every action and every historical event is an exercise in the exchange of power" (Rao & Wasserman, 2007, p. 35). Yet, at the core of this typology is the tension between global cultural homogenization and heterogenization, which the media, whether private or state-owned, buttress. For in messages punctuating the global mediascape, there is an embedded message of 'ethnic' incompetencies of 'the other. It is by acknowledging the multifaceted nature of these never-ending nascent global manifestations that discussing global normative standard of media and information ethics must be examined.

A cursory overview of the malleability informing global relations is significant before we interrogate the challenges of formulating a global media and information ethics. As Appadurai (2007, p.219) notes,

global relationship between ethnoscapes, technoscapres and finacescapes is deeply disjunctive and profoundly unpredictable, since each of these landscapes is subject to its own constraints and incentives (some political, some informational, and some techno-environmental), at the same time as acts as a constraint and a parameter for movements for each other.

A model on any global issue - not just media and information ethics - must take into account the shifting relationships characterizing human migration, ideological inclination, technological flow and media coverage (and representation). The relationship among these variables is itself disjunctive and revolves around, among other things, unprecedented rapid and voluminous flow of people, money and information. It is a more complex relationship than the one conceived by modernization revisionists who suggest "that aspects of political modernization could positively invigorate... traditional elements, albeit in a changed form, and also that these elements would invariably influence in some measure the form and pace of political change" (Randall, 2008, 20).

Appadurai (2009), by expanding on Benedict Anderson notion of global imaginary, imbues the global terrain, which emerge from this complex relations with discrete "flows," which he calls "scapes" – ethnoscapes, technoscapes, financescapes, mediascapes and ideoscapes. Despite their distinctiveness, these 'scapes' are in continuous motion, even physically. Global relations flow through these 'scapes,' and coalesce into complicated prism through which global media and information ethics must be perceived. In a sense, global media and information ethics seek to interrogate the intersection of global politics, power struggle, cultural identity and the role of the ideological hegemony in a world of isolated connectivities. But as Randall (2008, p.31) notes, there is an "assumption that the developing countries have to be made to change, which helps to rationalize continuing intervention of outside interests, experts, perspectives, and so forth."

The Appaduraian "scapes" dissect the dynamics of global imaginary by laying bare how its constitutional make up constantly mutates, deterritorializes or physically transverses "scapes," penetrating national borders in search of cosmopolitanism and/or globality. The media, in their copious manifestations, are at the nucleus of these changes. It is not surprising how the mediascape, as Appadurai illustrates, possesses the power to form scripts about "reality" informing how different parts of the globe can be formed. " These scripts can and do get disaggregated into complex sets of metaphors by which people live as they help to constitute narratives of the "other" and proto-narratives of possible lives, fantasies which could become prolegomena to the desire for acquisition and movement" (Appadurai, 1990, pg. 220). From this horizon, we can catch a glimpse on the impact manipulation of images by global advertising, marketing and public relations strategies where the "consumer is consistently helped to believe that he or she is an actor where in fact he or she is at best a chooser" (225).

The identity of the emergent citizenry in this global imaginary, therefore, can be framed by two dimensional axes, which are operational concurrently: the axis of similarity and continuity; and the axis of difference and rupture. While these axes may not necessarily be diametrically opposed to each other, their variability shifts their relation, constantly altering the perception of the global imaginary.

The global imaginary identity can, arguably, be, then, understood in terms of a dialogic relationship between these axes (See also Hall, 1996). Still, the intersection of relations informing the resultant perceptual identity is vital at the construction of the "other," – not just in Edward Said's "Orientalist" way, but also in a sense that makes the "other" experience himself/herself as constructed. Hall (1996) shows how the mutability of the identity that emerges from the intersection of 'small imagined worlds,' fails to "stand in the same relation of the 'otherness' to the metropolitan centers. Each has negotiated its economic, political and cultural dependency differently. And this 'difference,' whether we like it or not, is already inscribed in our cultural identities" (114). Hall's viewpoint must be read alongside Spivak's (2007, p. 231) re-alteration of the new subaltern who is "no longer cut off from lines of access to the center." If this maps the emerging global imaginary, then, its ethics, however provisional, must accommodate practices of consumption and production bolstering it. Friedman (1990, p. 327) explains:

The practice of identity encompasses a practice of consumption and even production. If we further assume a global historical frame of reference it is possible to detect and even to account for the differences among broad classes of strategies of identity and therefore of consumption and production as well as their transformations in time. This is the case, at least, to the extent that the different strategies of identity, which are always local, just as their subsumed forms of consumption and production, have emerged in interaction with one another in the global arena.

A reading of Entman and Rojecki's (2000) *The Black Image in the White Mind: Media and Race in America* by Mudimbe (2009, p. 313) is illustrative of inestimable ways of gaining entry and existing in the complexity of the modern global imaginary.

Robert M. Entman and Andrew Rojecki unveil interacting forces in ordinary conversations as they are reclaimed in American mainstream culture, their influences in terms of needs and limitations, and how all this is related to evolving economics of media industries, themselves shaped by new technology, global market competition and government policy decisions. Thus, against any dream of authenticity in assertions of "I ought," one faces the objective fascinations of the communication industry as well as the exigencies for personal growth in conflict with economic expectations as orchestrated through media content, both news and entertainment.

It would be fascinating to re-examine imagined worlds through the eyes of Elfreda Chatman (1985, 1991, 1999) an information scientist who coined the term "small world" in her human information seeking behavior research. Chatman developed this information theory from ethnographic studies she conducted when studying the information poor. Most of her studies focused on janitors, imprisoned women, and geophagists (dirt- eaters). She coined the term small world, which she defines as a world with a high predictability of information occurrence, with most of the information that binds this world together being relevant, justifiable, and occupying a sanctioned place in the general scheme of things. Activities such as information seeking and how information content is packaged in these small words are perceived as normative. According to Chatman people belong to particular small worlds because of a number of factors, important among them being shared life experiences and social norms. According to Chatman, sharing of sanctioned social norms makes life in a particular small world predictable.

To have a clear understanding of the small world concept, it is important to note that the small world thrives because it allows its members to "share a similar cultural and intellectual space" (Chatman & Huotari, 2001, p. 353). Like in Benedict Anderson's sense, Chatman's small worlds are also imagined communities. Among the integral things that hold any small world together include:

a. Common assessment of information worthy of attention
b. Social norms that allow its members to approach or ignore information
c. Behaviors that are deemed by other inhabitants to be appropriate for this world

Chatman's small world information theory comprises four key elements: social norms, worldview, social types, and human information behavior. Social norms are the core determinants of what is right or wrong in a small world and sanction not only information to be sought, but also information seeking behavior within a small world. The various ways members of a small world understand their world determine the importance of information they seek and the reality they construct. Worldview, therefore, becomes the "collective perception members of a social world hold in common regarding those things which are important" (Chatman, 2000. p.11).

Chatman does not interrogate the intersection of small worlds change. Unlike Chatman, Appadurai espouses a terrain which is not only symptomatic of Friedman's (2005) "ten forces that flattened the world," but also one that refuses to be defined as "flat" because it is fraught with ever-changing and interacting "scapes" or "flatteners." Friedman (2005) in *The World is Flat: A brief history of the twenty-first century,* consider the following as the major events that have not just shaped, but flattened the 21st century world:

1. The collapse of the Berlin Wall
2. Public use of Netscape
3. Work flow software
4. Open-sourcing
5. Outsourcing
6. Offshoring
7. Supply-Chaining
8. Insourcing
9. In-forming
10. The Steroids

Freidman's Universalist approach is grounded on both global political dynamics and global information flow. Using the collapse of the Berlin Wall to illustrate the flattening of the world, Friedman meticulously intersects global politics and information flow. He writes (49)

The fall of the Berlin Wall on 11/9/89 unleashed forces that ultimately liberated all the captive peoples of the Soviet Empire. But it actually did so much more. It tipped the balance of power across the world toward those advocating democratic, consensual, free-market-oriented governance, and away from those advocating authoritarian rule with centrally planned economies. The Cold War had been a struggle between two economic systems – capitalism and communism- and with the fall of the wall, there was only one system left and everyone had to orient himself or herself to it one way or another

In terms of information explosion, Friedman calls the period when the Berlin Wall collapsed to the mid-1990s, the age of "Me and my machine can now talk to each other better and faster, so that I personally can do more tasks" and the age of "Me and my machine can now talk to a few friends and some other people in my company better and faster, so we can become more productive." The walls had fallen and the Windows had opened ... but the age of seamless global communication had not dawned" (54/55) yet.

It is no surprise that the ubiquitous usage of the Internet – as an unprecedented form of instantaneous and voluminous dispenser of information in the global imaginary - inspired considerable writings and vibrant debates in academic and professional circles on several fronts with regard to information gathering, dissemination, storage, retrieval and security. Whether Information and Communication Technologies (ICT) would homogenize local cultures became a question worthy intellectual pursuit among cultural, media, literary and information scholars.

Also, there emerged an intense focus on the role of ethics in what had become a globalized world. Although scholars have suggested the need to be cognizant of both celebratory and counter-celebratory narratives of globalization, there has been a corollary epistemological discussion in search for universal media ethics in a globalized world (see also Gikandi, 2005).

Scholarly interest on global media and information ethics can be traced in the drafting of the 1948 UN Declaration of Human Rights, which, unfortunately, failed to craft a global agreement on freedom of information. But it was not until more than two decades after the UN Declaration of Human Rights that another debate regarding global media structures emerged. This time, 1970s, the UN General Assembly and UNESCO sought to establish what became the New World Information and Communication Order (NWCIO).

The establishment of NWCIO was not devoid of politics. Although Alleyne (2009, p. 383) problematizes how NWCIO was advanced "as a necessary corollary because it would attack global inequality at the ideological level," he emphasizes the need to pay attention to the operations of the United Nations as serving sovereign states "with primary legal standing in the international system" and "resolutions, declarations and conventions reflected the interests of states." Using the UN as a referent not only broadens the focus of discussing ethics from a mere parochial and sometimes selfish utilitarian priority of media owners and workers, but also acts a refraction of the antagonistic relationship between governments and the media as an institution.

The discursive space opened by using UN as a referent, presumably accommodates global socio-political, cultural and economic relations. The UN was formed, among other things, to maintain global peace in a landscape whose political terrain was shifting from the dynamics of waning colonialism to an environment supportive of self determination.

We do not necessarily endorse UN as an ideal benchmark in discussing global ethics because the body was formed to endorse concepts underlying the Enlightenment vocabulary. Still, a discussion on NWCIO only fomented old eco-political and ideological conflicts among members as evidenced through withdrawal of the United States, United Kingdom and Singapore from UNESCO in 1984 and 1985. To justify their withdrawal, these key countries cited how this UN agency was threatening, among other fundamental issues, press freedom. But the "politics" that surrounded UNESCO and NWCIO clearly pointed not so much to global concern for media and information ethics and freedom, but to political and ideological differences. So colossal were these differences that UNESCO became a field where Cold War was fought. It is telling that even recent UN initiatives have not been much successful. A case in point is the 2001 World Summit on the Information Society (WSIS), which sought a common global media vision but its deliberations were muddled in confusion. In spite meticulous language that punctuated the

WSIS proceedings, Hamelink (2004, p. 289) was not optimistic about its outcome. His reasons:

- The fundamental human right to free speech is universally violated through forms of political and commercial censorship.
- The Internet – in particular – has become the focus of censorship initiatives.
- The rights to corporate ownership of intellectual property are greatly extended.
- The access to information and knowledge is increasingly dependent upon the access to purchasing power.
- The consolidation of power on information and knowledge markets is consolidated in the hands of only a few conglomerates.
- There is minimal public accountability from the corporate actors controlling most of the technologies and the contents of the information society.
- Profitability more than human security drives ICT developments.
- The public sphere is increasingly limited.

Hamelink's frustrations point to the convolution of media companies and institutions, which must operate on the logic of capitalism to survive and how the very nature of their symbolic and ideological potential forces make these enterprises to confront complex realities – even virtual realities such as extremist religious groups such as jihadists or racist regimes. The Janusian, double-faced – an information disseminating and an economic entity - nature of the media is, broadly, the humus that nurtures the tension between utilitarianism and communitarianism as media ethics philosophical foundations.

Borrowing heavily from the Enlightenment philosophers, utilitarianism has the Western libertarianism as its basis. Guided by views of libertarian philosophers such as Immanuel Kant, John Stuart Mill and Jeremy Bentham, utilitarianism encompassed views of both deontologists and teleologists. It is consequence-driven just as it is duty- bound.

According to Christians (2007, p. 119) "utilitarian rationalism has served as the prevailing paradigm in communications for more than a century. Consistent with philosophical ethics generally since 1890, communications ethics has presumed that individual rationality marks all legitimate claims about moral obligations, so that the truth of those claims can be settled by formal examination of their logical structure." Although scholars have questioned human rationality after soldiers died in their millions in pitiable ditches in World War I, German holocaust, genocides in Cambodia, Rwanda and Darfur, Christians' (2007, p. 119) rebuttal is directed to utilitarianism shortcoming in addressing "dramatic technological innovation and the negative side of the market-driven global commerce have pulled the news profession away from its traditional role in facilitating democratic life." Proponents of utilitarianism (See John Merrill, 2006 for example) consider it to be the hallmark of individual rights and freedom "where the light shines brighter on consequences as predicted by the individual decision maker" (Land, 2006, p. 24). In praxis, utilitarianism is undoubtedly the overriding approach of moral reasoning among the media practitioners, especially in libertarian societies (See Etzioni, 1996). As a philosophical foundation, it places the individual at the centre of moral reasoning, the value-determinant. But despite being the "centerpiece of libertarianism," individuals are often guided by utilitarian ethics when they "weigh self-interest (ego) with competing interests in light of internalized codes of moral behavior and on the basis of perceived outcomes" (Land, 2006, p. 21).

Although utilitarianism recognizes "collectives" as aggregates of rational individuals, communitarianism focuses on the community, a concept that utilitarianism finds rather icy to define – is a community a village, state, nation, world? Along the same line, Wiredu (1996) acknowledges differences among cultures, but notes a similarity in the human use of language. Also, Wiredu argues, all languages – no matter their linguistic differences – exhibit similar phonemic complexity, serving practical roles just as they are markers of cultural configuration. Christians (2007, pg. 7), agrees with Wiredu that "languages everywhere are communal and give all peoples an identity as a species. Through the commonness we share as lingual beings, we can believe that there are universals at the same time as we live in our particular communities."

As an ethical guideline, communitarianism "seeks to strike a balance between individual freedom and the greater social order, based on shared virtues. Community writ large becomes a major starting point or source for moral reasoning, which seeks a judicious balance of rights and responsibilities" (Land, 2006, p. 24). Cognizant of individual rights and freedom, communitarian ethics seeks to understand the individual with the context of the *communitas*, because a "person is nothing without a community, a society, and that he or she must merge self with the group – the *Gatuung* – the entity that alone exists" (Merrill, 2006, p. 45). Community, in this view, is not limited to local communities, which reflect cultural and/or religious codes and beliefs that run counter to other cultures' moral philosophies. This views assumes accountability beyond the individual and the local community as the final arbiter of moral conflict; it must be able to appeal to the community of humankind, based on the principles inscribes in the United Nations' Declaration of Human Rights, for example. However, this UN system was "established after WWII on the assumption that a functional network of international organizations would promote universality as the preferred version of internationalism" (Alleyne, 2009, p. 386).

Communitarian agenda resonates with the 1947 Hutchins Commission report, which challenged individual-centered, libertarian ethics. This Commission's report, although ostracized by the American media owners and practitioners, highlighted five general responsibilities, which impacted on the media's philosophical foundations:

- Present a "truthful, comprehensive account of the day's events in a context which gives them meaning."
- Serve as a "form for the exchange of comment and criticism."
- Project a "representative picture of the constituent groups of the society."
- Present and clarify the "goals and values of society."
- Provide" full access to the day's intelligence."

Even as the Hutchins Commission report was being released, the world was yet to witness decades of socio-political and eco-cultural changes of mammoth proportions. These changes, as partly outlined by Land (2006, p.22), are indicative of the complexity that permeated the globe, making universalism untenable:

- The population explosion of post-Worlds War II and the subsequent migration toward sprawling suburbia in the United States, the revving up of the Cold War and the Civil Rights movement and the emerging nuclear threat in 1950s.
- A presidential assassination, the climax of the Civil Right Movement, the so-called counter cultural revolution, the Vietnam War and the continuing threat of nuclear holocaust in the 1960s.
- The women's movement, the Watergate scandal, the revelations of The Pentagon Papers, the ignoble end of the Vietnam War, and the continuing Cold War and the nuclear threat during the 1970s.

The monumentality of these changes and their impact, especially in the West, slowly led to conspicuous institutional mistrust. There was an, arguably, noticeable up-shoot of public or citizen journalism isomorphic with the growing hiatus between the media and the public. This generated a vibrant debate among scholars seeking to salvage the media profession.

ALTERNATIVE MODELS

With the deficiencies of Enlightenment now prominent, scholars sought for guidance from the moral universe. Grounded on universalism, Benhabib (1992) borrowed both from feminism and communitarian ethics. In her synthesis, she accommodates cultural polyphonies. Hamelink (2000) proposes the UN, especially the Human Rights Charter, as a referent for global ethical guidance. Land (2006, p. 25) identifies five principles, "which efficiently organize the majority of moral principles the world over: truth, justice, freedom, humaneness and stewardship." Land is careful in his delineation of the term "principles" and "values," which although can be used interchangeably, "principles serve as guideposts for what is right and wrong, while values, which may be thought of as "principle applied," define what is good and bad" (25-26).

Kùng (1998) sought a global common ground in similarities of values among religions. There have been also attempt corroborate ideas by examining non-Western societies. In India, Mahatma Gandhi developed *ahimsa*, which is derived from Hinduism, in reference to non-violence. Ethical obligation of showing compassion to the innocent, of not harming them, is a derivative of *ahimsa*. Gandhi (1942, p. 18) saw ahimsa as "absence of malice or hostility to all living beings in every way and at all times."

East Asia, Confucianism as a non-Western viewpoint from which a discussion on global ethics can be placed. First, Confucius believed in the power of *li* (禮), which translates to rites, morals, and social norms, to restore an orderly society. Confucianism is regenerative in orientation. Like the Judeo-Christian maxim of "do unto other as you would have them do unto you," Confucianism is rooted on a similar golden rule of reciprocity: "Do not impose on others what you yourself do not desire" (Confucius, trans. 1979, book 15, para. 24). But placing East Asian experience in a global relations with the rest of the world, especially the West, we fail to hear the echo of reciprocity. This dissonance has been captured in Kok-Chor Tan's *Tolerance, Diversity, and Global Justice* (2001, p. 145), who acknowledges the futility of the Cartesian plane of West vs. the Rest in global politics, but who nonetheless notices obstinate observance of Confucian golden rule both by East Asia and the West:

But while the student of international relations ignores issues of power, domination, and nation-interests at his or her own peril, it is just as distorting and misleading to squeeze entire debates into "West" against the "Rest" framework, as many defenders of the East Asian position tend to do. This realist paradigm does not do justice to the complexity of international relations; it fails to see that foreign policies are motivated by different competing realist considerations as well as idealist (normative) ones. A constructive cross-cultural human rights dialogue can commence only if both parties set aside "absolutist" frameworks – The West as imperialistic, the East as socially unprogressive – that have tended to poison the dialogue to date.

Let us briefly discuss African-based perspectives, which have been mentioned cursorily in media ethics debates, but which by and large require further interrogation.

Palaver Tree

The concept of the traditional "palaver tree" is especially common in West Africa and symbolizes a forum where controversies are resolved through lengthy discussions. Its dynamics are reminiscent of the Greek *Areopagus*, the Athenian hill of the highest judicial court. John Milton, the English wordsmith and a defiant fighter for freedom of expression speech, borrowed the title of his paean to free speech, *Areopagitica*, from this hill.

Under the shadow of the palaver tree, elders debated every dilemma until they reached a resolution through consensus. Land (1992, p.10) has extrapolated the importance of the palaver tree:

For centuries, sub Saharan Africans engaged in collective communication in the local shade of the sacred baobab or mango trees in their villages. Beneath the palaver tree misunderstandings were resolved and critical community issues were discussed under the leadership of village elders. Villagers explained their points of view and together, through group consensus, reached a final decision. But the palaver tree stood for more than group discussions and problems solving; it beckoned the villagers by means of the sacred talking drums to join in harvest celebrations and other festivals and rites of passage.

Before Western cultural influences were introduced in Africa, Land (1992, p.10) notes how "inhabitants of Cote d'Ivoire understood the meaning of communication – of myths, rituals, drama, and interpersonal and group interaction – because they lived communication within the context of community." But colonial domination, besides its econo-political subjugation, promoted Western-influenced communication innovations, which were geared toward endorsing Western cultural influence, "especially its journalistic practices and conventions that come with inherent assumptions, including philosophical (Land, 1992, p.10). Although Appadurai (2009, p. 217) suggests we consider how influences from other societies "tend to become indigenized in one way or another: this is true of music and housing styles as much as it is true of science and terrorism, spectacles and constitutions," we witness a struggling relationship between Enlightenment-crafted libertarian principles and dilemma-resolving principles of the indigenous people, especially when seeking to extricate themselves from repressive political regimes. Winner (1986 as quoted in Slack and Wise, 2002, p. 144) has extrapolated this concept in his examination of how technologies are "become woven into the texture of everyday existence," they "shed their tool-like qualities to become part of our very humanity." They become "deeply insinuated into people's perceptions, thoughts, and behavior." Technology historians have showed how technology, preceding its usage, mirrors as much as it enforces a social order (for example the transition from an agrarian order to one industrial). But facilitating a social order is political in itself. This assertion, however, must be laced with an attentiveness of technology's limitation to instrumental value, much less its ideological value.

Resonating with the metaphor of palaver tree is the Ethiopian *debo,* (Sopoya, 1999), the Malagasy's *fokolonoma* (Soahangy Mamisoa Rangers, 1999), *mugumo* (fig tree) among the Agîkùyù of Kenya (Kenyatta, 1962). Rangers illustrates how *fokolonoma* invoked a sense of community among the people of Malagasy. The term underlay their commitment to community development and sustainability, a concept violently disrupted by later metropolitan development agencies, including non-governmental organizations.

Ubuntu

Deriving its orientation from an assumption that "the community is prior to persons," (Christians, 2004, 235), ubuntu, "based on African thoughts, provides a cultural interpretation of an ethical principle such as human dignity as understood within the parameters of Western and liberal theories (Rao & Wasserman, 2007, p. 40). The term is coined from the expression "*umuntu ngumuntu ngabantu*" (I am a person through other persons' or "I am because we are' (Blankenberg as quoted in Rao and Wasserman, 2007, p. 40). Ubuntu echoes communitarian ethos by using the community as the guiding foci on which ethical decision-making should be placed. In fact Christians (2004) is so fervent about ubuntu that its focus on humans and "insistence on the moral dimension of society produce an ubuntu communitarianism that is the most mature version of nay to date."

GLOBAL MEDIA AND INFORMATION ETHICS?

We can use the 2005 publication of editorial cartoons of Mohammed by the *Jyllands-Postens* of Denmark to illustrate the problematic of having a global media and information ethics. The 12 cartoons depicting Prophet Mohammed were "testing whether fear of Muslim retaliation had constricted press freedom" (Alleyne, 200, p. 385). While publishing the cartoon was seen as an attempt at focusing the ravages of Islamic extremism, there were accusations of unmerited association of Islam with terrorism. Critics and Islam adherents considered the act as blasphemous as it was Islamaphobic. Some argued that the cartoons disparaged a Danish minority. There were demonstrations, which some slipped to violent riots in various countries around the world. Dozens of people were killed and hundreds injured, and property destroyed.

There were those who came in defense of the cartoons by invoking the principles of the First amendment – freedom of speech and expression, protection of debate and scrutiny of ideas in an *Areopagus*, marketplace of ideas, which is a fundamental trait in Western democracies. Why not publish the cartoons, some asked, while similar cartoons about other religions are printed frequently?

We may pause to speculate how much thought was given to the act of publishing these cartoons. Had the editors known the consequences, would they have changed their decision? How much should they be concerned about unintended consequences? Defending their right to publish the cartoons, the editors of *Jyllands-Posten* said that they "cannot apologize for our right to publish material, even offensive material" because one "cannot edit a newspaper if… paralyzed by worries about every possible insult" (Rose, 2006, p.x).

What was intriguing about this controversy was not so much the debate on freedom of expression and how the humiliation of the Muslim community. The repercussions of the publication of the cartoon precipitated economic and diplomatic consequences that would adversely affect Denmark. The Danish cartoon controversy ties this discussion back to Appadurai's disjunctive notion of global relationships and their attendant effect on the operation of nation-states where "the hyphen" linking "them is now less an icon of conjuncture than an index of disjuncture" (223). Alleyne (2009, p. 386) eloquently captures this disjuncture:

In the context of an international environment where the "War on Terror" had seemed to many a war on Islam, many Muslims felt that the central issue was respect for their religious identities. For this definitional reason, they attempted to retaliate by the only means at their disposals: formal written protests to the Danish government, mass protest, boycotts of Danish companies and products. The actions of a few Danish media workers had set in motion a series of events with profound consequences for the Danish economy and the Danish diplomatic services. Although the protestors sought and got an apology from the editor of Jyllands-Posten, the major responsibility for responding to the complaints was the domain of the country's diplomats- the primary representatives of a nation-state's national identity in an international order where governments have established for themselves the role of chief official transnational actors.

CONCLUSION

The discussion provided seeks to enhance literature on the discussion of global media and information ethics buy placing the discussion within a context of changes informing our globe. We argue how a discussion of global media and information ethics can only be discussed alongside global economic and political institutions.

Whether, as Alleyne (2009, 386) notes, that "a universal standard is…unlikely in the foreseeable future due to the deep primordial investment of different social systems in values very affected by the media," we invite media scholars to re-interrogate the leading Enlightenment philosophies that bolster media practices and their ethical dilemmas.

We encourage a critical examination of emerging dynamics of globalization and worldwide flow and counter-flow of cultures and sub-cultures, and information, capital and ideologies as pertinent reasons scholars should be seeking a more relevant philosophical approach. Global media and information ethics cannot be discussed in the absence of global economic and political structures.

REFERENCES

Alleyne, M. D. (2009). Global media ecology: Why there is no global media ethics standard. In Lee, W., & Clifford, G. C. (Eds.), *The handbook of mass media ethics*. New York: Routledge.

Appadurai, A. (1990). Disjuncture and difference in the global cultural economy. []. New York: Sage.]. *Theory, Culture & Society*, *7*, 295–310. doi:10.1177/026327690007002017

Benhabib, S. (1992). *Situating the self: Gender, community and postmodernism in contemporary ethics*. Cambridge, UK: Polity Press.

Chatman, E. (2000). Framing social life in theory and research. *New Review of Information Behavior Research*, *200*(1), 3–17.

Chatman, E. A., & Huotari, M. (2001). Using everyday life information seeking to explain organizational behavior. *Library & Information Science Research*, *23*, 351–366. doi:10.1016/S0740-8188(01)00093-7

Christians, C. (2006). The case for communitarian ethics. In Land, M., & Hornaday, B. W. (Eds.), *Contemporary media ethics: A practical guide for students, scholars and professionals* (pp. 57–69). Spokane, WA: Marquette Books.

Christians, C. G. (2004). Ubuntu and communitarianism in media ethics. *Ecquid Novi*, *25*(2), 235–256. doi:10.3368/ajs.25.2.235

Christians, C. G. (2005). Ethical theory in communication research. *Journalism Studies*, *6*(1), 3–14. doi:10.1080/1461670052000328168

Christians, C. G. (2007). Utilitarianism in Media Ethics and Its Discontent. *Journal of Mass Media Ethics*, *22*(2&3), 113–131.

Christians, C. G., Ferré, J. P., & Fackler, P. M. (1993). *Good news: Social ethics and the press*. New York: Oxford University Press.

Confucius,. (1979). *The analects* (Lau, D. C., Trans.). London: Penguin Books.

Cooper, T. (1996). *A time before deception: Truth in communication, culture, and ethics*. Santa Fe, NM: Clear Light Publishers.

Etzioni, A. (1993). *The spirit of community: The reinvention of American society*. New York: Touchstone Book.

Etzioni, A. (1996). *The New Golden Rule*. New York: Basic Books.

Friedman, J. (1990). Being in the world: Globalization and localization. In Featherstone, M. (Ed.), *Global culture: Nationalism, Globalization and modernity*. New York: Sage.

Friedman, T. L. (2005). *The World is Flat: A brief history of the twenty-first century*. New York: Farrar, Straus, Giroux.

Gikandi, S. (2005). Globalization and the Claims of Postcoloniality. In Desai, G., & Nair, S. (Eds.), *Postcolonialisms* (pp. 608–634). Oxford, UK: Berg.

Hall, S. (1996). Cultural identity and Diaspora. In Mongia, P. (Ed.), *Contemporary postcolonial theory: A reader* (pp. 110–121). New York: Arnold.

Hamelink, C. J. (2000). *The ethics of cyberspace*. Thousand Oaks, CA: Sage Publications.

Hamelink, C. J. (2004). Did WSIS Achieve anything At All? *Gazette: The International Journal for Communication Studies*, *66*(3-4), 281–290. doi:10.1177/0016549204043612

Hannerz, U. (1990). Cosmopolitans and locals in world culture. *Theory, Culture & Society*, (7): 237–251. doi:10.1177/026327690007002014

Kùng, H. (1998). *A global ethic for global politics and economics, (trans. John Bowden)*. Oxford, UK: Oxford University Press.

Land, M. (1992). Ivoirien television, willing vector of cultural imperialism. *The Howard Journal of Communications*, *4*, 10–27.

Land, M. (2006). Mass media ethics and the Point-of-Decision Pyramid. In Land, M., & Hornaday, B. W. (Eds.), *Contemporary media ethics: A practical guide for students, scholars and professionals* (pp. 15–38). Spokane, WA: Marquette Books.

Merrill, J. (2006). Utilitarianism Vs. communitarianism. In Land, M., & Hornaday, B. W. (Eds.), *Contemporary media ethics: A practical guide for students, scholars and professionals* (pp. 41–54).

Mudimbe, V. Y. (2009). In the name of similitude. In Njogu, K., & Middleton, J. (Eds.), *Media and identity in Africa, (International African Seminars)*. Edinburgh, UK: Edinburgh University Press.

Randall, V. (2008). Analytical approaches to the study of politics in the developing world. In Peter, B., & Vicky, R. (Eds.), *Politics in the developing world*. Oxford, UK: Oxford University Press.

Rangers, S. M. (1999, March). Madagascar: A waning tradition. *UNESCO Courier, 52,* 35. Retrieved December 23, 2008, from http://www.unesco.org/courier /1999_03/pdf/courier.pdf

Rao, S., & Wasserman, H. (2007). Global media ethics revisited: A postcolonial critique. *Global Media and Communication*, *3*(1), 29–50. doi:10.1177/1742766507074358

Rose, F. (2009). Why I published Those Cartoon. *Design*, *99*, 34–35.

Slack, J. D., & Wise, J. M. (2006). Cultural studies and communication technologies. In Lievrouw, L., & Livingstone, S. (Eds.), *The handbook of new media*. Thousand Oaks, CA: Sage.

Spivak, G. (2009). The new subaltern: A silent interview. In During, S. (Ed.), *The cultural reader* (3rd ed.). New York: Routledge.

Tar, K. (2001). *Toleration, diversity, and global justice*. University Park, PA: Pennsylvania State University.

Chapter 5
Public Discourses on the Korean Wave in Taiwan

Lihyun Lin
National Taiwan University, Taiwan

ABSTRACT

Recently, Korean dramas have won popularity in East Asia and provoked public discussions in newspaper forums in Taiwan. This chapter analyzes how the interpretive communities have formed their interpretive frameworks, what they have conceived from the Korean Wave and what policy implications have been drawn. According to the analysis, the dominant discourse is economic nationalism – while Korea has become a developed country, Taiwan should learn from Koran and catch up in this global competition. Imbedded in internal politics, different interpretative communities have learned different lessons from Korea. The pro-unification (with China) camp claimed that Taiwan should reduce local colors and open trade with China. The pro-(Taiwan) independence camp claimed that Taiwan should enhance national identity. The media reform group claimed that Taiwan should establish a strong public service broadcaster like the Korean counterparts. Based on this study, this chapter suggests that public discourses should address media problems in order to enhance communication at home and abroad.

INTRODUCTION

The spread of the Korean Wave has caught the attention of the press in Taiwan. The term "Korean Wave" was coined by the media to describe the popularity of Korea dramas in East Asia (Yang, 2008). In Taiwan, after Korean dramas enjoyed high ratings, discussions about the Korean Wave emerged in newspaper forums. The discussants tried to derive meanings from the Wave and draw implications that concern Taiwan.

The dominant discourse is economic nationalism. That is, to the discussants, Korea and Taiwan were once members of the Four Little Tigers (the newly industrializing countries in Asia) together. Yet, Korea has demonstrated itself as a leading country in the culture industry, while Taiwan has

DOI: 10.4018/978-1-60960-037-2.ch005

Copyright © 2011, IGI Global. Copying or distributing in print or electronic forms without written permission of IGI Global is prohibited.

been lagging behind and should learn from Korea in order to survive in this global competition.

However, imbedded in Taiwan's internal politics, different groups focus on different discourses and have learned different lessons from the Wave. They expressed different viewpoints in policy debates. In 2003, the government proposed to mandate terrestrial broadcasters to televise domestic programs during primetime, claiming that this is a lesson learned from Korea to boost local productions. This proposal was described by the media as an attempt to ban foreign programs including Korean dramas (Kao, January 11, 2006). In the policy debates, different groups, based on different discourses, expressed their perceived pros and cons.

The pan-blue camp, based on the discourse of economic (Chinese) nationalism, attacked the ban. This camp consists of the Chinese Nationalist Party (KMT) and related factions. After World War II, the KMT ruled Taiwan and imposed Chinese nationalism. Because of historical legacy, the pan-blue camp prefers unification with China to Taiwan independence (Chu and Lin, 2001; Chang, 2005). The pan-blue camp attacked the ban, claiming that, while Korean cultural products have transcended cultural differences, Taiwan should overcome "parochial" character and open up to the world, including China.

The pan-green camp, based on the discourse of economic (Taiwanese) nationalism, supported the ban. This camp consists of the Democratic People's Party (the DPP) and related factions. To challenge KMT's authoritarian rule, the DPP argued for the strengthening of Taiwanese identity (Chu and Lin, 2001; Chang, 2005). The pan-green camp claimed that as Korea has taken protection measures, Taiwan should also protect and support a culture industry that reflects Taiwan's own culture.

The other two discourses are not related to nationalism, but to the ideology concerning media policy. The media reform group, focused on the discourse of public intervention, supported the ban. Based on the belief that the media should play a role of public service in society, they have been concerned about solving the malaise in the media system after deregulation. To achieve this goal, they argued that Taiwan should follow Korea in adopting more public interventions. By contrast, the media industries focused on the discourse of *laissez-faire*. They argued that while Korea and other countries have been adopting de-regulation measures, further interventions would distort the market mechanism and affect the survival of domestic industries.

The above discourses can be read in newspaper forums. In Taiwan's political transition, the press has become implicitly partisan, paralleling to the ideologies in the political arena (Feng, 2003). The *United Daily* (*the UD*) has been aligned with the pan-blue camp, the *Liberty Times* (*the LT*), with the pan-green one. Another newspaper, the *Apple Daily* (*the AD*), launched by a Hong Kong newspaper group in 2003, is market-driven. The media reform group has delivered their ideas in all newspapers.

Thus, in drawing the meanings of the Korean Wave, different interpretive communities have formed. According to Fish (1980: 172-173), readers of certain cultural contexts would develop certain interpretive strategies and derive certain meanings from the texts. In this case, in newspaper forums, different interpretive communities have formed different interpretive strategies and derived different meanings from the Korean Wave.

This chapter aims to explore how different interpretive communities have formed their interpretive frameworks, what they conceived from the Korean Wave and what policy implications were drawn. In doing so, this chapter hopes to illuminate the public discourses in the wider contexts of political and inter-cultural communication.

BACKGROUND

The Korean Wave has fuelled research interest among the academia. Three analytical approaches have been adopted to study the Korean Wave: reception analysis, textual analysis and political economy analysis. In this section, we will discuss their main concerns, assumptions and implications regarding this study.

The reception analysis analyzes the experiences of the viewers of Korean dramas. It is assumed that in watching Korean dramas, audiences of different places gain new experiences but create meanings according to their local contexts. According to reception analysis, watching Korean dramas has become a source of pleasure for Asian audiences; the audience obtains a new image of Korea as a developed country, but reads texts through pre-established cultural repertories and sets of cultural meanings of their own (Chen, 2004; Chiang, 2004; Kim, 2005; Lin & Tong, 2008). Reception analysis has helped in the understanding of how cultural contexts can influence the interpretations of the Korean Wave.

Reception analysis focuses on viewers of Korean dramas, and considers the Korean Wave mainly as a part of popular culture. However, in this case study, the main concern is how the general public interpret the Korean Wave; further, the interpretations have also entered the policy arena. Thus, this case study is different from ordinary reception analysis, and will need to review other analytical approaches.

Textual analysis analyzes how the media represent the Korea Wave. It is assumed that texts contain news frames, which reflect the dominant ideology of a given society. According to textual analysis, the media representations of the Korean Wave are based on the discourse of economic nationalism; that is, the Korean Wave is seen to accumulate national wealth in international competition and to build stronger cultural identity both at home and abroad. The mainstream media in Korea consider the Korean Wave as boosting national economy and bringing national pride (Lee, 2008). In covering the Korean Wave, the media in other Asian countries consider Korea as developing into a media empire in Asia (Ju, 2007). Further, in these countries, some extreme nationalist groups consider the Korea Wave as a power invading and threatening local culture (Yang, 2008). According to textual analysis, the factor of nationalism is still an important one in trans-boarder communication, and could function effectively in this case when the Taiwanese public encounters the Korean Wave.

Textual analysis mainly focuses on texts. It pays less attention to the social contexts of the texts. However, not looking into the contexts, textual analysis cannot explain how the discourses were formed in a given society. To offer more plausible explanations about discourses, we need to look into a wider context of the texts.

The political economy approach analyzes the development of the media industry of import countries. According to Shim (2008:28), the Korean Wave was indebted to the media deregulation in Asian countries; in these countries, an increase in TV channels boosted the need for programs, and the channel providers were eager to introduce foreign programs, including Korean dramas. In the Taiwanese case, in the multi-channel environment after deregulation, the profit-seeking channel providers were important agents in promoting Korean dramas (Yeah, 2002; Kim, 2005). The political economy analysis helps to explain the spread of the Korean Wave in an import-led country.

The political economy approach mainly focuses on the structure of the media environment. It pays less attention to the cultural logics of a given society; the logics are revealed in the discourses and practices of actors. Thus, it is important to analyze the logics of the discourse and practices in order to obtain deeper meanings of human societies.

After reviewing the above approaches, this paper has developed its own analytical framework. Reception analysis shows that the audience would interpret Korean dramas based on pre-established cultural repertories. Textual analysis reveals that texts consist of the dominant discourse about the Korean Wave. The political economy analysis indicates that the media environment of the import-led country is an important condition for the rise of the Korean Wave. Based on the review, this chapter firstly analyzes the political economy of Taiwan's media environments. It will then secondly analyze how the interpretive communities formed their discourses, drew meanings from the Korean Wave and engaged themselves in policy debates.

ANALYSIS: FROM CONTEXTS TO TEXTS

Based on the above analytical framework, this section 1) looks into Taiwan's media milieu after deregulation, 2) analyzes how the interpretive communities have formed their interpretive frameworks and 3) examines how the communities, based on their interpretive frameworks, expressed their viewpoints in policy debates.

Deregulation in the 1990s

Taiwan experienced KMT's authoritarian rule until 1987. Taiwan was a Japanese colony between 1895 and 1945 and was returned to China at the end of World War II. In 1949, lost in the civil war against the Chinese Communists, the KMT regime moved from the Chinese mainland to the island of Formosa. In the cold war climate, the KMT imposed Chinese nationalism and justified its authoritarian rule. That is, the KMT claimed that the Taiwanese should shoulder the responsibilities of recovering the mainland and fight against the Chinese Communists. By claiming war-time necessities, the KMT imposed war-time regulations and froze the rights of the people guaranteed by the Constitution. Thus, the KMT party-state was able to control the media. The government, the party and the military respectively owned most shares of a terrestrial television station.

In the political liberation of the 1990s, the KMT did not reform the media. In 1988, the KMT declared to restore the rights guaranteed by the Constitution. However, the KMT still controlled the three terrestrial TV stations. As the official television stations could not satisfy people's needs for information and entertainment in a post-authoritarian society, underground cable and radio stations flourished and provided foreign programs (*e.g.*, US movies) illegally. To appease the people's resentment, the KMT did not impose harsh control over illegal cable television systems, except those involving political speech.

Under pressure from the US, in 1993, the KMT legalized the cable system. At that time, in order to protect the copyright of its culture industry, the US exerted pressure on Taiwan to legalize the cable system (Lee, 2000). The KMT government legalized the cable system with minimum restrictions on ownership, the number of channels and the percentage of domestic production (Chen, 1999).

Since the legalization, the cable system has become dominant and concentrated in Taiwan's media landscape. As there were no restrictions on ownership and the number of channels, two conglomerates entered the cable market and bought up existing cable systems; further, together with other channel providers, they launched several channels. Consequently, the number of channels in the cable system increased to more than 80. The penetration rate was boosted from 60 percent in 1997, to 80 percent in 2000. As a result, the two system operators controlled more than two-thirds of the market (Chen, 2001).

In this multi-channel environment, there is a strong demand for programs at low costs. On the one hand, the demand for programs greatly increased. In 1995, the three terrestrial television stations broadcasted approximately 21,000 hours of programming a year, while the cable channels combined televised a total of more than 360,000 hours of programming a year—16 times more than the total broadcasting hours of the three television stations (Government Information Office, 1995). On the other hand, the market share for each channel was dramatically reduced; the programs listed in the top 10 only had four percent ratings. Consequently, to gain profit, the channel providers had to trim the cost of production by resorting to sensational contents and re-runs (Examination Yuan of Taiwan Government, 2005). At the time, the channel providers imported programs mainly from Japan and Hong Kong.

To rectify the media system, a media reform movement was launched in 1999. Campaign for Media Reform (CMR) was established by progressive media scholars and professionals. The campaigners were based on the belief that the media should serve public interests and enhance the social welfares and cultural rights of the citizens. CMR argued for public interventions, including building a strong public television system and mandating private interests to fulfill public responsibilities (Feng, 2008). The aims of the movement included: 1) to restructure Taiwan's media with preliminary focus on issues such as curbing privatization; 2) commercialization of media ownership; and 3) lobbying for policies benevolent to the public interest of media access. The ultimate goal was to advocate for the communication rights of the citizens and to reform Taiwan's media ecology (CMR, 2008).

At this time, CMR mainly used Western European models to justify their claims. In 2000, the CMR demanded that the government transform the two state-controlled terrestrial television stations into public ones. They publicized articles in newspaper forums. In their articles, they used the public service broadcasting systems in Western Europe as examples to persuade the public that public service broadcasting would provide better services to democratic societies (Feng, December 20, 2001).

The discourse of public interventions was criticized by those who questioned that the European experiences did not fit Taiwan. In the authoritarian past, public television did not exist in Taiwan; main stakeholders (political parties and the media) did not have a clear vision of public media. In post-authoritarian politics, the dominant discourse was "free market." The private press argued that in the US public television only played a minor role and that the European model with different social and historical origins would not be appropriate for Taiwan (*the UD*, January 2, 2002).

Thus, in the debates, both camps used western models to justify their claims. The media reform campaigners used the European model to justify their claims for public interventions, while the free market camp used the US model. They did not use the models of other Asian countries until the arrival of the Korean Wave.

Learning Different Lessons from the Korea Wave

In the 2000s, the Korean Wave arrived at Taiwan, mainly driven by the demand for a larger variety of programs. As discussed previously, in this highly competitive multi-channel environment, the channel providers are always hungry for programs at low costs and of good quality. In 1999, while the cost of domestic dramas was as high as 20,000 US dollars (per hour), a channel provider, GTV, whose target audience is middle-aged housewives, began to import Korean dramas, which were sold at 10,000 US dollars (per hour), only half of the cost of domestic dramas (Hwang and Chang, 1999:117; Lieu, 2002: 52). In July 2000, the ratings of *Fireworks* climbed up and were further boosted to a historical high of 4.9 percent after the arrival of the Korean actors and actress on Taiwan soil. Thus, GTV continued to import more Korean dramas, and other channel providers began to follow suit (Lieu, 2000). Even the terrestrial television broadcasters began to broadcast Korean dramas during primetime, though given public airwaves, they were supposed to produce domestic programs. As a result, since 2004, Taiwan has become one of the largest importers of Korean dramas in Asia (Kim, 2004).

In order to boost ratings further, the channel providers also launched publicity campaigns. The channel providers put focus on the actors and actress as idols. The programmers also organized tours to Korea for journalists to conduct personal interviews with the stars and to experience the scenes of the popular dramas.

Consequently, all major newspapers devoted the bulk of their entertainment pages to cover the Korean Wave. In the entertainment pages, new sections on Korean dramas and stars appeared. These sections covered the stories of the dramas, the profiles and pictures of the stars and their schedules in Taipei (*i.e.*, their arrival times and hotels). Later in media coverage, thousands of fans gathered at Taipei airport for a glimpse of their idols (*Ettoday*, August 19, 2005).

The Korean Wave then rolled on from entertainment pages to forum pages. As the Korean Wave surged, the number of articles related to the Korean Wave increased. They were written by editorial writers, culture critics, scholars, and those who began to feel concern for Taiwan's culture industry and attempted to draw meanings of the Korean Wave in Taiwan.

The discussions were mainly based on the discourse of economic nationalism. The discussants addressed the issue from the perspective of economic competition between nations, after realizing the value of the culture industry in the process. Compared with the remote European media systems, the success of Korea has provided a great stimulus to the Taiwanese public. In the articles, it is stated that while Korea and Taiwan had similar historical development—both were Japanese colonies before 1945, marching toward modernization after World War II, and known as members of the Four Little Tigers in Asia—by establishing a thriving culture industry, Korea has taken a giant leap, while Taiwan was left lagging behind. The articles suggest that the Taiwan government should take active measures in order to catch up in this global competition. A culture critic put:

The Korean Wave is rising steadily. Demonstrating her economic capabilities, Korea is marketing her cultural goods globally...The Korean Wave, though not absolutely perfect, is to be learned and respected. The successful transformation of Korea in the past five years, allows us to reflect upon Taiwan's present predicaments and long stagnation (Wang, June 24, 2002).

However, in Taiwan's internal cultural politics, the pan-blue *UD* gradually formed the discourse of economic (Chinese) nationalism. After the rise of the Korean Wave, the *UD* began to advocate the intervention of the government to support the culture industry. In the first half of 2005, the *UD* was uncertain which policy would be better. Its editorial on March 20 analyzes that Korea and Taiwan have taken different paths of development; yet this editorial was uncertain which path would be right for Taiwan to build a culture industry. This editorial puts:

For the past decades, Korea has adopted a top-down approach, allocating national resources to certain industries...Taiwan, a de-centralization one...Our government should realize that Korea is becoming more and more competitive. We can not judge which one is good for Taiwan's culture industry (the UD, March 20, 2005).

However, a few months later, after the mobilization of a pro-independence group, the *UD* formed its discourse. Back to the authoritarian period, the KMT advocated Chinese nationalism, elevated the importance of Chinese culture (including Mandarin songs and language), debased Taiwanese culture (including Taiwanese songs and dialects) and gave lower-class Taiwanese the name of Taike. To challenge KMT's ruling ideology and later to find a niche within the currant music market, some Taiwanese signers called themselves Taike and publicly expressed their disdain for Mandarin popular music (Yang, 2008). In August, 2005, the Taikes held a festival, strongly advocating Taiwanese identity, mocking the mainstream consumption culture, including the Korean Wave. To fight against this force of Taiwanese nationalism, *the UD* argued that Taikes should avoid ethnic conflict and overcome its parochial character, because cultural products should reduce local colors to travel internationally. To strengthen its argument, *the UD* used Korean stars as examples. According to their argument, the Taike singers should learn from Bae yong-joon, a then popular star who has the charms to transcend all ethnic differences. *The UD* concluded that Taiwan should establish a culture industry based on universal values, not on Taiwanese local culture.

Why is the handsome Bae (yong-joon) able to succeed in Asia, gaining the hearts of men and women, young and old? Except his handsome outlooks, his gentleness, politeness and smiles make him irresistible…If we compare the Taike culture and the Korean Wave that Bae (yong-joon) represents, we can see the truth even clearly…The Korean Wave has conquered Asia, because Korea has the forward-looking national character…By contrast, except being indulged in the tragic past, what have we given to the so-called "Taiwanese"?

Try to make Bae yong-joon our teacher! … Appreciate his charms that transcend ethnic and cultural differences. Taike should not cause a rift among us (*the UD*, August 22, 2005).

By contrast, the pan-green camp argued that Taiwan should boost the culture industry to enhance Taiwan's culture identity. While the pan-green camp values the economic benefits of the culture industry, they also stress the importance of infusing Taiwanese elements into cultural products. The articles in *the LT* utilize the positive effects of the Korean Wave to strengthen this argument (Chang, July 12, 2005). A Taiwanese student in the US puts:

A group of Chinese students came to dinner the other day. They all talked about the Taiwanese idol dramas…They listened to the songs by Taiwanese singers. Apparently culture industry has far-reaching impacts…The Korean Wave has helped to establish the airlines between Taiwan and Korea… Hence, Taiwan should protect and promote her culture industry… (Chien, October 23, 2005).

While the Korean Wave has been widely discussed, the CMR began to use the Korean case to strengthen the argument. As discussed before, the CMR has aimed to build a strong pubic television system. Thanks to their efforts, in 2005 the government promised to transform one of two government-controlled terrestrial television stations into a public one. To urge the government to carry out the transformation project, the CMR argued that the Korean public broadcasters with a firm material base and stable investment in production are the powerhouses of the Korean Wave. As CMR states:

Though claiming to support the production of quality dramas, like Dae Jang Geum, the (Taiwanese) government will fail to deliver this promise. Dae Jang Geum was made by MBC, a strong public broadcaster. The Taiwanese government has procrastinated in establishing the infrastructure of the public television system (CMR, December 23, 2005).

Thus, because of the internal cultural politics, different camps formed their own discourses on the Korean Wave. The pan-blue camp, based on economic Chinese nationalism, argued that Taiwan's culture industry should reduce local colors. The pan-green camp, based on economic Taiwanese nationalism, argued that Taiwan's culture industry should reflect Taiwanese culture. The media reform group based their discourse on public intervention. Because of the different interpretive frameworks, their positions in related policy debates were shaped.

Policy Actions and Debates

In the second half of 2005, the DPP government translated the discourse of the pan-green camp into an action of policy. In February 2005, Premier Frank Hsieh came to office. As a DPP politician, Hsieh had been concerned about cultural affairs. When Hsieh came to office, the Korean Wave was hotly discussed. He claimed to follow the step of Korea in establishing the Taiwanese culture industry:

Our neighboring countries in Asia have put audio-visual industry as a key item in national development project. Their governments are sparing no efforts in promoting this industry. The achievement of Korea is the most impressive...Korea's audiovisual industry, together with her electronics industry, are two pillars in Korea's economy. Korea's success has given us the implications that we should pursue the creative industries (Hsieh, September 6, 2005).

To achieve his ideal, Hsieh appointed his long-term follower, Pasuya Yao, as Director of the Government Information Office (the government agency in charge of the audio-visual industry). Among Taiwanese politicians, Yao has been exceptional in taking an active role in his posts in cultural affairs (Lin, March 13, 2005). Following Hsieh from Kaohsiung city government to the Central government, Yao also followed Hsieh's footsteps, claiming to learn from the Korean policy actions and therefore taking several measures.

Firstly, claiming to learn from Korea, Yao proposed to promote Taiwan's film industry. According to Yao, the government would provide subsidies to directors and producers who have won international film awards: a total of 0.33 million US dollars (10 million NT dollars) for each award-winner. The government would also follow Korea in establishing a funding agent, providing funds, talents and assistance in promotion for those who are interested in the culture industry (Lee, August 23, 2006).

Secondly, responding to the demands of the media reform group, Yao fastened the pace to establish a strong public television system. In September 2005, Yao introduced the Statute on the Disposition of Government Shareholdings in the Terrestrial Television Industry. This Statute allowed the state-controlled terrestrial broadcasting stations to become public entities and merge with each other. Later in 2006, Yao promised to make a budget of 3 billion US dollars (92 billion NT dollars) for Taiwan's public service television to establish new channels.

Thirdly, considering that there were too many channels in Taiwan's cable system, Yao attempted to reduce the number of channels. In renewing the licenses of program providers, Yao suspended the licenses of seven program providers which were considered to be poor-managed (*The Epoch Times*, July 31, 2005). This measure was harshly criticized by the media industries and the KMT, who claimed that Yao attempted to control the media.

In January 2006, Yao proposed another controversial policy action—banning foreign dramas on primetime in terrestrial television stations. Yao claimed that, as Korea had imposed a screen quota system effectively, Taiwan should follow and enforce related regulations to protect local productions (Ko, January 12, 2006). According to Yao, in Taiwan's Broadcasting Law, "Locally produced programs shall not be less than 70 percent of the total radio/television programs" (Article 19), but the government has never seriously enforced this regulation. Yao's proposal soon triggered wide debates. The interpretive communities expressed their pros and cons, based on their interpretive frameworks.

The pan-blue camp and press, based on economic Chinese nationalism, attacked the policy action of the pan-green government. At this point, the pan-blue camp has realized the importance and the potentials of the culture industry in accumulating national wealth. However, with stronger Chinese nationalism, the pan-blue camp does not support the policy that advocated Taiwanese nationalism. Thus, in this controversy, the pan-blue camp did not support the protection of Taiwanese dramas. The KMT politicians attacked Yao as a Fascist, who was trying to control people's thoughts and ideas (Kao, January 11, 2006). An article in *the UD* argued that the pan-green government should not be blinded by Taiwanese nationalism, and that further protection would not improve the quality of programs. The *UD* published an article by a college student.

The policy-makers only deem the "Taiwanese subjectivity" as a Holy thing. The ban will not improve the quality of the programs. (The Taiwanese) audience will launch a "turn-off" campaign in the primetime (Cheng, January 11, 2006).

Conversely, the pan-green, based on economic Taiwanese nationalism, supported the ban on foreign programs. The pan-green camp has realized that the government should promote Taiwan's culture industry both for economic wealth and cultural identity. Thus, in this controversy, the pan-green camp supported the quota system. According to the articles in *the LT*, other countries (*e.g*., Japan, Korea and China) have imposed quota systems to protect their culture industry, and Taiwan should follow their example (Lo, January 11, 2006; *the LT*, January 13, 2006).

Both Japan and South Korea have long restricted foreign soap operas on primetime television to protect their own industries. While other countries do not want to play fair, Taiwan needs to reconsider its game rule (Chiu, January 12, 2006).

The CMR, based on the discourse of pubic intervention, supported the ban. Unlike the pan-blue and the pan-green camps, the CMR has been concerned about establishing the pubic character of the media, but not nationalism. To build a strong public service broadcasting system and to regulate commercial media, in this controversy, the CMR considered it necessary to mandate the local television industries to broadcast local programs (Chen, January 14, 2006). Furthermore, the CMR argued that in addition to the ban, the government should take additional measures to rectify the hyper-commercial market. A CMR member argues:

If Yao really wants to promote the local audio-visual industry...he should do more than just increase the proportion of local production...He should solve the problems of the distorted market. Not adjusting the market structure, not strengthening the material base of the industry and not supporting public television, the government will be not able to further constructive media policies (Kwan, January 13, 2006).

The media industry was based on the discourse of *laissez-faire*. The industry argued that the government should not intervene with the market. The television industries (including the cable program providers and terrestrial broadcasters) were against the ban, because they wanted to import more Korean dramas in order to trim the costs of production. The *AD* was also on the same front; it was based in Hong Kong and came to Taiwan only in 2003. The owner has not been involved in Taiwan's politics and adopted a market-driven approach. Thus, the *AD* argued that the ban would damage the market mechanism, and thus violating the interests of the industry and the audience (The *AD*, January 11, 2006).

Under the protest of the pan-blue camp and the media industry, the ban was never implemented. The pan-blue politicians continued to criticize Yao not only in Congress, but also in talk show programs on cable channels. Under the mounting pressures, Yao had to announce that the policy was only at the proposal stage. After Hsieh and Yao stepped down from the post in a cabinet reshuffle in March 2006, none of his successors ever adopted an interventional approach. Consequently, Taiwan's media landscape remained largely unchanged. The channel providers and the terrestrial television stations continued to import foreign programs, including Korean dramas.

After the turmoil, different interpretive communities have continued to be engaged in the policy debates and to take the Korean case as an important example to justify their claims. The pan-blue camp, based on economic (Chinese) nationalism, has adopted a pro-China policy. After the pan-blue candidate won the Presidential election in May 2008, the KMT government claimed to follow Korea in building a culture industry, mainly by strengthening cooperation with China through co-productions and common platforms. The KMT government states:

Taiwan's biggest market is mainland China... [The government] plans to strengthen the mechanism for cross-straits creative cooperation... The Chinese side has made a friendly decision..., to allow TV shows produced jointly by the two sides to be broadcast and published in the mainland, and to take part in exhibitions there. Joint productions now may enjoy equal treatment with TV shows produced by the mainland." (The Executive Yuan of Taiwan government, November 11, 2009)

The pan-green camp attacked KMT's pro-China policy. The pan-green camp, based on economic (Taiwanese) nationalism, considers that it is vital for Taiwan to build her own culture industry and to strengthen national identity. To challenge KMT's pro-China policy, the pan-green camp has accused the KMT of misinterpreting the Korean case. According to an article in *the LT* by a DPP politician, Korea succeeded by developing her own brand and national industry, not by relying on China; thus, in learning from Korea, Taiwan should create her own brand, instead of relying on China (Lin, February 16, 2008). This former congressman puts:

KMT politicians have delivered a myth to us: For the development of Taiwan's economy, we need to advance free trade with China and to treat China as our home market. The KMT has attributed Korea's success to the trade with China... However, the rise of Korea lies in innovation, research and developing her own brand and culture industry.

The CMR, based on the discourse of public interventions, exerted pressures on the two politician camps to implement media reforms. After Yao, later GIO chiefs made little progress in media policies. The CMR has protested against the procrastination on media reform. To strengthen arguments, the CMR held annual Citizens' Media Reform Conferences in which the participants mapped out policy actions based on the studies of different countries, including Korea. For example, the CMR analyzed Korea's public broadcasting systems, the quota system and the fund for promoting culture industry promotion (Feng, 2008). The CMR also observed media reform movements in Korea and their relevant strategies (Lin, 2008).

Because of internal cultural politics, the debates between the different interpretive communities have continued, all claiming to have learned from Korea. Claiming to learn from Korea in opening up the world market, the KMT government put the discourse of economic Chinese nationalism into practice by pushing free trade with China. The pan-green camp also claimed to learn from Korea by building national identity. The media reform campaigners continued to study Korea's television system. Thus, the interpretative communities, based on their discourses, "learned" different aspects of the Korean Wave to justify their claims.

Solutions and Recommendations

According to the above analysis, in internal cultural politics, main political camps in Taiwan have been mainly concerned about the issue of national identity but not media policy. Based on economic Chinese nationalism, the pan-blue camp has been interested in furthering relationships with China. The pan-green camp has been focused on building Taiwan's national identity. They have utilized different aspects of the Korean Wave to justify their claims and political aims. Though the media reform group has continued to stress the need for reforms, both camps have not yet focused on the media problem.

Not addressing the media problem, the public discourses have two limitations. Firstly, both political camps have not yet seriously faced media problems in Taiwan. The media problems have never been tackled and settled. Second, because of the different national imagination of the two camps, the policy discussions have been often stuck in unproductive debates and have not reached any consensus. Thus, any policy reform would have been impossible.

It is vital for the public to focus on media problems. Those who are concerned with the media environment should address media problems and propose media policies on two levels. On a micro-level, Taiwan should consider how to build a sound and sustainable media system. On a macro-level, in trans-border communication, Taiwan should consider how to have constructive trans-border communication, particularly with the neighboring East Asian countries, including China, Korea and Japan. If the two camps can focus on these issues, more productive dialogues can then be expected.

In proposing plausible options, the public should have full understanding of the experiences of other countries. At present, different interpretative communities have used only a part of the Korean experience to justify their claims and even serve their political purposes. To opportunely seize the valuable parts of the Korea Wave, the public should conduct a holistic and historical approach to understand the systems and histories of Korea.

Moreover, alternative discourse concerning trans-border communication should be proposed. According to the study, in Taiwan as well as in other countries, the dominant discourse is economic nationalism. However, in this line of thinking, cross-border communications have been mainly channeled through commercial agents and thus narrowed by the market logic of the import country. To enable further communication, alternative discourse should be proposed. Public broadcasters and independent producers of the region should endeavor to open up more public spaces not mediated by commercial interests. For example, they could exchange programs with those of more diverse perspectives. Further, they could co-produce programs (including dramas) based on the historical pasts, current situations and future scenarios of the region. For example, some Korean and Japanese producers have cooperated to produce a documentary program *Annyong, Sayonara*. In this film, a Korean woman launched a lawsuit against the Yasukuni Shrine and the Japanese government to withdraw the enshrinement of her father, which was supported by a Japanese man. Through this connection, the people of the two countries faced their painful history together and found a way for peace and conciliation. Communicating in this way, ordinary people of different countries will be able to encounter each other (not only the stars), untangle their real problems, acknowledge their commonalities and respect the differences.

FUTURE RESEARCH DIRECTIONS

The main concern of the chapter is: how the public interprets the trans-border communications in East Asia. In this chapter, we mainly analyzed newspaper forums as an important public discourse. The pubic discourses were mainly focused in wider social contexts. Based on this analytical framework, there are at least three future research directions.

Firstly, future research can focus on the public discourses on the Korean Wave in other countries. The Korean Wave has covered not only Taiwan, but also many other Asian countries. Different countries, because of their internal dynamics, could have different discourses on the Korean Wave. With more case studies, future research can compare the public discourses on the Korean Wave in different countries.

Secondly, future research can analyze the public discourses in addition to newspaper forums. The Korean Wave has provoked discussions not only in the newspaper forums but also in other public forums, including public hearings in policy debates, bulletin board systems and scholarly works. With different interpretive communities, these forums might have different discourses on the Korean Wave. With more case studies, future research can compare the public discourses on the Korean Wave in different public spaces.

Thirdly, future researchers can analyze the discourses on other cultural goods in trans-border communication. With enhanced global communication, more cultural goods have been invented and promoted to cater to people from different areas of the world. The cultural goods include both hardware (from digital television to mobile telephones) and software (from Hollywood, Bollywood to Hallyu-wood). It will be fruitful to analyze how the public interprets these cultural goods and what kind of cultural logics and values can be projected through these interpretations.

CONCLUSION

After Korean dramas enjoyed high ratings, discussions about the Korean Wave emerged in newspaper forums in Taiwan. However, different interpretive communities have different interpretive frameworks. This chapter has aimed to explore: How different interpretive communities have formed their interpretive frameworks, conceived the Korean Wave and drawn policy implications.

To answer this question, this chapter has reviewed main analytical approaches and developed its own analytical framework. The reception analysis showed that local audiences interpret Korean dramas based on the pre-established cultural repertories; but the reception analysis has not analyzed the reception of the general public and the public discourses. The textual analysis revealed that texts consist of the dominant discourse, but has paid less attention to the contexts of the texts. The political economy analysis showed that the media environment is an important condition for the rise of the Korean Wave, but it has paid less attention to the cultural logics of the discourses and practice.

Based on this review, this chapter has focused on public discourses in wider contexts. It has analyzed the political economy of Taiwan's media environments. It then analyzed, in these contexts, how the interpretive communities have formed their discourses, drawn meanings from the Korean Wave and engaged themselves in policy debates.

According to the analysis, the arrival of the Korean Wave has, to some extent, opened up discussions concerning media policy. Before the arrival of the Korean Wave, the dominant discourse was de-regulation, only the media reform group argued for public interventions and used the European model to strengthen their argument. Compared to the remote case of the European media systems, the Korean Wave has provided a great stimulus to the Taiwanese public. The pubic began to realize the economic and cultural values of the culture industry in international competition and demanded the government to take action. Thus, the Korean Wave has served as mirror for Taiwanese to reflect upon Taiwan's stagnation in recent years.

However, because of internal cultural politics, different camps have formed their own discourses concerning the Korean Wave. In the authoritarian past, the KMT imposed Chinese nationalism to justify its rule. To challenge this ruling ideology, the DPP has struggled for Taiwanese identities. Because of historical legacies, the pan-blue camp has formed the discourse of economic Chinese nationalism, arguing that Taiwan's culture industry should reduce local colors, and that Taiwan should open up trade with China. The pan-green camp has formed the discourse of economic Taiwanese nationalism, arguing that Taiwan's culture industry should reflect Taiwan's culture. The media reform group has formed the discourse of public interventions. In their discourses, different interpretative communities have learned different aspects of the Korean Wave to justify their claims.

Based on the discourse analysis, this chapter suggests that future public discourses should address media problems. In this case, the main political camps have been concerned about political problems (*e.g.*, national identity). They have used different parts of the Korean case to strengthen their arguments and to serve political purposes. They were unable to propose media policy, to have productive debates, or to take policy actions. Thus, the chapter suggests that the interpretive communities should address the media problems of a given society, fully understand the media systems of other societies, and propose alternative discourses in addition to the dominant discourse of economic nationalism.

REFERENCES

Chang, M. (2005). Taiwan's nationalistic politics and its difficult 'status quo'. *The Copenhagen Journal of Asian Studies*, *21*, 91–124.

Chang, Y. (July 12, 2005). Taiwan wave! (in Chinese) *Liberty Times*. Retrieved April 17, 2010, from http://www.libertytimes.com.tw /2005/new/jul/12/today-o4.htm.

Chen, P. H. (2002a). The role of the state in shaping Taiwan's cable industry. *Media Asia*, *29*(1), 37–41.

Chen, P. H. (2002b). Who owns cable television? Media ownership concentration in Taiwan. *Journal of Media Economics*, *15*(1), 41–55. doi:10.1207/S15327736ME1501_4

Chen, P. H. (January 14, 2006). *Ban foreign programs not close-door policy* (in Chinese). Retrieved April 17, 2010, from http://www.twmedia.org/modules/ news/article.php?storyid=228

Chen, Y. (2004). *The formation of television audience: The market of Korean dramas* (in Chinese). Unpublished master dissertation, Tankang University.

Cheng, Y. (January 11, 2006). Audience to watch Korean dramas (in Chinese). *United Daily*, p. A15.

Chiang, P. (2004). *The imagined cultural landscape: The Korean wave and the fans in Taiwan* (in Chinese). Unpublished master dissertation, National Chengchi University.

Chien, C. (January 11, 2006). Support the ban on foreign programs (in Chinese). *Liberty Times*. Retrieved April 17, 2010, from http://www.libertytimes.com.tw /2006/new/jan/11/today-o10.htm

Chien, C. H. (October 23, 2005). Defensive effect of culture (in Chinese). *Liberty Times*. Retrieved April 17, 2010, from http://www.libertytimes.com.tw/ 2005/new/oct/23/today-o9.htm

Chiu, D. (January 12, 2006). Taiwanese soaps need protection (in Chinese). *Taipei Times*. Retrieved April 17, 2010, from http://www.taipeitimes.com/News/ editorials/archives/2006/01/12/2003288638/wiki

Chu, Y.-H., & Lin, J.-W. (2001). Political development in 20th-Century Taiwan: State-building, regime transformation and the construction of national identity. *The China Quarterly*, *165*, 102–129. doi:10.1017/S0009443901000067

CMR. (December 23, 2005). Our anger at betrayal of government (in Chinese). Retrieved April 17, 2010, from http://www.mediawatch.org.tw/ modules/news/print.php?storyid=172

CMR. (2008). About the campaign for media reform. In Y. -P. Lin (Ed.), *A 21th century anthology of media criticism*. Retrieved April 17, 2010, from http://www.twmedia.org/

Ettoday (August 19, 2005). At the storm of handsome Bae, fans form "a united nations" (in Chinese). Retrieved April 17, 2010, from http://www.nownews.com/ 2005/08/19/91-1833232.htm

Examination Yuan of Taiwan Government. (2005). *Review of the audiovisual policy and practices in Taiwan*. Taipei: Examination Yuan of Taiwan Government. (in Chinese)

Executive Yuan (November 11, 2009). Cultural and creative industries: New engine for wider economic growth. Retrieved April 17, 2010, from http://www.ey.gov.tw/ct.asp?xItem=64011&ctNode=1334&mp=1

Feng, C. (December 20, 2001). Private ownership unable to solve problems (in Chinese). *United Daily,* A15.

Feng, C. (2003). Press freedom, to do or not to do [in Chinese]. *Journalist, 34*, 23–25.

Feng, C. (2006). Constructing and expanding Taiwan's public television from 1990 to 2006: Roles of academic intellectuals, social campaign and political authority [in Chinese]. *Journal of Communication and Society, 1*, 47–67.

Feng, C. (2008). Culture & economy: Comparing Taiwan and Korea. In Peng, H. (Ed.), *Two Asian dragons: Comparing Taiwan and Korea* (pp. 212–239). Taipei: Asia and Pacific Culture and Education Foundation.

Feng, C. (January, 26, 2008). Cultural policy in election (in Chinese). *Liberty Times*. Retrieved April 17, 2010, from http://www.libertytimes.com.tw/ 2008/new/jan/26/today-o1.htm

Fish, S. E. (1980). *Is there a text in this class?: The authority of interpretive communities*. Cambridge, MA: Harvard University Press.

Government Information Office. (1995). *The case analysis for the development of the Asian-Pacific Media Center*. Taipei, Taiwan: Government Information Office. (in Chinese)

Hsieh, C. (September 6, 2005). *First Coordinating Meeting under the Executive Yuan*. Retrieved April 17, 2010, from http://info.gio.gov.tw/ct.asp?xItem=26821&ctNode=3105&mp=1

Hwang, Y., & Chang, L. (1999). Overview of satellite and cable channels [in Chinese]. *Advertising Magazine, 95*, 15–119.

Ju, H. (November 14, 2007). *The nature of nationalism in the "Korean wave": A framing analysis of news coverage about Korean pop culture*. Paper presented at the annual meeting of the NCA 93rd Annual Convention. Retrieved April 17, 2010, from http://www.allacademic.com/ meta/p187925_index.html

Kao, L. (January 11, 2006). Lawmakers: Yao as Facist (in Chinese). *United Daily,* A4.

Kim, M. (May, 2004). *Cultural proximity and the type of Korean television programs in Asian market*. Paper presented at the 6th Media Economic Conference, Montreal, Canada.

Kim, H. M. (2005). Korean TV dramas in Taiwan: With an emphasis on the localization process. *Korea Journal,* (winter 2005), 183-205.

Ko, S. (January 11, 2006). GIO looking to take foreign soap operas off primetime TV. *Taipei Times*. Retrieved April 17, 2010, from http://www.taipeitimes.com/News/fron t/archives/2006/01/11/2003288408

Ko, S. (January 12, 2006). Soap opera plan's critics are ignorant, Yao retorts. *Taipei Times*. Retrieved April 17, 2010, from http://www.taipeitimes.com/News/tai wan/archives/2006/01/12/2003288583

Kwan, C. (January 13, 2006). *Dae Jang Geum a local dramas* (in Chinese). Retrieved April 17, 2010, from http://www.twmedia.org/modules / news/article.php?storyid=231

Lee, C. (2000). State, capital and the media. In Curran, J., & Park, M.-J. (Eds.), *De-Westernizing media studies* (pp. 110–120). New York: Routledge.

Lee, K. (2008). Mapping out the cultural politics of"the Korean Wave" in contemporary South Korea. In B. H. Chua & K. I. Iwabuchi (Eds.), *East Asian pop culture: Analysing the Korean wave* (pp.175-189). Hong Kong: Hong Kong University.

Lee, S. (August 23, 2006). The government spends 200 billion in promoting audio-visual industry (in Chinese). *United Daily. Retrieved* April 17, 2010. Retrieved from http://udndata.com/ndapp/Story?no=2...news_id=3069674

Lieu, P. (2002). The DNA of the Korean dramas [in Chinese]. *The Journalist (Weekly)*, *800*, 50–53.

Lieu, S. (August 24, 2009). Where is Taiwan's culture industry? Kim Dae Chung has led the miracle of the Korean wave (in Chinese). *United Daily*, A13.

Lin, A., & Tong, A. (2008). Re-imagining a cosmopolitan 'Asian Us': Korean media flows and imaginaries of Asian modern femininities. In Chua, B. H., & Iwabuchi, K. I. (Eds.), *East Asian pop culture: Analysing the Korean wave* (pp. 91–125). London: Eurospan.

Lin, L. (2005). Authoritarian states and television: A comparative study between Taiwan and South Korea [in Chinese]. *Mass Communication Research*, *85*, 1–30.

Lin, L. (2008). The press reform in Korea. *Mass Communication Research*, *95*, 257–282.

Lin, S. (March13, 2005). Aggressive Yao with dreams of films (in Chinese). *Economic Daily*, A7.

Lin, S., & Yang, W. (May 25,2005). Hsieh: Taiwan dangerous, if small and rigid (in Chinese). *Economic Daily*, A1.

Lin, T. (February 16, 2008). Taiwan's brand on the global stage (in Chinese). *Liberty Times*. Retrieved April 17, 2010, from http://www.libertytimes.com.tw /2008/new/feb/16/today-o1.htm

Lo, S. (January 15, 2006). Support the quota system! (in Chinese) *Liberty Times*. Retrieved April 17, 2010, from http://www.twmedia.org/mod ules/news/article.php?storyid=229

Lo, T. (January 11, 2006). Korean dramas in Japan (in Chinese). *Liberty Times*, Retrieved April 17, 2010, from http://www.libertytimes.com/20 06/new/jan/11/today-o11.htm.

Shim, D. (2006). Hybridity and the rise of Korean popular culture in Asia. *Media Culture & Society*, *28*(1), 25–44. doi:10.1177/0163443706059278

Shim, D. (2008). The growth of Korean cultural industries and the Korean wave. In Chua, B. H., & Iwabuchi, K. I. (Eds.), *East Asian pop culture: Analyzing the Korean wave*. London: Eurospan. [distributor]

The Apple Daily (January 11, 2006). Editorial: Yao mad (in Chinese). *Apple Daily*. Retrieved April 17, 2010, from http://tw.nextmedia.com/applenews/art icle/art_id/2329059/IssueID/20060111

The Epoch Times (July 31, 2005). Seven cable channels failed in the review (in Chinese). *The Epoch Times*. Retrieved April 17, 2010, from http://www.epochtimes.com /b5/5/8/1/n1003832.htm

The Liberty Times. (January 13, 2006). Editorial: Damn on the pro-China group (in Chinese). *Liberty Times*. Retrieved April 17, 2010, from http://www.libertytimes.com.tw/ 2006/new/jan/13/today-f1.htm

The United Daily (August 22, 2005). Taike, Hanru (in Chinese). *United Daily*, p. A2.

The United Daily (January 2, 2002). Editorial: Liberalization, best for state TV (in Chinese). *United Daily,* p. A2.

The United Daily (March 20, 2005). Editorial: Korea wins over Taiwan? (in Chinese).*United Daily,* p. A2.

Wang, B. (January 12, 2006). Yao: policy distorted (in Chinese). *Liberty Times.* Retrieved April 17, 2010, from http://www.libertytimes.com.tw/2006/new/jan/12/today-p12.htm

Wang, C. (August 20, 2005). Not boosting culture industry, Taiwan lags behind (in Chinese). *United Daily,* p. A15.

Wang, S. (1998). Where are the audience of domestic dramas? [in Chinese]. *Advertising Magazine, 85,* 127–130.

Wang, S. (June 24, 2002). Behind the Korean wave (in Chinese). *United Daily,* p. A15.

Yang, F. C. (2008). Rap(p)ing Korean wave: National identity in question. In Chua, B. H., & Iwabuchi, K. I. (Eds.), *East Asian pop culture: Analysing the Korean wave* (pp. 191–216). Hong Kong: Hong Kong University.

Yeah, W. (2002). *The management strategies of importing foreign programs: The example of vedioland* (in Chinese).Unpublished master dissertation, Minchuang University.

ADDITIONAL READING

Chen, P. H. (2002, May). The role of the state in shaping Taiwan's cable industry. *Media Asia, 29*(1), 37–41.

Chu, Y. H. (2004). Taiwan's National Identity Politics and the Prospect of Cross-Strait Relations. *Asian Survey, 44*(4), 484–512. doi:10.1525/as.2004.44.4.484

Jin, D. (2006). Cultural politics in Korea's contemporary films under neoliberal globalization. *Media Culture & Society, 28*(1), 5–23. doi:10.1177/0163443706059274

Kang, M.-K. (2000). Discourse politics toward neoliberal globalization. *Inter-Asia Cultural Studies, 1*(3), 443–456. doi:10.1080/14649370020009933

Livingstone, S., Lunt, P., & Miller, L. (2007). Citizens and consumers: Discursive debates during and after the Communications Act 2003. *Media Culture & Society, 29*(4), 613–638. doi:10.1177/0163443707078423

Ryoo, W. (2005). The role of the state in the national mediascape: The case of South Korea. *Global Media Journal, 4*(6), 1–32.

Shim, D. (2002). Hybridity and the rise of Korean popular culture in Asia. *Media Culture & Society, 28*(1), 25–44. doi:10.1177/0163443706059278

Shim, D. (2005). Globalization and cinema regionalization in East Asia. *Korea Journal, 45*(4), 233–260.

Shim, D., & Jin, D. Y. (2007). Transformations and development of the Korean broadcasting media. In Murphy, P., & Blankson, I. A. (Eds.), *Globalization and media transformation in new and emerging democracies* (pp. 161–176). New York: Suny Press.

Tiffen, R., & Kwak, K.-S. (2005). Cable television and democratization in Taiwan and South Korea. In Romano, A. R., & Bromley, M. (Eds.), *Journalism and democracy in Asia* (pp. 135–144). London, New York: RoutledgeCurzon.

Wang, V. (2004). Broadcasting: PSB in Taiwan a bright future. *Intermedia, 32*(4), 6–9.

Young, D. (2003). Discourses on communication technologies in Canadian and European broadcasting policy debates. *European Journal of Communication, 18*(2), 209–240. doi:10.1177/0267323103018002003

KEY TERMS AND DEFINITIONS

Korean Wave: The popularity of the South Korean cultural/media products and images in Asia

Interpretative Community: The collectivity of people who share strategies for interpreting, using, and engaging in communication about a media text or technology

Discourse Analysis: The analysis of written, spoken or signed language, focusing on the ways social and political domination are reproduced by text and talk

Public Discourse: A particular cultural form in which members of the public imagine themselves as belonging to a particular public

Political Parallelism: The extent to which the media are linked, formally or informally, with parties and other political institutions

Chapter 6
The Emerging Media Exchange in the Cultural Regionalization of Asia

Peichi Chung
National University of Singapore, Singapore

ABSTRACT

This chapter focuses on the emerging media regionalization that takes place in Asia in 2000s. Japan and Hong Kong used to be the dominant cultural exporters commercializing their national media products to the nearby Asian markets. The recent market success of Korean wave and the gradual opening of Chinese market bring media regionalization to a different level. The chapter selects three cases to present the detailed image of cultural standardization in Asia's media regionalization. The first centers on the circulation of media text in television drama, emphasizing on Korean wave and the particular TV series, Boys Over Flowers. The second case discusses Taiwanese popular music and its influence on Mandopop in the Chinese communities. The last case studies the regionalization of online game from China. This case examines the localization of Chinese online game, Westward Journey Online II. Chinese online games initially begin with the imitation of Korean game but later form their national branding based upon a mixture of global and local cultural elements that speak to the largest group of online game consumers in the pan-Asian market.

INTRODUCTION

In recent years we have seen an increase of regional media exchange emerging in Asia. As the media systems in most Asian countries become liberalized and privatized, more media companies are able to export media products to nearby countries. Part of the reason for this growth in media regionalization is a developed system for media content delivery resulting from decades of cultural trade in the region. Another reason speeding up the cultural exchange in Asia is the advancement of new media technology, as satellite and the internet have provided convenient access for audiences to receive diverse programming from various sources (Iwabuchi, 2005). The other factor

DOI: 10.4018/978-1-60960-037-2.ch006

Copyright © 2011, IGI Global. Copying or distributing in print or electronic forms without written permission of IGI Global is prohibited.

benefiting the frequency of media exchange is the improvement of Asian economies in the 2000s. As the size of the international media market in developed economies shrinks during the economic downturn, international advertisers begin to look for consumers in emerging markets. These factors altogether contribute to an active production and consumption environment that allows a self-sustainable economic community to form in Asia.

This chapter provides an analysis of emerging trends in Asian regionalization. The chapter examines the complex relationship that shapes the current expansion of popular culture in Asia. According to a multinational survey on Asian soft power in 2008, United States, Japan, China and South Korea are the four countries that have prevalent popular culture in Asia (East Asia Institute, 2008). On the one hand, as the United States benefits from the free market and open competition in the region, many Asian countries are strongly influenced by the US at the levels of policy, politics, economy and culture. On the other hand, we also witness Asian regionalism on the rise. The report shows that Japan, South Korea and China have increasingly established active interdependence that leads to a greater integration among nations in the Northeast Asia. A shifting hierarchy of cultural order in the region also emerges in the area of popular culture. Japan and Hong Kong used to be the dominant cultural exporters commercializing their national media products in the Asian markets. For instance, Japanese comic books have been routinely translated into the local languages in South Korea, Thailand, Indonesia and Taiwan (Otmazgin, 2008). A variety of Japanese TV programs are also widely apparent in many cities in East Asia (Keane & Moran, 2005). In addition to Japan, Hong Kong film and music also achieve wide circulation, in particular with Hong Kong film talents successfully crossing to Hollywood (Lo, 2001). The recent market success of *Korean wave* and the gradual opening of the Chinese market further bring this structure of media regionalization to a different level. There is a shift of power in media production and the obvious is the case of *Korean wave*. The increasing popularity of Taiwanese music in China and the gradual commercialization of Chinese online games in Southeast Asia reveal an evolving state of Asian media culture that does not purely rest on any particular national territory in forming the collective identity of the region

The chapter looks into the hierarchy of this emerging media trend in the 2000s. The purpose is to use the theoretical concepts of cultural adaptation and hybrid regionalism to explain the evolving process of cultural convergence in the popular media of Asia. I argue that media regionalization in Asia should move beyond the discourse of national media in which global information flow means an encounter of the global to the local. The cultural phenomenon in Asia shows that certain Asian media industries are emerging in the region. These industries have formed strategies to increase the internationalization of their media content. Their Asia-centered approach demonstrates a new regional identity based upon power negotiation among national media at the regional level.

The chapter examines three cases to present a detailed image of cultural regionalization in Asia. Firstly, the paper focuses on the circulation of media text in television drama – with emphasis on *Korean wave* and the particular TV series, *Boys Over Flowers*. This television series is selected for analysis because the show provides a unique example on the marketability of Korean television drama in Asian markets. Secondly, I discuss Taiwanese popular music and its influence on Chinese Mandopop in China. This case focuses on the connectivity of Taiwanese popular music to Chinese market. Taiwanese popular music also reflects the complexity of cultural exchange in various Chinese communities. Thirdly, I study the regionalization of online games from China. This section examines the adaptation of foreign content in Chinese online games. The section uses the case of *Westward Journey Online II* to show a collective trend of cultural standardization that encourages the consumption of local games among Chinese gamer. Many Chinese games initially begin with the imitation of Korean games but later form their national branding based upon a mixture of global and local cultural elements. The strategy of cultural adaptation in *Westward Journey Online II* shows the diversity of cultural element mixed in online game of China. *Westward Journey Online II* shows the influence of Hong Kong film culture in the addition of Chinese theme in local Chinese games. The game was popular among Chinese gamers due to its connectivity to a larger group of consumers that follow *Kung Fu* culture in the pan-Asian market.

LITERATURE REVIEW

Most of the literature that discusses Asia popular culture is framed within the theory of globalization. The underlining theme in this line of scholarship is to argue against the simplicity of one way information order in media imperialism theory (Shim, 2005). Globalization reflects the multiple dimensions of information exchange that is under the global spread of neoliberal deregulation. With the World Trade Organization and transnational corporations playing key roles in mastering capital and commodities, the global information order has demonstrated an extraordinary growth in direction in the increasingly networked global society. Thussu (2006) argues that while global information is still dominated by American media, contra information flows also exist in the transnational space. The internationalization of telenovela and Bollywood shows that subaltern flows are hybrid and circulating in global communication.

Responding to the transnational effect of hybrid media contents from Latin America and India, globalization scholars continue to complicate the concept of cultural globalization by examining the intensification of regional media flows in Asia (Chua, 2008; Iwabuchi, 2002; Shim, 2006). Iwabuchi (2005) shows that the power relations operated in cultural globalization are processed in a dispersed and decentered manner. He states that "cultural globalization does not just mean the spread of the same products of Western (mostly American) origin all over the world through these media conglomerates" (Iwabuchi, 2004, p.1). Rather, since the early 1990s, Japanese media industries have developed successful strategies to enter the booming Asian audiovisual market. The Japanese media industries appropriate a hyperactive strategy that "indigenizes America" in their creation of media content. This indigenization of American popular culture leads to the rise of non-US television culture that is "culturally odorless" but made-in-Japan (Iwabuchi, 2004). The most actively developed localization is portrayed by the popularity of Japanese pop idols across Asia. "Japanese music industries produced local pop stars in China, Indonesia and other countries, who were marketed to other countries by learning the Japanese know-how of the cultural appropriation of American popular culture" (Iwabuchi, 2008, p. 146).

Being the media center that inspires transnational Asian flow has turned Japanese popular culture into a leading source for a regional popular cultural trend among Asian countries (Lee, 2004; Liu & Chen, 2004; Shim, 2005). Korean television culture, for instance, internalizes Western culture into its local context through Japanese media culture (Lee, 2004). Despite the fact that the South Korean government used to ban cultural products from Japan, Korean cultural industries use Japanese media as a reference of imitation and cultivated Japanese tastes for Korean audiences (Shim, 2005). In Taiwan, Japanese media culture also assisted the Taiwanese media industry to survive the hegemony of American media (Liu & Chen, 2004). The local producers revive Taiwan's television entertainment industry by heavily adapting the Japanese quiz show format in variety shows. This cultural borrowing also offers Taiwanese idol dramas a distinctive Taiwan look that results in popular reception of Taiwanese idol dramas, including *Meteor Garden*, among the Southeast Asian audience communities. Overall, the Japan-based media regionalization shows a transnational flow of East Asian pop culture achieving cultural economy integration among various Asian audience communities (Chua, 2008). Further development of this integration is the similar indigenization process of Korean dramas, where dubbing and subtitles allow Korean dramas such as *Winter Sonota* (겨울연가), *Dae Jang Geum*(대장금), The 1st Shop of Coffee Prince(커피프린스 1호점), *My Lovely Sam Soon* (내 이름은 김삼순) and *Boys Over Flowers* (꽃보다男子), to become popular in the Chinese locations of Hong Kong, Taiwan, Singapore and the People's Republic of China.

Two concepts can help to explain the significance of the continuous expansion of cultural economies from East Asia: cultural adaptation and hybrid regionalism. According to Keane and Moran (2005), adaptation has become the dynamism of today's circulation of global popular media culture. The adaptation of format from reality shows such as *American Idol* and *Who Wants to be a Millionaire* indicates the willingness of TV producers to extend the life cycles of their media products. The tendency to adapt global programs that are successful comes from the economic concern that media producers "adapt successful media content because these materials offer some chance of duplicating past and existing successes" (ibid, p. 2). In Asia, adaptation of popular formats helps media producers to anticipate how media markets work with consumer expectation across media systems in Asia (Keane, 2004). There are two major kinds of adaptation. One is adaptation at the institution level as film becomes television series (or vice versa). The other is adaptation at the content level in the case of serialization through sequels (Keane & Moran, 2005). In Asia, adaptation often works at the content level that ideas and know-how are transferred culturally without involving any licensing agreement (Liu & Chen, 2004). Some media producers even directly copy and imitate the original programs simply for the purpose of the convenience of applying a successful formula to secure audiences in the local markets.

Another concept that further explains the expansion of East Asian popular culture further is media regionalization. I refer to two terms, hybrid regionalism, proposed by Katzenstein (2006) and strategic regionalism, proposed Sinclair and Wilken (2009) in this paper to indicate the complex regionalization process that East Asian popular culture currently encompasses. As Katzenstein (2006) states, Asia is going through a hybrid form of regionalization that goes beyond any one national model. The shared desire to achieve economic efficiency has led governments to form clustering nations under the regionalization process of three countries. Japanization, Americanization and Sinicization are the three cultural flows that take place at the same time in Asia's forming of a regional identity (Munakata, 2006). The East Asian economy appears as a result of intraregional competitive dynamics that involve industrial nations and less developed nations. In media industries, Sinclair and Wilken (2009) identify similar market dynamics that international companies utilize in their advertising campaigns when they enter regional markets. In Asia, leading brands such as Coca-Cola and McDonald have applied strategic regionalization in their marketing campaigns. On the one hand, these international companies use standardization strategies to maintain quality control of their global brands. However, on the other hand, these companies also diversify their marketing strategies based upon regional differences. The recognition of regional market dynamics further leads international companies to recognize the needs of strategic management in their local adaptation. McDonald's effective use of standardization and localization strategies can be seen in the example of the "I'm Loving It" global advertising campaign when in Asia McDonald replaces Justin Timberlake with Chinese singer

Wang Li Hong as part of its regional promotion strategy (ibid, p. 152).

HYBRID REGIONALISM IN EAST ASIAN POPULAR CULTURE

Korean Television Drama: Boys Over Flowers

Korean popular culture has been an emerging popular culture in Asia since the 1990s. The spread of Korean drama shows the transnational mobility of Korean cultural products. Kim (2007) argues that *Korean wave* is the consequence of active government support in the culture industry of Korea. Politically, the Korean government promotes its popular culture as the "soft power" of the country and uses it to increase their national image abroad. Economically, *Korean wave* also contributes to the national economy because domestic conglomerates internationalize Korean popular culture by exporting films and television dramas to other Asian markets. Iwabuchi (2004b) identifies this emerging success of Korean wave an uneven cultural flow embedded in the intraregional cultural flow in Asia. Shim (2006) considers this new shift of cultural location as a form of cultural hybridization based upon globalization trends and democratic reforms in the Korean media market. In all, the asymmetric power relation reflects a niche market that the Korean wave occupies in the region. Korean television and dramas work on an alternative value system that combines an up-to-date fashion style and slick production values with "Confucian" or "traditional East Asia" cultural norms (Armstrong, 2008). This Confucian theme resonates with cultural values in many Asian societies, in particular, Vietnam. Korean media content provides a fresh and new feel that is different from the dominant culture. American and Japanese cultures are already well-known in the region.

This section applies the concepts of media adaptation and hybrid regionalism to discuss the changing structure of intraregional cultural flows in Asia. By analyzing the factors contributing to the success of Korean television drama – *Boys Over Flowers*, this section demonstrates the complexity of cultural flow in *Korean Wave*. *Boys Over Flowers* is a Korean television drama that was released on KBS2 in Korea early in 2009. The show created a hit success during its premiere, enjoying an average of 30% viewership rating during the Monday and Tuesday timeslots (Ki, 2009). Immediate regional success followed after the release of the show in Korea, selling rights to eleven countries in Asia including the Philippines, Taiwan, Indonesia, Japan and Singapore. The show brought the actors and actresses to immediate stardom. Lead characters such as Koo Hye Sun (구혜선), Lee Ming Ho (이민호), Kim Hyun Joong (김현중), Kim Beom (김범) and Kim Jun (김준) and even the K-pop boy band SS501 received immediate international attention through the popularity of the F4 drama. The DVD release in the US market began with a US-based Korean publishing company, YA Entertainment, in November 2009.

The discussion on the factors contributing to the success of *Boys Over Flowers* should start with the successful adaptation strategy using the Japanese comic book narrative for Korean television drama. The original story is based upon the Japanese best selling manga, *Hana Yori Dango*, by Yoko Kamio. This comic book series had 37 volumes, ran continuously from October 1992 to September 2003, and was one of the best selling Japanese manga at all times. The comic book reached a total sale of 54 million copies by 2005. In 2001, Taiwanese television producers began the first TV adaptation with producing the Taiwanese version of *Meteor Garden*. The show was popular among Taiwanese audiences and enabled Taiwanese dramas to become popular in Southeast Asian markets (Liu & Chen, 2004). For instance, *Meteor Garden* is listed as one of the most popular television dramas in the Philippines. In 2005, Tokyo Broadcasting System produced the Japanese version of *Hana Yori Dango*. The television series had the most popular J-pop idol, Jun Matsumoto, cast in the lead role, Tsukasa Domyoji (道明寺司). Both the Taiwanese and Japanese series became hit TV dramas in the Asian markets. The Taiwanese version even brought international fames to new talents such as Jerry Yang (言承旭), Vic Zhou (周渝民) and Barbie Hsu (徐熙媛) whom later became heavyweight actors and actress in Taiwan's idol dramas. After the successes of the Taiwanese and Japanese version of *Hana Yori Dango* in the regional market, the success of the Korean version reaffirms the popularity of the formula that adapts popular literary narrative for pan-Asian audience consumption. One related cultural phenomenon is that, the Philippines will produce its own version of *Boys Over Flowers* and release the TV series on the local television channel, ABS-CBN, in 2011.

The second factor contributing to the success of *Boys Over Flowers* shows a strategic production formula that is new and atypical to traditional Korean wave dramas. This formula involves an alternative financing method in gaining capital from overseas investors in the regional market. While the regular Korean television drama remains a local production that emphasizes stories taking place in Korean society, *Boys Over Flowers* shows a hybrid storytelling style that mixes an international cultural element (the Japanese comic narrative) and a local cultural element (the fairy tale identity in Korean wave drama). For instance, the main plotline of *Boys Over Flowers* has a Japanese presence that both producers and audiences share in terms of originality as a *Hana Yori Dango* story. While the director has remained truthful to represent characters in the development of the story, he also implements the traditional fairy tale formula in typical Korean wave dramas by adding more storyline to connect most of the characters (ie. the relationship between So Yi Jung and Chu Ga Eul). In addition, the Korean version also demonstrates a specific social setting that allows audiences to gaze the life style of the economic and political elites in the Korean society. On the production side, the financing process reflects a hybrid regional production approach as part of the production budget comes from pre-sales with foreign investors for the total amount of 6.5 billion Korean won (US$ 5.4 million). The director adds more sensational scenes in order to compare competitively with the previous Japanese and Taiwanese versions. This heavy investment in production shooting involves foreign location shootings and frequent product placement in the show. The hybrid financing approach also shows that while KBS only subsided 150 million Korean won (US$ 0.126 million) for an episode, the actual budget required 280 million Korean won (US$ 0.236 million). The production has to move toward developments that strongly appeal to regional sales so that the director can maximize profit in pan-Asian markets.

Mandopop in Taiwan

Another emerging cultural trend is the gradual rise of Mandopop in Taiwan. Popular Chinese music in 2000s has shown a shift in consumption from Cantopop to Mandopop (Fung, 2007). This rise of Chinese popular music from Taiwan is similar to the increasing popularity of Taiwanese idol drama that benefits from the massive Chinese market. There is a refashioning of Taiwanese idol drama after the success of *Meteor Garden* in Taiwan's television export market (Chen, 2008). Similarly, the music industry in Taiwan functions as a springboard for overseas ethnic Chinese singers to gain fame on the global Mandopop scene. Overseas Southeast Asian singers such as Stephanie Sun and JJ Lin both started their music careers in Taiwan before they entered the Chinese market. Malaysian Mandopop singers such as Leung Jing Ru and Guang Liang also moved to Taiwan in order to connect to the Chinese market outside of Malaysia. In Taiwan, Mandopop serves as a base for both local and international Chinese artists.

The music industry in Taiwan produces popular Chinese music that reaches the Mandarin-speaking populations of the world. According to a report from the International Federation of Phonogram and Videogram Producers (IFPI), the trade revenue of music industry in Taiwan totaled $68.9 million in 2007. This market is comparable to China, an emerging market that suffers from piracy. China's total market revenue was $69.4 million in the same year (Worth, 2009).

Iwabuchi (2002) argues that the music industry in Taiwan was influenced by Japanese idol systems in the 1980s, but the industry later moved away from Japan to form an identity of its own. Ho (2006) reviews Taiwan's music history and finds that political changes shaped different stages of development in Taiwan's popular music. Fung (2007) further argues that the circulation of music products in Hong Kong, Taiwan and China reflects the changing ideologies of the People's Republic of China. In all, Taiwan's music industry first was influenced by Japanese colonization, but then shifted to focus on music production under the influence of single party control by the Kuomintang. The lifting of martial law in 1987 led Taiwan to democratization and also further shapes the dynamism of Taiwan's music scene today. Political factors lead Taiwan's Mandarin popular music to absorb influences from other Chinese music such as Shanghai and Hong Kong. The Mandopop music mixes these elements with the local style (Shim & Ho, 2009). Up to the 2000s, the Mandopop in Taiwan formalized a style of music production and songwriting that mixes American hip hop, Japanese pop and local ballad. The popular music scene in Taiwan has a combination of Taiwanese, Mandarin and non-mainstream singers. It also contains major international record companies such as Sony, EMI, BMG, Warner and Polygram. During the 1980s, Taiwanese Mandopop queen, Teresa Teng, became extensively popular among listeners in the Mandarin-speaking communities. Cassettes of Teng were illegally carried in luggage into China, as Teng represented an icon of modernity and for imagining a foreign lifestyle for common Chinese listeners (Fung, 2007). Taiwanese Mandopop stars started to receive fame in China and triumphed in creating music across all genres. Current notable Taiwanese artists include singers, A-mei, Jay Chou (周杰倫), David Tao (陶喆), Jolin Tsai (蔡依林), Lihorn Wang (王力宏) and girl band, S.H.E. (女朋友). Their music contains a mixture of western "R & B" and ballads in Chinese art form. These singers tend to establish a large scale fan base in Chinese communities despite the fact that digital piracy prevails in the consumption of Chinese Mandopop music. One notable example is the completion of more than 40 concerts in the world tour of Taiwanese mandopop band, Super Band (縱貫線). The band consists of four veteran heavyweight singers, Lo Ta-Yu (羅大佑), Jonathan Lee (李宗盛), Wakin Chu (周華健) and Chang Chen-Yue (張震嶽). Concerts of Super Band were held in many cities in China, while the band performed with over 70 albums published by the four since 1970s in their tour from 2008 to 2010.

Unlike the *Korean wave*, the adaptation of the Mandopop music genre demonstrates a trend of standardization in the making of music for Chinese-speaking markets from the perspective of international music companies. The gradual rise of political and economic power in People's Republic of China brought to the influx of Taiwan music that inevitably replaced Cantopop (Fung, 2007). Most of the music industry in Taiwan is controlled by international music companies including Universal Music Asia Pacific, Warner Music Chinese Mandarin Group and EMI Taiwan. These companies consider Taiwanese music talents having the most qualified production technique to produce music among the current Chinese speaking communities. The companies rely on Taiwanese know-how to create Mandopop because the young generation in China is believed to have cultivated a taste of consumption that sees popular music in Taiwan cooler than home-grown (Worth, 2009). As Mandopop is still in the process of transformation in China, major label companies use the strategy of standardization to control a stable regional growth as popular music is developing in China. The strategy has proved to be effective as the market revenue of digital music continues to increase in China. One example is the profits resulting from digital download in China. According to the *Digital Music Report 2009*, China is potentially a huge music market with a serious problem of online piracy (IFPI, 2009). However, despite rampant piracy, online companies such as Baidu still earn an annual advertising earnings of RMB 30 million (US$5 m) simply from the illegal download service with the majority of the illegal download coming from popular music in both Chinese and English languages.

Chinese Online Game: The Westward Journey Online Series

This section discusses the most recent regionalization in online gaming, and focuses on the gradual internationalization of online game distribution in China. Chinese games have gradually followed the expansion pattern of Korean games, penetrating into the markets of South Korea, Taiwan and the emerging markets of Southeast Asia. In 2006, Chinese game company sold its self-developed game, *Perfect World* (完美世界), to Japan and South Korea (Kshetri, 2009). In 2007, one of China's most profitable games, *Zheng Tu* (征途), was sold to Malaysia. In 2009, as the global game market became much accessible to players whom played games at the internet terminals and cellphones, top tier game companies from China such as Changyou, Perfect World and Shanda also exported games to the global market. China is expected to have more than 40 percent of the global market by 2011 (Reuters, 2009).

From 2000 to 2004, Korean games used to dominate the market of China. The domination did not last until 2005 when the Chinese government employed a protectionist policy to restrict the number of foreign games imported into China. In 2004, Korean games occupied 70 percent of China's game market. After 2005 Chinese games topped the list of popular games in China (Chung & Yuan, 2009). The policy has cultivated a market with gamer culture that prefers Chinese games that have historical themes and *xuxia* genre. Despite the fact that most Chinese games have specific local cultural appeal and sometimes reflect similar design concept to foreign (mostly Korean) game products, certain Chinese game companies are still able to overcome these local market barriers and enter the global market. For instance, in 2009, 64 online games were exported to overseas markets (Xin, 2010). These games reached the countries of Japan, South Korea, Southeast Asia, Hong Kong, and Taiwan. One online game company, Perfect World, successfully entered the Korean market by working with the major game publisher, CJ Internet, in Korea. The company drew on Chinese culture to produce games of *Wulin* and *Perfect World* for which led to the distribution Chinese games to Malaysia, Singapore, Thailand, Taiwan and Japan (KPMG, 2007).

Table 1 shows the top five money-making MMO games in 2008. Except for the titles of *World of Warcraft* and *Runescape* that were produced in the US and UK, games ranked from the second to the fourth on the list are Asian MMORPGs that are developed by Chinese and South Korean game companies. This table reflects a notable development regarding to the economic performance of online games in Asia. As the online game, *Fantasy Westward Journey* (梦幻西游) is a game circulated only in the Chinese market, the other two games, *Maple Story* and *Lineages I & II*, demonstrate regional appeal and have proved to be successful in many Asian markets. NetEase is a Chinese game developer and publisher that designed the game, *Fantasy Westward Journey*. The company creates a successful genre that reflects a typical development strategy used by local game companies in China.

In contrast to previous two cases of *Korean wave* and *Mandopop*, the case of online game

Table 1. The Top Five Money-Making MMOs of 2008

Rank	Title	Genre	Country of Development	Estimated 2008 Revenue
1	World of Warcraft	MMORPG	USA	$500 million
2	Fantasy Westward Journey	MMORPG	China	$150 to $500 million
3	Maple Story	MMORPG	South Korea	$150 to $500 million
4	Lineage 1 & Lineage 2	MMORPG	South Korea	$150 to $500 million
5	Runescape	MMORPG	United Kingdom	$150 to $500 million

Source: The Top 10 Money-Making MMOs of 2008 at <http://www.salon.com/mc/giga_om/print.mc?URL=/tech/giga_om/tech_insider/2009/0_10_money_making_mmos_2008 >

shows that Chinese game companies use the strategy of localizing foreign media content to increase the marketability of their games in China. Keane (2005) emphasizes the element of "imitation" in China's creative media industry. The success of *Fantasy Westward Journey* is based upon the popularity of game series -- *Westward Journey Online I & II* (大话西游) that NetEase produced in 2001 and 2002. *Fantasy Westward Journey* and the *Westward Journey Online Series (*大话西游系列) were listed in the top ten online games in 2007 (Chung & Yuan, 2009). Both games reflect the industry development process that local game companies use to mix local and foreign cultural elements to create effective branding to Chinese consumers.

This section focuses on *Westward Journey Online Series* for analysis. In 2001, NetEase developed China's first MMORPG game, *Westward Journey Online I* (大话西游1). The company published *Westward Journey Online II* (大话西游2) in 2002. The game is an adaptation of ancient Chinese story, *Journey to the West* (西遊記). However, unlike other Chinese online games that tried to adopt Korean style of game development, NetEase differentiated its product by conceptualizing the game based upon a historical narrative in traditional Chinese folk stories. The series focused on a theme that was commonly reflected in Hong Kong cinema. The game series attracted attention from local gamers in China that by 2003 *Westward Journey Online II* reached around 3 million users (Baidu Encyclopedia, 2010). In 2007, the company developed the upgraded version of *Westward Journey 3* (大话西游3). This game was a 2D MMORPG that used improved graphic style and advanced animation technology. The company also expanded its market scale by developing *Westward Journey Wai Zhuan* (大话西游外传) and the 3D version of *Chuan Shi Westward Journey* (创世西游). The overall performance of NetEase's adoption of Hong Kong formula was the success of *Westward Journey Online II* and *Fantasy Westward Journey.* Both games became top ten popular online games in China in 2007 (Chung & Yuan, 2009). The popularity continued in 2009 as the former game had 96 million and the later game attracted 99 million registered users in China (Baidu Encyclopedia, 2010).

Westward Journey Online II reflected a Hong Kong artistic concept that allowed players to enjoy humorous dialogues and gang competition when they constructed their game narrative. Most Chinese game developers emphasized on the importance of developing games based upon ancient Chinese stories. The game *Westward Journal Online II* concentrated on the classic Chinese folk story of the Monkey King in Tang Dynasty. While *Journey to the West* is a story well known among common households in China, NetEase created a storytelling style that went beyond classic Chinese literary traditions. The game heavily built its narrative upon modern version of *Journey to the West* – a literary text that was a remade comedy by Hong Kong filmmaker, Stephen Chow. Chow's movies were popular among the young generation in China. NetEase adopted the comic film style to create game that contained humorous action and slapstick. For instance, in the movie of *Journey to the West* Stephen Chow named the leading actress, Lin Zixia. The name shared similar pronunciation to famous Taiwanese actress, Lin Chin Hsia (林青霞), a reputable actress who has been popular in Chinese movies since 1970s. Similarly, naming in *Westward Journey Online II* also shared this comedic result to the gamers, as gamers followed Chow's naming style to engage in social relationships online. In additions, *Westward Journey Online II* also mixed Chow's comedic style with traditional cultural elements that showed a virtual space about gods, human beings and monsters in ancient Chinese society. The adaptation of Hong Kong –based media content showed the influence of non-local culture in the creation of competitive local games in China. This game therefore showed a media strategy of cultural regionalism that NetEase used to create brand effect and achieve product differentiation in the company's competition with game developers from Korea and America.

FUTURE RESEARCH DIRECTION

Future research direction of this paper should show a comprehensive analysis on the emerging cultural exchange in the media regionalization of Asia. As the chapter argues that the advancement of media technology increases the frequency of cross border media flow within Asia, further studies should trace the complexity of cultural exchange among various media production locations in Asia. For instance, Korea, Taiwan and China are the three emerging production locations that follow the media production style populated by Japan and Hong Kong. This potential power shift within new and old geographic locations discloses a complex power negotiation process among Asian audiences. While new cultural producers are emerging into the media scenes of television drama, film, music and online game, what type of power restructuring does this new development change our understanding of collective identity in Asian popular culture? As media regionalization represents a fluid process of media production and consumption, there is also a need to critically examine the influence of this emerging media flow to popular consumption areas including Japan, Hong Kong, Singapore, the Philippines, Vietnam and other South East Asian countries.

CONCLUSION

A review of three emerging cultural locations in 2000s shows that cultural exchanges have become frequent and complex in the Asian context. While the popular culture in Asia at the globalization level indicates a focus on the decentralization of the west through Japanese and Hong Kong media, the emerging cultural presence of *Korean wave*, Taiwanese Mandopop and Chinese online game further points out new directions of cultural convergence that goes beyond Japan and Hong Kong. In the case of Korean television, *Boys Over Flowers* provides an alternative model to evaluate a new approach of content production in television drama. A production strategy that combines branding of *Korean wave* and a popular story plotline that is already popular in Asia guarantees market success across various Asian communities. In the case of Mandopop, Taiwan being the epicenter of Chinese popular music points to the gradual rise of Chinese music market after the demise of Cantopop in the Chinese music market after 1990s. The presence of transnational music companies reveals sources of corporate strategies as these companies use the know-how in Taiwanese music to control the uncertainty of an emerging market. Taiwanese Mandopop has gained a strong presence in many Chinese speaking communities including Singapore and other Chinese communities in Southeast Asia and China. The Mandopop reaches the young generation in China that is the largest consumption group in any form of popular entertainment. In the case of Chinese online game, local game companies follow the game development style from either Korea or the United States. The internationalization of *Perfect World* demonstrates an effort to "denationalize" the Chinese identity through representing cultural images that resemble game products from Korea. Another example is *Westward Journey Online II* that echoes this denationalization process by locating regional cultural element from Hong Kong. It is through their relocation to a different regional locale that local games can become "local" in the game development process. Consequently, adaptation becomes a standardization strategy for the local media producers to reach economic success in media regionalization. As the intraregional cultural flow goes beyond a fixed location, it is the constant negotiation among different cultural trends that creates the influential presence of East Asian popular culture in Asia.

The three cases in all provide two new directions to rethink the concept of media regionalization in Asia. First, media regionalization in Asia is in an evolving process due to the emerging trend of cultural standardization in popular culture. Korean television dramas show that regional success can be achieved through the local appropriation of Japanese literary content. Taiwanese Mandopop shows an effort of standardization from international music companies. The case of game development in China also sees the localization of foreign media content as the strategy for Chinese online game companies to achieve production standard in their local markets. While some Chinese games incorporate Korean and American influences in their content creation, an alternative is to remix media cultures from other overseas Chinese media industries. Hong Kong cinema in the case of *Westward Journey Online II* has thus become a cultural resource for local game companies to produce games that both reflects Chinese themes but contains foreign content to guarantee success in the local market.

Secondly, the chapter complicates the concept of media regionalization in Asia. The decentralization of "America" and "Japan" in the globalization debate leads to a further argument about media regionalization and its various different forms in Europe, Latin America and Asia. In Europe, media regionalization concentrates on the various media cultures contributing to the pluralism that shapes the collective regional identity of Europe. Be it in the form of film, television, music and advertising, European media lacks a common ground to reach cultural standardization at both the global and regional levels. However, in the American continent, media regionalization means power struggle under a strong presence of American popular culture. There is an imbalance of power between Hollywood and the media production locations within the free trade region of North American Free Trade Agreement (NAFTA). For instance, Canada and Latin America are closely connected to the media network in the United States. The close integration of Canada as Hollywood North and Latin America as Hollywood South illustrates a cultural dilemma that both Canada and Latin American countries face in their resistance to assimilate to American popular culture. Media regionalization in Latin America therefore represents an important concept of cultural autonomy that the audience communities in Latin America embrace and celebrate. Differently, in Asia, the constitution of media regionalization is fragmented, diverse and still emerging. The economic development of Asia after the global economic crisis will soon speed up the process of a changing collective identity of Asian regionalism. Media regionalization in Asia will be a continuously transforming concept influenced by various political, economic and cultural forces that work toward the goal of regional integration among participating countries.

REFERENCES

Armstrong, C. (2008). Korean wave: A view from America's shores. *Korea Herald*, 5 February. LexisNexus. Consulted on October 31, 2009.

Baidu Encyclopedia. (2010). *NetEase company*. Retrieved April 3, 2010 from http://baike.baidu.com/view/149363.html (In Chinese).

Chen, Y. (2008). Looking for Taiwan's competitive edge: The production and circulation of Taiwanese TV Drama. In Zhu, Y., Keane, M., & Bai, R. (Eds.), *TV drama in China* (pp. 175–187). Hong Kong: Hong Kong University Press.

Chua, B. (2008). East Asian pop culture: Layers of communities. In Kim, Y. (Ed.), *Media consumption in everyday life in Asia* (pp. 99–113). New York: Routledge.

Chung, P., & Yuan, J. (2009). Dynamics in the online game industry of China: A political economic analysis of its competitveness. *Journal of Eptic On Line, 11*(2). Retrieved March 29, 2010 from http://www.eptic.com.br/arqu ivos/Revistas/vol. XI,n2,2009/11- PeichiChungJiangPingYuan.pdf

Fung, A. (2007). The emerging (national) popular music culture in China. *Journal of Inter-Asia cultural studies, 8*(3), 425-437.

Ho, W. (2006). A historical review of popular music and social change in Taiwan. *Asian Journal of Social Science, 34*(1), 120–147. doi:10.1163/156853106776150216

IFPI. (2009). *Digital music report 2009: New business models for a changing environment*. Retrieved October 31, 2009 from http://www.ifpi.org/content /library/DMR2009.pdf

Iwabuchi, K. (2002). *Recentering globalization: Popular culture and Japanese transnationalism*. Durham, NC: Duke university press.

Iwabuchi, K. (2004a). *Cultural globalization and Asian media connections*. Hong Kong: Hong Kong University Press.

Iwabuchi, K. (2004b). Discrepant intimacy: Popular culture flows in East Asia. In Erni, J., & Chua, S. (Eds.), *Asian media studies* (pp. 19–36). Malden, MA: Blackwell Publishing.

Iwabuchi, K. (2005). Introduction: Cultural globalization and Asian media connection. In Iwabuchi, K. (Ed.), *Feeling Asian modernities: Transnational consumption of Japanese TV dramas* (pp. 1–22). Hong Kong: Hong Kong University Press.

Iwabuchi, K. (2008). Cultures of empire: Transnational media flows and cultural (dis)connections in East Asia. In Chakravartty, P. & Y. Zhao, (eds.), *Global communications: Toward a transcultural political economy* (pp. 143-162). Lanham, MD: Rowman & Littlefield publishers, Inc.

Katzenstein, P. (2006). East Asia – Beyond Japan. In Katzenstein, J., & Shiraishi, T. (Eds.), *Beyond Japan: The dynamics of East Asian regionalism* (pp. 1–36). Ithaca, NY: Cornell university press.

Keane, M. (2004). Asia: New growth areas. In Moran, A., & Keane, M. (Eds.), *Television across Asia: Television industries, programme formats and globalization* (pp. 9–20). London: Routledge.

Keane, M. & Moran, A. (2005). (Re)Presenting local content: Programme adaptation in Asia and the Pacific. *Media international Australia: Incorporating culture and policy, 116*, 88-99.

Ki, S. (2009). Boys over flowers gains attention overseas. *Korea Herald*, 9 February, 2009. LexisNexus. Consulted on October 31, 2009.

Kim, J. (2009). *"Boys over flowers" planning director Bae Jong-Byung*. Retrieved October 31, 2009 from http://www.asiae.co.kr/news/view.htm?sec=ent7&idxno=2009102015351081887

Kim, Y. (2007). The rising East Asian 'Wave': Korean media go global. In Thussu, D. (Ed.), *Media on the move: Global flow and contra-flow* (pp. 135–152). London: Routledge.

KPMG. (2007). *The video games market in China: Moving online*. Retrieved April 3, 2010 from http://www.kpmg.com/CN/en/IssuesAndIns ights/ArticlesPublications/Docum ents/video_games_china_0712.pdf

Kshetri, N. (2009). The evolution of the Chinese online gaming industry. *Journal of technology management in China, 4*(2), 158-179.

Lee, D. H. (2004). A local mode of programme adaptation: South Korea in the global television format business. In Moran, A., & Keane, M. (Eds.), *Television across Asia: Television industries, programme formats and globalization* (pp. 36–53). London: Routledge.

Liu, Y., & Chen, Y. (2004). Cloning, adaptation, import and originality: Taiwan in the global television format business. In Moran, A., & Keane, M. (Eds.), *Television across Asia: Television industries, programme formats and globalization* (pp. 54–73). London: Routledge.

Lo, K. C. (2001). Double negations: Hong Kong cultural identity in Hollywood's transnational representations. *Cultural Studies, 15*(34), 464–485. doi:10.1080/09502380110046669

Ministry of culture, sports and tourism Korea game industry agency (2008). *2008 White paper on Korean games: Guide to Korean game*. Seoul, South Korea: Korean Game Industry and Promotion Institute.

Munakata, N. (2006). Has politics caught up with markets? In search of East Asian economic regionalism. In Katzenstein, J., & Shiraishi, T. (Eds.), *Beyond Japan: The Dynamics of East Asian Regionalism* (pp. 130–160). Ithaca, NY: Cornell University Press.

Otmazgin, N. (2008). Contesting soft power: Japanese popular culture in East and Souteast Asia. *International relationships of the Asia-Pacific, 8*, 73-101.

PricewatehouseCoopers. (2007). *Global entertainment and media outlook: 2007-2011*. New York: PricewaterhouseCoopers LLP.

Reuters. (2009). Makers of online games in China look overseas. *New York Times*, October 29, 2009. Retrieved March 26, 2010 from http://www.nytimes.com/2009/10/27/business/global/27iht-game.html?_r=1

Shim, D. (2005). Globalization and cinema regionalization in East Asia. *Korea Journal, 45*(4), 233–260.

Shim, D. (2006). Hybridity and the rise of Korean popular culture in Asia. *Journal of media culture and society, 28*(1), 25-44.

Shim, H., & Ho, T. (2009). Translation of 'America' during the early Cold War period: A Comparative study on the history of popular music in South Korea and Taiwan. *Inter-Asia Cultural Studies, 10*(1), 83–102. doi:10.1080/14649370802605274

Thussu, D. (2006). Mapping global media flow and contra-flow. In Thussu, D. (Ed.), *Media on the move: Global flow and contra-flow* (pp. 11–32). London: Routledge.

Worth, T. (2009). Best bets, global: Strait to stardom. *Billboard.com*, January 24. Retrieved from NexusNexus on October 31, 2009

Xin, H. (2010). Domestic online game industrial chain getting mature. *China Economic Net*, March 5. Retrieved April 3, 2010 from http://en.ce.cn/Insight/201003/05/ t20100305_21064568.shtml

Chapter 7
New Imagined Community by Cultural Participation: A Study on Taiwan's Digital Cultural Heritage

Pi-Chun Chang
National Taiwan Normal University, Taiwan

ABSTRACT

Although the preservation of cultural heritage has always been a primary task of cultural policy in many countries, the idea of combining digital technology and cultural heritage was almost entirely unknown as recently as 1990. It is undeniable that digital technologies have played an important part in our lives. In the case of Taiwan, the government has been working on digitizing cultural heritage by launching National Digital Archives Program since 2002. Most scholarship has focused either on technical practices or the economic value of such practices. Scanty attention has been paid to the relationship between digital cultural heritage, cultural citizenship, and one's imagined community. In other words, the application of digital technology onto cultural heritage has been largely unmapped in terms of identity formation. This study explores the social and cultural implication of the combination of technology and heritage. When heritage meet contemporary technology, how does it shape and what does it implicate for one's cultural identity and imagined community?

INTRODUCTION

To speak of the digital is to engage with the impressive array of virtual simulacra, instantaneous communication, ubiquitous media, global interconnectivity, and all their multifarious applications. It is only recently that the UNESCO responded to the ongoing development of applying digital technologies to cultural heritage sector by articulating the *UNESCO Charter on the Preservation of Digital Heritage*.[1]

In the case of Taiwan, cultural heritage preservation becomes an important issue started from the late 1980 and the need for documenting cultural heritages digitally is increasing recently (Chiu, 2007: 230). Taiwan government has been working on digitizing cultural heritage by launching National Digital Archives Program since 1999.

DOI: 10.4018/978-1-60960-037-2.ch007

Copyright © 2011, IGI Global. Copying or distributing in print or electronic forms without written permission of IGI Global is prohibited.

The purpose is to provide users access to museum database via internet browser, making it easier for users to utilize information and resources and to learn, understand, and share history and culture.[2] Although the cultural heritage sectors acknowledge that digital technology poses new challenges, many philosophical, social, historical and cultural issues have not yet been critically explored.

When digital technology has been applied to cultural development or cultural preservation, most scholarship has focused either on the economic value of this practice or on technical considerations of such practices. Scanty attention has been paid to the relationship between digital cultural heritage, cultural citizenship, and one's imagined community. In other words, the application of digital technology onto cultural heritage has been discussed from economists and information scientists' perspectives, while how it implicates for one's identity formation left largely unmapped. If previous discussions illuminate the economic and technical dimensions of the use of digital media in the (re)production of art works and in the acquisition, representation and conservation of heritage, this study explores the social and cultural implication of the combination of technology and heritage. Heritage on display or in exhibition is regarded to consolidate national identity (Dicks, 2003). Yet when heritage meet contemporary technology, how does it shape and what does it implicate for one's cultural identity and imagined community?

Heritage was once regarded as heredity, a form of inheritance, and possessions bequeathed from the heir, today it implies roots, identity, and sense of place and belonging (Hoelscher, 2006: 200). Despite its aged lineage, it is described as "a mode of cultural production in the present that has recourse to the past" (Kirshenblatt-Gimblett, 1998: 7) and has become a specific way of interpreting bygone times that links individuals with a larger collective. Heritage is not just a thing displayed in the museum, but is also bound up with the understanding of objects outside the museum and of the nature of the subject and subjectivity. It informs not just how we see what is in them, but also how we see what is outside and how we see ourselves. In this sense, it can be said that heritage holds the nexus of identity, and cultural politics.

Material representation of national lineage is deeply related to one's identity formation. The display of heritage objects is not only the staging of history, but also helps to know one's own history and thus confirms one's own "roots" (Beier-de Hann, 2006). But when traditional exhibitions are transformed into digital form, the visitors are freed from given definitions and interpretational frameworks. That is, the practice of displaying objects in showcases along with written information ceases to dominate how something is to be interpreted. How does the connection between the individual and a larger common ground change when individuals are encouraged to view things and history in their own way?

This question will be analyzed with Taiwan's National Digital Archives Program executed by National Palace Museum and National Historical Museum in which artifacts, paintings and calligraphy, documents, publications and films are primary objects and users who have used this program. By examining the characteristics of the digital project, this study argues that when these artifacts are displayed in public museums, their implication for imagined community is based on ethnicity. Yet when cultural heritage is digitized, viewed through technological lens, using digitized heritage program offers a new way of imagining one's own community by cultural participation.

This study will firstly introduce Benedict Anderson's *Imagined Community* which combines technology and cultural transformation in terms of identity. The study will proceed under this theoretical framework with the discussion of more advanced technology--digitization. Secondly, it will introduce the history of Taiwan's National Palace Museum and National History Museum, and delineate how the roles of these museums are changing partly in accordance with the policy of digitization. Thirdly, this study analyzes the characteristics of National Digital Archives Program and how these characteristics promote a sense of participation. Then, this study will discuss the convergence of new media theory and histories of Taiwan's public museums and conclude that as digitized, web-based display practices are invented, the task of transforming the basis of cultural identity from ethnicity to participation is also invested.

IMAGINED COMMUNITY AND TECHNOLOGY

In the book *Imagined Community*, Benedict Anderson tried to explain the origin and spread of nationalism. He begins with a provocative postulate that a nation is an imagined community. All communities are imagined, but what distinguished them is the style in which they are imagined. As language, always important in creating a sense of community, was not necessarily decisive, new way of imagining national community is print capitalism. That is, capitalism's print technology made it possible for people to "imagine" large communities that had previously enjoyed no special form of togetherness. When print capitalism turn to publishing novels and newspapers in the vernacular and brought this style to elites and then to large numbers of people, the nature of the medium itself created the possibility of a new form of imagined community. In this way, it sheds light on the relationship between technology and socio-cultural transformation.

Anderson's approaches to national identity as a socio-cognitive construct also points out that the "nature" of collective identity is not a fixed concept rooted in the similarities or shared attributes around which group members coalesce. Early literature approached these attributes as "natural" or "essential" characteristics—qualities emerging from physiological traits, psychological predispositions, regional features, or the properties of structural locations. A collective's members were believed to internalize these qualities, suggesting a unified, singular social experience. Anderson delineates key moments of identity construction, times during which cultural (language) and social factors (capitalism, print technology) convene in a particular historical moment, effectively remaking collective images of the national self.

In the present environment, new communication technologies have contributed to the construction of the self and the establishment of communities. Anderson's work reminds us that technologies have changed the backdrop against which identity is constructed. Along with Anderson, Joshua Meyrowitz was among the first to explore the relationship between technology and identity (1985, 1989, 1997). He examines the ways in which electronic media reorganize the sites of social interaction. That is, new communication technologies weaken the connections between physical and social "place," and locate the self in a new hybrid arenas of action. For example, when the public and the private were meshed, what distinguishes home from work may be the online/offline border.

In a similar vein, David Altheide (1995) argues that new communication technologies create new environments for self-development and identification for they enable new communication formats—new modes of selecting, organizing and presenting information. In turn, these new formats reshape social activity. Altheide takes keyboard as an example to illustrate that the technology limits and directs the form and substance of the social interaction it enables. For example, keyboard technology homogenizes work and play sites; merge adult work and play worlds; link adult workspaces and children's play spaces and reconfigure children's play via adult's tools.

Some others explore the impact of media technology on community formation and resulting collective identity, demonstrating concrete effects of technologies as social factors on the self and the community (Cerulo et al 1992, Cerulo 1997,

Purcell 1997, Steuer 1992). These works hold implications for the connection between identity and technology and are, in one way or another, premised on the innovative elaborations of Anderson's *Imagined Community* whose national identity construction socially, historically and culturally relates to the emergence of new technology. As cultural heritage in exhibition gradually change changes from a physical culture to a digital new media culture, it is important to continue to theorize, study, and historicize emerging relationships between new media, identity and cultural heritage.

TECHNOLOGY AND MATERIALITY: (MATERIAL) ORIGINAL VS. (DIGITAL) COPY

During the 1980s and 1990s, media theorists began to examine the epistemological and ontological effects of information and communication technologies. Much of this scholarship focused on using new media to collaborate in the construction of knowledge through hypermedia (Barrett, 1994), to create constructionist learning communities (Bruckman, 1998), to develop user-guided narrative experiences (Davenport, 1994), to explore how cyberspace changes storytelling (Murray, 1997), and to show the influence of new media on how individuals construct and perform computer-mediated identities (Turkle, 1995). Much of this initial scholarship on new media portrayed cyberspace as a realm where embodiment, time, and space were made irrelevant (Negroponte, 1996). In recent years, new media scholars have further contextualized the materiality of new media as a site for the development, construction, and performance of knowledge and culture.

Since the introduction of digital media into the field of heritage through such things as Web sites, computer interactives, and digitized film and sound, museum practitioners and culture critics have either lamented or celebrated these developments. The critical thinking that has centered on the status of the (digital) copy as inferior to its non-technological original, and the potential of the former to undermine the values and meanings of the original can be traced back to Walter Benjamin. In his often-quoted essay "The Work of Art in the Age of Mechanical Reproduction," Benjamin expresses the value of material authenticity and strong repugnance for reproductions. He claims that original art works have an auraic presence defined as the "essence of all that is transmutable from its beginning, ranging from its substantive duration to its testimony to the history which it has experienced" but that "withers in age of mechanical reproduction" (1968: 3).

Similarly, Jean Baudrillard argues that mechanical reproduction and more recently simulations pose a threat to the "real" object and works of art. This is because he regards media as an instrument that destabilizes the border between the original and the copy. Virtual reproductions, for example, are usually thought and sold as a perfect analogon of the "real," and therefore intimately intricate with the real in terms of the meaning, values, and the rhetoric that the "real" object carries. If viewers are unable to distinguish the replica from the real, the auraic and ritualistic qualities of the "real" object will be lost.

The assumption of both Benjamin and Baudrillard is based on an original/copy binary in which the copy is viewed as a disturbing force that undermines the ideas of authenticity, originality, and aura. More recently, culture critics who focus on museum and heritage studies have pointed out that the view real historical objects attributed as "original," "material," and "objective" whereas digital historical objects as copies, immaterial, inferior, surrogate has its roots in nineteenth-century empiricist and evolutionist ways of thinking (Cameron, 2006: 52-53). Some media theorists, museological scholars or sociologists argue that when digital technologies are applied to the museum collections, the original/copy distinction becomes blurred, and thus advocate liberating the digital cultural heritage from a materialist epistemology.

The empiricist tradition, according to Fiona Cameron and Helena Robinson, privileges classification and descriptions of the physical attributes and the history of an object rather than its meaning and interpretation (2006: 169-170). They recognize the validity of "empirical" records and "verifiable" details and facts, such as date, maker, technique, style, etc., but they also contend that alternative forms of analysis, documentation (e.g., 3D imaging, sound recordings, etc.) and different perspectives of object should be included (2006: 170-171).

From a different view Bolter and Grusin point out the "double logic" of new media. They define a medium as "that which appropriates the techniques, forms, and social significance of other media and attempts to rival or refashion them in the name of real" (1999: 66), and argue that the evolution of media is driven by a consistent motivation to create more accurate representations of sensory reality. Thus the "double logic" of new media is the process of media appropriation to be refashioned and improved versions of older media. Through "immediacy," Bolter and Grusin claim that new media offers the experience of "really being there" as the presence of the media becomes transparent. As the process of media appropriation involves the intersection of technical, social, economic and cultural forces, the materiality of digital representation should not be denied.

On discussions about the breakdown of the distinction between real-world and virtual experiences through electronic communication, N. Katherine Hayles introduces the concept of "material metaphors" to describe the interaction between words and physical artifacts, particularly "inscription technologies"—devices that "initiate material changes that can be read as marks" (2002: 24). She uses the idea of a "technotext" to present a class of literary artifact that reveals its own means of production and exposes the relationships between the production and reception of a text. Through inscription technologies, the interpretation of a text cannot be separated from the meanings of its semiotic elements, including visual interfaces, words, images, video and sound. A technotext illuminates the connections between the imaginative world that it generates and the material apparatus embodying that generation as a physical presence. Technotexts invite critical evaluation, discussion and innovative action in response to the embodiment of both the form and content of a text (1999). By considering the positionality of producer and audience, Hayles argues that each text exists in its own type of materiality.

From the perspective of shifting nature of knowledge and its relationship to museum practices, Eilean Hooper-Greenhill argues that the provision of standard and universal descriptive taxonomy for objects resulted from strict materialist epistemology fails to acknowledge the polysemy of objects. Facing the poststructuralist and postmodernist turn of the practices of museum documentation, objects and their meanings should now be seen as contingent, fluid, and can be interpreted in a variety of ways. User needs, behaviors, and wider social issues should be taken into account (1992). From this perspective, heritage as museum objects should be viewed in a new way, in that the meaning and significance of it should not be seen as fixed, but rather as inherently polysemic, and can be interpreted in different ways at any given time and space and dependent on various factors, be it cultural, social, institutional, or individual.

Examining the contrast between the digital copy and the material real with the aim to explore how digital technology implicates people's cultural lives, activities and identities does not mean this study is technology-determinism oriented. Nor is it saying that digital cultural heritage can replace the real. Yet such debate sheds light on the possibilities that technology may heighten the potential of diversifying the interpretation of objects, push museums more centrally into our lives or unshackle that were once the preserve of museums.

THE HISTORY OF TAIWAN'S CULTURAL POLICY AND CHANGING ROLES OF PUBLIC MUSEUMS

In 1949, after fighting with the Chinese Communist Party, the Chinese Nationalist Party (also known as Kuomintang, KMT) lost the mainland and fled to Taiwan. From 1949 to 1965, the Nationalist faced the threats of the Chinese Communists and challenges from Taiwan people's daily lives, cultural development at this time was not national priority. However, in 1954, the Nationalist built National History Museum for educational functions. From 1966 to 1975, as the Chinese Communist started "Cultural Revolution" in the mainland, the Nationalist in Taiwan launched a state cultural program "Chinese Cultural Renaissance Movement" as the main anti-communist propaganda (Winckler, 1994). In 1967, the government established "Chinese Cultural Renaissance Promotion Committee" with the intention to become the spokesman of "authentic" Chinese culture. During this period, the consolidation of National Palace Museum had been of importance of preserving and displaying the long genealogy of Chinese cultural heritage, and thus legitimized Nationalist regime and embodied the substantial existence of Chinese culture in Taiwan (Chang, 1997: 29).

The origination and development of National Palace Museum quite corresponds to Tony Bennett's description of the public museums of the nineteenth century. As Bennett (1995) shows, the artefacts were arranged according to a historical developmental sequence while at the same time the disciplines of history, art history, archaeology, anthropology, ethnology, public museums embodied a new, universalizing, post-Darwinian principle—that all peoples and civilizations developed through time in a series of consecutive stages. As Bennett points out, museums annexed these universal histories to national rhetorics, positing the national collections as the culmination of the universal story of civilization's development (1995: 75-76). It was this similar imperialist mindset that gave rise to the establishment and consolidation of National Palace Museum.

Past scholarship has analyzed the role of museums in nation-building and in the construction of national identity (Kaplan, 1994; Henning, 2003). As Michelle Henning suggests, critical writing today tends to see museums, especially public museum, as having been instrumental in the construction of concepts on nationhood. The country had gained an identity that was bound up entirely with their glorious past through the museums (2004: 39). In this light, the cultural heritage that is sediment with historical knowledge and displayed in National Palace Museum implies that the visiting citizens are the "natural" offspring of thousands-year-old civilization. Visiting itself constructs a relationship with one's national identity, which such an identity is based on the notion of "ethnicity" and Chinese culture. In other words, Taiwan as a Chinese society and as a polity was synonymous with that of an ethnic group. The very idea of national identity was based on the notion of Chinese culture invoked. And, at the same time, "culture is not just imagined but authorized and institutionalized as well" (Chun, 1996: 114). As Flora Kaplan says, "Material representations of traditionality and age help to legitimate an ethnic group's claims to a unique identity and political power and to their attempts to create a sense of unity among themselves" (2006: 153).

On the other hand, the way the object in museum is revealed, accessed and consumed affects the meaning of this object and thus dominates the visitors' response to the past, and to the community the objects represent (Fisher, 1991). Taiwan public museums' rules of access limit what people do and what they may touch, the past may seem stifling and enfeebling. As the museum also functions as a ritual space that aims governing people through the attempt to manage the conduct of visitors and through the social routines and performances they elicit (Duncan, 1995; Bennett, 1995), one could say that the idea of "identity" or "community" is disciplined by displaying shared cultural heritage objects. Identity based on ethnicity is internalized without fully understanding. Yet such bounded notion of identity that public museums imply would change in accordance with the advance of technology.

CHARACTERISTICS OF DIGITIZATION PROGRAM

Objects Being Better Consumed

With digitization becomes a hot issue over the past few years, the National Palace Museum and National History Museum have initiated several important core projects in steps: digital archives, digital museum, and digital learning.[3] The digitization program of National Palace Museum manifested itself in a number of ways. The first is a comfortable evolutionary extension of museum activity. That is, interacting with an online resource in the comfort of one's own places, the chance of a successful searching/learning is heightened. In contrast, during a museum visit, one is not inclined to read labels and soon get tired of walking, standing and staring and perhaps also tired of the curator's singular approach to interpretation. In National History Museum's and National Palace Museum's exhibitions of Digital Archives Program,[4] the collections can be viewed with more detailed description and explanation and yet with more leisure and thus enhances the potential that the object moves from art historical icon preserved in the glass to something very clearly of human manufacture.

Digital technologies do open up the possibilities for sophisticated interpretation. This does not mean that digital exhibit transcends the power of the real thing or the image should replace the object. Yet certainly access is extended as online content is not constrained by the linearity of the publication, delicate objects become visible, the storehouse is opened up and collections become accessible to remote communities. Digitization offers the potential to dissolve the barrier of a traditional museum and gives a sense of a tangible reality. A virtual visitor may understand the thing better and be better prepared to interpret it.

Widening and Indiscriminate Access

The second is that people unfamiliar with the basic tenets of the museum profession are welcomed to take part in museum activity. In its previous material state, the archive is available only to specialists or researchers. Once in digital format, however, the archive is expected to attract an audience that is very diverse. For example, the program of National Palace Museum's e-learning leads people to appreciate ceramic, porcelain manufactured in the 18th century, calligraphy and ancient paintings which people consider to be remote from their daily lives and hard to understand. In addition, this e-learning program offers people new understanding of antiques which emphasizes using scientific ways to observe and analyze ancient objects.[5] It also provides knowledge of antiques caring with the starting point of teaches people how to care their valuable objects, including clothes, photos, diaries and furniture.[6] A digital media environment that includes a master plan for storing, classifying and searching the materials, and for aggregating them into meaningful wholes from parts, can provide strategies and adequate solutions to the issue of accessibility. As such, digital technology changes the museum/heritage-audience relationship, and the direction of change is to bring these two poles much close together.

Interactivity

The third is interactivity. The main characteristic of the program E-learning is interactivity. As interactivity has mainly characterized digitization program, the philosopher Hilde Hein points out that such dimension also characterizes museums' tendency to be more "people-centered," and attentive to visitors' own experiences and values. That is, museums in general have moved from focusing on objects to an emphasis on the subject (2000: 66). Hein argues that the tendency toward the use of objects in simulations and reconstructions to elicit subjective sensations, emotions and feelings, rather than as objects of knowledge that might "speak for themselves" (2000: 79). In other words, museums are increasingly emphasizing "experience," which Hein refers to a concept of subjective aesthetic experience and attitudes that entail an increased appeal to people's emotional and sensory responses (2000: 67). For instance the program of Exploring Chinese Ceramic with Dr. Dragon elicits empathy and identification through the use of a fictional character Dr. Dragon to link visitors to the development history of ceramics, whose experiences are narrated to individual visitors in stages as they move through the exhibit.[7]

Similarly, Andrew Barry echoes Hein's view by saying that interactivity has become "a dominant model of how objects can be used to produce subjects," intended to turn "unfocussed visitor-consumers" into "interested, engaged and informed technological citizens" (2001: 147-8, 129). Barry uses interactive science exhibit as an example to delineate that visitor's role has undergone a significant change. The interactive devices shifted the focus of the museum away from objects and toward the demonstration of scientific principles, processes and phenomena. According to Barry, this process involves changes in the relationship between the exhibition and the visitors in which the visitors are conceived of as political subjects (2001).

As visitors are becoming more centered in the digital exhibit environment, with a new sense of empowerment, they receive information through participatory discovery. The game design of e-learning program is a telling example.[8] In her discussion of virtual heritage and users' senses a virtual environment may evoke, Bernadette Flynn (2007) points out that digitized programming may provide a user-centered model of interaction with the past. That is because digital simulation undermines the indexical relationship to the object, but emphasizing on spatial modeling. As Flynn notes,

Virtual movement has the potential to create a simulated spatiality that extends the real to a more imaginative, enchanted, lyrical relationship with spatial immersion. As users pass through and inhabit virtual space, the creation of a simulated reality moves beyond the purely visual. As the player inhabits screen worlds—takes up space—the relationships between mastery of terrain, kinetic engagement, and experiences of space transform data space into place.... This structure demands kinetic journeys across a fragmented terrain, allowing users to apprehend the world and use it for their own ends. (2007:363)

The chance for the user to navigate acquires a sense of discovery through exploration and a sense of being there. The role of the visitors as merely a knowledge recipient is thus replaced by an active and interactive agency.

This may be also why media theorist Lev Manovich argues that interactivity of new media involves both imaginative and conceptual engagement, and more importantly, "plays to notion of the experiential" (2001: 182). Manovich thus argues that digital technologies and the Internet are tools that are culturally created and informed by current epistemologies. The technical possibilities they offer are that information can be organized manipulated, segmented, reworked, and delivered in multifarious ways (2001: 131). To extend from his view, digital technologies offer the potential to liberate heritage/museum objects from limited empiricist understanding as well as from a museum-based context. Digital historical objects should be seen as objects in their own right rather than as an inferior representation of our reality. With the characteristic of concerning users' experiences and sensibilities, digital technologies heighten the interpretive potential of heritage/museum objects by utilizing a range of senses and thus bring a new relationship between the viewer and the heritage.

Social Implications

This new relationship, in Dipesh Chakrabarty's eyes, is revolutionary in affecting the display of museum objects in that technological application to cultural heritage plays to notions of experience and brings experiential forms of knowledge production, which he positions as an important part to play in contemporary democracies (2002). He differentiates this experiential form of democracy from pedagogical democracy, which is based on the notion that to be a citizen, a person first needs to educated in the values of democracy in institutions like schools, universities, museums, and libraries. Chakrabarty calls "pedagogical" because such places privilege a form of knowledge based on analytical skills or the ability to abstract from the immediate example. As he puts it, "Abstract reasoning made it possible for the citizen to conceptualize such imaginary entities as 'class,' 'public,' or 'national' interests, and adjudicate between competing claims" (2002, 5).

As opposed to pedagogical model of democracy, Chakrabarty suggests experiential model of democracy is emerging and becoming prevalent. It is the knowledge production embedded in the senses which provides an increasingly important information base to citizens of contemporary mass and media democracies. This kind of knowledge production privileges the experienced, rather than the analytic and is embodied in a material way of sharing experiences, empathy, and memory.

In a similar vein, George Landow (1997) sees the new media technology as essentially democratizing as our notions of the meaning of the text, the authority of curatorial practices are destabilized and people, granted free access to information, become less passive recipients than active learners. He views the formation of voluntary, electronic associations as the horizontal counterweight to any vertical dimensions of power. The construction and reconstruction of one's own intellectual world through internet is inherently empowering.

The digitization program extends the role of museum to a place not bound by material culture, but "a place for inquiry into the memories of the past, a forum for consideration of the present, and a site from which to inspire, instruct, and inform" (Weil, 2002: 111). Digitization establishes an unprecedented relationship between museum, heritage and audience. It reconfigures the exhibition space to enable visitors to interact with the display, using a number of different strategies, provides egalitarian principles of access to culture for everyone and a significant platform for cultural activity and has become a factor in the shaping of cultural interests.

CONCLUSION

Museums have long played a critical role in meaning making, relying not only on the intellect, but also on different ways to form valuable historical continuity and cultural connections. On the other hand, the main function of heritage is the preservation and exhibition of ancient civilization, with objects as tangible expressions of permanence and historical authority. Digitization has provided people with indiscriminate access to culture, history and visual information beyond anyone's past expectations or predictions.

The project of digitization seeks to investigate new forms of representation of heritage, as well as to achieve an understanding of the changing role of the museum as cultural and educational institution. Surveying historical narratives can make us aware of the differences between cultures and ways of living throughout history. New technologies facilitate the museum's contribution to the corpus of knowledge by suggesting and enabling new ways of understanding heritage. Digital Archives Program is a step in this direction.

Notions of repository, evidence, authenticity, authority, preservation and access have changed.

With this change, one forms his/her historical narratives rather than being informed by messages from any kind of cultural institutions. The power of cultural institutions, such as public museums, used to be hierarchical and disciplinary (Bennett, 1995). This was especially true in Taiwan where cultural values were, more often than not, subject to politics and cultural values were shaped by ruling party's attitudes and beliefs. With Taiwan's political democratization force grown up, this force has also taken place in cultural institutions. National Archival Digitization Program as an indexical cultural policy has showed this tendency. When user behavior and experience become key defining principles, which acknowledge the particular characteristics of digital medium (Witcomb, 2007), digital cultural heritage is more than an object, but cultural constructs that have the power to implicate the shaping of cultural identities, engage emotions, perceptions, and values. And this is why digitization of cultural heritage promotes a sense of participation.

Cultural values were arbitrarily subject to politics, but now they can be enhanced by new media technology. This goes parallel with the character of NPM that was anchored in terms of ethnicity. The exhibitions and concurrent programs validated grand narratives and had created a strong sense of continuity of Chinese descent. National identities represented in museums had been rooted in the imagined community on display in which objects bespoke civilization and ethnicity. Yet, in responding to political demands to be more democratized, museums have had to move to displays which privilege experience (Witcomb, 2007: 46). Digitization of cultural heritage is one of the main ways that experience can be communicated through affect. The result is to empower the visitors to undergo a process, that is, "the process of coming to know (that) is not framed by a didactic top-down approach characteristic of conventional museological interpretations of the material world, but (that) rather is more open to the process of dialogue and interactivity" (ibid. 47). This dialogue and interactivity requires engaging emotions and perceptions to actively participate cultural activities and thus implicates one's re-conceptualization of his/her new imagined community.

REFERENCES

Altheide, D. (1995). *An ecology of communication: Cultural formats of control*. Hawthorne, UK: Aldine de Gruyter.

Anderson, B. (1991). *Imagined communities* (2nd ed.). London: Verso.

Barry, A. (2001). *Political machines: Governing a technological society*. New York: Athlone Press.

Baudrillard, J. (1996). *The system of objects, (Trans. J. Benedict)*. London: Verso.

Beier-de Hann, R. (2006). Re-staging histories and identities. In Macdonald, S. (Ed.), *A companion to museum studies* (pp. 186–197). Malden, MA: Blackwell. doi:10.1002/9780470996836.ch12

Beier-de Hann, R. (2006). Re-staging histories and identities. In Macdonald, S. (Ed.), *A companion to museum studies* (pp. 186–197). Malden, MA: Blackwell. doi:10.1002/9780470996836.ch12

Benjamin, W. (1968). *The work of art in the age of mechanical reproduction, (Trans. H. Zohn)*. London: Jonathon Cape.

Bennett, T. (1995). *The birth of the museum: History, theory, politics*. London: Routledge.

Bolter, J., & Grusin, R. (1999). *Remediation: Understanding new media*. Cambridge, MA: MIT Press.

Cameron, F. (2006). Beyond the cult of the replicant: Museums and historical digital objects—traditional concerns, new discourses. In Cameron, F., & Kenderdine, S. (Eds.), *Theorizing digital cultural heritage: A critical discourse* (pp. 49–76). Cambridge, MA: MIT Press.

Cameron, F., & Robinson, H. (2006). Digital knowledgescapes: Cultural, theoretical, practical, and usage issues facing museum collection databases in a digital epoch. In Cameron, F., & Kenderdine, S. (Eds.), *Theorizing digital cultural heritage: A critical discourse* (pp. 165–192). Cambridge, MA: MIT Press.

Cerulo, K. (1997). Re-framing sociological concepts for a brave new (virtual?) world. *Sociological Inquiry*, *67*(1), 48–58. doi:10.1111/j.1475-682X.1997.tb00428.x

Cerulo, K., Ruane, J., & Chayko, M. (1992). Technological ties that bind: media generated primary groups. *Communication Research*, *19*(1), 109–129. doi:10.1177/009365092019001005

Chakrabarty, D. (2002). Museums in late democracies. *Humanities Report*, *9*(1), 5–12.

Chiu, M. L. (2007). Digital services for preserving cultural heritages: Digital Tainan experience. In L. H. Chang, Y-T. Liu, & J-H. Hou (Eds.), *The Proceedings of DACH 2007: The 2007 international conference on digital applications in cultural heritage* (Vol. 2, pp.229-242). Tainan, Taiwan: National Center for Research and Preservation of Cultural Properties.

Chun, A. (1996). Fuck Chineseness: On the ambiguities of ethnicity as culture as identity. *boundary 2 23*(2), 111-138.

Dicks, B. (2003). *Culture on display: The production of contemporary visitability*. London: Open University Press.

Duncan, C. (1995). *Civilizing rituals: Inside public art museums*. London: Routledge.

Fisher, P. (1991). *Making and effacing art: Modern American art in a culture of museums*. Cambridge, UK: Cambridge University Press.

Flynn, B. (2007). The morphology of space in virtual heritage. In Cameron, F., & Kenderdine, S. (Eds.), *Theorizing digital cultural heritage: A critical discourse* (pp. 349–368). Cambridge, MA: MIT Press.

Hayles, N. (1999). *How we became posthuman: Virtual bodies in cybernetics, literature, and informatics*. Chicago: University of Chicago Press.

Hayles, N. (2002). *Writing machines*. Cambridge, MA: MIT press.

Hein, H. (2000). *The museum in transition: A philosophical perspective*. Washington, DC: Smithsonian Institution Press.

Hoelscher, S. (2006). Heritage. In Macdonald, S. (Ed.), *A companion to museum studies* (pp. 198–218). Malden, MA: Blackwell. doi:10.1002/9780470996836.ch13

Hooper-Greenhill, E. (1992). *Museums and shaping of knowledge*. London: Routledge.

Kaplan, F. E. S. (Ed.). (1994). *Museums and the making of "ourselves": The role of objects in national identity*. London: Leicester University Press.

Kaplan, F. E. S. (2006). Making and remaking national identities. In Mcdonald, S. (Ed.), *A companion to museum studies* (pp. 152–169). Malden, MA: Blackwell. doi:10.1002/9780470996836.ch10

Kirshenblatt-Gimblett, B. (1998). *Destination culture: Tourism, museums, heritage*. Berkeley, CA: University of California Press.

Landow, G. (1997). *Hypertext 2.0: The convergence of contemporary critical theory and technology*. Baltimore, MD: Johns Hopkins University Press.

Manovich, L. (2001). *The language of new media*. Cambridge, MA: MIT Press.

Meyrowitz, J. (1985). *No sense of place*. New York: Oxford University Press.

Meyrowitz, J. (1985). The generalized elsewhere. *Critical Studies in Mass Communication*, *6*(3), 323–334.

Meyrowitz, J. (1997). Shifting worlds of strangers: Medium theory and changes in "them" versus "us.". *Sociological Inquiry*, *67*(1), 59–71. doi:10.1111/j.1475-682X.1997.tb00429.x

Purcell, K. (1997). Towards a communication dialectic: Embedded technology and the enhancement of place. *Sociological Inquiry*, *67*(1), 101–112. doi:10.1111/j.1475-682X.1997.tb00432.x

Steuer, J. (1992). Defining virtual reality: dimensions determining telepresence. *The Journal of Communication*, *42*(4), 73–93. doi:10.1111/j.1460-2466.1992.tb00812.x

Weil, S. (2002). *Making museums matter*. Washington, DC: Smithsonian Institution Press.

Winckler, E. (1994). Cultural policy on postwar Taiwan. In Harrell, S., & Huang, C. (Eds.), *Cultural change in postwar Taiwan* (pp. 111–134). Oxford, UK: Westview Press.

Witcomb, A. (2007). The materiality of virtual technologies: A new approach to thinking about the impact of multimedia in museums. In Cameron, F., & Kenderdine, S. (Eds.), *Theorizing digital cultural heritage: A critical discourse*. Cambridge, MA: MIT Press.

ENDNOTES

1 Please see http://portal.unesco.org/en/ev.php-URL_ID=17721&URL_DO=DO_PRINTPAGE&URL_SECTION=201.html (accessed 5 October 2009)

2 Please see http://digital.nmh.gov.tw/ndap/1_a.htm (accessed 5 October 2009)

3 Due to the volume of material to be digitized is huge, problems caused by digitization to cultural institutions such as integration of metadata, storage of huge amounts of data, and vast costs related to these tasks are not yet solved, digitization executed by NPM is still an ongoing, rather than finished project.

4 Please see http://digital.nmh.gov.tw/ndap/en/exhibitions_en_02.htm (accessed 13 October 2009) and http://www.npm.gov.tw/da/ch-htm/about02_8.html# (accessed 13 October 2009)

5 The scientific analyses applies to the investigation of science history, the foundation of conservation, the scientific identification of materials, and further research of ancient objects. Please see http://elearning.npm.gov.tw/sci_analysis/index.html (accessed 13 October 2009)

6 Please see http://elearning.npm.gov.tw/protection_en/index.html and http://elearning.npm.gov.tw/digital_2-4-2-3-en.htm (accessed 13 October 2009)

7 Please see http://elearning.npm.gov.tw/courses/digital_2-4-2/digital_2-4-2-1/lesson_e/index.html (accessed 15 October 2009)

8 For example, the game of make-up of a beauty in Tang Dynasty and the making of bronzes. See http://digital.nmh.gov.tw/ndap/3_b.htm (accessed 20 October 2009)

Section 2
Convergence of New Media

Chapter 8
New-Media-Based Attention Implosion and Television Publicness in the Era of Internet-TV Convergence

Francis L. F. Lee
Chinese University of Hong Kong, Hong Kong

ABSTRACT

This chapter discusses how television captures and organizes the attention of the mass public in the age of media convergence. It is argued that media proliferation and technological convergence have led to a fragmented audience. Paradoxically, audience fragmentation also provides the condition for the powerful return of collective attention in the case of new-media-based attention implosion. This chapter uses the case of the extraordinary (online) popularity of television drama character Brother Laughing in Hong Kong to illustrate the phenomenon of new-media-based attention implosion and the dynamics behind it. The analysis shows that attention implosion is generated by audience members' time-space shifting practices and key individuals' organizational efforts, both facilitated by the new media. It led to the formation of an interpretive public surrounding the fictional character. Implications of the phenomenon on our understanding of the relationship between television, social life, and collective public attention are discussed.

INTRODUCTION

On March 18, 2009, an unprecedented public mourning service was held in Hong Kong. According to media reports, more than 100,000 people participated in the mourning. The person being commemorated was nicknamed Laughing Gor (which means Brother Laughing), a police officer who died during his undercover mission as a gangster. The mourning service was unprecedented not because of the number of people joining it; it was special because Laughing Gor was a television drama character. The service was conducted online at the Facebook group created for the character. On the day when television broadcast the episode in which Laughing Gor died, the Facebook group already had more than 140,000 members.

DOI: 10.4018/978-1-60960-037-2.ch008

Copyright © 2011, IGI Global. Copying or distributing in print or electronic forms without written permission of IGI Global is prohibited.

As many people flocked to the Internet to lament the death of Laughing, others urged the television station to alter the story so that the character did not have to die. Responding to the "online uprising," the production team at Television Broadcasting Ltd. (TVB) shot a brief additional scene and put it at the end of the whole drama series. The scene shows that Laughing Gor is still alive and is ready to have his police identity back as his undercover mission finally ends.

The event, in one sense, was highly illustrative of many new developments in the era of Internet-television convergence. As Jenkins (2006a) points out, convergence does not mean the reduction of various communication and media technologies into a single magic box. In contrast, "boxes" continue to pile up around people. Delivery systems diverge. What converges is the content delivered. Convergence in this sense implies new opportunities and new practices of television watching. Certainly, television watching has long been embedded in domestic life in numerous ways such that it never simply means sitting in front of the television set to watch a program (Morley, 1980; Ang, 1985; Fiske, 1987). Yet new technologies are often associated with new practices. In the 1980s and 1990s, the remote control device facilitated zipping and zapping (Bellamy & Walker, 1996), and the video recording machine facilitated time-shifting (Cubitt, 1991). Nowadays, with the Internet, watching television can mean watching missed episodes on the official websites of TV stations, downloading pirated files of foreign television dramas, or watching clips of old or new programs on YouTube. TV stations, in the U.S. at least, discover that Internet viewing of their programs might not depress conventional television viewing substantially. It may even tend to generate higher levels of overall viewing (Waldfogel, 2009). Hence they develop formats to "encourage viewers to embrace the computer through television" (Parks, 2004, p. 144). Some stations may also offer extensions of the television text in their websites, e.g., by providing additional information about the personal background and/or everyday life of drama characters (Brooker, 2004; Caldwell, 2004).

Meanwhile, the audience members themselves also contribute to the extension and elaboration of the television text by their various "user-generated-contents" (UGC). Convergence culture thus also raises questions of audience power and agency. While cultural studies scholars have long argued that audience members are active in appropriating and interpreting the meanings of popular cultural products (Fiske, 1987), the emergence of the Internet, especially the UGC sites, provides not only new means of "tele-participation" (Ross, 2008), but also new means for the wide distribution of the audiences' views and productions.

Traces of all the above developments can be found in the television scene in Hong Kong. However, besides being illustrative of convergence culture, the Laughing Gor phenomenon is also a case of what we may call new-media-based attention implosion. The notion of attention implosion presumes the presence of a fragmented audience produced by the proliferation of television channels and media technologies since the 1990s (Katz, 1996; Webster, 2005). Decades ago, a very limited number of channels would have captured the attention of the bulk of the citizenry. Nowadays, audience attention tends to spread more evenly and thinly over a much larger number of outlets. Of course, the actual extent and characteristics of audience fragmentation vary from country to country. Yet generally speaking, mass television broadcasting around the world has lost at least some of its power in capturing and coordinating the collective attention of the mass public.

Paradoxically, audience fragmentation also provides the condition for the powerful return of collective attention in the case of attention implosion. Besides the Laughing Gor phenomenon, another case in point is Susan Boyle, the middle-aged British woman who captured the world's imagination through her performance on *Britain's Got Talent*. Within a single weekend, tens of millions of viewers around the world went to YouTube and other Internet sites to watch Boyle's rendition of the song I Dreamed a Dream. If television, since its heyday, has never been so powerless in coordinating the collective attention of people in a given society on a regular and daily basis, it has never been so powerful in generating global collective attention with such a scale, scope, and speed.

Theoretically, the significance of new-media-based attention implosion can be examined in relation to Dayan's (2009) conceptualization of models of television publicness. For Dayan, schedule television "orchestrates the temporality of collective attention on a regular basis" (p. 22). Conventionally, television schedules are "in sync" with the dominant pattern of everyday lives in a society. Different groups of people are expected to be watching television at different times of the day, and everyone is expected to be watching television at the prime-time, family hours. In contrast, multichannel television services, coupled with the VCR, led to the emergence of what he calls carousel television, which suspends the coordination of the public's collective attention. With the proliferation of choices and channels, television viewing patterns become much more individualized. Meanwhile, media event "reintroduces the coordination of collective attention with a vengeance," but on an exceptional basis and with the additional temporality of live-ness.

The following analysis builds on Dayan's (2009) discussion. But while Dayan emphasizes the significance of the live broadcasting of prescheduled, historic events in reintroducing collective attention, this article argues that new-media-based attention implosion represents another type of occasion which refocuses collective attention in the contemporary mediascape (Appadurai, 1991). In other words, the most fundamental concern behind this study is the relationship between television and the organization of social life and collective attention. And an underlying presumption of this chapter is that an understanding of new-media-based attention implosion would help us understand the social role of broadcast television in the age of media convergence. Given the fundamental concern and the presumption, this chapter asks: What are the social and technological conditions for attention implosion? What are the characteristics of the "publicness" involved in this kind of events? How does this publicness differ from that associated with the "live broadcasting of history" analyzed by Dayan and Katz (1992)? What is the significance of the new-media-based attention implosion to contemporary culture?

This chapter discusses these questions by using the Laughing Gor phenomenon in Hong Kong as a case study. The following analysis aims at reconstructing the phenomenon in a way such that the social and cultural significance of the phenomenon can be highlighted. Methodologically, the empirical materials analyzed included 17 in-depth interviews with local and overseas television audience members who have joined the Facebook group created for the drama character. The interviewees included the Facebook group's administrator and three of the group's "officers" at the time of the interview study. One of the officers, Frank, was also a worker at TVB.com. Besides in-depth interviews, this study also systematically analyzed the contents of the Laughing Gor Facebook group, 100 blog articles discussing the phenomenon, and news coverage of the phenomenon. Lastly, the article will discuss the characteristics and significance of the Laughing Gor phenomenon within the specific context of television and society in Hong Kong, as well as the broad theoretical concern of television and organization of public attention in the Internet age.

SOCIAL AND TECHNOLOGICAL BASES OF ATTENTION IMPLOSION

As mentioned above, new-media-based attention implosion presumes the presence of a television audience which has already been fragmented to a certain extent. Therefore, some background discussion about television audience fragmentation in Hong Kong is crucial to a reconstruction of the Laughing Gor phenomenon.

Commercial television broadcasting in Hong Kong began in the 1950s.The economic take-off of the city in the 1970s further fueled the growth of the industry (Ma, 1999). By 1986, 98% of Hong Kong households have already owned at least one television set.[1] Throughout the decade of the 1980s, there were only two free-to-air commercial broadcast television stations in Hong Kong,[2] with TVB being the overwhelmingly dominant player, whereas Asia Television (ATV) often found it hard to capture even 10% to 20% of the market share. Each station offered only one Chinese and one English channels for the then five million Hong Kong people. It was arguably the heyday of mass broadcast television in the city. As Ma (1999) stated, in that period of time it was "not unusual for the final episode for major [TV drama] series to achieve ratings of 60 to 70 percent, which meant it could empty streets or restaurants" (p. 34). Besides, people often enthusiastically anticipated the coming of the annual "big shows" produced by TVB, such as the Miss Hong Kong Pageant. In other words, it was a time when commercial television, especially TVB, had few difficulties in organizing and dominating the collective attention of the public on a regular basis.

It was also under this historical context that the typical program schedule of commercial television was stabilized. Most pertinent to the present discussion is the practice, continued to today, for TV stations to broadcast their serial dramas five days a week. This practice is in clear contrast to the often one-episode-per-week practice in countries such as the U.S. and Japan. Moreover, the typical television dramas in Hong Kong are dominated by the serial element instead of by episodic narratives. The scheduling and program characteristics, added together, assume an audience who is available in front of the television set on every weekday evening.

If this assumption of the audience's everyday life was not too far away from the social reality in the 1980s, changes in the television industry and the Hong Kong society at large have undoubtedly rendered the assumption problematic since the 1990s. First, in the late 1980s, the Hong Kong government decided to introduce subscription-based television services as a way to enhance the city's information infrastructure (Lau, 1988, 1992). Cable TV, the first multi-channel television service provider in the city, began broadcasting in 1994. By August 2006, three pay television service providers were operating their businesses, offering a total of 239 channels and 39,640 hours of programs on a weekly basis.[3]

Admittedly, the impact of multi-channel television services on broadcast television in Hong Kong is limited by various factors, including audience habits, channel loyalty, and the continual significance of television as a "family medium" (Lee, 2007, 2010). Nevertheless, the emergence of multi-channel television can be considered as a part of the broader proliferation of entertainment choices for Hong Kong people. Meanwhile, the lifestyle of the city's population continued to evolve. Although there is a lack of relevant survey data, it should be safe to state that, for the majority of the working population and especially for the younger generation, going home directly after work and turning on the TV set immediately has become less and less frequent practices.

In any case, into the late 2000s, weekday primetime (8:00pm to 11:00pm) television dramas aired by TVB could usually obtain between 25 to 30 ratings points. It is not uncommon, however, for ratings for a drama series to go up during its broadcast period (usually four to six weeks) if the drama received good popular acclaim. These days, a drama series obtaining a rating point of about 40 in its last week of broadcast would be regarded as a significant success. These figures, of course, still represent a relatively high level of audience concentration when compared to the U.S television scene. But it is also a far cry from the television scene of the 1980s in Hong Kong as described by Ma (1999) and others.

The situation described above does not mean that the other 60% to 70% of Hong Kong people no longer watch television. Rather, it means that their television consumption is no longer falling within the model of what Dayan (2009) calls schedule television. New media technologies, especially the rise of the Internet, have provided ample opportunities for various time-shifting practices among the population. TVB.com itself, for example, usually provides an episode of a drama to be viewed online just a few hours after its conventional television broadcast. But appearing even quicker would be other pirated electronic files circulating via file-sharing software such as Foxy or available through websites such as todou.com. Such online distribution is especially important for the overseas audience who want to follow TVB dramas at the same time as they are being shown in Hong Kong. As many of our overseas interviewees pointed out, while TVB dramas used to be available to them with a time lag of weeks or even months, nowadays they can follow TVB dramas with a minimal time difference. For example, Samuel, a Chinese immigrant and a college student living in New York, would nowadays download computer files of TVB dramas from the Internet at around 11:30am New York time (which means around 11:30pm Hong Kong time). He would then watch the drama with his mother and cousin together in the evening of the day, using a computer cable to show the drama on the television set in the living room.

Returning to the local audience, time-shifting afforded by the Internet allows many to follow the drama series even if they cannot stay at home every evening. Moreover, while time-shifting was historically made possible by the VCR in the early 1980s (Cubitt, 1991; Gauntlett and Hill, 1999), the Internet has certainly broadened the range and altered the characteristics of people's time-shifting practices. Most importantly, if the VCR allowed a person to record a program for later consumption, the person nonetheless would still need to know beforehand what the television schedule is and be aware of what programs are currently available. Moreover, with the VCR, even though a person does not have to watch a show as it is broadcast, he or she would still have to record the show as it is being broadcast. In these senses, the VCR did not completely free the television audience from the broadcast schedule.

The Internet, in contrast, allows people to catch up even if he or she does not have any prior information about the program and broadcast schedule. For example, Amy Ho, a local interviewee who would later become an active member of the Laughing Gor Facebook group, stated that she actually did not have the habit of watching television dramas. As she recalled, she learned about the drama *Emergency Unit 3* – i.e., the drama which featured Laughing Gor – only accidentally one evening as her husband was watching the show: she was working on other things but overheard the television drama's dialogue and found one character (who turns out to be Laughing) to be particularly "noisy." After finding out that the show was quite interesting, she downloaded a whole set of earlier episodes to catch up. Notably, even after finding out about the show, she did not follow it every evening because her husband went to a business trip later on. She waited her husband to return to Hong Kong and then watched the remaining episodes of the series with him

On the whole, at least in Hong Kong, we can describe the social condition for attention implosion as one in which a sizable number of people have turned away from habitual, daily consumption of broadcast television. Yet at the same time they retain an interest in television entertainment and would return to it when something interesting to them is available. Their occasional return to television entertainment is facilitated by new media technologies which allow people to engage in different practices of time-shifting.

Of course, the significance and role of the new media goes beyond how it facilitates time-shifting. As the platform for attention implosion, websites such as YouTube and Facebook also made attention implosion something which can be seen and recorded. In fact, it is important at this point to further elaborate upon the notion of attention implosion. Up until this point, attention implosion is discussed in terms of being an occasion when the attention of large number of people suddenly converges to the same television object. However, it is not the same as having the attention of large number of people converging onto the television screen. In fact, in the case being discussed here, the rating of the drama *Emergency Unit 3* has remained largely stable over the period of broadcasting – the average ratings of the first five weeks of broadcast were 29 to 30 points. Only in the last week of the broadcast did the rating go further up, probably due to the buzz generated by the Laughing Gor phenomenon. But the increase was still only up to an average of 34 rating points.

Therefore, attention implosion does not entail a surge in television ratings. Rather, attention implosion is largely mediated and represented by other new and conventional media. New media are particularly important in recording the attention implosion. For instance, Figure 1 shows the estimated number of wall posts on the Laughing Gor Facebook group from March 13 to April 12. The Figure shows that the number of wall posts indeed suddenly surged on March 18, i.e., when the episode in which Laughing died was broadcast.[4] The number of wall posts declined sharply immediately afterwards, despite another smaller peak on the day when the final episode of the whole drama series was broadcast.

For our discussion here, the important point is that the Facebook group can "record" audience reaction such that the latter becomes objectified and observable. Napoli (2003), in *Audience Economics*, argues that television stations and advertisers do not respond to the actual audience. Instead, they respond to a *measured audience* produced through specific measurement techniques. Similarly, we may contend that the significance of websites such as Facebook and YouTube resides partly in their capability of providing their own "audience figures": number of people watching a video clip, number of people joining a group, or number of wall posts. In the case of Laughing Gor, the membership size of the Facebook group is the most important figure "proving" the existence of a television phenomenon.

This recorded and objectified audience response also provided the bases for the mainstream media to cover and discuss the phenomenon. As Lee (2009) discusses, events generated by and through new media communications are often amplified and further promoted by mainstream media coverage. What happens in "new media events" (Qiu and Chan, 2009) is often the creation of a "media loop" (Manning, 1997), as information and images are circulated and re-iterated in a converged mediascape composed of both old and new media. In the case under study, Figure 2 shows the number of newspaper articles mentioning "Laughing Gor" from March to May, 2009.[5]

There are two important differences between Figure 2 and Figure 1. First, media mentioning of Laughing Gor reached its peak not on March 18; it came about a week later on March 25. Second and more important, mainstream media's mentioning of Laughing Gor did not just surge on and decline rapidly after specific dates. Rather, although media mentioning fluctuated from day to day, occasional coverage related to Laughing Gor persisted into April. Such media coverage partly reflected the increased publicity the actor attracted after the character's success. But it also partly signified the transformation of Laughing Gor from a television phenomenon to a social phenomenon. That is, the fact that many people were enthusiastic about the drama character became a phenomenon calling for attention, interpretation, and explanation. Into this stage, paying attention to Laughing Gor no longer only

Figure 1. Daily number of wall posts at the Laughing Gor Facebook group, Mar 13-Mar 31

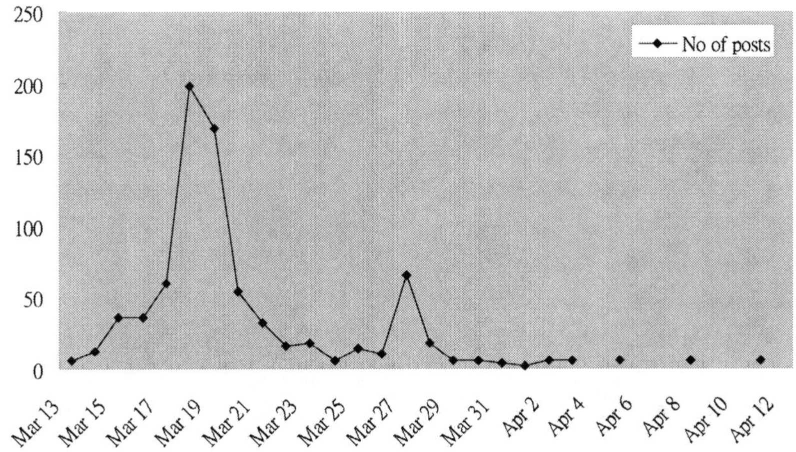

meant watching the drama on television and/or the Internet; it could also mean paying attention to the broader social phenomenon of the drama character's popularity.

ATTENTION IMPLOSION AS MOMENT OF COLLECTIVE INTERPRETATION

As Laughing Gor became firstly a television phenomenon and then a social phenomenon, what kind of publicness did it reintroduce into the society? In this particular case, one can argue that what the phenomenon led to was the formation of an "interpreting public." That is, the phenomenon posed questions for interested people to ponder on: why did Laughing Gor capture so much popular support? What did the phenomenon tell us about Hong Kong television, its audience, and the society in general?

Of course, as many cultural studies of television consumption since the 1980s have pointed out, the television audience has always been actively

Figure 2. Daily number of newspaper articles mentioning Laughing Gor, Mar 13-May 31

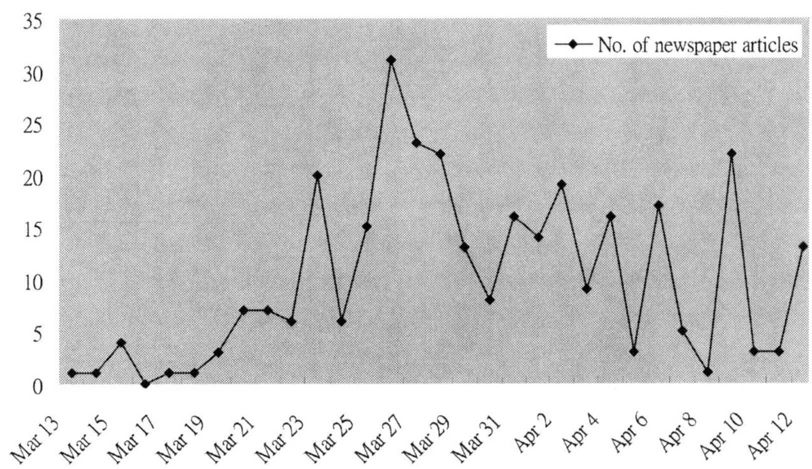

interpreting the meanings of texts. The point here, however, is that with the existence of new media platforms, the Laughing Gor phenomenon has generated a moment for Hong Kong people to engage in a process of *collective* and *public* interpretation: people did not only discuss the questions with their acquaintances in relatively private environment; they expressed their views by participating in online forums, writing blogs, posting Facebook messages, etc. Many of these online writings have the public at large as their addressees.

So how did Hong Kong people interpret the meanings of Laughing? We employed online search engines to look for blog articles which have mentioned the term "Laughing Gor." We selected 100 articles, mostly published in late March, for a textual analysis. What we see is that bloggers have adopted a few major modes of interpretations. Many bloggers explained the popularity of Laughing Gor in terms of how the character embodied certain social experiences, cultural values, or ideal images upheld by people. For example, some interpreted Laughing Gor as embodying many Hong Kong people's personal struggles in their working environment. Others regarded Laughing as an ideal hero for Hong Kongers, whereas some adopted a more "gendered" perspective and interpreted Laughing as representing a "good man" for many Hong Kong women.

The economic situation right now is not good. All people are afraid. They fear that they could lose their jobs if they say something wrong. Very few people dare to express their true thoughts. Laughing is different. He fears neither the heaven nor the earth. He says what he has to say. He has courage and strategies. People find the kind of courage they lack in him, so naturally they come to like him and worship him.[6]

The core of the question is the image of hero. Laughing Gor is an undercover police. He is misunderstood by others. When the truth comes out, people find that the unjust deeds were for the bigger justice…...What people want is not a hero who is from the start and always remaining on the bright side; people want a hero who surprises. More concretely people hope that all others who bring misfortune to them are also Laughing Gor – their apparent lack of conscience is for the purpose of building a better home.[7]

The most important thing is that he is not selfish. He is willing to sacrifice his life to protect his colleague. The last scene on the car was touching. What is so bloody romantic is that he did not just get shot and die. Instead, he asked Man: "Fire! Add a shot!" Are there really men like this in the world?[8]

These character and/or narrative based interpretations of the popular reaction toward Laughing Gor are essentially interpretations of the social values and desires of the Hong Kong people. Besides, some bloggers may adopt a somewhat different mode of interpretation, focusing on the actor Michael Tse and his life, instead of focusing on the character and the drama narrative. Nevertheless, such actor-centered interpretations are still likely to presume or uphold certain basic social values.

The key of his success is "try the best and never say die." All successful people have done a lot of hardwork. You have to find a person who recognizes your ability, you have to find the right time, the right place, the right people around you. [But] if you want all these, you have to prepare yourself…...luck is only for those who are prepared.[9]

His acting makes us realize that hardwork may not give you immediate reward. But when the time comes, the audience will appreciate. From his role "Big Heaven Junior" in Young and Dangerous in the 1990s to his role "Cheap Man" in Desperate Women a few years ago, Tse's acting has always impressed the audience deeply.[10]

Michael Tse has been in the entertainment industry for nearly two decades. Originally a singer, Tse's career did not take off and has remained a second or third-tier movie and television actor. Many blog articles such as the two quoted above, however, insisted that Tse's talents have long been appreciated by the more discerning group of audience. It is just that his "time" had not yet come. Such interpretations thus reaffirm not only the cultural values of diligence and determination, but also the idea that luck – in terms of timing – is also essential to one's career success.

Beyond interpretations pertinent to the cultural values of the society, some bloggers celebrated the "online uprising." They treated TVB's response to online requests to change the drama script as a sign showing the power of Hong Kong people when they are united together and exhibit their solidarity. Others emphasized the power of the Internet, seeing the event as a formal announcement of the coming of the era of web 2.0.

It again proves that, no matter how impossible it is, we can turn the rotten into magic as long as Hong Kong people are united! Within a few days, Hong Kong people stood up together to change the ending of a television drama series![11]

It is time for marketing people to wake up! Laughing proved that the mass TV drama audience has already entered the lifestyle of web 2.0. The clicking rates of the drama blogs and forums on TVB.com are not bad. The noise created by the Facebook mourning activities triggered coverage by the traditional media. The coverage came from common people's self-initiated participation, and this is exactly the hallmark of the web 2.0 era.[12]

Of course, not everyone viewed the Laughing Gor phenomenon in a positive light. There were bloggers who criticized the phenomena from various perspectives. In contrast to the celebration of "audience power," some argued that asking for the "resurrection" of Laughing in the drama would only destroy his appeal. Underlying this argument is a criticism of the audiences as people who do not understand the artistic aspects of cultural production. More broadly, some bloggers treated the whole Laughing Gor phenomenon as a symptom of Hong Kong people's preference for the trivial and meaningless over serious matters.

However, Hong Kong people are interested only in Laughing Gor. I saw that all the mini-feeds on the Facebook accounts of my friends are about the death of Laughing. The commentaries on newspapers are also about Laughing. Why would it arouse so much interest? When it comes to medical reform, the raising of cigarette tax, the plan to enhance employment of university graduates; I only see the public's inattention.[13]

It really proves that Hong Kong people are extremely/extraordinarily nonsensical. Hong Kong is really too small......Such a small matter can create a flock effect......This is the problem of over-concentration of media and popular culture in Hong Kong.[14]

The above excerpts do not exhaust the full range of perspectives offered by people in their blogs. And it is not important for us to evaluate which blogger's interpretation is more valid. The point is that the Laughing Gor phenomenon provided an occasion for people to engage in "public" interpretations. The interpretive activities are public in three senses. First, through new media, common people's interpretations are circulated publicly. Second, conventional reception analysis in cultural studies has emphasized the ways people connect popular cultural texts to their personal, everyday lives. Although this kind of "personalized interpretations" did exist online in the case under study, the interpretations quoted above all attempted to address the "public" significance of the phenomenon. Third and related to the second point, many bloggers were not fans of Laughing Gor. They might not be frequent television watchers at all. They cast themselves as a member of the Hong Kong public commenting on a "public" phenomenon.

The peculiarity of the interpreting public generated through attention implosion can be further elaborated by comparing it to the publicness associated with the type of live-broadcast, historic media events analyzed by Dayan and Katz (1992). As mentioned at the beginning of this article, Dayan (2009) regards media events as capable of reintroducing collective attention. However, in a typical media event as conceptualized by Dayan and Katz (1992), the meanings of the event are largely fixated by how the event is being narrated – as a conquest, a contest, or a coronation. The public is invited to attend the event as witnesses/ spectators and to take up the "dominant reading" proffered by the narration. Even the journalists are expected to act deferentially and respect the authority of the organizers. Certainly, other scholars working on the concept of media events have pointed to the possibility of contestation of meanings (see Liebes and Curran, 1998), and Dayan and Katz (1992) themselves also talked about "event pathology." But it remains the case that a prototypical media event is not expected to generate an actively interpreting public.

In contrast, in the case of Laughing Gor, the attention implosion came unexpectedly. As a matter of fact, Laughing Gor was not the most important character in the drama *Emergency Unit 3*. Even TVB was caught by surprise about the character's success. Hence there was no "dominant reading" of the phenomenon. The success of Laughing Gor thus invited interested people to play the role of the cultural critic. The publicness associated with attention implosion was therefore a more active and participatory one, even though, as the next section discusses, audience participation is a complicated concept. There is no simple equation between audience participation and activities with audience power.

AUDIENCE PARTICIPATION AND AUDIENCE POWER?

The question of audience agency and power is central to many discussions of the impact of new media technologies on the relationship between people and popular culture. Some scholars, such as Jenkins (2006a, 2006b) and Ross (2008), emphasized that new media technologies offer new opportunities for tele-participation. On the surface, the discussion of the interpretive public in the previous section falls in line with such arguments of increased audience agency. In fact, as we have seen, the success of the television audience to "force" TVB to alter the ending of the drama was taken by some as a sign of audience power.

However, other scholars have rightly pointed to the need of caution when discussing audience agency. Van Dijck (2009), for instance, problematizes the notion of user agency in relation to UGC sites. She argues that the claim of user participation and power neglects the complexities of the notion. Among her critique of the myth of audience agency, two arguments are particularly pertinent to our discussion here. First, Van Dijck (2009) points out that there are many different types and levels of user participation. While some people do create contents, others mainly distribute contents, whereas some others act merely as spectators. In the case under study, there were organizers of the online group, active and sustained participants of the group, people who joined the online mourning activities and were active for a few days only, non-television-viewers who nonetheless commented on the phenomenon, and people who simply remained conventional television watchers. The distinction among these participants could be important if we want to fully understand how an online phenomenon such as the Laughing Gor group can come into being. More specifically, much public discussion in the local media missed the fundamental point that the Laughing Gor Facebook group was successful not just because of the popularity of the character, but also because of a proper set up and management of the group.

Here, the story of Ronald, the Laughing Gor Facebook group administrator, is worth telling. A young person in his early 30s who operates his small businesses, Ronald has long been active in various kinds of "new media participation." He is an officer in an online radio station in Hong Kong. Regarding Facebook, he opened his account in late 2006, which was about one year before Facebook became a popular website in the city. At the time of the in-depth interview, Ronald had more than 2,000 "friends" on his main Facebook account. Moreover, he has been active in creating different types of Facebook group targeting the larger "public" as members/participants. For instance, just two months before the Laughing Gor phenomenon, he created a Facebook group called "the Strange and the Weird," which invites people to post photos concerning supernatural or mythological objects, ranging from ghosts to UFOs. The group has successfully attracted 10,000 members by early March, and the group had already had 20,000 members by early September 2009.

Ronald, in other words, is an experienced Facebook group organizer, and he did use his experience in organizing the Laughing Gor group. As he recalled, he joined the Laughing Gor group very early, at a time when it had only a few hundred members. When he witnessed the rise of membership beyond 10,000, he noticed that "junk members" and "junk posts" started to appear, i.e., people who joined the group just to post promotional wall posts. Hence he proactively contacted the original creator of the group and asked the creator to let him become the administrator. Since then, he would "police" the group every day, removing those promotional posts. He would warn those people posting promotional materials to stop, and he would kick them out of the group if the warning was not heeded. In terms of further promoting the Facebook group, Ronald made sure that the group has new materials for people to consume on a constant basis. He also opined that there are various minor things which can be related to the success of a Facebook group. For instance, he deliberately chose a picture of the actor Michael Tse which looks "ordinary" to serve as the profile picture of the Facebook group. He also pointed out that an administrator of the Facebook group has to constantly provide new materials but without giving people the feeling that the group has become the administrator's property (e.g., by talking too much about the administrator himself or herself).

When the group began to attract public attention, he also had to deal with various parties interested in using the group to advance their own interests. For instance, Ronald recalled that several bars and restaurants have contacted him and claimed that they could organize events with the participation of Michael Tse. The bars and restaurants wanted Ronald to mobilize members of the Laughing Gor group to participate in those events. But Ronald refused to cooperate as he doubted – and rightly so as it turned out- - whether the bars and restaurants could really invite Tse to attend the events.

Certainly, the effort by the Facebook group organizers does not explain every aspect of the online phenomenon. In fact, another reason for us to adopt a more differentiating view of the participating audience is that the views and actions attributed by public discourse to an undifferentiated group of "online participants" may actually be the views and actions of only a vocal minority within the group. In our in-depth interviews with both the local and overseas audience-participants, we found that only a couple of them were actually supportive to the idea of rewriting the drama script to keep Laughing Gor alive. Most of the interviewees expressed indifference toward whether Laughing would be alive or dead at the end, whereas a number even readily agreed with the critique that forcing the change of the drama script would undermine the appeal and integrity of the character. Although we do not have statistics showing how many members in the Laughing

Gor Facebook group were for or against altering the drama script, our in-depth interviews at least pointed to the possibility that the call for Laughing to remain alive was only issued by a vocal minority who happened to have their voice picked up and amplified by the mainstream media.

Besides the need to differentiate types and levels of participation, van Dijck (2009) also argues that, in the case of UGC sites, users are laboring for the commercial interests of the sites as they create contents to be shared with others. In the case of Laughing Gor, it is important to note that many aspects of the relationship between the audience and the television station remained unchanged. In fact, all the active Laughing Gor Facebook group members interviewed expressed some dissatisfaction regarding how TVB treated the online fan group. According to Ronald, he tried to contact TVB to organize activities for the fans since late March, but TVB did not respond positively and immediately. Rather, Ronald felt that TVB was highly suspicious of the Laughing Gor Facebook group at the beginning.

Later, after Ronald successfully established contacts, TVB would occasionally ask him and the Facebook group officers to see if they could mobilize fans to attend events, such as studio recordings for game shows in which Tse participated or press events involving Tse. As Ronald and others pointed out, it is a common practice for TVB and other entertainment companies in Hong Kong to mobilize "fan presence" when an actor or singer attends public events. When the actor or singer does not have enough popularity to guarantee spontaneous fan presence, sometimes entertainment companies would pay young people to act as fans in those occasions. Given this background, Ronald and others understood that TVB was only using them as free labor, and TVB had yet to devise new models of how the station can relate to fans. The more critical ones, such as Amy, even explicitly stated that TVB does not respect the fan group. Nevertheless, the interviewees also understood that they had no choice but have to rely on TVB if they have to see the actor Michael Tse.

As Miller (2004) points out, being active does not mean being critical. And in the case of Laughing Gor, being critical toward the power of mainstream media organizations does not mean being active in resisting such power. Although the interviewees were critical toward TVB and were aware of their own relative powerlessness, they tended to accept it as something inevitable. They focused on how to get the most of what they want given the existing system, instead of thinking about whether and how the system might be changed.

Therefore, while the attention implosion has generated a public actively participating in interpreting the meanings of the character and the event, more participation does not entail more power. The "success" in forcing TVB to alter the drama script belonged only to a section of the online participants, and it is also just about as much the audience participants succeeded to do. There is a lack of more meaningful and fundamental change in the relationship between the audience and the television company, at least for now.

CONCLUDING DISCUSSION

This article sets out to discuss whether Internet-television convergence has led to new models of television publicness. Using the Laughing Gor phenomenon as an example and a case study, this chapter argues that, in the age when the audience is normally fragmented and spread over the convergent mediascape, collective action is nonetheless reintroduced in moments of attention implosion. The reconstruction of the Laughing Gor phenomenon shows that attention implosion can be facilitated by new media technologies in a number of ways. Most importantly, new media technologies proffer the opportunities for new and more various practices of time-shifting. New media technologies also become a platform for the imploding audience attention to be recorded and made conspicuous. In the era of web 2.0, attention implosion is recorded not so much by television ratings but by Facebook group membership and YouTube video clip viewership.

It is then argued that attention implosion brought about a moment for collective interpretation. Different from "media events" (Dayan and Katz, 1992), attention implosion is unpredicted and unpredictable, and there is a lack of a dominant narrator and a dominant narrative for the event. While audience members have long been active interpreters of meanings of television texts, through new media platforms, the audience engages in "public interpretation" of television phenomena – their interpretations are made public and their interpretations focus on the "public" significance, instead of merely personal significance, of the phenomena. The formation of the interpreting public means that the publicness associated with new media based attention implosion is a participatory one.

Nevertheless, participation is a complex notion. "Participants" should not be seen as an undifferentiated group. Different types of participants play different roles in the generation of the phenomenon, with some being core organizers. Finally, audience participation does not entail audience power. Instead, at least in the case of Laughing Gor and the Hong Kong television industry in general, the relationship between the audience and the television companies has not been changed substantively.

Admittedly, the analysis is an exploratory one, and certain aspects of the phenomenon deserve more in-depth analysis in the future. For instance, scholars interested in fandom may further examine whether the changing mediascape has led to the formation of new fan practices. The analysis also raises a number of questions which are yet to be answered. For example, while we have seen the formation of an interpreting public, the publicness and collectiveness of such interpretive activities would be strong only to the extent that people do read and even discuss what others have written. Our analysis has not looked into whether and to what extent the interpreters and the net users at large actually interacted with each other. Besides, the discussion has pointed to the need to differentiate types of participants and their roles, and yet more systematic categorization of the participants and analysis of their relationships and interactions would be needed.

While this article employs the Laughing Gor phenomenon as a case, one may question if the phenomenon is an idiosyncratic, one-off event. In other words, one may question if what is called attention implosion is going to be a recurrent feature of contemporary culture. Only time will provide the answer. But as mentioned at the beginning of this article, the sudden fame of Susan Boyle through the circulation of video clips about her performance in *Britain's Got Talent* is another case in point, whereas the global commemoration of Michael Jackson following his unexpected death in June 2009 can be considered as another instance of new media based attention implosion.

There are, certainly, similarities and differences among these different cases. The Laughing Gor phenomenon occurred in a context in which the television industry is marked by an oligopoly, and the dominant player in the industry hitherto had little incentives to explore ways to appropriate the potential of the Internet. Laughing Gor was also primarily a local phenomenon – its transnational dimension is limited to the participation of overseas Chinese. Yet certain aspects of the Laughing Gor phenomenon should also be applicable to other cases. For example, the point that new media technologies contribute to attention implosion through facilitating time-shifting practices would be applicable to other cases as well. And we may be able to discover new meanings and practices of "time-shifting" when we look at other occasions, such as when a teenager catches up with his knowledge about Michael Jackson through the huge amount of online materials.

Better understanding of the formation, characteristics, and consequences of new-media-based attention implosion can be derived only when more case and/or comparative analyses are conducted. But as concluding remarks, it would be suitable for us to return to the most fundamental concern underlying the study, i.e., the relationship between television and the organization of social life and collective attention.

As mentioned at the beginning of the article, the trend toward convergence has led some scholars to develop new notions, such as flexible microcasting (Parks, 2004), to describe the contemporary television scene. Gripsrud (2004), however, draws upon the analysis of television by Raymond Williams (1975) and argues that broadcasting is likely to continue to exist and play important social roles. He argues that Raymond Williams situated the rise of broadcasting onto two social conditions: 1) the centralization of resources, power, and knowledge in general, and 2) mobile privatization of people's lives, i.e., the increased social and geographical mobility of individuals and nuclear families at the expense of the stability of older, traditional social communities" (p. 212). With these two conditions, television and mass media in general thus become important channels for the efficient distribution of information on a broad scale and the production of a shared national or regional identity. Gripsrud contends that these social conditions are still pertinent to the contemporary society, and hence the technological developments of channel proliferation and Internet-television convergence would not lead to the complete demise of broadcasting.

The case of Laughing Gor suggests that Gripsrud (2004) is likely to be correct in saying that broadcasting is unlikely to become obsolete. Yet it is also true that the contemporary television audience is more fragmented, spreading thinly over not only a larger number of television channels, but also a larger number of media and non-media entertainment and information choices. In this condition, we may need to re-imagine the place of broadcasting in our social lives. In *Ambient Television*, McCarthy (2001) has illustrated the cultural and social significance of the numerous television screens which exist in shopping malls, at the airport, and other public locations. However, conventionally, television as a conspicuous piece of furniture in the living rooms of people's homes has long been the most prominent image people have when thinking about television broadcasting. Broadcasting supposedly brings the public world into people's private homes.

Nevertheless, in the contemporary convergence era, the television screen which can best serve as the image of broadcasting may no longer be the living room television screen. Instead, the large television screen on the wall in front of a shopping mall may become more and more pertinent: It is a highly conspicuous object, but it is also usually "unseen." People passing under and in front of the huge television screen do not often pay attention. What the screen is showing is just a background against which they go to work, go to shop, and go to hang out with friends. But occasionally, when something special appears on the screen, people would stop and pay attention. A crowd would come into being in front of the television screen, with their attention converged onto the television object. The consumption of the television object is more "public" than the usual private viewing at home. Television broadcasting in this sense not only remains alive; it arguably has become even more prominent in the era of Internet-television convergence.

REFERENCES

Ang, I. (1985). *Watching Dallas*. London: Routledge.

Appadurai, A. (1991). *Modernity at large*. Minneapolis: University of Minnesota Press.

Bellamy, R., & Walker, J. (1996). *Television and the remote control*. New York: Guilford Press.

Brooker, W. (2004). Living on *Dawson's Creek*: Teen viewers, cultural convergence, and television overflow. In Allen, R. C., & Hill, A. (Eds.), *The Television Studies Reader* (pp. 569–580). London: Routledge.

Caldwell, J. (2004). Convergence television: Aggregating form and repurposing content in the culture of convergence. In Spigel, L., & Olson, J. (Eds.), *Television after TV* (pp. 41–74). Durham, NC: Duke University Press.

Cubitt, S. (1991). *Time-shift: On video culture*. London: Routledge. doi:10.4324/9780203358955

Dayan, D. (2009). Sharing and showing: Television as monstration. *The Annals of the American Academy of Political and Social Science*, *625*, 19–31. doi:10.1177/0002716209338364

Dayan, D., & Katz, E. (1992). *Media events: The live broadcasting of history*. Cambridge, MA: Harvard University Press.

Fiske, J. (1987). *Television culture*. New York: Routledge.

Gauntlett, D., & Hill, A. (1999). *TV living: Television, culture, and everyday Life*. London: Routledge.

Gripsrud, J. (2004). Broadcast television: The chances of its survival in a digital age. In Spigel, L., & Olson, J. (Eds.), *Television after TV* (pp. 210–223). Durham, NC: Duke University Press.

Jenkins, H. (2006a). *Convergence culture: Where old and new media collide*. New York: New York University Press.

Jenkins, H. (2006b). *Fans, gamers, bloggers: Exploring participatory culture*. New York: New York University Press.

Katz, E. (1996). And deliver us from segmentation. *The Annals of the American Academy of Political and Social Science*, *546*, 22–33. doi:10.1177/0002716296546001003

Lau, T. Y. (1988). Introducing cable television into Hong Kong: Political and economic implications. *Telecommunications Policy*, *12*(4), 379–392. doi:10.1016/0308-5961(88)90060-2

Lau, T. Y. (1992). From cable television to direct broadcast satellite: Emerging policy issues in the Asia-Pacific Region. *Telecommunications Policy*, *16*(7), 576–590. doi:10.1016/0308-5961(92)90023-I

Lee, F. L. F. (2007). Development of pay television in Hong Kong: History, audience consumption, and challenges. *Journal of Comparative Asian Development*, *4*(2), 247–272.

Lee, F. L. F. (2009). Paradigm defining events and event routinization: Hong Kong newspapers' coverage of YouTube video clips. *The Chinese Journal of Communication and Society*, *9*, 181–202.

Lee, F. L. F. (2010). The influence of family viewing preferences on television consumption in the era of multichannel services. *Asian Journal of Communication*, *20*(3), 281–298. doi:10.1080/01292981003802176

Liebe, T., & Curran, J. (Eds.). (1998). *Media, ritual, and identity*. London: Routledge.

Ma, E. K. W. (1999). *Culture, politics, and television in Hong Kong*. London: Routledge.

Manning, P. (1997). Media loops. In Bailey, F., & Hale, D. (Eds.), *Popular culture, crime & justice* (pp. 25–39). Belmont, CA: Wadsworth.

Miller, T. (2004). Trainspotting the *Avengers*. In Gwenllian-Jones, S., & Pearson, R. (Eds.), *Cult Television* (pp. 187–197). Minneapolis: University of Minnesota Press.

Morley, D. (1980). *The Nationwide audience*. London: British Film Institute.

Napoli, P. (2003). *Audience economics*. New York: Columbia University Press.

Parks, L. (2004). Flexible microcasting: Gender, generation, and television-Internet convergence. In Spigel, L., & Olson, J. (Eds.), *Television after TV* (pp. 133–156). Durham, NC: Duke University Press.

Qiu, J. L., & Chan, J. M. (2009). Toward the study of new media events. *The Chinese Journal of Communication and Society*, *9*, 19–38.

Ross, S. M. (2008). *Beyond the box: Television and the Internet*. Malden, MA: Blackwell.

Van Dijck, J. (2009). Users like you? Theorizing agency in user-generated content. *Media Culture & Society*, *31*(1), 41–60. doi:10.1177/0163443708098245

Waldfogel, J. (2009). Lost on the web: Does web distribution stimulate or depress television viewing? *Information Economics and Policy*, *21*(2), 158–168. doi:10.1016/j.infoecopol.2008.11.002

Webster, J. G. (2005). Beneath the veneer of fragmentation: Television audience polarization in a multichannel world. *The Journal of Communication*, *55*(2), 366–382. doi:10.1111/j.1460-2466.2005.tb02677.x

Williams, R. (1975). *Television: Technology and cultural form*. New York: Shocken. doi:10.4324/9780203450277

ENDNOTES

1 The figure is from the Hong Kong Annual Report published by the government.

2 Radio Television Hong Kong (RTHK), the public broadcaster in the city, did not have its own television channels. Rather, government regulations requested commercial broadcasters to spare a certain amount of airtime to broadcast RTHK programs.

3 The three service providers obtained the domestic pay television service licenses from the Hong Kong government. A fourth service provider, Hong Kong Broadband, operated a television content service only through the Internet. It is not classified as a domestic pay television service provider by the government.

4 Figure 1 was produced based on a simple content analysis of the wall posts in the Laughing Gor Facebook group. Adopting a systematic sampling procedure, every 10th wall post was coded in terms of a few simple variables. From the results of the content analysis, it is possible to estimate the number of wall posts appearing each day.

5 Figure 2 was produced using the electronic news archive Wise News. "Laughing Gor" was used in the keyword search to generate numbers of articles per day.

6 The article appeared originally at: http://309668875.qzone.qq.com/blog/1238314009.

7 The article appeared originally at: http://www.inmediahk.net/node/1002758

8 The article appeared originally at: http://yipyipsir.blogspot.com/

9 The article appeared originally at: http://blog.24reader.com/vincentlam/

10 The article appeared originally at: http://hk.myblog.yahoo.com/iama-cs/article?mid=31308

11 The article appeared originally at: http://ebenezer-ebenezer.blogspot.com/.

12 The article appeared originally at: http://ad101hk.blogspot.com/2009/03/12laughing.html

13 The article appeared originally at: http://eugeneyhc.wordpress.com/

14 The article appeared originally at: http://viennalam2006.spaces.live.com/blog/

Chapter 9
Bearing Witness through Technology:
War and Visual Culture in a Converged World

Yasmin Ibrahim
Queen Mary, University of London, UK

ABSTRACT

The convergence of technologies brings a two-fold phenomenon into our technologically mediated world. Firstly, mobile technologies have enabled the recording of war and conflict by bystanders who can then transmit their recordings to the wider world through the Internet. Secondly, this form of bearing witness creates a proliferation of images of suffering leading to a new visual economy that invites new types of spectatorship which reconfigure events and integrate them into mainstream and niche media platforms. This chapter argues that the aestheticization of suffering is a dominant and integral part of our culture and new media technologies illuminate and offer new ways to both engage with and commodify suffering. The commodification of suffering on video-streaming platforms on the Internet results in spaces of perverse obsession with pain and suffering but also performative counter-sites which offer different ways to record and narrate human suffering beyond the vantage point of mainstream media.

INTRODUCTION

Kevin Carter's photograph of a starving emaciated Sudanese toddler stalked by a vulture whilst trying to make her way to a feeding camp won him a Pulitzer Prize in 1994. The picture caused immense controversy when it was published in the *New York Times* and subsequently the world over. As an iconic image conveying the depths of poverty in Africa it invited global spectatorship but equally it ignited debates on the ethics of photography. Ironically, the vulture came to symbolize the photographer aestheticizing a horrific image of suffering in the practice of his craft. The worldwide gaze and criticism of the image may have contributed to Carter's suicide a few months after the release of the picture. Nevertheless the image provided a platform for raising awareness of the poverty in Africa and the vulnerability of the children in such conditions. Inevitably it also

DOI: 10.4018/978-1-60960-037-2.ch009

Copyright © 2011, IGI Global. Copying or distributing in print or electronic forms without written permission of IGI Global is prohibited.

reopened traditional debates about the 'ethics of gaze' and the morality of spectatorship in conditions of suffering.

The aesthetics of suffering occupies a tenuous ethical terrain both in art history and in philosophical discussions of the human condition in relation to our relentless obsession with and pull towards images which depict suffering. Undeniably, the history of human civilization can be depicted by images of death, disaster, war, terror, starvation, disease and countless other tragedies which often become the vehicle to authenticate history and memory. The visual image occupies a space of double subjugation. It denotes both the vantage point of the person who captures the image and equally the secondary 'gaze' of those who consume it and ascribe to it new meanings and values. This double subjugation, along with the problematics of representation and its association with memory and history, often politicizes the image as a value-laden artefact that can thwart reality and realism through the mediation of technology and the ethics of gaze.

Both mobile technologies and the convergence of technologies, which incorporate visuals, sounds and the ability to communicate, interact and publish on a global platform like the Internet, have created the possibilities for new forms of consumption. The ability to record and store images and sound and to upload them on the Internet enables civilians to become image creators when events happen across the globe. The act of recording through mobile technologies and the ability to circulate images across the world craft new rituals in event creation where the bystander, through the act of recording, bears witness through technology. This material act of bearing witness through technology unleashes a new form of moral economy where the images of suffering captured through convergent technologies convey new visibilities where the civilian gaze can narrate events, without the mediation of newsmakers, onto multimedia platforms which can potentially invite non-stop viewing from audiences around the globe.

The civilian bystander equipped with mobile technology becomes an instrument of intervention in representing suffering away from the vantage point of the entrenched media and established newsmakers. The proximity of the gaze enabled through mobile and convergent technologies reconfigures the well-established relationship between news producers and audience. The appropriation of suffering away from the established storytellers heralds the formation of a new image economy where the proliferation of user-generated content is enabling new engagement with suffering and different ways to narrate and publicize it to the world. Multimedia platforms (which enable the upload of images and sounds) beyond creating new forms of visibility elicit new ways to engage with suffering. Often images of suffering become commodified with video streaming portals such as YouTube and additionally become integrated as data into search engines which organize images into a coded archive of information enabling audiences to search and consume through the click of a mouse. This commodification of visuals (whether still or moving) has consequences for both the aestheticization of suffering and our moral engagements with consuming suffering as new forms of visual entertainment often disembedded from its context.

This paper analyses this politics of creation and consumption of suffering where the omnipresent gaze of the technologically mediated mobile body captures events on the move and stages them for worldwide consumption. The consumption of war, conflict and suffering then occupy a liminal world between pleasure and taboo which invite consideration of both the moral and ethical paradigms as well as the political interventions we can make through the consumption of these images. The multimedia platforms which enable new ways of bearing witness through technology and the ability to create 'media events' raise issues of aesthetics, cognitive distance or proximity as well as our ability to empathize and relate to human suffering.

This paper argues that the co-presence of the mobile and connected body becomes an integral construct in event-making as its non-stop omnipresent gaze enables the capture of events beyond the reach of media institutions. The increasing reliance of mainstream media on this 'omnipresent gaze' to augment news stories, whether in the case of the Indian Ocean Tsunami in 2004, the 7/7 bombings in London or the massacre of monks in the 2007 anti-government protests in Burma, reiterates the growing importance of mobile bodies and technologies in re-negotiating event creation and memory-making in a globalized world. This can reconfigure the ways in which we narrate, understand and archive significant events and can be instrumental for the ways in which we imagine the nation, seek communion, but mostly for the overarching project of historical memory and historicity.

MEDIATED MODERNITY AND MORALITY

Modernity as mediated, accelerated and sensitized through technology is a well-worn trope in social sciences. The renegotiation of cognitive distance and meaning-making through the processes of technological mediation needs to be revisited in the context of mobile technologies and their ability to bear witness and represent reality. Martin Heidegger (1971: 165) locates technology as the defining characteristic of modernity where technology can shrink time and space without necessarily bringing any nearness. For Heidegger, technology can aid in the perishing of senses to offer new ways of revealing and relating to unseen objects. Equally, technology can bring new forms of existence in spatio-temporal terms and reconfigure the horizon in which human interactions are placed and unfolded (Lattas 2006:31). In tandem with this, Paul Virilio's (2000) notion of hypermodernity positions the 'accident' and 'disaster' as culturally fundamental to a technologically mediated society. Here the notion of 'catastrophilia' is integral to and embedded in both technocratic and perceptive realities of everyday life (cf. Feldman 2005:223) such that technologies shape our environment through non-stop images which induce anxiety and awe.

Virilio's hypermodernity captures both the acceleration of speed and pervasiveness of technology. In his critique of modern art and media, Virilio alludes to a modern visuality characterized by 'pitilessness and excess' and inundated with 'tortured forms' which thwart humane engagement. The aesthetics of this new visuality in modernity amplified by technology then has political and ethical consequences for humanity. The thematization of accidents as central to modernity places information and communication technologies (ICTs) as central in proliferating mediatized images of events where there is an issue of separating the reality from the simulacrum of images. For Virilio the real world is substituted by another reality produced by the conjoining of information technology and mediatized images. This results in an 'overexposed' world where there is nowhere to hide (cf. Newman 2005:2-3). In the same vein, Ulrich Beck's (1992) notion of modernity as a risk society filled with mediatized images and risk discourses represents the interconnectivities of a globalized world as creating a pervasive environment of risk mediated through ICTs.

In considering the question of aesthetics in relation to the new media ecology there is a need to integrate the dynamics of image seeking and profiling on the Internet. The proliferation of images of suffering, communication of war and conflict through visual narratives and instant access to videos disembedded out of their context into video streaming platforms creates a politics of pity that is decentred and mediated through a 'download and upload culture' where looking, sharing, archiving, publishing uploading denote new forms of engagement, politicization and aestheticization of war, conflict and human suffering.

Timothy Jackson (1999) points out that the term 'aesthetics' has come to mean a discourse focusing on the critique of dominant conceptions of taste and beauty in the arts. He contends that the philosophical field of aesthetics has a complex political history and is linked to how we engage with our world and articulate meaning through sensory input and processing. By focusing on how ideas are formed, shared and contested through our senses it provides a way of delving into both the ethical and the political. Jackson (1999: 314) reasons that it is important to ponder the future role of radically new aesthetics in a world in which senses are manipulated by new technology. New media present new challenges for negotiating meaning through sensory input by providing new types of experiences and forms of art production and consumption. New media inaugurate a new reality which is re-shaping our world and world-view where there is an interplay between the natural and the prosthetic. This creates new forms of thinking, viewing and engaging with multiple realities. The Internet, with its video-streaming facility, enables spectatorship at a mouse click, bringing the horrific and gory, the traumatic and tragic onto our screens through new forms of news seeking, archiving and publishing. Here the imagination of communion with unseen societies (Anderson 1991) or detachment through 'compassion fatigue' (Sontag 2003) can be equally representative of modernity where technology can induce moral apathy or empathy regarding the suffering of others.

Daniel Dayan and Ellul Katz (1992) categorize the 'media event' as a construct which pauses the everyday, compartmentalizing it into the unusual through the narration of events such as the death of Princess Diana where these are mediated by the production codes of broadcasting and the temporality of live coverage. In contrast with broadcasting, the Internet does not hold audiences worldwide captive from a temporal perspective. The disembedding of both temporality and context reconfigure the consumption practices on the Internet where a media event reported through the broadcasting medium could trigger people's need to augment television coverage through extra visuals and civilian narratives of the event on the Internet. The conjoining of these different temporalities presented by broadcasting and the Internet is a distinct aspect of viewing 'trauma' in postmodernity.

A case in point is the recent assassination of the former president of Pakistan, Benazir Bhutto. The assassination of Bhutto, like other world events, was also about the exploitation of new visibilities. Within hours of Bhutto's assassination malware makers capitalized on the news to dupe users into downloading viruses which can compromise the PC with a wide range of other malware. These malware makers duped users into believing that they needed to download software in order to view video clips in high definition and in the process PCs became infected with new viruses (Keizer 2007). Beyond the questions of authenticity, truth and simulation of reality, this new visuality is textured through a paradigm of risks, duplicity and opportunities crafting it as a space that can both empower and deceive users. The Internet as an intertextual space which binds the ideologies and realities of the physical world as well as the attributes and communication practices of the virtual world becomes a contested terrain where counter sites can emerge to challenge dominant paradigms and present new forms of visibilities. On the other hand, it can equally manipulate and simulate reality, making it difficult to verify or authenticate information.

DISTANT PROXIMITY OF MEDIATED REALITY

Beyond the notion of technological mediation, acceleration and the pervasiveness of risk construction in our conceptions of the world, there is a need to consider the embeddedness of information and communication technologies and media in our everyday lives where the wider world is conceptualized through sound bites and image-laden narratives. As discussed in the last section, the proliferation of ICTs is seen as a major force in decoupling time and space (Giddens 1984:113; Moores 2004). Distance and temporality are negotiated through new interactive technologies which not only connect the world but create new visibilities by highlighting remote parts of the world. Our intrusion into these hitherto inaccessible spaces constructs the globe as one that is wired and open to media intrusion enacting a pervasive gaze that can capture the human condition when it is faced with new travesties or traumas, where new media events and cultural tropes can be created through these technological visibilities and connectivity.

Roger Silverstone (1994), in his theorization of the domestication of technology, postulates that the meaning and significance of all our media and information technologies depend entirely on the engagement of the user where the cognitive participation can both blur the spatio-temporal spaces while reinforcing them. Media technologies and their domestication are seen as playing a crucial role in the ontology of everyday life. Life is experienced through both the formal (rituals) and informal (mundane) structures of the everyday (Silverstone 1994: 169). Technology enables the translation of the new into the familiar and the unfamiliar to be objectified, integrated or distanced through the meanings imposed in the private realm. The world around us is streamed through broadcast and online platforms, engaging us beyond the realms of our situated domesticity.

Clive Barnett (2004: 63) alludes to the ability of media environments, particularly of print media, to 'detach symbolic forms from local contexts and re-inscribe them in new contexts depending on the physical transportation of tangible objects over material infrastructures'. Hence media technologies enable new forms of signification and integration over expanded spatialities. Barnett (2004) emphasizes John Carey's contention that telecommunications and broadcasting signal a departure from previous media such as print. Their interrelationship with communication, infrastructure and transportation mean that they can be detached from the transportation of tangible objects over space.

Similarly, John Thompson (cf. Barnett 1004:64) points out that electronic communication technologies give rise to 'despatialized simultaneity' where 'the experience of "now" is detached from shared locales and the sense of "distance" is detached from physical movements and travel'. The dissolution of distance, temporality and the inscribing of new values through the representation of images signify a global media economy where the transportation and dissemination of materials happens non-stop in our media-saturated environment.

Paddy Scannell's (See 1996a; 1996b) seminal writings implicate the mass media (i.e. radio and television) in the emergence of a mass consciousness shaped by technology, the rituals of broadcasting and its incumbent temporality. Broadcasting, through its 'dailiness', provides a 'continuity, reliability and a familiarity'. The entwining of the personal with the mediated broadcast space signifies the coalescing of imagination to form our cognitive realities of the wider world. The media space is the stage where the global is narrated through the dailiness of news and documentary, where our secure 'imagined communities' (Anderson 1991) are constantly juxtaposed through the presentation of the 'imagined others' who also inhabit the globe. Benedict Anderson qualified such communities not through the notion of authenticity of their presence but in accordance to the style in which they are imagined (cf. Barnett 62). This in some senses implicates the biases in communication technologies where, for instance, the eye may be privileged over the ear or visuality over orality. However, moving beyond the arguments of bias instituted by Harold Innis (See 1950; 1964), Anderson stresses the rituals of communion where media artefacts and consumption provide the means to base this imagination.

Hence imagination is both limited and governed through the different types of medium.

Media such as television and the Internet provide forms of voyeurism where one can 'travel' the globe and imagine it through both the non-stop narratives that are streamed in via broadcasting or hyperlinks and networks in the World Wide Web which constructs the globe through one's active engagement with content and technology. ICTs provide 'imaginary mobility' to witness the world and, in particular, satellite broadcasting provides an 'instantaneity' to the rest of the globe. The domestic space becomes a site for global consumerism as well as spectatorship of the global.

Raymond Williams (1974:171), in his book *Television: Technology and Cultural Form,* employs the term 'mobile privatization' to describe how information and communication technologies enable the experiencing of the world from the domestic sphere where there is movement from public engagement to private consumption. Williams highlighted two dialectical strands which emerged in urban industrial society: mobility through the development of transportation, especially through the construction of railways and mass migration by steamship, and the emergence of the atomized self-sufficient family home (cf. Mcguigan 2007). These 'mobile and home-centred ways of living' constituted a 'mobile privatization' for Williams and within this framework of mobility and atomization of the family unit, broadcasting fitted neatly into the arrangement providing both a site for family communion and private consumption as well as access to events happening in distant locales (cf. Mcguigan 2007).

Space is a recurrent metaphor in media theory for it signifies the construction of cognitive realities such as public spheres where we debate and scrutinize issues consequential to our public and private lives. Thus throughout the trajectory of media studies, media is theorized as a 'doubling' (Scannell 1996b) or a multiplying of places (Couldry 2000) or even as a sense of 'placelessness' (Meyrowitz 1985). Paddy Scannell (1996b: 76), in proposing a phenomenological approach to studying radio and television, argues that media are concerned with the 'ways of being in the world, providing audiences in different settings and localities "the possibilities of being in two places at once"'. In setting forth his argument, Scannell (1996b:167) alludes to the doubling of space where the audience share a co-presence in witnessing a 'live' event in remote places through broadcasting without being there materially, bringing them 'close' or 'within range' of the broadcast event and removing the 'farness'. In particular, Scannell discusses the ways in which media, particularly broadcasting, arrange or perhaps rearrange temporality and in the process construct a 'simultaneity of experience.'

Lisa Parks (2004: 38), in reviewing newer media forms, observes that the 'meanings, knowledges and experiences of time/space and movement have themselves shifted with different technologies, geographies, users and sociohistoric conditions'. In critiquing the 'annihilation of time-space discourse' against the Web environment, she points out that it has led to the creation of a fantasy of 'digital nomadism' where the user in the Internet environment is free to roam through the electronic terrain and assume different identities. This fantasy she contends is the 'extension of the fantasy of global presence which emerged with the first live international television broadcasts in the 1960s'. Parks (2004:38) is critical of such subjectivities 'privileging and centralizing a transcendent Western subject' that she proclaims as 'existing above and beyond technology rather than in relation to it.' With specific reference to the network infrastructures of the Internet, Parks contends that our cognitive realities are limited through a combination of geographic, artistic, linguistic and visual systems of signification. Our power economies within the home and the outer world as well as our cultural systems of signification both define and constrain our imagination of the global and our sense of 'otherness' that can emerge in our postmodern condition.

The media's role in shaping our cognitive reality, temporality and notions of distance has been well documented in media theory. The socialization of mass consciousness through the broadcasting space and its role in creating mass spectatorship of media events is an important element in the popularization of the 'tragic' and the aestheticization of trauma. The mainstreaming of tragic narratives through the broadcasting space creates acceptability in integrating suffering as a narrative of mass appeal in our everyday lives. Additionally, this image-manufacturing culture of postmodernity is inevitably entwined with the political ideologies and power economies of any given society. This crafts modern visuality as a problematic device amenable to the inherent hegemony of existing socio-economic and political ideologies, but equally resistant to a dominant reading and accommodative of alternative paradigms. While images can reiterate dominant ideologies they can also become counter sites. Kari Andén-Papadopoulos (2008: 5), commenting on the infamous photographs of abuse at Abu Ghraib, asserts that images don't always become an instrument of existing ideologies. In contrast, they can become sites of protest and opposition to the very deeds they represent. For Andén-Papadopoulos the Abu Ghraib photographs 'function as a critical prism through which elite and popular views on US foreign policy are refracted and the Bush administration's "war on terror" is understood.'

VISUALITY AND THE COMMUNICATION OF SUFFERING

Kathryn Moore (2003: 26) reasons that the concept of visual thinking is deeply embedded in our culture and an intrinsic part of our lives. The role of images in structuring or perhaps restructuring reality has consumed much of the debates both in art history as well as in the field of photography (See Barthes 1977; Sontag 1977; Zelizer 1998, 2001; Sekula 1984; Mitchell 1994). Relevant to this discussion is the notion of the photograph as a referent to an event or incident, Roland Barthes (1981: 76-77) contending that photography certifies and corroborates reality as one 'can never deny the thing has been there'. Often the image is a short-hand for both objectivity and subjectivity and hence creates 'a crisis of faith in optical empiricism' (Tagg 1992). The capacity to forge a personal connection with a past public trauma depends first on the materiality of photographs, whereby photographic images stand in for the larger event, issue or setting they are called to represent (Zelizer 2002: 699).

In C. Wright Mills' apt description of our 'second-hand world', this constant circulation and repetition of images constitutes a terrain of mediation and as such images are not innocent of cultural formation nor are they innocent of forming culture. Annette Kuhn (1985) calls this the ideology of the 'visible as evidence' where the 'illustrative function of photographs leaves opinions, prejudices, fantasies, misinformation untouched' (Sontag 2003:75). Mark Danner (cf. Bauman 2006: 106) ascertains that the most powerful weapon of the 9/11 terrorists was 'that most American of technological creations: the television set.' The TV can push universal fears of vulnerability and the sense of ubiquitous danger far beyond the limits of the terrorists' own capacity (Bauman 2006: 106). In our media-saturated world 'repetition is instrumental in sustaining a sense of the real' (Deleuze 1994).

Dudley Andrew (1997: viii), in writing on the works of Walter Benjamin, comments that Benjamin, in relation to other theorists, 'strove to relate this new image culture to a modernity that involves new forms of subjectivity and social organization.' Similarly, Jean Paul Sartre's (1956:340-400) conceptualization of 'The Look' again emphasizes the power economies of gazing where there is an objectification of the subject gazed at. For Sartre this 'look' disempowers people of their agency to narrate themselves. In addition to this, with reference to photography, Paul Petrovic contends that in the digital age it enables the ability to reproduce like never before and augments the additional charge of photographic images enabling trickery, deception and manipulation.

Susan Sontag (1977: 7-10) situates the camera lens as mediating reality and conveying participation in the situation of an event. The camera nevertheless adds an element of detachment, creating a measure of distance to the event. But more importantly, the act of being photographed turns the image into 'a part of a system of information fitted into schemes of classification and storage' (Sontag 1977:156; Twomey 2007: 297). For Sontag (1977:41), 'our ability to stomach this rising grotesqueness in images' creates a pseudo-familiarity which in the long run reinforces alienation thus making one less amenable to react in real life. Images of atrocity or the repulsive become 'ordinary, remote or inevitable'. Keith Tester (1994:130), in commenting on televisual images of suffering, makes a similar argument where the stream of non-stop images of suffering becomes commonplace and incapable of evoking moral significance. In this sense, Frederic Jameson (1992: 1) categorizes the visual as 'essentially pornographic' for it unleashes rapt mindless fascination. John Taylor (2000: 135) concedes that the intellectual movement against the proliferation of photographic realism is driven by the belief that visuals result in mindless fascination. The sensory appeal of visuals, for Jean Baudrillard (1988: 109-111), leads to audiences being seduced and held hostage by media images of violence.

Julie Kristeva (1982), in writing about the 'power of horror' in a myriad scenarios of complex human existence, finds the abject in the law, prophetic literature and the gospels, but the ultimate abjection is in the fiction of the end of the world or the apocalypse (cf. Pippin 1997:2). The concept of abjection illuminates the difficult connection between horror and the human psyche, highlighting both our attraction and revulsion to images of horror. The television as a communication medium further distorts the imaginary distance and the closeness of the depictions of horror creating archives of evidence through this visibility. Sontag (2003: 91) points out that our ability to feel sympathy means that 'we feel we are not accomplices to what caused the suffering for our sympathy proclaims our innocence as well as our impotence.'

Ernst Junger (1989:207-208) argues that media technology enables the freezing of trauma and suffering into a manageable format characterized by an aesthetics of detachment. This photographic detachment neutralizes pain and suffering as it is cast outside of the 'realm of sensibility.' The mediatization of reality through technology creates a secondary consciousness where the subject sees himself as standing outside of the realm of pain.

With broadcasting and the availability of multimedia facilities on the Internet, there has been a privileging of the eye and an emphasis on visuality. Individual experiences of pain and suffering become public spectacles when portrayed through the mass media (Yamada et al. 2006) and become a 'cultural appropriation of suffering' (Kleinman & Kleinman 1997) en masse. Anita Biressi (2004) asserts that in our post-documentary culture the consumption of pain and suffering is facilitated through new television formats and genres whether it be reality television or the uploading of particular scenes of suffering on the Internet.

The commodification of suffering through clips and hyperlinks creates a ritual of segmenting suffering from its context to be consumed on demand. This mouse-click aestheticization and consumption of suffering can invite global spectatorship into unknown events of suffering in remote spaces. In the process it can work to raise awareness of events which mainstream media may not have access to. On the other hand, the commodification of suffering can enter an economy of circulation where it serves to assuage our human obsession with suffering where everything can be uploaded as a global public spectacle without any consideration for human suffering or dignity. This pervasive visuality is dialectical as it can be trigger communal agency and empathy and equally work to anaesthetize and numb our ability to connect and feel the pain of others. Unlike broadcasting which can push information to its audience, the user-led environment of the Internet can initiate selective amnesia and a culture of fetishistic image seeking where images of suffering and pain are decontextualized and dismembered from the events and circulated in the context of communication and interaction between people and online communities. They then become artefacts or cultural commodities for aesthetic consumption, exchange and reciprocity.

THE VIRAL APPEAL OF SUFFERING

The individual acts of co-presence in an event often have the potential for collective viewing and witnessing when they are published on the Internet. Images of the protest which left up to 200 dead in the first few days of a violent crackdown in Burma on 23 September 2007 were captured on cell phones, satellite phones and digital cameras and circulated around the world (ONI 2007, Stillman 2008; Meyer 2008). The human rights abuses by Chinese forces in the Himalayas were also broadcast to the world through clips on YouTube (Naim 2007). Mobile technologies and the Internet became instruments of intervention in illuminating the plight of the protesters in Burma. In contrast, during the 1988 uprising in Burma it was relatively easy for the authorities to cut phone lines and impose a news blackout. Similarly, the anti-China protest in Tibet in 1989 took a considerable time to reach the world community (Mellgren 2008; AFP 2008).

The convergence of technologies has enabled new forms of news gathering and dissemination, for example in the case of microblogging in China among the more Internet-savvy citizens. Microblogging entails publishing short messages from blogging sites on mobile phones and vice versa (AFP 2008). This form of communication has been useful in disseminating information about events which may not be reported by the mainstream media as in the case of China. The role of image in transmitting suffering has also been crucial in gathering public support and in building public sentiment. An image's ability to corroborate the occurrence of events where solid eyewitness narratives may not be evident has been vital in repressive or authoritarian regimes. For example, YouTube showed one recording of a huge protest near Labrang Monastery in Sichuan, China, which validated the eyewitness reports of 4,000 people (AFP 2008).

The YouTube revolution is symbolic of the modern visuality where endless clips ranging from the significant to the mundane form a repository of images which can be accessed and viewed disembedded from their context. They can include clips from mainstream media as well as those captured on mobile technologies by individuals. The lack of any hierarchy within this repository, the decentring of what is political or significant, and the culture of image-seeking through individual preference and choice is an integral characteristic of this visuality of demand. YouTube, founded in February 2005, receives 20 million visitors every day. On a typical day 100 million clips are viewed and over 650,000 clips are posted showcasing everything from the politically sensitive to spoofs of politicians (Naim 2007). An endless array of images from the tragic to the comic, and from the banal to the salacious, vie for audience attention in these spaces. Naim (2007) points out that multimedia platforms such as YouTube create a 'double echo chamber'. When content is first posted on the Web it is re-aired by mainstream TV networks while memorable clips from television are given a permanent presence by bloggers or activists who redistribute them through websites. Naim contends that the power of the gaze has not gone unnoticed as the notion of non-stop gaze is used as a weapon to hold authorities and powers to account. For example, the human rights group Witness hands out video cameras to activists and individuals in conflict zones. In other instances, electoral watchdogs are taping elections. In the US, for example, the military has asked soldiers to stop posting videos unless they are vetted. On the other hand, these multimedia platforms have been used for the purposes of duping and deceiving people and for publicly displaying people's private lives for the benefit of voyeurs.

These new visibilities enabled through mobile technologies are inevitably crafting new relationships with established media institutions who are adopting mobile visual images into their newsmaking. People's co-presence in media events and the expansion of mainstream media narratives through the augmentation of images captured through mobile technology is becoming an embedded feature in mainstream media. The communication of suffering becomes an open-ended phenomenon continuously fed and expanded through the convergence of technologies and co-presence of people bearing witness through technology. When the former president of Iraq Saddam Hussein's execution was captured on a mobile phone by one of the witnesses to the event, the mainstream media were much more restrained in broadcasting it while audiences were able to access the whole footage online.

Even after the end of a media event and its extensive coverage in mainstream media, images of suffering can be endlessly circulated on the Internet where they enter an information economy of being retrieved, archived, searched, shared, exchanged and hyperlinked to diverse websites. The consequences of our shifting communication practices through new media platforms and the manipulation of images as well as concerns of deception and trickery become important issues in our image-filled world. Equally, the proliferation of images and our need to consume and engage with images of trauma and suffering raise ethical questions about how truth can be possible in a global economy which has a pervasive image-manufacturing culture (Jackson 1999:315).

According to Allan Feldman (2005: 203-20), since the 'first Gulf War there has been a global repositioning of visual communication practices with the intensification and proliferation of threatening images of violence, terror, and social suffering in our public culture.' This re-positioning of communication practices of terror symbolized through iconic images of the World Trade Center attacks or abuses at Abu Ghraib are indicative of both the power economies of war and terror but equally of the 'material reproductive' efficacy of imaging technologies. Feldman (2005:218) argues that the first series of photographs of Abu Ghraib were comparable to the 'celebratory and horrific carnivalesque atmosphere of the pictures postcards that were sold as souvenirs of the lynching and mutilation of African-Americans in the 1920s.'

Photography and videography also become tools to 'mobilize shame' (cf. Beckman 2007). For Thomas Keegan (2004) this notion of mobilizing shame is rooted in the belief that where law is absent reason works when it exposes and reveals. In the images of torture at Abu Ghraib and of Saddam Hussein's execution, Keegan moots the notion of 'intervention' through the aesthetics of consumption. He contends that the 'aesthetic finds itself in extreme proximity to the ethico-political now; the proximity is perhaps discomforting to some but it is also the condition of any serious intervention' (2004:446). The concept of aesthetics then becomes a point of intervention and a means to connect with an image (cf. Beckman 2007: 65).

Hal Foster 1996 (354-358), in citing Lacan, propounds that the traumatic is a missed encounter with the real and as such while the missed real cannot be entirely represented, it can be repeated as a form of technique to instil trauma. In commenting on the phenomena of a spectacular society, Foster (1996:366) contends that it is in the role of the witness that the mass subject reveals the sadomasochistic aspect of the human condition. In consuming suffering the mass subject 'is often split in relation to a disaster; even as he or she may mourn the victims, even identify with them masochistically, he or she may also be thrilled sadistically by the victims of whom he or she is not one.' According to Peter Sloterdijk (1987), mass spectatorship of modernity is sustained through technological possession of modern consciousness and equally conveys the political passage from the unconscious to automatic consciousness which also incorporates the politics of suggestion. Thus a mechanized consciousness is enabled through the detachment of perception from the event for the individual spectator and where a memory-making institution takes over the collective notion of perception and witnessing (Feldman 2005: 215).

CONCLUSION

The omnipresent gaze facilitated through mobile bodies and technology crafts the notion of the 'media event' as an open-ended exercise that can be expanded and narrated from different vantage points. The ability of mobile bodies to create 'media events' which can court the attention of audiences worldwide as well as mainstream media institutions signifies a new visibility where non-stop visuals highlight both our technologically mediated modernity as well as the notion of visual thinking as an intrinsic aspect of our cognitive reality. The commodification of suffering through new media platforms is symbolic of both our fascination with the morbid and the need for our individual and collective aestheticization of suffering. Equally, these spaces function as counter-sites for eliciting human agency and mobilization of support and empathy for the suffering of others. Additionally, the commodification of suffering, the disassociation of video clips from their political and social contexts on the Internet, and their placement into a repository of disparate visuals craft them as cultural and communication artefacts which can be shared and exchanged not only for their content but as cultural capital between users and producers on the Internet.

REFERENCES

Agence France Presse (AFP). New technology triggers battle for information from Tibet. (2008, March 18). Retrieved from http://afp.google.com/article/Aleq M5gyuONfqd4fnQB61MMrrD-PyjMDjw

Andén-Papadopoulos, K. (2008). The Abu Ghraib torture photographs: News frames, visual culture, and the power of images. *Journalism*, *9*(1), 5–30. doi:10.1177/1464884907084337

Anderson, B. (1991). *Imagined communities: Reflections on the origin and spread of nationalism*. London: Verso.

Bann, S. (1984). *The clothing of Clio: A study of the representation of history in nineteenth-century Britain and France*. New York: Cambridge University Press.

Barnett, C. (2004). Neither poison nor cure: Space, scale and public life in media history. In Couldry, N., & McCarthy, A. (Eds.), *Mediaspace: Place, Scale and Culture in a Media Age* (pp. 58–74). London: Routledge.

Barthes, R. (1977). *The photographic message. In Image-Music-Text*. New York: Hill and Wang.

Barthes, R. (1981). *Camera Lucida: Reflections on photography, (Trans. R. Howard)*. New York: Hill and Wang.

Baudrillard, J. (1988). *America*. London: Verso.

Bauman, Z. (2006). *Liquid fear*. Cambridge, UK: Polity.

Beck, U. (1992). *Risk society, (Trans. M. A. Ritter)*. London: Sage.

Beckman, K. (2007). Telescopes, transparency, and torture: Trevor Paglen and the politics of exposure. *Art Journal*, *2007*(Fall), 62–67.

Biressi, A. (2004). "Above the below": Body trauma as spectacle in social/media space. *Journal for Cultural Research, 8*(3), 335–352. doi:10.1080/1479758042000264975

Couldry, N. (2000). *The Place of media power: Pilgrims and witnesses of the media age*. London: Routledge.

Dayan, D., & Katz, E. (1992). *Media events: The live broadcasting of history*. Cambridge, MA: Harvard University Press.

Dudley, A. (1997). A preface to disputation. In Dudley, A., & Shafto, S. (Eds.), *The image in dispute: Art and cinema in the age of photography* (pp. vii–xiv). Austin, TX: University of Texas Publication.

Feldman, A. (2005). On the actuarial gaze: From 9/11 to Abu Ghraib. *Cultural Studies, 19*(2), 203–226. doi:10.1080/09502380500077763

Foster, H. (1996). Death in America: Shocked subjectivity and compulsive visual repetition. In A. Michelson (Ed.), *October: The second decade, 1986-1996* (349-370). Cambridge, MA: MIT Press.

Giddens, A. (1984). *The constitution of society*. Cambridge, UK: Polity Press.

Heidegger, M. (1971). *'The Thing.' in Poetry, language, thought, (Trans. J. G. Gray)*. New York: Harper & Row.

Heidegger, M. (1977). *'The age of the world Picture'. The question concerning technology and other essays, (Trans. W. Lovitt)*. New York: Harper Torchbooks.

Innis, H. (1950). *Empire and communication*. Oxford, UK: Oxford University Press.

Innis, H. (1964). *The bias of communication*. Toronto: University of Toronto Press.

Jackson, T. (1999). Taste and aesthetics of simulation: Art, technology and the future. *Journal of Arts Management. Law and Society, 28*(4), 314–319.

Jameson, F. (1992). *Signatures of the visible*. London: Routledge.

Junger, E. (1989). Photography and the second consciousness. In Philips, C. (Ed.), *Photography in the modern era: European documents and critical writings, 1913-1940* (pp. 207–210). New York: Aperture.

Keegan, T. (2004, Spring/Summer). Mobilizing shame. *The South Atlantic Quarterly, 103*(2-3), 436–447.

Keizer, G. (2007, Dec). Hackers quickly move to exploit Bhutto assassination. *Computerworld, 28*, Retrieved from http://www.netoworkworld.com/cgibin/m ailto/x.cgi?pagetosend+/export/home/httpd/h.

Kleinman, A., & Kleinman, J. (1997). The appeal of experience, the dismay of images: Cultural appropriations of suffering in our times. In Kleinman, A., Das, V., & Lock, M. (Eds.), *Social Suffering*. Berkeley: University of California Press.

Kristeva, J. (1982). *Powers of horror: An Essay on abjection, (Trans. L. S. Roudiez)*. New York: Columbia UP.

Lattas, K. (2006). Technology of visibility: The utopian politics of cameras, videos and dreams in New Britain. *The Australian Journal of Anthropology, 17*(1), 15–32. doi:10.1111/j.1835-9310.2006.tb00045.x

Mcguigan, J. (2007, June 18). Mobility: Comment. *Fifth-Estate-Online – International Journal of Radical Mass*. Retrieved from http://www.fifth-estate-online.co.uk/ comment/mobility.html

Mellgren, D. (2007, Sept 26). Cell Phones, web spreads news of Myanmar. *ABC News*. Retrieved from http://abcnews.go.com/Intern ational/wireStory?id=3656364

Meyer, K. (2008, Jan-Mar). Yin and Yang: The Media's Role in the Burmese Peace Movement. *Peace Magazine*, 9. Retrieved from http://archive.peacemagazine. org/v24n1p09.htm

Meyrowitz, J. (1985). *No sense of place: The impact of electronic media on social behaviour*. New York: Oxford University Press.

Mitchell, J. T. (1994). *Picture theory: Essays on verbal and visual representation*. Chicago: University of Chicago Press.

Moore, K. (2003). Overlooking the visual. *The Journal of Architecture*, *8*, 26–40. doi:10.1080/1360236032000068497

Moores, S. (2004). The doubling of place: Electronic media, time-space arrangements and social relationships. In Couldry, N., & McCarthy, A. (Eds.), *Mediaspace: Place, scale and culture in a media Age*. London: Routledge.

Naim, M. (2007, Jan-Feb). The YouTube effect: How a technology for teenagers became a force for political and economic change. *Foreign Policy*, *158*, 104–106.

Newman, S. (2005). Virtual virillo. *IM Issue,* 1. Retrieved from http://nass.murdoch.edu.au/docs /S%20%20Newman.pdf

Open Net Initiative Bulletin. Pulling the plug; A technical review of the Internet shutdown in Burma. (n.d.). Retrieved April, 12, 2008, from http://opennet.net/sites/opennet.net/files /ONI_ Bulletin_Burma_2007.pdf

Parks, L. (2004). Kinetic screens: Epistemologies of movement at the interface. In Couldry, N., & McCarthy, A. (Eds.), *Mediaspace: Place, scale and culture in a media age*. London: Routledge.

Peters, J. D. (2001). Witnessing. *Media Culture & Society*, *23*, 707–723. doi:10.1177/016344301023006002

Petrovic, P. (2005). Between visibility and invisibility: Baudrillard, Jean-Luc Marion, and Lance Olsen's gild imagined by chance. *Extrapolation*, *46*(2), 249–258.

Pippin, T. (1999). *Apocalyptic bodies: The Biblical End of the World in Text and Image*. New York: Routledge.

Pisani, F. (2006, Winter). Journalism and Web 2.0: Tomorrow's potential readers are using the web in ways we can hardly imagine, and we want to remain significant for them. We need to understand how. *Nieman Reports*, *60*(4), 42–45.

Sartre, J. (1956). *Being and Nothingness. (Trans. H. Barnes)*. New York: Washington Square Press.

Scannell, P. (1996a). For a phenomenology of radio and television. *The Journal of Communication*, *45*(3), 4–19. doi:10.1111/j.1460-2466.1995. tb00741.x

Scannell, P. (1996b). *Radio, television & modern life; A phenomenological approach*. Oxford, UK: Blackwell.

Sekula, A. (1984). *Photography against the grain: Essays and photo works 1973-1983*. Halifax: The Press of Nova Scotia College of Art and Design.

Silverstone, R. (1994). *Television and everyday life*. London: Routledge.

Sloterdijk, P. (1987). *Critique of cynical reason*. Minneapolis: University of Minnesota Press.

Sontag, S. (1977). *On photography*. New York: Farrar, Struas & Giroux.

Sontag, S. (2003). *Regarding the pain of others*. London: Penguin.

Stillman, S. (2008, Jan 22). The whole world was watching. *Truthdig.com*. Retrieved from http:// www.truthout.org/docs_ 2006/012308I.shtml

Taylor, J. (2000). Problems of photojournalism: Realism, the nature of news and the humanitarian narrative. *Journalism Studies*, *1*(1), 129–143. doi:10.1080/146167000361212

Tester, K. (1994). *Media, culture and morality*. London: Routledge.

Twomey, C. (2007). Emaciation or emasculation: Photographic images, white masculinity and captivity by the Japanese in World War Two. *Journal of Men's Studies*, *15*(3), 295–310. doi:10.3149/jms.1503.295

Virilio, P. (2000). *The information bomb*. London: Verso.

Williams, R. (1974). *Television: Technology and cultural form*. London: Fontana. doi:10.4324/9780203426647

Yamada, (2006). Casualties: Narrative and images of the war on Iraq. *International Journal of Health Services*, *36*(2), 410–415. doi:10.2190/6PXW-LQ3B-DWN6-XD97

Zelizer, B. (1998). *Remembering to forget: Holocaust memory through the camera's eye*. Chicago: University of Chicago Press.

Zelizer, B. (Ed.). (2001). *Visual Culture and the Holocaust*. New Jersey: Rutgers University Press.

ADDITIONAL READING

Bal, M. (2005). 'The Commitment to Look'. *Journal of Visual Culture*, *4*(2), 145–162. doi:10.1177/1470412905054671

Boltanski, L. (1999). *Distant Suffering: Morality, Media and Politics*. Cambridge: Cambridge University Press. doi:10.1017/CBO9780511489402

Capurro, R., & Pingel, C. (2002). Ethical Issues of Online Communication Research. *Ethics and Information Technology*, *4*, 189–194. doi:10.1023/A:1021372527024

Coman, M. (2005). Cultural Anthropology and Mass Media: A Processual Approach. In Rothenbuhler, E., & Coman, M. (Eds.), *Media Anthropology*. London: Sage.

Dant, T., & Gilloch, G. (2002). Pictures of the Past: Benjamin and Barthes on Photography and History. *European Journal of Cultural Studies*, *5*(1), 5–25. doi:10.1177/1364942002005001153

Debrix, F. (2006). The Sublime Spectatorship of War: The Erasure of the Event in America's Politics of Terror and Aesthetics of Violence. *Millennium: Journal of International Studies*, *43*(3), 767–791. doi:10.1177/03058298060340031401

Dekel, I. (2009). 'Ways of Looking: Observation and Transformation at the Holocaust Memorial, Berlin'. *Memory Studies*, *2*(1), 71–86. doi:10.1177/1750698008097396

Ellis, J. (2000). *Seeing Things: Television in the Age of Uncertainty*. London: I. B. Tauris.

Fortunati, L. (2001). The Mobile Phone: An Identity on the Move. *Personal and Ubiquitous Computing*, *5*, 85–98. doi:10.1007/PL00000017

Friedlander, S. (1995). *Probing the Limits of Representation*. Los Angeles: University of California Press.

Good, M.-J. D. V., Bodwin, P. E., Good, B. J., & Kleinman, A. (Eds.). (1994). *Pain as Human Experience: An Anthropological Perspective*. Berkeley: University of California Press.

Hirsch, M., & Spitzer, L. (2009). The Witness in the Archive: Holocaust Studies/Memory Studies. *Memory Studies*, *2*(2), 151–170. doi:10.1177/1750698008102050

Hoijer, B. (2004). The Discourse of Global Compassion: the audience and media reporting of human suffering.'. *Media Culture & Society*, *26*(4), 513–531. doi:10.1177/0163443704044215

Kitch, C. (2000). 'Mourning in America': Ritual and Redemption, and Recovery in News Narrative after September 11. *Journalism Studies*, *4*(2), 213–224. doi:10.1080/1461670032000074801

Kuipiers, G. (2006). The Social Construction of Digital Danger: Debating, Defusing and Inflating the Moral Dangers of Online Humour, Pornography in the Netherlands and the US. *New Media & Society*, *13*(3), 379–400. doi:10.1177/1461444806061949

Licoppe, C., & Heurtin, J. P. (2001). Managing One's Availability to Telephone Communication Through Mobile Phones: a French Case Study of the Development Dynamics of Mobile Phone Use. *Personal and Ubiquitous Computing*, *5*(2), 99–108. doi:10.1007/s007790170013

Marx, G. (2001). Murky Conceptual Waters: The Public and the Private. *Ethics and Information Technology*, *3*, 157–169. doi:10.1023/A:1012456832336

Mirzoeff, N. (2002). 'Ghostwriting: Working Out Visual Culture'. *Journal of Visual Culture*, *1*(2), 239–254. doi:10.1177/147041290200100206

Mitchell, W. (2002). Showing Seeing: A Critique of Visual Culture. *Journal of Visual Culture*, *1*(2), 165–181. doi:10.1177/147041290200100202

Morrison, W. (2004). 'Reflections with Memories: Everyday Photography Capturing Genocide'. *Theoretical Criminology*, *8*, 341–358. doi:10.1177/1362480604044613

Neale, S. (1986). Sexual Difference in the Cinema: Issues of Fantasy, Narrative and the Look. *Oxford Literary Review*, *8*, 123–132.

Palen, L., Marilyn, S., & Youngs, E. (2001). Discovery and Integration of Mobile Communications in Everyday Life. *Personal and Ubiquitous Computing*, *5*, 109–122. doi:10.1007/s007790170014

Slater, D. (2002). Making Things Real. Ethics and Order on the Internet. *Theory, Culture & Society*, *19*(5/6), 227–245.

Thompson, J. (1995). *Media and Modernity: A Social Theory of the Media*. Cambridge: Polity.

Thompson, J. (2000). *Political Scandal. Power and Visibility in the Media Age*. Cambridge: Polity.

Thompson, J. (2005). The New Visibility. [London: Sage.]. *Theory, Culture & Society*, *22*(6), 31–51. doi:10.1177/0263276405059413

Vasterman, P., Yzermans, J., & Dirkzwager, A. (2005). The Role of Media and Media Hypes in the Aftermath of Disasters. *Epidemiologic Reviews*, *27*(1), 107–114. doi:10.1093/epirev/mxi002

Walters, Tony, & Littlewoood, Jane., and Pickering, Michael. (1995). Death in the News: the Public Invigilation of Private Moment. *Sociology*, *29*(4), 579–69. doi:10.1177/0038038595029004002

Zelizer, B. (1990). Achieving Journalistic Authority Through Narrative. *Critical Studies in Mass Communication*, *7*(4), 366–376. doi:10.1080/15295039009360185

KEY TERMS AND DEFINITIONS

Mobile Technologies: Communication gadgets and equipment that can be accessed and used to communicate and disseminate information and engage in a multitude of activities whilst on the move.

Aestheticization of Suffering: The depiction and consumption of trauma that dissects the event from the context by inserting different forms of gaze.

Commodification: The act of ascribing material formats to processes or acts which may not be tangible.

Counter-Sites: Spaces which offer alternative ideas and views to mainstream ideology and content.

Bearing Witness: Partaking in an event through the act of watching or gazing.

Technologically-Mediated World: The world as viewed and understood through the vantage point of communication and media technologies.

Visual Economy: Processes, visual formats and acts of consumption that emerge in the new media environment which enable new forms of empowerment and vulnerabilities.

Chapter 10
Mum's the Word: A Case Study of Students' Intergenerational New Media Literacy in Shanghai

Larissa Hjorth
RMIT University, Australia

ABSTRACT

In 2009, for the first time, Internet penetration rates in China have surpassed the global average level with over 298 million users (CINIC 2009). In this burgeoning of use, one of the largest and oldest social networking systems (SNS) in China, QQ, has become part of the daily diet of many Chinese both in cities and big towns. In this chapter I explore the role of QQ — and SNS more generally — through a case study of university students studying in Shanghai. As I posit in this chapter, QQ has become a rite of passage for the migrating youth leaving home to study in another city or country. As the first generation to grow up within the IT educational reforms that saw a growing importance placed on new media and Internet literacy, these ba ling hou students are passing on their knowledge to their family and friends back home, demonstrating that SNS is far from the prerogative of Western, middle-class youth. In particular, SNS is creating new ways for intergenerational connectivity and contact in the face of growing geographic mobility.

INTRODUCTION

'If I was a fish then QQ would be my water.'

(Bao, 25 year old, male postgraduate student)

In 2009 Chinese Internet usage surpassed the global average with over 298 million users (279 million of these being broadband users), to become the number one user of the Internet (CINIC 2009). This phenomenon is particularly evident in China's voracious use of social media such as blogs, where they have more users than any other nation. Just as Internet take-up rates in the developing world are increasing faster than in the first world, all-be-it from a lower base, so take-up rates in this the largest of the developing countries

DOI: 10.4018/978-1-60960-037-2.ch010

Copyright © 2011, IGI Global. Copying or distributing in print or electronic forms without written permission of IGI Global is prohibited.

are now increasing sharply among lower-income and less-educated people.

And just as the order and rate of take-up differs among divergent socioeconomic groups, so too do the software applications and the hardware access routes of choice. Many millions of Chinese make use of the Internet through personal computers and social networking applications that would be familiar across the globe. But for many millions outside of big cities (like many countries in Southeast Asia and Africa), the only entry route to the Internet is via the mobile phone — not the Personal Computer — employing applications that are particular to that group of users and their mobile platforms. This diversity of users, platforms and socio-cultural demographics is reflected in China's diverse Social Networking Sites (SNS), and in particular, is reflected in QQ, China's oldest SNS, where many different services cater to diverse platforms (many of which are free) and modes of access.

So while technological infrastructure and access to the Internet are still predominantly an urban phenomena in China, changes have occurred in rural areas with the number of rural Internet users reaching nearly 85 million (CINIC 2009). In short, the growth rate of rural Internet usage now far exceeds that of urban Internet usage. While the Internet is mainly accessed via personal computer in urban city areas by the middle and upper class, the hundred million odd migrant working class — along with the rest of China's predominantly working class demography — deploy GPRS (General Packet Radio Service) for Internet access via the mobile phone. In China's undulating landscape of the twenty-first century, class (as a socioeconomic category) continues to be a dominant factor influencing the types of emergent technocultures. This technocultural template is marked by both class and geo-spatial (rural and urban) features; the diverse characteristics of socioeconomic class shape, and are shaped by, the numerous types of technocultures that on a different scale come together in shaping and reshaping contemporary China.

In this rise of Internet access among these dynamic economic and geographic groupings, the increased uptake of social media is notable. In China's varied technoscape numerous types of SNS — representing different classes, communities and lifestyle clusters — can be observed. Broadly speaking, these SNS fall into two techno-cultural trajectories; sites such as Renren (akin to Facebook), Kaixin (used by female 'white-collars' [workers]) and MSN are accessed via a personal computer and are often all simultaneously opened and used on someone's desktop; whereas Fetion and the aforementioned QQ differ in the important respect that they can be accessed both via mobile phones or via computers.

In this chapter I explore the role of QQ — and SNS more generally — through a case study of university students studying in Shanghai. As I posit in this chapter, QQ has become a rite of passage for the youth leaving home to study in another city or country to keep in contact with friends and family at home. As the first generation to grow up within the IT educational reforms (such as the ten-year Chinese Education Research Network called CERNET) that saw a growing importance placed on new media and Internet literacy, these *ba ling hou* students are passing on their knowledge to their family and friends back home, demonstrating that SNS is far from the prerogative of Western, middle-class youth. For many of the parents of the *ba ling hou* have never used the Internet, let alone SNS. Though discursive routes such as mobile phone Internet via GPRS, these older generations — often from lower socioeconomic backgrounds living in country towns and occasionally illiterate — are able to gain entrance into twenty-first century media practice.

In this chapter I will firstly discuss SNS literature in general, moving onto the particular technoculture of China — and specifically Shanghai. The context of Shanghai presents a very particular 'cosmopolitan' technocultural with many local and global SNS — a phenomenon not replicated elsewhere in China. From there I move onto the sample study of university students at Fudan University in Shanghai. Far from demonstrative of all university students, this sample study focuses upon university students who have come from elsewhere in China — often from small towns and families that would be socio-economically defined as working class. Through education and attendant IT policies initiated by national and state government over ten years ago, this new generation of media literates are gaining access into emerging twenty-first century technocultures whilst guiding their friends and family back home.

This chapter is a preliminary study exploring these new forms of mobility through discursive media literacy in which young adults impart their media knowledge to their parents and grandparents. In this study I specifically focus upon the students' discussion of the media usage highlights that as social media evolves it is far from a western or youth prerogative as suggested by the focus of SNS literature to date. A future study will involve interviewing parents and connecting their new media literacy and experience with that of their 'teachers', the children.

LOGGING ONTO THE SOCIAL: CONTEXTUALIZING CHINA AND SOCIAL MEDIA

Within Internet Studies, the rise of SNS as a dominant area of study has been notable. One glance at the program of key annual conferences such as AoIR (Associated of Internet Researchers) and we can see the various methods and approaches in the burgeoning SNS. As part of the phenomenon of so-called Web 2.0 (Allen 2009; O'Reilly 2005; DiNucci 1999), social media — and the role of the locality in informing the practices and experiences — has become a pivotal feature within Internet Studies (Goggin & McLelland 2009). Within the contestation, Web '2.0' characteristics include 'participatory', 'dynamic' and 'user-centred' (Munster & Murphie 2009) — key features of the social architecture of SNS. Some researchers of SNS have used qualitative approaches (boyd 2003, 2004) which deploy revised models of virtual ethnography (Hine 1998, 2000; Miller & Slater 2001) to focus upon micro and localized understandings of intimacy and sociality, whilst others have been more broad sweeping in their analysis and criticism of social media (Shirky 2008). Studies around SNS — as part of broader forms of twenty-first century media literacy (Ito et al. 2008) — have begun to emerge; as has discussion within game studies about the increasing role of SNS to recruit non-traditional demographics of users and transform them into players can be seen as part of the 'casual revolution' (Juul 2009). Online gaming, once the prerogative of niche 'hardcore' players has now become the preoccupation of casual and mainstream audiences as games expand their ability to be social.

Indeed, SNS literature is replete with references to the role of SNS in establishing and maintaining weak links, bridging and bonding ties, cultural movements, learning communities, communities of practice, local and global connectedness, systems of trust, wider access to education and to employment opportunities, facilitating civic engagement and social participation, providing more efficient access to government services and to a more participatory form of democratic involvement. SNS are both a set of cultural practices as well as cultural artifacts; they are both commercial *and* cultural; they manifest the identity of a network (Castells 1996) *and* a community (Arnold 2007); they are present in our online relations *and* structure our offline relations, and they manifest both hierarchical *and* horizontal structures.

Along with the rise of SNS literature has also been the awareness and need to address the gap in non-Anglocentric models of the Internet (Ho, Kluver & Yang 2003; Gottlieb & McLelland 2003; Goggin & McLelland 2009). Increasingly studies are starting to focus upon locations in regions such as the Asia-Pacific — a region that boasts some of the most sophisticated technological infrastructures and highest broadband rates, along with important centres for hardware production and consumption — namely Japan, Korea, Taiwan and China. The region is home to some of the oldest Social Network Services such as Cyworld (Hjorth & Kim 2005), and demonstrates examples of new social media that merge gaming, socialising and work all in one (Hjorth & Chan 2009). As both a site for the global production and consumption of new forms of social networked media (evidenced for example, in Korea's global grip on online multiplayer games), the region is a fascinating site to analyze the divergences of emerging Web 2.0 social networked media spaces.

Across the plethora of different SNS, modes of access (via PC or mobile), and online communities in the region, there are commonalities as well as distinctions that mark the Asia-Pacific's embrace of Web 2.0. For youth across the region, SNS is not only a fundamental part of everyday life and social capital, it is also a space that allows them to maintain intergenerational contact when geographic distance might be involved. It is no longer just "youth" that are using SNS as adult on adult and intergenerational forms of dialogue and literacy expand with the increased net accessibility. Moreover, the demographics are shifting too as Internet access becomes not only a middle class prerogative but also an integral part of the new mobile working class (Qiu 2008). For example, in China, it is the working class that is growing exponentially as the main users of the Internet (CINIC 2009).

In particular, China — as part of what Arif Dirlik has called the 'Global South' (2007) — has continued to gain much focus. The unique features of China within the global chains of ICT production and consumption over the last couple of decades (Goodman & Robison 1996; Donald et al. 2002) has been most evident within the twenty-first century through the deployment of the Internet as a site for contestation and technonationalism (Damm & Thomas 2006; Hughes 2003; Tai 2006). However much of the literature on China's Internet has focused on its role as a contested site between public opinion and government policy. Like South Korea, the Internet in China has provided a way in which to conceptualise struggles of democracy with social media such as blogs attracting much focus (Bruns & Jacobs 2006; Lovink 2007; Yu 2007), while seemingly 'less' political media such as SNS have been relatively overlooked (Koch et al. 2009; Yu 2007) despite often being deemed threatening and thus banned by the government (in 2009 Facebook and Twitter were barred).

China's SNS technoculture features numerous examples both local and global. Local SNS include QQ, Kaixin, Renren (previously Xiaonei), and Fetion; while MySpace, Facebook, Twitter and MSN feature as global examples. However, the use of both local and global SNS is only apparent in 'cosmopolitan' cities such as Shanghai. For the general population of China, few local SNS are used or even known apart from QQ. Part of QQ's success has been ensured by its inclusion and convergence of various platforms from chatting and SNS (QQzone) to online, localized news. Via Internet portals the service is free, and its various platforms such as QQ chat can be adapted to devices such as mobile phones. This is important given that while most of the population outside of the big towns and cities do not have the Internet they all have a mobile phone. QQ represents the largest cluster of SNS services in China; resulting in QQ being among the largest in the world. Moreover, as we shall see in the case study, QQ represented a type of Chinese nationalism and many expressed a strong sense of sentimentalism towards QQ as their first entrance with SNS. Despite the significance of QQ in emphasizing the vernacular and local of Chinese sociality — and its history as the first SNS and IM that ushered a new generation of digital natives through to Web 2.0 — studies of QQ have been limited.

As Pamela Koch, Bradley Koch, Kun Huang and Wei Chen (2009) note in their study 'Beauty is in the Eye of the QQ User: Instant Messaging in China', 'the emergence of this local homegrown QQ software indicates that it plays an important role in Chinese society' (2009: 265). It is for this reason that QQ can help to 'understand the influence of this specific Internet platform as situated within Chinese culture' (2009: 265). As they note the various services and platforms of QQ mean that it differs greatly from the 'Internet' in China in general. Rather, it is the portal *between* media, between platforms and generations, between media experiences, between classes.

Thus, through the lens of such SNS as QQ, we can chart this transitory period in which China's technocultures and the emergent forms of intergenerational new media literacy. In this case study I focus upon the current generation of university students in Shanghai (who have come from elsewhere in China to study) to explore some of their practices and feelings experienced as they transfer their new media literacy skills to family and friends back home.

In particular, by focusing upon students in Shanghai who have migrated from elsewhere in China to study, this chapter considers some of the experiences of the unique technocultural fabric of the city through the portal of SNS, and specifically QQ, between being *home and away.*

HOME AND AWAY: A CASE STUDY OF FUDAN UNIVERSITY STUDENTS

As the first generation to grow up in China's emerging net culture, the current university students are a product of the first education IT initiative (beginning in 1994). This initiative, in the form of the first national network, was called CERNET. Many of these students were the first group of students to be affected by the EISS (an acronym for 'Electronic Information Service System' or in Chinese 'xiaoxiaotong' [校校通]) policies in which the government orchestrated, over a ten-year period (2000-2010), the implementation of computers and online education within primary and secondarily schools.

The EISS was initiated in November 2000, as a project spanning ten years to enable 90% of the independent middle schools and primary schools throughout the country to have access to the Internet, along with increasingly the deployment of online resources to be shared amongst teachers and students. For the average current university student, they are the EISS generation, growing up with the benefits of these policies informing their high media literacy whereby the Internet is viewed as pivotal part of everyday life. The EISS, along with the general rise of Web 2.0 and its participatory entertainment industries, has ensured that for most university students in China today, the Internet and mobile technologies are an integral part of their daily life. Current Fudan University students are a good example of these new media literates and their attendant forms of mobility and co-presence.

During June 2009 three sets of focus group discussions were conducted at Fudan University in Shanghai. Postings were placed on the university BBS as well as community online sites. As this period of recruitment was the end of the semester, many of the students still on campus where those whose families lived outside Shanghai. The focus groups consisted of students from various disciplines, levels of education, and a cross-section of socioeconomic backgrounds. The first two focus groups consisted of ten participants with roughly 70% females and 30% males. Half of these participants were not from Shanghai and had family living elsewhere — especially in towns.

The third and final focus group consisted of ten female students, nine of whom were not from Shanghai, provided the most insight into understanding changes (and challenges) experienced by migrating students as they moved onto socioeconomic mobility through education. Education has long been viewed as a pathway for mobility

(Hau & Salili 1996), although previously it was predominantly the activity of the middle-class (Donald et al. 2002; Goodman & Robison 1996). Here, the role of SNS, and specifically QQ, provides both a window and a portal into this mobility across geographic, generation and socioeconomic differences. This final group was indicative of a new rite of passage in which working class young people were now entering university education, whilst often their parents remained illiterate back in their hometown. However, far from these young students trying to distance themselves from their lower socioeconomic backgrounds, instead, through the deployment of various forms of SNS, they were reconnecting and engaging regularly with old friends and family.

In addition to the focus groups I also conducted one-to-one in-depth interviews (one and half hours long) with 12 respondents, half male and half female. In the interviews I discussed the particular daily habits of SNS usage and how it functioned in the context of other activities such as schoolwork and part time employment. The discussion of home and family was mainly instigated by respondents rather than lead-by the interview questions. In both the interviews and focus groups respondents provided examples of their various SNS accounts and we discussed customization and also types of etiquette and networks and how these differed between the SNS. Amongst all the variables, every respondent had a QQ account and many used it predominantly for keeping in contact with family and friends back at home.

This sample study is not meant to be indicative of all university students' experience in Shanghai. Each university, like province in China, has a different technoculture and associated demography of students. Rather, this chapter marks a preliminary study into the lives of the new emerging media literates — and their attendant forms of unofficial intergenerational literacy. By no means exhaustive, this qualitative study aims to provide insight into how some of the individual media literates are grappling with their new mobility both home and away.

QQ in a Time of Mobility

It became clear through the discussions by respondents that SNS occupies an important place in daily life, that QQ is special among the SNS alternatives, and that the reasons for this are the product of our informants' various forms of mobility — socioeconomic, geographic, personal, interpersonal and intergenerational. In this discussion of the case study I focus upon the prevailing intergenerational usage that has emerged within the students' socioeconomic and geographic mobility.

Socioeconomic and Geographic Mobility

Education in China, as elsewhere, is associated with socioeconomic mobility, and educational mobility and socioeconomic mobility are both associated with geographic mobility as young people move to University. Through the experience of these three interrelated dimensions runs the thread of SNS and QQ in particular.

When asked to discuss how migration had affected their use of SNS, the focus group discussed the various SNS and the types of deployment and associated social networks. Many were in agreement about the different usages of MSN and QQ and their attendant class, socioeconomic and lifestyle images. While many in their workplace environments (either part time work or last year internships) were prohibited from using QQ (given its "social" tool implication), employers strongly supported the use of "professional" media such as MSN. QQ, MSN and other technologies have qualities or characteristics that go beyond their technical functionality. They also have identifiable ideological, class, or socioeconomic qualities that make them welcome in some environments and not others. MSN is welcome in many white-collar work and university environments, QQ is less so, and so for many of our informants the first MSN experience was in Shanghai via university or employment. Lieei sums it up:

QQ is used by everyone, everywhere. Before I came to university I had not heard of MSN, Shanghai people like MSN very much. Many of my Shanghainese friends use MSN.

Zi, another female respondent, aged 26, observes:

I noticed when I came to Shanghai people were using MSN and Skype which I hadn't seen used elsewhere... In other parts of China there are a lot of people who don't have a computer in their home, let alone Internet. So QQ is much more popular with these people as it can be accessed via the phone mobile — something that everyone has. MSN is popular here in Shanghai, unlike the rest of China. A lot of children who don't have a computer or Internet access at home go to an Internet café. At those places, they don't have MSN downloaded onto the desktop and it requires people to be able to use English language, so many just use the readily available QQ. Young students, especially college and elementary use QQ, and in developing areas there would be more people using QQ than anything else. But in big cities like Shanghai and Beijing, there are so many companies and so many white collars so they use MSN a lot because they are older (authors' italicization).

The interplay of SNS, geography, and socioeconomics, and the fluid, mobile variations of practice produced by this interplay is well illustrated in the following conversation:

Interviewer: Do you think there is a digital ("technological") divide between different classes, generations or locations (rural/urban)? So how many people think class is an issue?

[All put up their hands].

Interviewer: Do you think class is a more important factor than generational or location?

Ji: I think class is just a normal factor.

Interviewer: So would people say that all of the aforementioned factors influence usage.

[The group nods].

Ji: While the white collar may use MSN more often, she will also have a QQ account. Different geographies will influence what people use. Geography influences the hardware — some places don't have computers and Internet access, they just have a mobile phone. Or they may have the hardware and not the software.

Ling: In the countryside computers are not very common

Interviewer: Or relevant.

Ling: So often if they want to use the Internet they must go to a café. Some villages only have one Internet café. Whereas every person has a mobile phone. In rural areas, they can only use the mobile phone. Also, many older generations use QQ — parents, grandparents. They use it on the computer. On the phone they feel they don't want to leave a message because it might be too troublesome.

Genji: In other parts of China there are a lot of people who don't have a computer in their home, let alone Internet. So QQ is much more popular with these people as it can be accessed via the phone mobile — something that everyone has. MSN is popular here in Shanghai, unlike the rest of China. A lot of children who don't have a computer or Internet access at home go to an Internet café. At those places, they don't have MSN downloaded onto the desktop and it requires people to be able to use English language, so many just use the already available QQ.

Tao stresses the relationship between SNS choice and socio-economics:

Interviewer: You mentioned the issue of class there in terms of white-collar workers; do you think class informs who uses QQ and who doesn't?

Tao: MSN is used by the white collared.

Interviewer: Is this because it seems more professional and cosmopolitan?

Genji: Some bosses if they see you using QQ at work might think you are not professional.

And Lina provides us with a useful summary:

Xiaonei (now Renren) is mainly university students while Kaixin is for office people. MSN is more international. When my classmates go overseas to study in the US they use MSN. They can't use QQ. MSN is more official and international.

Concomitant with this mobility along a socioeconomic axis, is geographic mobility. The young people moving to Shanghai for a University education are geographically mobile as they move around the city on a day-to-day basis, and they are on the move across the country through a lifetime. A dual-edged implication of contemporary communications technologies for this mobility is that they make the geographic distances between places less relevant without diluting the significant differences between those places. For Daiyu, aged 27, this dual-edge is evident and QQ's success was not only its easy accessibility across platforms and technologies (facilitating geographic flexibility) but also its abilities to localize its services (thus foregrounding place). Features such as the QQ news service that focused on the region of the user's access means that the service was very important in defining a sense of place. Reiterating points made by Genji, she notes,

When you open QQ the news pops up, for example the earthquake in Sichaun province, it is news that Chinese people are more interested in... While the white collar may use MSN more often, she will also have a QQ account. Geography will influence what people use. Geography influences the hardware — some places don't have computers and Internet access, they just have a mobile phone. Or they may have the hardware and not the software. In the countryside computers are not very common so often if they want to use the Internet they must go to a café. Some villages only have one Internet café. But every person has a mobile phone. In rural areas, they can only use the mobile phone. Also, many older generations use QQ — parents, grandparents. They use it on the computer. On the phone they feel they don't want to leave a message because it might be too troublesome (italicization for emphasis).

Cultural Mobility between Generations

Education has long been viewed as a pathway for mobility in China (Hau & Salili 1996), although from the 1980s and 1990s it was predominantly the prerogative of the middle-class (Donald et al. 2002; Goodman & Robison 1996). Our mobile students are indicative of a new rite of passage in which working class young people are now entering university education in cities, whilst often their parents remained illiterate (educational backgrounds for parents varied from no or little schooling) back home. In the West a move away from home to University is often an opportunity for a young person to establish an independent subjectivity and identity, facilitated by constructing a certain critical distance from family and home. However, far from these young students trying to distance themselves from their family, hometown, and lower socioeconomic backgrounds, various forms of SNS were deployed to connect and engage regularly with old friends and family, maintaining continuity and connection rather than establishing distance and independence.

One way of doing this is through the exchange of images:

Qiao: ...Because QQ is for me and my friends and family, I don't need to put a real photo of me. But I often upload my QQ photo to send to my mother.

Interviewer: Another thing you do for your mother

Qiao: I like her very much

[Laughter]

Another is through gaming. For example, Chyou, a female respondent aged 22, frequently played online games such as Mahjong (online version of a traditional Chinese board game), because her mother liked to play with her. She said,

'... the more I play, the happier my mother is. I like to make her happy'.

Luk concurs:

Luk: I play games to make my mother happy. (laughter) I play poker because my mum likes to play it. We play together.

Interviewer: and when would you do this?

Luk: any free time.

Hong, another female respondent aged 20, complained that she believed her father to be addicted to playing games. She said, 'he has so much time on his hands, he just wastes it on gaming. Our generation doesn't have time.' Here we see an interesting role reversal in the stereotypes associated with age and media practice. Rather than it being just the youth that are wasting their time with new media instead we see it is the parents.

Given that most of the focus group participants were from elsewhere than Shanghai, issues around family, home and how the SNS functioned to facilitate this within the various forms of mobility dominated the discussion. For Xun, a female postgraduate student aged 24, one's origin strongly influenced what media they used. She noted changes of SNS dependent on life stages (work etc) and where one is living, and she noted that QQ continued to play a special role in maintaining her familial and kinship ties. In order to do so, she had taught her family to use QQ. This was a common pattern found in almost all of the participants — regardless of different SNS usage, QQ continued to function to maintain kinship ties and geo-social and economic connections with their past. Shuang, a female respondent, aged 27, notes

For me, Xiaonei (now Renren) is not a place to make new friends, just to reconnect to already known friends and contacts... But I continue to use QQ for my parents. Last year they didn't know how to use a computer, let alone QQ, so I showed them. Now my family has moved to a new place so we use QQ video so we can see each other whilst chatting.

Others have reservations in which they felt that their parents could watch their movements and perhaps misunderstand some of the contexts...

Interviewer: Oh, so your mother is Xiaonei (now Renren)? You must find that difficult and it must make you censor a lot of things.

Dan: Yes. My mother often goes to my website to see what I'm up to. If I could choose I would hope that she didn't have an account, but she does.

For Rong, a, female aged 22, the usage of QQ everyday provided a familiar experience in an unfamiliar environment (as she moved away from home to attend university). It also helped to maintain her contact with her family, reducing homesick feelings.

I use QQ almost every day. I use both my computer and mobile phone all day, every morning and every night. I use various services of QQ. This is the first year I have lived away from home and so I use QQ mainly to keep in contact with my family at home as I live away from home. I use QQ almost every day to make contact with my family. I get very homesick and so QQ helps with that. Last month my mother had an operation and I couldn't come home because of exams so instead I used QQ to keep in contact with her continuously.

And...

Lifen: I think these tools are very useful for those living a long way away from home and family. It's cheap and convenient. Actually my grandparents also use QQ.

Interviewer: Did you have to introduce them to it and teach them?

Lifen: Yes. And as they can't type the words we communicate through pictures and spoken words. They know they can use QQ to contact me. Telephone calls cost whereas QQ is cheap.

And...

Fu: My great memorable experience with QQ was when I was abroad and I used QQ every day to chat with my mother. Just before I left China I applied with QQ id for my mother. QQ is easier (for the mother) because it's all in Chinese. I think it is quite convenient because something like MSN my mum won't use it. The Chinese writing makes my mum feel more familiar with it.

Dongnei: I have a similar situation to Olivia as when I was in America I had my PC and the only thing I really needed was QQ. It doesn't cost money. Everyday I chatted with my parents. I used the webcam so I could also see them.

And...

Kang: I always use QQ to keep in contact with my parents and cousins.

STILL MOBILE: CONCLUSION ON CHINA'S MOBILE TECHNOCULTURES

As the young move away from home in the quest for a university education, often there are many feelings of homesickness. The mobility of the lower socioeconomic children through social media is a growing phenomenon given the new media mobility afforded through policies such as EISS implemented in 2000. For the youth studying in another place, SNS offers these youth efficient and convenient ways in which to keep connected with family and friends at home, helping to alleviate some of the feelings of loneliness and sadness.

In this emerging landscape, social networked media is an integral part of everyday life. As evidenced in the case study group of university students, as well as their families and friends, there is a new generation of networked mobility. As they have grown, so too has the technological infrastructure, access and education burgeoned. They are the first generation to grow with the rise of Web 2.0. This growth has been accompanied by the oldest SNS in China, QQ.

In particular, through QQ, university students away from home are teaching their parents how to navigate online in order to communicate frequently. QQ is not only first SNS used for the young, it is now the portal that introduces older generations to SNS. Moreover, for many, this introduction to SNS is *also* an introduction to the Internet, and, in this way, could be seen as an important rite of passage in beckoning a new breed of intergenerational new media literacy. For illiterate parents or grandparents, features such as QQ video and games like *Happy Farm* allow them to see each other and talk for free. Thus far from SNS being vehicles for endorsing their new mobility in which the past is forgotten, it provided a portal for reconnection and re-rooting with one's past. While SNS like Renren are predominantly used by university students to connect with new and old 'non-close' friends and networks, many of the other SNS, especially QQ, is being used to maintain older, more intimate, relationships.

This case study of university students in Shanghai is the first of many that need to be conducted if we are to understand the relationship between these emerging forms of intergenerational mobility and China's twenty-first century technoculture. This initial study into this emerging phenomenon not only explores intergenerational literacy but also contests the notion that SNS is for Western, middle-class youth. As noted earlier, a follow-on study should focus upon the impressions and experiences of both parents and children as they accompany each other through this emerging rite of passage. Whilst Shanghai has long been a location in which working class university students from elsewhere come to study and transgress their socioeconomic background, the role of ubiquitous technology in this equation is new. As the first generation to grow up with the Internet, this generation — and their relationship to other generations — can provide some new models for media literacy. While these new media literates are demonstrating new forms of mobility across geographic and socioeconomic differences, they are highlighting the importance of maintaining a connection with their past.

Through early precursors such as QQ, we can gain much insight into the ways in which mobility — socio-economic, geographic, cultural, intergenerational — are shaping China's burgeoning technoscape. Rather than this new media reinforcing generational, class and geographic divides, SNS such as QQ are functioning as the glue between these once disparate areas, illustrating that SNS — both in China and globally — are shifting away from the prerogative of the western middle-class and becoming reflective of the diversity of experiences within contemporary global technocultures. As a lens onto emerging forms of intimacy, kinship and media literacy, SNS demonstrate that in the increase of various modes of mobility, home (or a sense) is never far away.

ACKNOWLEDGMENT

This research was kindly supported by The State Innovative Institute for the Studies of Journalism & Communication and Media Society at Fudan University, Shanghai. Larissa Hjorth would also like to thank Chu Yuewen. This study is part of two broader research projects — one South China grant with Jack Qiu, Ding Wei and Baohua Zhou looking at migrant workers and QQ in China and secondly an Australia Council Research grant with Michael Arnold exploring online communities in the Asia-Pacific.

REFERENCES

Arnold, M. (2007). The Concept of Community and the Character of Networks, *Journal of Community Informatics*, 3 (2). Retrieved 20 October 2008 from http://www.ci-journal.net/index.php/ciej/article/view/327/355

boyd, D. (2003). *Reflections on Friendster, Trust and Intimacy.* Presented at School of Information Management & Systems (SIMS), University of California, Berkeley.

boyd, D. (2004). *Revenge of the User: Lessons from Creator/ User Battles*. Presented in the O'Reilly Emerging Technology Conference, February, http://www.danah.org/papers/Etech2004.html

Bruns, A., & Jacobs, J. (Eds.). (2006). *Uses of Blogs*. New York: Peter Lang.

Burgess, J. (2008). All Your Chocolate Rain Are Belong to Us? Viral Video, YouTube and the Dynamics of Participatory Culture. In Lovink, G. (Eds.), *The VideoVortex Reader*. Amsterdam: Institute of Network Cultures.

Castells, M. (1996). *The Rise of the Network Society, The Information Age: Economy, Society and Culture* (*Vol. I*). Cambridge, MA: Oxford University Press.

CERNET. (n.d.). Retrieved from http://www.edu.cn/CERNET_1377/index.shtml

CINIC [China Internet Network Information Center]. (2009). *Statistical Survey Report on the Internet Development in China* (January 2009). Retrieved 14th June 2009 from http://www.cnnic.cn/uploadfiles /pdf/2009/3/23/131303.pdf

Damm, J., & Thomas, S. (Eds.). (2006). *Chinese Cyberspaces Technological Changes and Political Effects*. New York: Routledge.

DiNucci, D. (1999). Fragmented Future. *Print 53*(4), 32. http://www.cdinucci.com/Darcy 2/articles/Print/Printarticle7.html

Dirlik, A. (2007). Global South: Predicament and Promise. *The Global South, Winter*, 1(1), 12-23.

Donald, S. H., Keane, M., & Hong, Y. (Eds.). (2002). *Media in China: Consumption, content and crisis*. London: Routledge.

Dong, D. (2003). *Resistance within Contexts: A Study of University BBS Youth Culture in China*. Paper presented at the annual meeting of the International Communication Association, Marriott Hotel, San Diego, CA, May 27.

Editorial. (2009). Shanghai relaxes residency rules. *BBC News online*. Retrieved 14th June 2009 from http://news.bbc.co.uk/go/pr/fr/-/2/hi / asia-pacific/8107518.stm

Goggin, G., & McLelland, M. (Eds.). (2008). *Internationalizing Internet Studies: Beyond Anglophone Paradigms*. New York: Routledge.

Goggin, G., McLelland, M., & Hjorth, L. (2009). Waiting to participate: Introduction. *Communication. Politics & Culture*, *42*(2), 1–5.

Gottlieb, N., & McLelland, M. (Eds.), *(n.d.). Japanese Cybercultures*. New York: Routledge.

Hau, K. T., & Salili, F. (1996). Achievement goals and causal attributions of Chinese students. In Lau, S. (Ed.), *Growing up the Chinese way* (pp. 121–146). Hong Kong: The Chinese University Press.

Hine, C. (1998). *Virtual Ethnography*. Paper from Internet Research and Information for Social Scientists Conference, University of Bristol, UK, March 25-27.

Hine, C. (2000). *Virtual Ethnography*. London: Sage.

Hjorth, L., & Kim, H. (2005). Being there and being here: gendered customising of mobile 3G practices through a case study in Seoul. *Convergence*, *11*, 49–55.

Hjorth, L., Kim, H., & Chan, D. (Eds.). (2009). *Gaming Cultures and Place in Asia-Pacific*. London: Routledge.

Ho, K. C., Kluver, R., & Yang, K. C. C. (2003). *Asia@com*. London: Routledge.

http://journal.fibreculture.org/issue14/

http://oreilly.com/web2/archi ve/what-is-web-20.html

Hughes, C. (2003). *China and the Internet: Politics of the Digital Leap Forward (Politics in Asia Series)*. New York: RoutledgeCurzon.

Ito, M. boyd, d. & H. Horst (2008). *Digital Youth Research*. Retrieved from http://digitalyouth.ischool.berkeley.edu/

Jenkins, H. (2006). *Fans, Bloggers, and Gamers: Essays on Participatory Culture*. New York: New York University Press.

Juul, J. (2009). *A Casual Revolution: Reinventing Video Games and Their Players*. Cambridge, MA: MIT Press.

Koch, P., Koch, B., Huang, K., & Chen, W. (2009). Beauty is in the Eye of the QQ User: Instant Messaging in China. In Goggin, G., & McLelland, M. (Eds.), *Internationalizing Internet Studies* (pp. 265–284). London: Routledge.

Kücklich, J. (2005). Precarious Playbour: Modders and the Digital Games Industry. *Fibreculture Journal, 5*. Retrieved 23 July 2008 from http://journal.fibreculture.org/issue5/kucklich.html

Lovink, G. (2007). *Zero Comments: Blogging and Critical Internet Culture*. New York: Routledge.

Miller, D., & Slater, D. (2001). *The Internet: An Ethnographic Approach*. London: Berg.

Munster, A. & Murphie, A. (2009). Editorial — Web 2.0: Before, during and after the event. *Fibreculture journal, 14*. Retrieved 10 October 2009 from

O'Reilly, T. (2005). *What Is Web 2.0: Design Patterns and Business Models for the Next Generation of Software*. Retrieved 10 September 2009 from

Qiu, J. (2008). Wireless Working-Class ICTs and the Chinese Informational City. *Journal of Urban Technology, 3*, 57–77. doi:10.1080/10630730802677954

Robison, R., & Goodman, D. S. G. (Eds.). (1996). *The new rich in Asia: Mobile phones, McDonalds and middle-class revolution*. London: Routledge.

Shirky, C. (2008). *Here comes everybody*. Keynote presented at the Aspen Ideas Festival, June 30-July 8th, Aspen Coldrado. Retrieved 20 December 2008 from http://fora.tv/2008/07/06/Clay_Shirky_o n_Social_Networks_like_Facebook_and_MySpace#chapter_01

Tai, Z. (2006). *The Internet in China: Cyberspace and Civil Society*. New York: Routledge.

Yu, H. (2007). Blogging Everyday Life in Chinese Internet Culture. *Asian Studies Review, 31*(4), 423–433. doi:10.1080/10357820701710724

Chapter 11
The Electronic Pontifices Maximi:
Social Networking and the Hope of a Right to Communicate

Aliaa Dakroury
University of Ottawa, Canada

ABSTRACT

Drawing from the work of Canadian political economist Harold Innis, as well as the French activist, Jean d'Arcy—the father of the right to communicate, this chapter argues that it is almost impossible to have (and maintain) a right to communicate with the presence of a modern form of Pontifice Maximum: the media conglomerates. More particularly, it highlights the various possibilities of enforcing a human right to communicate with the introduction of the new social networking sites –such as Myspace, Facebook, Twitter, etc. Yet, it also underscores the challenges posed by global corporations that tried to capitalizing some popular social networking sites and user-generated content, and hence threatening a full and equitable participation of citizens in the democratic discourse, which is key to a fuller participation by the communicative citizen in democratic dialogue.

INTRODUCTION

More than forty years ago, Jean d'Arcy—the father of the right to communicate—admitted that controlling the different means of communication is imminently threatening the practice of democracy and the human right to communicate arguing that historically, "control of bridges gives control of roads and thus communications" (1977, p. 49). Drawing from this argument, this chapter aims to analyse from a political economy perspective the obstacles that might face a real practice of free communication using the case study of social networking sites that have arguably enabled more free and unrestrained communication than conventional forms of electronic media.

DOI: 10.4018/978-1-60960-037-2.ch011

Harold Innis, one of the prominent figures of the political economy of communication in Canada, underscored the historical relationship between power, political organizations and communication technology, asserting the importance of the different powers in facilitating or impeding the

Copyright © 2011, IGI Global. Copying or distributing in print or electronic forms without written permission of IGI Global is prohibited.

human communication. In such cases, the ideas of access to the different means of communication; and, the right of citizens to participate effectively in society via means of communication would be in a real danger.

Drawing from the work of Innis, this chapter argues that it is almost impossible to have (and maintain) a right to communicate with the presence of a modern form of *Pontifice Maximum*: the conglomerate media giant. Their existence transformed the use of information from a tool of human communication to a pure commodity available to the highest bid and a means of social control. Most recently, French President, Nicolas Sarkozy fires many criticism to the state of capitalism in the international realm, arguing that it "has gone 'badly wrong'", and calling "reforms to restore [the] 'more dimension' to world economy" (Akin, 2010, p. d-3). For that, this chapter underscores the possibility of a human right to communicate with the introduction of the new communications innovation, particularly social networking sites –such as Myspace, Facebook, Twitter, etc. Not surprisingly, global corporations showed great interest in capitalizing and dominating the trends of social networking and user-generated content. Such corporate concentration reinforces a mass media mentality thereby limiting the full participation of citizens in the democratic discourse, which is key to a fuller participation by the communicative citizen in democratic dialogue.

However, the trend to greater corporate concentration of the media is in tension with the growing impetus of citizens seeking greater opportunities to access and participate in global electronic, interactive communication. Early efforts to promote a right to communicate were hindered by the lack of experience most people had with electronic modes of interactive communication. However, this situation dramatically changed in the 1990s with the convergence of the Internet, the World Wide Web, and the widespread use of the personal computer. As well, there has been during the closing decades of the twentieth century a growing attainment of basic human rights throughout the world. These developments have profoundly changed the technological and political landscape surrounding the concept of a right to communicate. Consequently, there is a renewed interest in achieving such a right for everyone and in overcoming the socio-economic constraints limiting access. Indeed, as the chapter will articulate, and as envisioned by the early advocates of a right to communicate, over the last few years, users of communication technologies, particularly internet users are creating a "space-less" public sphere, an arena where the utopian Habermassian conceptualization of public discussion is created. Of equal importance in this discussion is the pioneering efforts of Jean d'Arcy who predicted in the late 1960s of the great promise and extraordinary possibilities of new technologies to disseminate both images and sounds and ultimately to facilitate communication between human beings. In its 2008 editorial, *VSD*—one of the French news Magazines—published a rare interview with d'Arcy on November 12th, 1969, titled "L'homme qui avait vraiment prédit l'Internet" in which he thought that communication should ultimately aim to "serve" human beings: "Un outil de communication au service des hommes" (as cited in Dakroury, 2009, p. 15). Although this statement was made more than thirty years ago, to a great extend it reflects the hope and firm belief in the freedom of communication between human beings that d'Arcy held.

THE RIGHT TO COMMUNICATE: A CONCEPTUAL FRAMEWORK

Many scholars argue that the concept of the right to communicate is the legal child of the twentieth century's various languages and legislations starting from the Post World War II period. They consider Article (19) of the *Universal Declaration of Human Rights* adopted by the United Nations in December 1948, as the first launching of that concept, in which it is stated that "Everyone has the right to freedom of opinion and expression; this right includes freedom to hold opinions without interference and to seek, receive and impart information and ideas through any media and regardless of frontiers". Thus, it is evident that this article affirmed not only an individual freedom of expression but also a collective right that requires cooperation between human beings for it to be practiced in real life. This includes the right to send, receive, and impart any kind of information, ideas, and thoughts using any type of media without any barriers. It was then included, with slightly altered wording to emphasize the right of humans to choose their medium of information, in the *International Covenant on Civil and Political Rights* in 1966, which entered into force ten years later, i.e.,: "Everyone shall have the right to freedom of expression; the right shall include freedom to seek, receive and impart information and ideas of all kinds, regardless of frontiers, either orally, in writing or in print, in the form of art, or through any other media of his choice".

There is a notable difference in the wording of this article, eighteen years after the *Universal Declaration of Human Rights*: many new technologies emerged in the post-WWII era in the world of communications, and "any other media of his choice" implies the future developments in this area.[1] From a communication standpoint, this covenant "receives greater weight than the... declaration, especially because [it] introduces essential specification to the freedom of information concept as understood in the Universal Declaration" (Nordenstreng & Hannikainen, 1984, p. 134). However, the term "right to communicate" was not clearly realized in real life until a renewed debate was sparked by Jean d'Arcy's[2] 1969 article "Direct broadcast satellites and the right to communicate" in the European Broadcasting Union Review, where he claimed that[3]:

The time will come when the Universal Declaration of Human Rights will have to encompass a more extensive right than man's right to information, first laid down twenty-one years ago in Article 19. This is the right of man to communicate. This is the angle from which the future development of communications will have to be considered if it is to be fully understood (1977, p. 1).

However, and despite the preceding historical account, I have argued elsewhere (Dakroury, 2009) that the right to communicate is not a recent one, but rather is rooted in the ideas of certain thinkers of the Age of Reason beginning in the seventeenth century, such as John Milton, John Locke, Voltaire, Montesquieu, Jeremy Bentham, and John Stuart Mill; I suggest that the right to communicate begins with the work of the 17th century English writer and poet John Milton (1608-1674). In his speech addressed to the English Parliament, Milton defended several types of freedoms and liberties that combine a possible right to communicate; he said in his landmark *Areopagitica*:

If it be desired to know the immediate cause of all this free writing and free speaking... it is liberty, Lords and Commons... liberty which is the nurse of all great wits; this is that which hath rarefied and enlightened our spirits like the influence of heaven... give me the liberty to know, to utter, and to argue freely according to conscience, above all liberties (as cited in Dakroury, 2009, p. 85).

Many human rights theorists and thinkers believe that John Locke's work constitutes the foundation of the modern conception of human rights. Central to his philosophy is the concept of "freedom"; the main focus of any discussion of human rights that every human being is free to act reasonably according to his own will and belief. Included here is an adept example Locke's perception of the concept: "Freedom being the foundation of all the rest; as he that, in the state of society, would take away the freedom belonging to those of that society or commonwealth, must be supposed to design to take away from them every thing [sic] else" (Ibid, p. 93). Voltaire (1694-1778) is also considered to be one of the greatest advocates for an absolute right to communicate, evident in one of his famous dialogues "I disapprove of what you say but I will defend to the death your right to say it" (Ibid, pp. 99-104). Finally, John Stuart Mill (1806-1873) argued for the freedom of thought as an inseparable human right, and at the same time promotes the importance of individuality, emphasizing the treatment of one's opinion as one's own "personal property". This apparently unlimited freedom is not, in fact, the freedom to express "heretical" statements and opinions; rather, Mill suggests that freedom of thought, and opinion could be channelled into the well-being of society as a whole through the individual ability to criticize the "corrupt" government.

Furthermore, the legal language of eighteenth-century philosophy, the *Virginia Declaration*, June 12, 1776, the *Declaration of Independence*, July 4, 1776, and the *American Bill of Rights*, 1789, appeared to placed great emphasis on freedom through the use of phrases such as "by nature free," "inherent rights," "enjoyment of life and liberty," "no men be deprived of this liberty," "unalienable right," and "pursuit of happiness." For example, Article 1 of the *Virginia Declaration* stated:

That all men are by nature equally free and independent, and have certain inherent rights, of which, when they enter into a state of society, they cannot by any compact deprive or divest their posterity; namely, the enjoyment of life, liberty, with the means of acquiring and possessing property, and pursuing and obtaining happiness and safety.

Even if historically the American Revolution precedes the French, many scholars argue that, in many ways, the French Revolution represented a more radical social change than the American, especially for the realization of the ideals of democracy and freedom as well as for the realization of the right to communicate and its embodied freedoms of speech, expression and so on. This idea may be due to the radical changes effected by the French Revolution on the beliefs, structures, and policies of society. The Declarations spawned by the American Revolution are conceived as extensions of the *Magna Carta* and the English Bills of rights; as a former British colony, America's main aim in the Revolution was independence from British sovereignty. However, *La Déclaration des Droits de l'Homme et du Citoyen*, or the *Declaration of the Rights of the Man and of the Citizen* (1789) was undeniably a turning point – especially in the history of the idea of treating communication as a basic human right – as it synthesized from the Enlightenment thinkers the concepts of "freedom," "liberty," and "fraternity" in expressing opinions and attitudes. It was the first formal realization of the right to "unrestrained" communication as a "sacred" right of human beings. In addition, the Declaration formally states in Article (11) the importance of an "unrestrained" and "free" communication for every man and citizen. It also mentioned different formats of communication (opinions and attitudes, etc.) and different media (speaking, writing, etc.): "The unrestrained communication of thoughts and opinions being one of the most precious rights of man, every citizen may speak, write, and publish freely, provided he is responsible for the abuse of this liberty, in cases determined by the law".

MEDIA TECHNOLOGY, POWER AND SOCIETY

Canadian studies has a long tradition of studying the relationship between technology and society since the legacy of George Grant and his understanding of technology as a human fate, to the more recent interpretations that of Harold Innis, who related this relationship to the study of power relations in societies. It is undeniable that power "is so tied up with technology" (Goyder, 2005, p. 28), and as seen in various analysis, such as of R.A. Buchanan in his *The power of the machine*

(1992), is representing one of the "concentric circles representing 'elements of technological revolution'" (Ibid).

Harold Adams Innis (1894-1952), is one of the prominent Canadian theorists and one of the major Canadian figures not only in the field of political economy in which he started his academic research, rather as a communication and media pioneer as much of his work focused on the historical relationship between power, political organizations and communication technology. Innis represents one of the pioneering scholarly efforts to understand how empires (as symbols of power) were extended and transformed through social control of the structures of media and culture.[4] Nicholas Garnham asserts that Innis' most important contributions came from his study of the influence of technologies not only on their messages' reception, but rather on media technologies as "mode(s) of coercion and production." He noted that Innis's understood media as including "physical transport systems" and studied "their relationship to systems of economic and political power" (2000, p. 28). Viewed in this manner, Innis's political economy of communication was *prophetic* in that it influenced subsequent approaches to regard the development of communications systems as central to the economic structure and powers of the state that could be seen in contemporary Canadian communication scholars (such as the work of Vincent Mosco, and Robert Babe). Garnham clarifies that Innis's adoption of communication as a field of study had a significant impact on the direction of research in this field, influencing it to include the history of modernity in which arguments focus on state formation, bureaucracy, and modes of persuasion and production, rather than a perspective that merely focuses on the state as a mode of coercion (Ibid, p. 29).

Innis emphasizes the importance of studying the effects of media in different contexts as he explains in his *The Bias of Communication* "A medium of communication has an important influence on the dissemination of knowledge over space and over time and it becomes necessary to study its characteristic in order to appraise its influence in its cultural setting" (1964, p. 33). This Innisian pioneering perspective could be easily translated if correlated to our modern, advanced technologies and their global launching and their powerful effect on the field of communication. Technology has impacted the lives of individuals in many cultures throughout the world. Indeed, new technologies have removed many physical and geographical barriers between people living in different and distant places around the globe (Eid & Dakroury, 2008).

Harold Innis posed two key questions in his research; the first is to determine the reasons of change in societies. For him, the answer is related to the way the different media of communication transformed the social and human organization in society arguing that the rise and fall of human civilization rely on the degree of competition of the existed forms of media. Particularly, Innis asserted on the importance of the "monopoly of knowledge" and power domination of different means of communication technology. The second question that occupied Innis is finding the reasons behind social and cultural stability, which he argued that, it is related maintaining a balance between the bias of "time" and "space" such as the case of the Greek society for example. Innis genuinely studied media and culture adopting a "macro" level of analysis which was lately criticized of being too abstract and difficult to read (McPhail & McPhail, 1990, pp. 62-63).

Empire and Communications (1950) is another major contribution of Harold Innis where he tracks the uses of different media of communication in the ancient empires. He started his longitude research from the ancient Egyptian civilization. Particularly, he explained that with the use of papyrus in the ancient Egypt, there has been a move from carving into stone as a medium of communication (bias of time since these are heavy media) to writing on papyrus that resulted in a major shift from an absolute monarchy system to a more democratic system of governance. Innis explains the rise in power of religion, priests, and scribes as well as the increased demand for literacy (Eid & Dakroury, 2008).

Thus, Innis' research helped explaining the rise and fall of political empires (as the case of ancient Egypt, China, and Greece) by relating them to the *social changes* resulted in the uses of different media technologies. Such research has indeed contributed to the communication research generally as it took a historical analysis turn through Innis' analysis of power relations associated with the media technologies. In short, Harold Innis has, as Marshall McLuhan outlined that Innis' *The Bias of Communication*, has opened new windows for us as communication researchers to read and understand history:

Most writers are occupied in providing accounts of the content of philosophy, science, libraries, empires, and religions. Innis invites us instead to consider the formalities of power exerted by these structures in their mutual interaction. He approaches each of these forms of organized power as exercising a particular kind of force upon each of the other components in the complex (McLuhan, 1964, p. ix).

Viewing the relationship between media technologies and societies from a different angle, McLuhan's *Understanding Media* (1964) revealed to many the "magical" world: the new media of communication — radio, television, film, photography, satellites, and later computers — that were for many the means to "restructuring civilization" (Rosenthal, 1968, p. 19). McLuhan argued that, with new electronic technologies, human beings had returned to primitive, tribal ways of communicating. Particularly, he viewed media technology as an extension of the individuals arguing that each medium embodies a certain bias toward one of the human senses (sight, sound, touch..etc). He argues that media are transmitting our knowledge and experience embedded within the very structure of our society. He puts it in his famous equation: "the medium is the message", which means that we should not be concerned only with the tools of communication (the media) and ignore the (message) constructed. For McLuhan each medium requires us to get engaged using a different bias. For that, McLuhan argues that the dominant medium in a given time shaped the way we perceive and understand the world surrounded it.

Thus, McLuhan's main thesis is centered on the argument that with the new technological advances of the media (especially television during his time), societies are restoring the 'tribal" character that existed before the invention of print. For him, print asserted the senses of individualism and competition opposing to electronic media that constructed a "global village". He asserts that the world had become a "global village" where people send and receive messages instantly. The media reduced the spaces between, and the separation from, fellow humans and hence could alter their behaviour: "Our new electric technology that extends our senses and nerves in a global embrace has large implications for the future of language" (McLuhan, 1964, p. 80).[5] In relating he previous discussion to the core thesis of this chapter, one must assert the way in which global powers who "own" the different means of communication can accordingly own the message and leave no place to alternative voices in the mainstream media. More particularly, the problem of media ownership and control and its impact on the right to communicate has long been, and still is, one of the greatest barriers to the enjoyment of this right, and constitutes one of the important arguments against its universality in practice. This was Jean d'Arcy's[6] argument over two decades ago when complaints and problems arose due to media ownership and monopolies in the communication industry. As he pointed out, "any power, religious, political or private, always knew that control over communication meant control over society. One of the titles of Caesar was 'Pontifex Maximus,' the builder of bridges. Control of bridges gives control of roads and thus communications" (1977, p. 49). In highlighting the impact of the many levels of "powers" that can dominate and control the right of people to communicate freely, d'Arcy says clearly, that who controls the media of communications can certainly control society (Dakroury, 2005).

MEDIA, GLOBAL CAPITALISM AND THE THREAT OF THE RIGHT TO COMMUNICATE

Following Immanuel Wallerstein's[7] modern world system approach, many scholars, notably, Herbert Schiller, adopt the global capitalist market economy arguing that a hierarchy of power relations is built between the "core"; "peripheries"; and the "semi-periphery" countries countries. For example, the capitalist, Western, and developed countries are considered to be the "core" countries, while, the "periphery" countries are those that follow the core countries economically, politically, and also culturally, and finally, the 'semi-periphery' countries, which combine both situations. This body of literature deals with the idea of cultural imperialism through the assumption of dependency theory, in which a belief that the less-developed former colonial regions, although officially independent from their colonizers, arguably are still dependent on the developed world's move toward "modern global capitalism" (Tomlinson, 1991, pp. 36-7). Multinational media corporations are good examples of this perspective, in which these corporations that operate on a transnational scale, are producing, and distributing the media products which emphasize those values and consumeristic ideals that the capitalist imperial powers promote, thereby leading the world's tastes and culture in that direction. Cees Hamelink, in his *Cultural Autonomy in Global Communications: Planning National Information Policy* argues that many international practices reflect this discourse in daily life. For example, he says, "in a Mexican village the traditional ritual dance precedes a soccer match, but the performance features a gigantic Coca-Cola bottle" (1983, p. 2). [8] In other words, he believes that the various capitalist transnational corporations are the primary agents responsible for hegemony, particularly in the Third World, impacting even those cultural practices related to future generations:

replacing breast-feeding by bottle feeding [which is promoted by companies like Nestle for instance] has had disastrous effects in many Third World countries. An effective, adequate, and cheap method has been exchanged for an expensive, inadequate and dangerous product (Hamelink, 1983, p. 15).

Interestingly, Merrill argues that "*Concentration of ownership is dangerous for people and society*" (Dennis & Merrill, 2002, p.38, *italics in original*), asserting that we cannot deny the fact that media corporations are capitalist institutions, however, we cannot claim at the same time that they are advocating for diversity and plurality in opinions in practice which are the core concepts for democracy in society. He adds that with the spread of giant media corporations into local newspaper ownership, statistics show that readers became passive and are not involved in their local issues/matters, newspapers tend to use more advertising to attract the audience, and more importantly, the press is "losing a real sense of trust and respect for its contents" (Ibid, pp. 38-39). As for looking to media as only responding to the audience tastes and requests, Merrill argues that this is a one-sided view; he believes that media are both responding to the public but at the same time forming and constructing its choices, and tastes.

Relating to this understanding, Alger asserts the importance of looking to giant media as key players in the economic and political realms: "The main media fall increasingly under the control of a small number of giant corporations and extremely wealthy and wilful people, especially when such people are inclined to use the powerful media of mass communication for their own political and economic purposes" (1998, p. 2). In real life, giant media corporations who are vertically and horizontally integrated are playing a key role in shaping the individual, corporate, and specialist choices. Back again, to the AOL/Time Warner example: one can see that this merger impacted on other types of media corporations. In his study *Lost in Cyberspace*, Dwayne Winseck explains how multimedia owners play a very significant role in either allowing or restricting individuals' uses of the net. He describes this procedure in his notion of the walled garden concept:

The role of content, images and language in creating cybernetically enclosed worlds is also embodied in contracts between AOL/Time Warner and other service providers. For example, AOL's contract with Disney requires Disney to deter users from using Disney's site and related hyperlinks as departure points for larger forays into cyberspace as a condition of being granted entry into the "walled garden." Indeed, in this story of the 'magic kingdom' meets the "walled garden," AOL can cancel the agreement if more than 25% of users visiting Disney's site subsequently leave the AOL "space" (2002, p. 16).

One of the important studies in this field is that of Danah Boyd, and her focus on the role of networked publics in the social life of the teenagers and young users using new media of technologies.[9] In one of her recent work, she highlighted the growing attempts of giant media corporations to launch[10] (or attach their companies to) social networking sites. Examples of that include Google, which attach its name to Orkut—a social networking site created by its own Google employees. As well, Microsoft, Yahoo!, AOL, and Wal-Mart who tried to create their own social networking sites, yet, it has been proven not as successful as the other sites (2008, pp. 120-121). In 2005, MySpace was purchased by Fox Interactive Media (one body of Murdoch's News Corporation) US $580 million US dollars, an ownership that resulted to some changes in the site itself and the on-line participation which push MySpace to grow in different spaces and within different cultural and linguistic settings (Ibid, pp. 122-123).

SOCIAL NETWORKING AND THE RIGHT TO COMMUNICATE: IS THERE A POSSIBILITY?

Various studies have recently explored the extent to which communication on the internet, including blogs, conforms to Habermas's criteria for the public sphere (O Baoill, 2004, and Froomkin, 2003). For Habermas himself, the potential of the internet as a public sphere is decidedly ambivalent (Habermas, 2006). In this context, Habermas fails to appreciate the extent to which blog participants are countering the traditional dominance of a mass media mentality and public discourse elites as they construct a communicative consciousness within the new communication environment. The innovation of blogs has dramatically enhanced the ability of individuals to connect globally with each other in an instantaneous, interactive mode. Typically blogs, a web based mode of social networking, allow readers to leave comments in a conversational mode. Blogs are often devoted to particular topics or serve as online diaries. They have achieved so much prominence in the political arena that major TV networks report on the "chatter" on the blogs. As of December 2007, the blog search engine Technorati monitors almost 113 million blogs. It reports that over 175,000 new blogs are created each day and that blogs are updated at a rate of over 1.6 million posts a day (http://www.technorati.com/about/). Their prominence and growth raises a question about whether they can be characterized as a form of public sphere.

Ruggiero (2000) writes that new media like the Internet possess three attributes of data not commonly associated with traditional media: interactivity, demassification, and asynchroneity that were not evident in traditional media. Focusing on social networking sites, and according to the Australian social media statistics, more than 800,000 visitors have visited Twitter in one year (from June 2008 to June 2009) in Australia alone. While facebook ranked the most visited (more than 6 million) of the rest of social networking sites (www.switchedonmedia.com.au). Users of these websites, tend to use these new tolls of media to gratify specific needs, as shown in one of the studies, as Snijders proved in his study of the evaluation of social network dynamics arguing that actors evaluate the network structure to obtain a more "pleasant" configuration of relations (2001, p. 363). By using social networking sites, as another research showed, users tend "to present themselves, articulate their social networks, and establish or maintain connections with others" (Ellison, Steinfield, & Lampe, 2007). The uses varies from merely establishing social relationships (like Friendster, CyWorld), sharing interests in music or politics (MySpace), connecting through work-relates (LinkedIn.com), and the infamous Facebook which initially launched as an interaction between college students. Studies tracking the number of Facebook users, argue that it jumped from half million users in 2006 to more than 21 million users in 2007 (Orra, Sisic, Arseneault, Simmering, and Orra, 2009, p. 579).

Social networking has been a growing field of research, in particular the uses of these sites as a toll of free expression and voicing opinions and ideas, to the extent that scholars argue that the trend of social networking has created cyber communities, with cyber friends, and cyber chat, discussion, and forums. Since the introduction of the telegraph humans having been struggling to deal with the implications of "speaking into the air", as John Durham Peters observes: "Though humans were anciently dubbed the 'speaking animal by Aristotle, only since the late nineteenth century have we defined ourselves in terms of our ability to *communicate* with one another." With the introduction of each new mode of communication "The meaning of communicative connections, large scale and small, is an ongoing conundrum" (Peters, 1999).

For some, the interactivity offered by the Internet and blogs in particular holds the promise of creating a Habermasian public sphere. As media scholar Luke Goode notes: "What better news for advocates of the Habermasian public sphere than to be told that the era of mass, one-way communication flows is in terminal decline?" (Goode, 2005). Moreover, other studies have elaborated this understanding contending that in examining blogs as communicative connections and analyzing their importance to the development of a public sphere, it is necessary to consider the role of an expanding communicative consciousness as a necessary prerequisite to a genuine, "space-less" public sphere (Dakroury & Birdsall, 2008). In other words, it has been argued that "communicative consciousness" as "the self-reflection of an individual on the meaning and nature of his or her communicative relationships and others through interactive communication via global electronic networks" (Ibid, p. 5).

Although advocates of human communication, d'Arcy rejected McLuhan's infamous term "the global village", arguing that it would be more plausible to anticipate the revival of human scale as small "microsocieties" where information would be freely exchanged (d'Arcy, 1969). But to achieve this free exchange of information, the new social structures resulting from global, interactive communication would have to be within a framework of the universal right to communicate. With the innovations in satellites, telecommunications, and computing, individual would be able to participate in horizontal interactive communication through a global "fine-meshed nervous system" that provides an abundance of communication channels.

In trying to relate the uses of social networking sites to the free enjoyment of the right to communicate, it is important to recall Boyd's understanding of "networked publics" (2008, p. 125), where she explained that while there are different meanings of "public", yet, social networking sites have allowed the "public to gather" (Ibid) online, creating forums, discussing issues of interests, and important fighting censorship and oppression. In other words, social networking sites have affected their "mediated nature of interaction" between the public Particularly, the notion of networked publics is built through four different characteristics: persistence (where all communications are recorded and carried on); searchability (where users can digitally search different persons, spaces, and ideas); replicability (where communication can be copied and transferred easily and instantly); and invisible audiences (which is surely related to the quality of communication using technologies over time and space) (Ibid, pp. 125-126). Collectively, one can argue that the advances of information technologies in the recent years, particularly, the uses of social networking sites have indeed paved the way to a more plausible understanding of the practice of the human right to communicate. Evidently, and depending on original archival materials, Winter and Wedemeyer (2009) have related three main evolutionary stages of the Right to Communicate to the emergence of participatory, networked policy, and the emergence of an Internet-based Public Sphere.[11] First, the recognition of the right to information, found in Article (19) of the *Universal Declaration of Human Rights* (1948); secondly, the graduation of the concept that human beings have a right to communicate, launched by d'Arcy's 1969 article, "Direct Broadcast Satellites and the Right of Man to Communicate"; and finally, the online, networked era of the Right to Communicate, emerging in the context of the Internet, World Wide Web, and mobile communications. it is true to emphasize that:

For now, though, let us say that the networked information economy as it has developed to this date has the capability to take in, filter, and synthesize observations and opinions from a population that is orders of magnitude larger than the population that was capable of being captured by the mass media. It has done so without re-creating identifiable and reliable points of control and manipulation that would replicate the core limitation of the mass-media model of the public sphere – its susceptibility to the exertion of control by its regulators, owners, or those who pay them (as cited in Winter & Wedemeyer, 2009, p. 40).

As seen from the previous discussion, global capitalism is seen as one of the reasons behind hindering a human right to communicate, since it is concentrating on the colonization of space and the monopoly of resources and means of dissemination. For that scholars, notably McChesney, who admitted that capitalism encourages a culture tending to "commercialize every nook and cranny of social life in way that renders the development or survival of nonmarket political and cultural organizations more difficult" (as cited in Papacharissi, 2002, p. 11). This underscores the main thesis of this chapter that explains the importance of new media as a means of the human right to communicate.

REFERENCES

Akin, D. (28 January 2010). Capitalism has gone 'badly wrong'. *The Ottawa Citizen*, d-3.

Alger, D. (1998). *Megamedia: How giant corporations dominate mass media, distort competition, and endanger democracy*. New York: Rowman & Littlefield.

Bachen, C., Raphael, C., & Lynn, K. M. (2008). Civic engagement, pedagogy, and information technology on Web sites for youth. *Political Communication*, *25*(3), 290–310. doi:10.1080/10584600802197525

Baoill, O. A. (2004). Weblogs and the Public Sphere. In L.J. Gurak, S. Antonijevic, L. Johnson, C. Ratliff, & J. Reyman (Eds.), *Into the Blogosphere: Rhetoric, Community, and Culture of Weblogs*. Retrieved December 2009 from http://blog.lib.umn.edu/blogosphere/ weblogs_and_the_public_sphere.html

Benedetti, P., & DeHart, N. (Eds.). (1996). *Forward through the rearview mirror: Reflections on and by Marshall McLuhan*. Scarborough, UK: Prentice Hall Canada.

Boyd, D. (2008). Why Youth Social Network Sites: The Role of Networked Publics Teenage Social Life. In Buckingham, D. (Ed.), *Youth, Identity, and Digital Media* (pp. 119–142). Cambridge, MA: The MIT Press.

d'Arcy, J. (1969). Direct broadcast satellites and the right to communicate. *EBU Review, 118*, 14–18.

d'Arcy, J. (1977). The right of man to communicate. In Harms, L. S., Richstad, J., & Kie, K. A. (Eds.), *Right to Communicate: Collected papers* (pp. 45–52). Honolulu, HI: Social Sciences and Linguistics Institute, University of Hawaii, Manoa.

Dakroury, A. (2005). Who owns the medium owns the message? The ambiguity of the right to communicate in the age of convergence. *Reconstruction: Studies in Contemporary Culture, 5*(2).

Dakroury, A. (2009). The baron of the right to communicate: Jean d'Arcy (1913:1983). In Dakroury, A., Eid, M., & Kamalipour, Y. R. (Eds.), *The right to communicate: Historical hopes global debates, and future premises* (pp. 21–42). Dubuque, IA: Kendall/Hunt Publishing.

Dakroury, A. (2009). *Communication and Human Rights*. Dubuque, IA: Kendall/Hunt Publishing.

Dakroury, A., & Birdsall, W. (2008). Blogs and the right to communicate: Towards creating a space-less public sphere? In *International Symposium on Technology and Society proceedings*. Fredericton, New Brunswick: Institute of Electrical and Electronics Engineers (IEEE).

Dakroury, A., & Eid, M. (2008). Canadian Communication perspective. In Eid, M., & Dakroury, A. (Eds.), *Introduction to communication and media studies* (pp. 303–309). Boston: Pearson.

Dennis, E., & Merrill, J. (1984). *Basic Issues in Mass Communication: A Debate*. New York: Macmillan Publishing Company.

Dickin, J. (2002). The Internet as a site of citizenship. *Canadian Journal of Communication, 27*(4), 509–534.

Froomkin, A. M. (2003). Habermas@discourse net: Toward a Critical Theory of Cyberspace. [from http://osaka.law.miami.edu/~froomkin/discourse/ils.pdf]. *Harvard Law Review*, 751–873. Retrieved December 2009.

Garnham, N. (2000). *Emancipation, the media, and modernity: Arguments about the media and social theory*. Oxford, UK: Oxford University Press.

Goode, L. (2005). *Jürgen Habermas; Democracy and the Public Sphere*. London: Pluto Press.

Goyder, J. (2005). *Technology and society: A Canadian perspective*. Peterborough: Broadview Press Ltd.

Habermas, J. (2006). Does Democracy Still Enjoy An Epistemic Dimension? The Impact of Normative Theory on Empirical Research. In *Communication Theory* (pp. 411–426). Political Communication in Media Society.

Hamelink. C. (1983). *Cultural Autonomy in Global Communications: Planning National Information Policy*. New York: Longman.

Harms, L. S., & Richstad, J. (1977). Right to Communicate: Human rights, Major Communication Issues, Communication Policies and Planning. In Harms, L. S., Richstad, J., & Kie, K. A. (Eds.), *Right to Communicate: Collected Papers*. Hawaii, HI: The University Press of Hawaii.

Innis, H. A. (1964). *The bias of communication*. Toronto, Canada: University of Toronto.

McLuhan, M. (1964). *Understanding media: The Extensions of man*. Toronto, Canada: McGraw-Hill.

McPhail, T. L., & McPhail, B. M. (1990). *Communication: The Canadian experience*. Toronto, Canada: Copp Clark Pitman ltd. A Longman company.

Nordenstreng, K., & Hannikainen, L. (1984). *The Mass Media Declaration of UNESCO*. Norwood, NJ: Ablex Publishing Corporation.

Papacharissi, Z. (2002). The virtual sphere: The internet as a public sphere. *new media & society, 4*(1), 9-27.

Peters, J. D. (1999). *Speaking Into the Air: A History of the Idea of Communication*. Chicago: University of Chicago Press.

Rosenthal, R. (Ed.). (1968). *McLuhan: Pro & con*. New York: Funk & Wagnalls.

Ross, C., Orra, E., Sisica, M., Arseneaulta, J., Simmeringa, M., & Orra, R. (2009)... *Computers in Human Behavior, 25*(2), 578–586. doi:10.1016/j.chb.2008.12.024

Ruggiero, T. E. (2000). Uses and gratifications theory in the 21st century. *Mass Communication & Society, 3*, 3–37. doi:10.1207/S15327825MCS0301_02

Snijders, T. (2001). The Statistical Evaluation of Social Network Dynamics. *Sociological Methodology, 31*, 361–395. doi:10.1111/0081-1750.00099

Stein, L. (2008). Speech without rights: The status of public space on the Internet. *Communication Review, 11*(1), 1–23. doi:10.1080/10714420801888385

Tomlinson, J. (1991). *Cultural Imperialism: A Critical Introduction*. London: Printer Publishers.

Wallerstein, I. (1974). The Rise and Future Demise of the World Capitalist System: Concepts for Comparative Analysis. *Comparative Studies in Society and History, 16*(4), 387–415. doi:10.1017/S0010417500007520

Winseck, D. (2002). Lost in Cyberspace: Convergence, consolidation and power in the Canadian mediascape. *Media, culture and society*.

Winter, J., & Wedemeyer, D. (2009). The roots of the right to communicate and emerging participatory policy. In Dakroury, A., Eid, M., & Kamalipour, Y. R. (Eds.), *The right to communicate: Historical hopes, global debates, and future premises* (pp. 43–70). Dubuque, IA: Kendall/Hunt Publishing.

ENDNOTES

1 It is often argued that the two Covenants—the *International Covenant on Civil and Political Rights* and the *International Covenant on Economic, Social and Cultural Rights*—are two important international documents that complete the recognition of basic human rights. Their advantage lies in their statement of procedures for reporting and enforcing on human rights' status.

2 Jean d'Arcy was the Director of Radio and Visual Services during the 1960s in the Office of Public Information at the United Nations in New York. President of the International Institute of Communications until his death in the1980s, he was among the greatest advocates for the right to communicate.

3 It is worth noting here that although Richstad, Harms, & Kie argue that he "did not specify in detail what he meant by the Right to Communicate,... [his article] inspired others to attempt formulations over the following years" (1977, p. 114).

4 Particularly in his two landmark books: *Empire and Communications* (1950), and *The Bias of Communication* (1951).

5 For example, he believed that there was great potential in satellites, which for him exemplified how media could create a "global theatre." He told Ed Fitzgerald of CBC television in 1970: "You could say that with the satellite, the global village has become a global theatre ... [with] everybody on the planet simultaneously participating as actors" (Benedetti & DeHart, 1997, p. 66).

6 Jean d'Arcy was the Director of Radio and Visual Services during the 1960s in the Office of Public Information at the United Nations in New York. President of the International Institute of Communications until his death in the1980s, he was among the greatest advocates for the right to communicate.

7 See Immanuel Wallerstein in his "The Rise and Future Demise of the World Capitalist System: Concepts for Comparative Analysis" (1974), and his "Semi-Peripheral Countries and the Contemporary World Crisis" (1976), in which he argues that the World Economy became a fixed model as early as the 1640s because this is where unequal exchange between the core-peripheries countries first becomes evident, not only in terms of economic factors but also in power relations.

8 A relatively recent example of the application of this cultural homogenization of global capitalist powers is the work of Neo-Marxist, Armand Mattelart, in particular, his criticism of Disney comics and cartoons, which are widely distributed in the developing countries. He believes that these cultural products are "'carriers' of the American capitalist cultural values" (Tomlinson, 1991, p. 42).

9 In her "Why Youth Social Network Sites: The Role of Networked Publics in Teenage Social Life" Boyd tracks the timeline of the social networking sites worldwide, explaining that while MySpace was a very popular site for Americans by mi-2005, Friendster gains more popularly in Singapore, Philippines, Indonesia, and Malaysia, while Orkut and Hi5 in were equally used Brazil and India by both teenagers and adult users. While in United Kingdom, New Zealand, Australia, and Japan, social networking sites were mainly targeting youth (such as Tagworld, Bebo, Piczo, Faceparty, and Mixi) (Boyd, 2008, pp. 119-120).

10 Boyd cited Cyworld website as one of these examples, where it Cyworld was a property of SK Telecom, the largest mobile phone operator in South Korea (Boyd, 2008, p. 120).

11 In this important study, Winter & Wedemeyer use various examples to confirm the importance of new technologies and particularly social networking sites to assert the importance of the right to communicate. For example, the use of sites like MySpace, Facebook; the user-generated tags (as in Amazon or Flicker) that categorize information, news and photos; the Wikis (enabling collaborative authorship); blogs (both on individual and organizational levels); the RSS (Really Simple Syndication) for instant information and news; podcasts, and YouTube (2009, pp. 41-42).

Chapter 12
A Picture and a Thousand Words:
Visual Scaffolding for Mobile Communications in Developing Regions

Robert Farrell
IBM T J Watson Research Center, USA

Catalina Danis
IBM T J Watson Research Center, USA

Thomas Erickson
IBM T J Watson Research Center, USA

Jason Ellis
IBM T J Watson Research Center, USA

Jim Christensen
IBM T J Watson Research Center, USA

Mark Bailey
IBM T J Watson Research Center, USA

Wendy A. Kellogg
IBM T J Watson Research Center, USA

ABSTRACT

Mobile communication is a key enabler for economic, social and political change in developing regions of the world. The authors argue that engaging citizens in developing regions in information creation and information sharing leverages peoples' existing social networks to facilitate transmission of critical information, exchange of ideas, and distributed problem solving. All of these activities can support economic development. They describe Picture Talk, a mobile social computing application framework designed to facilitate local information sharing in regions with sparse Internet connectivity, low literacy

DOI: 10.4018/978-1-60960-037-2.ch012

Copyright © 2011, IGI Global. Copying or distributing in print or electronic forms without written permission of IGI Global is prohibited.

rates and having users with little prior experience with information technology. Picture Talk runs on today's internet-enabled smartphones as well as camera phones with multimedia messaging, now available in many of these regions.

INTRODUCTION

We are interested in designing applications that enable people at the base of the economic pyramid (BoP) to create, share, and discuss information as is commonly done on the World-Wide Web today, but through mobile technologies. The BoP includes over one billion people with little access to computer technology often living on less than $1US per day in some of the least developed countries in sub-Saharan Africa, South/Central America and the Indian Sub-continent, as well as several other parts of Asia. As others have recognized (Prahalad, 2004; Kumar et al, 2008), enabling connections among a wide spectrum of people can lead to the empowerment of the disenfranchised and enable people who have been largely excluded from modern technology and economic opportunity to express their entrepreneurial tendencies. This could result, for example, in the creation of broader markets for local goods and services. The global reach of mobileM communication networks offers, for the first time, a broad platform for delivering applications and software services that can realize this potential.

We have three design goals for the mobile applications we build. First, we want these applications to be usable by even the most disadvantaged users, many of whom are illiterate Second, we want to enable these users to document local needs, problems, and issues by creating, storing, and sharing digital artifacts (e.g., maps, photos, graphics, radio news reports, music, games, TV segments, informal news). Third, we want to enable these users to engage in conversation about these digital artifacts to offer solutions, share perspectives, or to engage in purely social exchanges.

Our initial implementation toward these goals is Picture Talk, a mobile social computing application framework that supports three major features: first, mobile phone users can start an inclusive discussion forum with other mobile phone users by asynchronously sharing short recordings of their voice over a wireless network; second, they can supplement and provide structure for the unfolding conversation by adding pictures; and third, they can discover and participate in discussions shared by other users, reaping the rewards of prior users' ideas, comments, and solutions. These features could enable Picture Talk to become a powerful tool to support existing social behavior and social connections between users (Danis et al., 2009).

This article first provides some background on some of the obstacles that BoP communities face in trying to access information technology, then introduces the problem of supporting information creation and dissemination in this environment, and then describes mobile social computing as a potential solution, including the Picture Talk application framework we have developed. We conclude with some future directions and conclusions.

BACKGROUND

BoP populations face a number of obstacles in becoming part of the global online community. In the economically developed world, access to information technology has been largely through Internet-connected computers. An important benefit of access to the Internet has been the potential for contact with the worldwide community of users. Through online communities, Internet users engage in discussions on topics of common interest, write comments that serve as a means of self-expression, and solve each others' problems. We argue here that through mobile phones, similar applications could be deployed to BoP communities to enable discussions on topics of local interest, provide a voice for individuals who would otherwise have no forum for their ideas, and enable solutions to communal problems through information exchange.

There remain at least three obstacles to the widespread use of information technology to support online communities and other applications in developing regions. First, despite initiatives such as One Laptop Per Child (OLPC, 2009), conventional computing technology continues to remain out of reach of the large majority of the BoP population. Lack of reliable networks to access the Internet further limits the ability of people in these regions to access information even where capable devices are available. BoP users are necessarily very cost conscious, driving a need for a low cost platform. Adapting applications designed for powerful computers with a keyboard and a large display for BoP users with mobile phones remains a challenge. Second, low literacy rates in some regions prevent significant portions of the BoP users from using the Internet's predominantly textual interaction mode. Despite vast quantities of online information, the plethora of native languages and alphabets in BoP regions render many of these resources useless even for people who are literate in their local language. Third, despite skills and experience in the social use of mobile phones, many BoP users may have little familiarity with or motivation to use the device to access information services, preferring face-to-face interaction. A radical rethinking of the human interface to information technology may be required.

The statistics for even basic access to electricity in developing regions are alarming. According to the Open Society Initiative for Southern Africa (OSISA, 2009), approximately 90% of Africa's one billion people have no regular access to electricity. Even where there are facilities, blackouts are common. The bulk of power plants and transmission facilities in West Africa, for example, are over 50 years old. With little maintenance or investment, the aging equipment fails or operates below capacity (Madamombe, 2005). Despite these obstacles, where power to homes is not available, people often travel to a centrally located solar-powered, wind-powered, or coin-operated charging station to maintain use of their mobile phone.

Global statistics for computer usage demonstrate huge differences between developed and developing countries. For example, the highest rates of access to the Internet in 2007 were in Sweden (82%), the US (81%), South Korea (81%), and other developed countries, whereas the lowest access rates included Tanzania (6%), Kenya (12%), and Uganda (11%). Similarly, computer ownership is lowest in Uganda and Tanzania, both at 2% (The Pew Research Center for the People and the Press, 2007).

The picture changes radically when considering mobile phones rather than computers. A recent survey by the International Telecommunications Union (ITU) found that while only one quarter of the earth's population of 6.7 billion uses the Internet, nearly two thirds of the population uses mobile phones (ITU, 2009). Wireless phone use is exploding in the developing world: sixty-eight percent of mobile phone subscribers worldwide are outside of North America and Europe. In Africa, regular users of mobile phone services have jumped from 10 million to 400 million in the five years from 2003 to 2008 and the growth is still accelerating (ITU, 2009). The rate of mobile phone ownership in the Ivory Coast, Mali, Nigeria and South Africa is over 60%, higher than in Canada (The Pew Research Center for the People and the Press, 2007). Mobile phone use in economically developing regions crosses the barriers of gender, age, and education (Samuel et al., 2005).

The rapid growth of mobile phone networks in BoP markets is fueled by the need for communication in environments where there are few alternatives. The lack of traditional wired infrastructure creates an opportunity: much of the developing world is a "green field" where new technologies can be deployed without being hampered by existing business models, infrastructures, or user expectations. For example, in many parts of Africa, wireless networks have leapfrogged the public switched telephone network in terms of installed base. In 2007, the African continent had 280 million total telephone subscribers, but 260 million of these were mobile cellular subscribers. Building a new wireless network is faster, easier, more reliable, and less expensive than putting in a whole new wired infrastructure.

Despite the growth of wireless networks, few developing countries yet have data communication channels sufficient to provide rural populations with access to the public Internet. In 2008, only 5% of Africa (Appfrica, 2008) had access to the Internet. Still, it is likely that mobile data services (e.g., Global Packet Radio Service, GPRS) and mobile broadband (e.g. WiMax) will be the first contact that most Africans have with the Internet.

The capabilities of mobile phones are also increasing rapidly. The first deployments of camera phones occurred in 2001 and by 2004, 370 million mobile phones with digital cameras were sold (InfoTrend/CAP Ventures, 2004). The total number of mobile camera phones is expected to exceed 1 billion by 2011 with fastest growth in emerging markets (Lyra Research, 2007).In the late 1980s and early 1990s, mobile phones were used for voice communication only and users typed on a numeric keypad. Today's smartphones have high resolution touch screen displays, miniature keyboards, and other flexible input methods and are used for mobile computing, multimedia messaging, information access, and more. Worldwide smartphone sales increased 12.7 percent in the first quarter of 2009 and sales are anticipated to grow at more than a 30% compound annual growth rate over the next five years (Gartner Group, 2009). We anticipate that smartphones connected to a wireless phone network will become a viable platform for delivering mobile ICT in Africa in the coming years.

Another major hurdle is literacy. Literacy is typically defined as the ability to read and write, however there is an inherent lack of precision that results from the methods of assessment and thus official figures often over-estimate functional literacy. For example, the commonly cited statistics, such as those compiled by UNESCO (2009), are based on census and other self-report methods, which are fundamentally inexact. Also, the definition of literacy can vary from 'the ability to write a simple sentence' to 'being able to freely communicate ideas in literate society.' Individuals as young as fifteen who may have been counted as literate because they were attending primary or secondary school may, because of limited continued use of written language, become functionally illiterate as adults (Seshagiri, Sagar & Joshi, 2007). According to UNESCO (2009), two-thirds of the world's 785 million illiterate adults are found in only eight countries (India, China, Bangladesh, Pakistan, Nigeria, Ethiopia, Indonesia, and Egypt). Low literacy rates are concentrated in South and West Asia, sub-Saharan Africa, and the Arab states, with percentages averaging in the 60s, though some countries like Mali and Niger report rates for 15 to 20 year olds of less than 30%. Men typically have higher rates of literacy than women in traditional societies (UNESCO, 2009).

While there are no generally accepted statistics on how much of the Internet is available in different languages, it is generally accepted that the dominant language on the Internet is English, making much of the Internet linguistically inaccessible to the large majority of the BoP. The large number of languages spoken in BoP countries is intertwined with literacy and access to written information. UNESCO (2009) estimates over 2000 languages are spoken on the African continent. An additional challenge is that individuals who may be literate in their native language may nevertheless be functionally illiterate if information is available only in one of the country's official languages (Plauché and Nallasamy, 2007).

A third major challenge in introducing technology in developing regions is being sensitive to the social and cultural context in which the technologies operate. To provide one example from the southeastern Indian state of Kerala, fishermen now use mobile phones to get market price information before deciding where to sell their fish (Abraham, 2007). About 40% report an increase in income and 50% report fewer losses due to unsold or spoiled fish when they start calling for prices. Interestingly, however, few of these fishermen consistently go to markets with the highest prices; instead many choose ports where their "commission agent" has a presence. Because commission agents invest in the fisherman's business (e.g., financing the purchase of a fishing vessel), the fisherman feels a social obligation to bow to the agent's wishes, even when doing so may prevent him from maximizing his income.

Several field reports illustrate specific ways in which trust in one's social network may provide a bridge to the acceptance of information from official, but unfamiliar sources of information through computing technology. For example, farmers in the southern state of Tamil Nadu use web-connected kiosks (telecenters) fielded by a local sugar factory to ask only "simple" (i.e., low-stakes) questions of a purported agricultural expert who is not known to them, saving "high-stakes" questions for successful farmers with whom they have some pre-existing relationship (Srinivasan, 2007). Gopakumar (2006) explains that local people play a critical intermediary role in the success of telecenters. For example, living in the same village led target users of the Akshaya telecenter to develop trust in the entrepreneurs and intermediaries who ran the centers. By extension, they also developed trust in the abstract systems of medicine and government that were the ultimate sources of the information.

To summarize, these studies demonstrate the power that access to information can have in improving people's lives, but also how the impact of information is gated by social factors like trust, accountability, and social and institutional pressures.

MOBILE SOCIAL COMPUTING FOR DEVELOPING REGIONS

Social computing makes the claim that people are fundamentally social; therefore their use of technologies cannot be understood separately from social considerations. In other words, people access, evaluate, and consume information in a social context. We believe that social computing is uniquely well suited to contribute to effective communication and information dissemination in developing regions.

Issues, Controversies, Problems

The widespread availability of mobile phone applications in developing regions is already having a broad impact at the BoP. It is giving individuals a way to share critical information in a timely fashion, for example. Furthermore, individual contributions can be aggregated to provide value to a broader audience. For example, Ushahidi, meaning "testimony" in Swahili, is a platform for crowdsourcing crisis information. Ushahidi allows anyone to transmit geo-coded data via Short Message Service (SMS) and the application aggregates this data (Ushahidi.com, 2009). We think that mobile social computing applications need to support an inclusive dialogue, enabling both the ability for citizens to report their thoughts and experiences and comment on the contributions of others. a

As we mentioned above, illiteracy is also a major challenge. Designers of applications geared towards illiterate users have focused on non-text modalities in order to design more generally accessible applications in the countries with low rates of literacy. For example, speech is a widely used modality. However, limitations on the generality of speech recognition technology in multi-lingual environments (Plauché and Nallasamy, 2007) demands the use of other modalities for a broad set of functions needed to complement spoken language interfaces. For example, Joshi, Welankar, Kanitkar and Sheikh (2008) developed and tested a phonebook they call Rangoli aimed at low literacy populations. Rather than entering phone numbers based on alphabetical order, users are able to use a combination of color, icon and spatial location. Similarly, Froehlich and colleagues (2009) proposed applying digital storytelling (for example, video or sequences of still photographs accompanied with spoken annotations) as ways of enabling low literacy individuals to participate in information creation and sharing. However, the mobile devices prevalent in developing regions of the world have limited support for images and almost no support for video.

As noted previously, even for literate users in the population, the large number of languages in BoP countries makes it unlikely that the user's preferred local language will be used in the user interface. Thus the use of pictures to augment spoken language may allow many more people to have meaningful access to information. However, when speech is transposed into digital settings, a number of well-known problems arise. In the type of applications we are envisioning, conversations would be asynchronous, that is, carried out between people in different places speaking at different times. This means that conversations would lack some characteristics that are important for establishing and maintaining common ground (Clark & Brennan, 1991) – "the knowledge that the participants have in common, and they are aware that they have it in common" (Olson & Olson, 2000, pp. 157) – which is critical for enabling communication among participants in a discussion. For example, it would mean that speakers would not be able to see one another, or share visual cues like glances, gestures and shrugs that in collocated speech enable interlocutors to control the conversation's flow, easily refer to objects, and verify that they are being understood (e.g., Yankelovich et al., 2004). It also potentially means that many more people and people from a wider geographic area might be able to engage in a conversation, something that could be desirable but which also could exacerbate these problems.

Solutions and Recommendations

Picture Talk is a mobile social computing application framework intended to support a wide range of social interactions that can be accomplished through asynchronous communication, including conversations with remote participants, question and answer exchanges, and peer production of localized content. We begin by laying out the rationale that underlies Picture Talk by describing the scenarios and design sketches that marked the beginning of our design process. After presenting the initial vision, it goes on to describe a working prototype.

The concept of Picture Talk arose out of consideration of the issues mentioned above and how they might be addressed in the context of a mobile phone-based communication system. The crux of the solution was to augment speech with three types of visual component: 'comment proxies', pictorial contexts and visual controls. 'Comment proxies' are visual representations of digital speech that depict various types of meta-information, such as the identity of the speaker, the length of the comment, and the relationship of the comment to other comments (e.g., a reply); in the Picture Talk user interface they also provide direct access to the comment they represent, thus mitigating the difficulty of navigating voice posts. Pictorial contexts are diagrams or photographs that provide a background for a particular conversation; pictorial contexts serve both to represent the conversation as a whole, and allow comment proxies to take on additional meaning by virtue of their location with respect to the pictorial background. Finally, visual controls are a variety of visual user interface components for controlling the system, for example, a message play button.

Figure 1. Three design sketches of the Picture Talk concept for applications set in rural India: (a) Rice Talk, for farmers to discuss problems with their Rice plants; (b) Health Talk, for villagers to discuss health problems; and (c) TinkerTalk, for people in a region to indicate that they need the services of a traveling tinker. Background images of rice plant © Ivan Kopylov | Dreamstime.com, of human body © Dannyphoto80 | Dreamstime.com, and of map © Robert Adrian Hillman | Dreamstime.com. Used with permission.

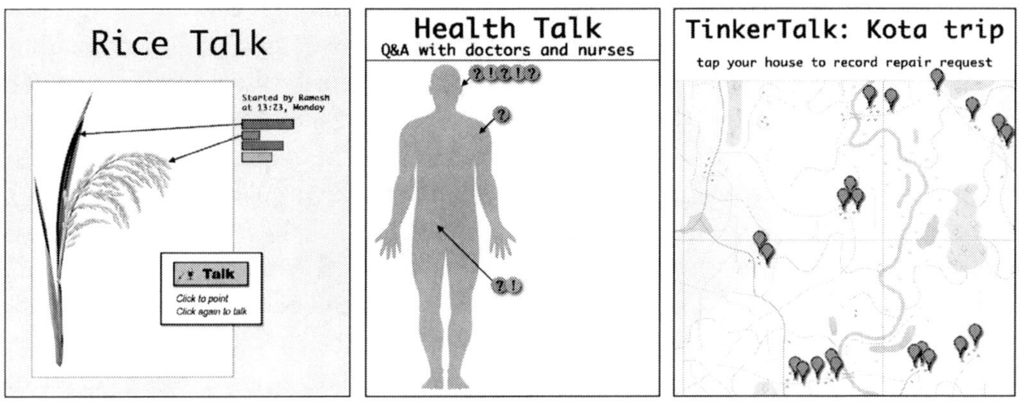

Figure 1 shows three early 'design sketches' of Picture Talk developed in the context of a scenario set in rural India. (By 'design sketches' we mean provisional concepts that are intended as conversation starters with stakeholders, rather than as depictions of well-considered solutions.) The first sketch, Rice Talk, envisions an asynchronous conversation among farmers about pests and diseases affecting their rice plants. It consists of (1) a diagram of a rice plant (the pictorial context); (2) a series of shaded bars (the comment proxies) that represent spoken comments, showing their durations, which of them have been made by the same speaker, and the part of the plant to which the comments refer; and (3) a floating 'talk' button (the visual control). The second sketch shows a health-oriented conversation with circles (the comment proxies) superimposed over a diagram of the human body (the pictorial context), the circles' positions indicating what aspect of the body or health they refer to and how they are related to other comments. The third sketch shows a conversation between a traveling tinker (i.e., a mender of pots) and potential customers, the pictorial context being a map of the region, and comment proxies (the balloons) indicating where the speaker is located.

Besides communicating the basic idea behind Picture Talk – using pictures, and simple visual representations of voice comments to provide scaffolding for asynchronous speech-based communication – the sketches serve other purposes. First of all, they illustrate the flexibility of the basic concepts. The pictorial contexts, and similarly the comment proxies, can represent a large range of topics, and even when depicted as simple geometric shapes, they can represent a considerable array of meta-information. Perhaps more importantly, the sketches are useful in raising a number of questions both within the design team, and with other audiences. How do the pictures get into the system? What sort of meta-information should comment proxies depict? Do different conversations benefit from the display of different comment meta-information? What sort of visual representations will be understandable by the envisioned user populations? How do users find their ways to particular conversations? As the aim of this paper is not to trace the trajectory of the design, it will not detail its evolution, but will instead move on to describe the user experience of the resulting working prototype.

Our implementation of Picture Talk consists of a client application running on the T-Mobile G1™ mobile phone, an Internet-enabled smartphone running the Google Android™ operating system, and a centralized data server running an application-specific Web service in the Ruby on Rails™ Web application server environment.

When users launch the client application on their mobile phone, their phone number is used to retrieve their user profile from the Web service. If this is the first time the user has accessed the service, he is prompted to record his name and take a picture of himself. The user is then presented with a menu that has four options: take a picture, view the gallery of shared pictures, view the profiles of other users, or update one's own profile.

Users can start a discussion by simply taking a picture using the phone's built-in camera and tapping anywhere on the photo. The system stores the picture in the gallery of shared pictures and records various metadata (e.g., who started the discussion, the time and date). Additional metadata could be stored, such as the location where the picture was taken. Subsequently, users can join an ongoing discussion by finding the picture in the gallery and tapping on it, and being led to the discussion screen.

The discussion screen (see Figure 2) implements the three user interface elements identified earlier: the pictorial context (in this case, a picture of a plant), 'comment proxies' (graphics on the upper right depicting spoken comments about the picture and associated speaker photo below), and visual controls (buttons at the bottom of the screen to control audio recording and playback). In the spoken comments area, to the right of the picture, each graphic represents a single comment from a user. Clicking on a graphic element displays the associated speaker's photo.

Clicking on a speaker's photo leads to the speaker's profile. The speaker's profile can help discussion participants determine their trust in the information provided. For example, a speaker may be a friend who is instantly recognizable from her photo or may be someone not known to the discussants but nonetheless reputable.

We designed the audio controls to allow the user to compose and review a recording before posting it to the discussion for others to hear. Before or after recording audio, we will allow the user to tap on the picture to point out something of interest in the picture, for example, the diseased part of a rice plant. The resulting visual annotation will consist of the comment in association with the particular location the speaker indicated on the picture. When a user posts a comment, the bar graphic is posted to the discussion area to the right of the picture. The bar graphic provides a visual "residue" of the comment recording (Hollan, Hutchins, & Kirsh, 2000) for subsequent users. The length of the bar reflects the length of the recording. Pressing anywhere on the bar graphic starts playing the recorded audio and displays any corresponding visual annotations on the picture. The same set of controls is used for both recording and playback, much like a music player. Users

Figure 2. A Picture Talk discussion

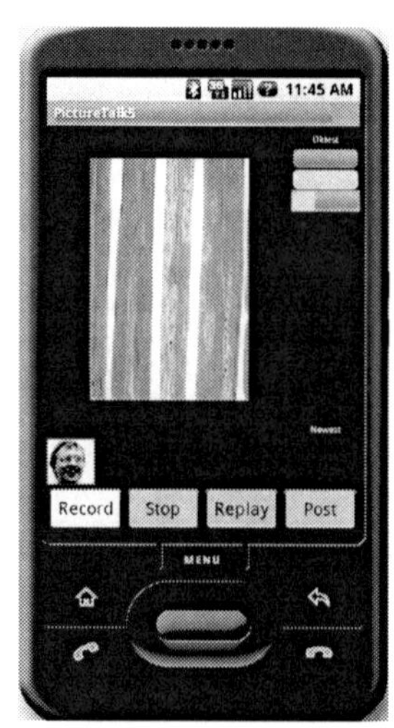

can pause the playback or replay the audio from the beginning, but they can only listen to one recorded comment at a time. The bar graphics in the spoken comments area are listed chronologically from top to bottom in a scrolling window with the most recent always visible.

Posting a comment stores the audio, and any visual annotations, with the discussion so that subsequent users accessing the picture can access the comment's audio and visual elements. It also stores metadata (who made the comment, the date and time of their comment). Picture Talk can also notify other participants in the discussion that there is a new comment. Individuals can block these notifications.

Picture Talk users can also browse through a scrolling gallery of all user photos. Touching a user's photo leads to that user's profile. The user profile includes a photo, contact information (telephone number) and a scrolling gallery of the pictures that anchor discussions the user has started. Touching a picture leads to a discussion screen with the given picture as the context. In this way, users can quickly find and engage in discussions started by other participants. This could be useful, for example, if a user has come across a farmer who has posted useful information about, for example, rice fungi and wants to see what other advice the farmer may have provided on similar topics. As users engage in discussions on a topic, their network of co-discussants grows. Photos of co-discussants can cue memory for relevant discussion contexts and serve as a visual index to organize the pictures anchoring discussions.

Picture Talk implements a concept we call 'blended synchrony' (Erickson et al., 2006), meaning that the same application supports (near) synchronous and asynchronous interaction among participants. Picture Talk discussions persist over time, with remarks separated by seconds, minutes, days, or even months. Some discussions will feel quite immediate and rapid-fire, whereas others may be slower paced, or might be more like announcements than a true conversation. It just depends on the pattern of participation. Blended synchrony is useful in environments where communication needs to be close to real-time in some cases but can be asynchronous in others. Cultural and societal as well as pragmatic factors may come into play in deciding when and how to communicate through ICT applications (Hudson, Christensen, Kellogg, & Erickson, 2002).

Picture Talk is architected as a client-server application (see Figure 3). The Picture Talk client is installed as a third-party Android application on the G1 mobile phone. The G1 has suitable

Figure 3. Picture Talk architecture

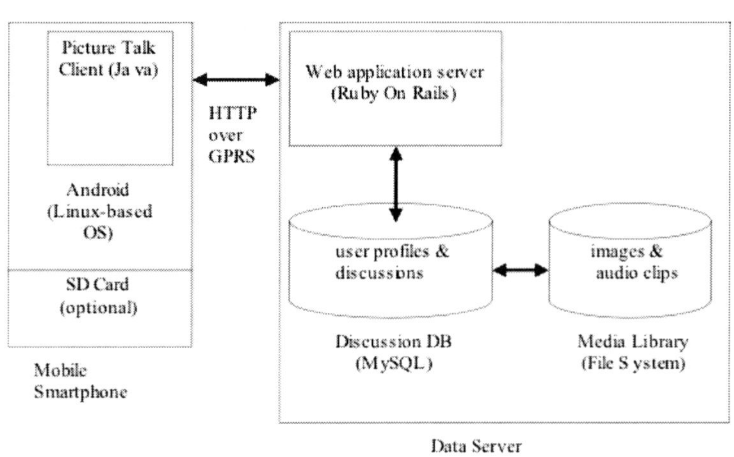

hardware for running the Picture Talk client: a 3.2-inch touch-screen display, wireless networking, a microphone, built-in speakers, a camera, and gigabytes of storage on a removable flash memory card. A number of smartphones provide similar functionality, but Picture Talk takes advantage of the Android operating system's (OS's) capability of accessing the phone's hardware, including detecting the presence of wireless network services, recording and playback of audio, controlling the built-in camera and storing pictures on the phone and in external storage.

The Picture Talk data server provides a persistent data model for the application's objects (discussions, pictures, comments, audio clips, users, etc.). To minimize the data exchanged between clients and server (and hence conserve wireless bandwidth), the server assigns version numbers to the data objects so that both client and server know when data object updates are needed to synchronize the data model. The server uses Rails' active record support to store and access the data objects in a MySQL® database. Pictures (images) and voice recordings (audio clips) are stored in files.

Communication with the Ruby on Rails server takes place over a wireless network. The Picture Talk client sends pictures captured with the phone's camera and audio captured with the phone's microphone to the Picture Talk server and automatically updates the currently displayed discussion. When the phone is not able to connect to the wireless network, new pictures from the camera are stored on its Secure Digital (SD) card, if available, or on the phone's local storage, thus enabling users to start discussions, make audio postings, listen to previously accessed audio postings, and update personal information. In this disconnected mode, data objects are stored and retrieved from a database local to the phone using Android's SQLite software library.

Given the current technology trajectory in developing nations, we expect to see increased adoption of smartphones in the next three to five years. But in order to get early feedback on our designs and provide benefit to the BoP, we are interested in deploying Picture Talk as widely as possible in the near term as well. Thus, we have developed a voice-only version of Picture Talk in order to make the application available to people with lower end mobile phones. We have also created a web version, suitable for kiosk or telecenter use.

The Picture Talk voice version (the "voice client ") allows people using basic mobile phones, commonly found in BoP environments, to listen to and record discussion comments. Users with camera phones can also exchange pictures with the Picture Talk data server via MMS. A Picture Talk server-side component, built using the open source Asterisk® Public Branch Exchange (PBX) telephony toolkit, provides voice and telephone keypad Interactive Voice Response (IVR) interfaces, and uses the Picture Talk data server to access application objects.

Both web and voice clients access the same data as the Android client, but display that data in a suitable way for the target platform. For example, the rice plant anchoring the discussion in Figure 1 could be sent to the Picture Talk data server using MMS. Subsequently when another user wants to participate in the discussion, the server can send the picture to their phone in another MMS message. Users listen to voice comments over a normal voice channel. We had to develop additional server-side functions to transform the audio and image objects into formats usable by and optimized for both mobile phones and desktop computers.

Searching and browsing discussions is difficult using basic mobile phones, so we have explored how to perform these operations using voice only. The first feature we implemented of this nature is 'voice tagging'. When users create a discussion, they are asked to add optional 'voice tags.' A voice tag is a spoken word, phrase, or other short descriptive audio recording that is then associated with the discussion. Voice tags play a similar role to textual tags on the Web (e.g., Youtube, Flickr). Users can press touchtone keys to listen to tags, navigate to more specific tags, and add a new tag. After choosing a voice tag, the user can listen to the description of discussions and select discussions using a touchtone key. We have also implemented a form of 'voice bookmarking'. After listening to a voice post, the user can use touchtone to bookmark the post. Like social bookmarks on the Web, users share bookmarks with one another. We are just starting to explore how to provide visual proxies for voice tags and bookmarks on smartphones.

While the user experience on basic mobile phones is necessarily more restrictive than with smartphone or web browser clients, having the option of using these inexpensive mobile devices makes Picture Talk discussions potentially available to a broad range of BoP users. Further research is needed on methods of effectively finding and navigating relevant content, sharing information, and tapping into discussion databases on mobile devices. We hope that future mobile phone users at the BoP will have access to information services that have heretofore been usable only from desktop computers in the hands of experienced users.

FUTURE RESEARCH DIRECTIONS

Many of Picture Talk's features represent general capabilities that could be applied in a variety of contexts. In future work, we would like to apply the concepts to a wider range of mobile social computing applications.

Another area of future work is improving user identification. Like many social software applications, Picture Talk helps users share photos, contact information, and so on, with each other. However, in many developing regions of the world, it is common for mobile phones to be shared amongst members of a family or even an entire village. Mobile social computing applications in many BoP contexts may need to allow users to identify themselves to the system explicitly and in innovative ways (e.g., by selecting their picture or identifying a vocal sample) in order to work in contexts of shared phone use.

One additional area of research is the propagation of information by social networks. Picture Talk currently promotes participation and inclusion by providing an open forum for discussion. Furthermore, the ability to notify and invite people to participate, even people who are not currently registered users of the system, supports the possibility of "viral" growth of the Picture Talk user population. In developing regions where the idea of using technology to access information beyond one's social network may be unfamiliar, this kind of "viral spread" can be a key bridging mechanism. However, supporting the application at a larger scale involves creating mechanisms whereby users can find each other easily. We are exploring how to organize information about people in ways that can be easily appropriated by the population, such as by village or family.

Another problem that has arisen in our work is efficient and effective navigation through large amounts of audiovisual content. As a conversational system, Picture Talk's audio postings could quickly grow to an unmanageable size as more users access the system. Interfaces for time-varying multimedia e.g., audio and video, do not offer the same navigational affordances as visual interfaces (Muller, Farrell, Cebulka & Smith, 1992). In Picture Talk, we have mitigated this problem by anchoring aural information to metadata that is made explicit through photos and graphics. Ultimately we would like users to be able to easily switch between visual or voice menus organized by author, topic, time period, location, and so on. A more complete solution will no doubt ultimately be needed.

CONCLUSION

We are at an exciting point in the history of mobile computing. For the first time, the billions of people in some of the world's most economically disadvantaged countries have the promise of participating in the information revolution through mobile phones. If successful, this could bring about positive social, political, and economic change in regions struggling with illiteracy, disease, poverty, natural disasters, oppression, and other challenges. Enabling ordinary citizens to become both producers and consumers of information could facilitate the spread of critical information from one person to the next during crises, encourage broad exchange of ideas, connect experts with those needing help, strengthen social networks, and enable people at the base of the economic pyramid to become full participants in society and world economic markets.

We introduced Picture Talk, a mobile social computing application framework we designed for use in environments with low literacy rates, limited Internet connectivity, and user populations with little familiarity with information services. Picture Talk can be accessed over a voice channel from mobile camera phones available to many BoP populations, but we have also implemented a smartphone version, anticipating further penetration of mobile touch screen devices in developing regions, and a Web version, suitable for kiosk or telecenter use. The limitations of using mobile devices to access rich structured multimedia content by users with limited literacy skills exposes human-computer interaction challenges that are key to enabling broad access to information by people in BoP populations.

REFERENCES

Abraham, R. (2007). Mobile phones and economic development: Evidence from the fishing industry in India. *MIT Press Journal*, *4*(1), 5–17.

Appfrica (2008). *The current state of Internet penetration in Africa.* Retrieved June 15, 2009, from http://appfrica.net/blog/archives/248

Clark, H. H., & Brennan, S. E. (1991). Grounding in Communication. In L. Resnick, J. Levine & S. Teasley (Eds.), *Perspectives on Socially Shared Cognition* (127-149). Hyattsville, MD: American Psychological Association.

Danis, C. Bailey, M., Christensen, J., Ellis, J., Erickson, T., Farrell, R., & Kellogg, W. A. (2009) Social Computing Applications for the Next Billion Users. In *Designing Future Mobile Software for Underserved Users Workshop* at CSCW 2008.

Erickson, T., Kellogg, W. A., Laff, M., Sussman, J., Wolf, T. V., Halverson, C. A., & Edwards, D. (2006). A persistent chat space for work groups: the design, evaluation and deployment of loops. In *Proceedings of the 6th Conference on Designing Interactive Systems 06* (pp. 331-340). New York: ACM Press.

Froehlich, D. M., Rachovides, D., Riga, K., Bhat, R., Frank, M., Edirisinghe, E., et al. (2009). StoryBank: mobile digital storytelling in a development context. In *Proceedings of CHI 2009* (pp. 1761-1770). New York: ACM.

Gartner Group. (2009). *Gartner Says Worldwide Mobile Phone Sales Declined 8.6 Per Cent and Smartphones Grew 12.7 Per Cent in First Quarter of 2009.* Press released dated May 20, 2009. Retrieved June 15, 2009, from http://www.gartner.com/it/pa ge.jsp?id=985912

Gopakumar, K. (2006). E-governance services through telecentres: Role of human intermediary and issues of trust. *Information Technologies and Development*, *4*(1), 19–35.

Hollan, J., Hutchins, E., & Kirsh, D. (2000). Distributed cognition: toward a new foundation for human-computer interaction research. *ACM Transactions on Computer-Human Interaction*, *7*(2), 174–196. doi:10.1145/353485.353487

Hudson, J. H., Christensen, J., Kellogg, W. A., & Erickson, T. (2002). I'd be overwhelmed, but it's just one more thing to do. In *Proceedings of the SIGCHI conference on Human Factors in Computing Systems*, (pp. 97-104). New York: ACM Press.

Infotrend/CAP Ventures. (2004). W*orldwide Camera Phone and Photo Messaging Forecast: 2004-2009.* London: Kluwer. Retrieved June 15, 2009, from http://store.infotrendsresearch.com/PhotoGall ery.asp?ProductCode=MobileImagingStudy10106

ITU. (2009). *New ITU ICT Development Index compares 154 countries.* (press release dated March 2, 2009).http://www.itu.int/newsroom / press_releases/2009/07.html

Joshi, A., Welankar, N., Bl, N., Kanitkar, K., & Sheikh, R. (2008). Rangoli: A Visual Phonebook for Low-literate Users. *MobileHCI 2008*, Amsterdam, The Netherlands.

Kumar, A., Rajput, N., Agarwal, S., Chakraborty, D., & Nanavati, A. A. (2008). Organizing the unorganized: Employing IT to empower the under-privileged. In *Proceedings of the International World-Wide Web Conference* (pp. 935-944). Beijing, China: ACM Press.

Lyra Research. (2007). *The Worldwide Camera Phone Forecast.* Retrieved March 3, 2010, from http://lyra.ecnext.com/coms2/ summary_0290-801_ITM

Madamombe, I. (2005). Energy key to Africa's prosperity: Challenges in West Africa's quest for electricity. *Africa Renewal*, *18*(4), 6.

Muller, M. J., Farrell, R., Cebulka, K. D., & Smith, J. G. (1992). Issues in the usability of time-varying multimedia. In Blattner, M. M., & Dannenberg, R. B. (Eds.), *Multimedia interface design* (pp. 7–38). New York: ACM Press.

OLPC. (2009). *One Laptop Per Child.* Retrieved June 15, 2009, from http://laptop.org

Olson, G. M., & Olson, J. S. (2000). Distance matters. *Human-Computer Interaction*, *15*(2), 139–178. doi:10.1207/S15327051HCI1523_4

OSISA. (2009). *Electricity for Africa?* Retrieved June 15, 2009, from http://www.osisa.org/node/4164

Plauché, M., & Nallasamy, U. (2007). Speech interfaces for equitable access to information technology. *Information Technologies and International Development*, *4*(1), 69–86. doi:10.1162/itid.2007.4.1.69

Prahalad, C. K. (2004). *The Fortune at the Bottom of the Pyramid: Eradicating Poverty through Profits*. Upper Saddle River, NJ: Wharton School Publishing.

Samuel, J., Shah, N., & Hadingham, W. (2005). Mobile communications in South Africa, Tanzania and Egypt: Results from Community and Business Surveys. *Africa: The Impact of Mobile Phones. The Vodafone Policy Paper Series*, *2*(March), 44–52.

Seshagiri, S., Sagar, A., & Joshi, D. (2007). Connecting the 'Bottom of the Pyramid' – An Exploratory Case Study of India's Rural Communication Environment. *WWW 2007, May 8-12, 2007*, Alberta, Canada.

Srinivasan, J. (2007). The role of trustworthiness in information service usage: The case of Parry information kiosks in Tamil Nadu, India. In *Proceedings of the International Conference on Information and Communication Technologies for Development (ICTD),* (pp. 345-352). New York: ACM Press.

The Pew Research Center for the People and the Press. (2007). *The Pew Global Attitudes Project.* Retrieved October 4, 2007, from http://www.pewglobal.org

UNESCO. (2009). *UNESCO Institute for Statistics*. Retrieved June 15, 2009, from http://www.uls.unesco.org

Ushahidi.com. (2009). *Crowdsourcing Crisis Information (FOSS)*. Retrieved June 15, 2009, from http://www.ushahidi.com/

Yankelovich, W. W., Roberts, P., Wessler, M., Kaplan, J., & Provino, J. (2004). Meeting Central: Making distributed meetings more effective. In *Proceedings of the Conference on Computer-supported Cooperative Work (pp.* 419-428). New York: ACM Press.

ADDITIONAL READING

Heeks, R. (2008). ICT4D 2.0: The Next Phase of Applying ICT for International Development. *Computer*, *41*(6), 26–33. doi:10.1109/MC.2008.192

Parameswaran, M., & Whinston, A. B. (2007). Social Computing: An Overview. *Communications of the Association for Information Systems*, *19*, 762–780.

Zaslavsky, A. B., & Tari, Z. (1998). Mobile computing: Overview and current status. *Australian Computer Journal*, *30*(2), 42–52.

KEY TERMS AND DEFINITIONS

Visual Scaffolding: Images, words, or other visible elements with the sole purpose of supporting some primary user interaction e.g., tactile and/or aural user interfaces.

BoP: Stands for 'Base of the Pyramid' aka 'Bottom of the Pyramid'; a term referring to the largest but poorest economic group in the world with billions of people living on a small number of dollars per day (e.g., less than $1US/day).

Developing Regions: Parts of the world with a relatively low average standard of living when compared with advanced economies including parts of sub-Saharan Africa, South/Central America, the Indian Sub-continent, and southeast Asia.

Social Computing: An area of computer science concerned with intersection of social behavior and computation. This includes social behavior using computer systems and computing through the social behavior of large numbers of people.

MMS: Stands for Multimedia Messaging Services; a standard way to send messages that include multimedia content to and from mobile phones. A popular use is to send photographs from camera-equipped handsets.

Smartphone: A mobile phone offering capabilities beyond text message and voice communication, often including functions often found on personal digital assistants (PDAs) or personal computers (PCs), for example, a way to connect to the Internet to send e-mail or documents.

ICT: Stands for Information and Communication Technology; hardware, software, or networks in services of the creation, transfer, or storage of information in all its forms (e.g., text, audio, images, video) including over voice over telecommunications networks; often used as an extension of IT (Information Technology) and

ENDNOTES

1. Android is a trademark of Google, Inc.
2. G1 is a trademark of T-Mobile USA, Inc.
3. Asterisk is a registered trademark of Digium, Inc.
4. MySQL is a registered trademark of MySQL AB in the United States, the European Union and other countries.

Section 3
Mobile Technology and Culture

Chapter 13
Going Online in the PC Graveyard:
The Sociocultural Evolution of Japan's Mobile Internet

Inkyu Kang
Penn State University, USA

ABSTRACT

This chapter explores how the keitai (mobile) Internet has come to dominate Japan, marginalizing the PC-based Internet. The discussion focuses on the country's cultural and ideological aspects that have worked as driving forces behind its mobile Internet boom. Special emphasis will be given to the validity of the common belief that the Japanese language has been a barrier to PC and Internet diffusion. This chapter argues that the Japanese language has actually encouraged rather than discouraged Internet adoption in Japan. Relying on media comparative analysis, this essay examines how Japan has developed a qualitatively different Internet use pattern that cannot be revealed in statistics like "number of Internet subscribers" or "household Internet penetration rates."

INTRODUCTION

Japan has been known for its technological innovations for a long time. Its pioneering adoption of the mobile Internet has been no exception, reinforcing Japan's reputation as a "techno-wonderland." For example, NTT DoCoMo's *i-Mode* was nothing short of revolutionary; the world's first 3G network service turned Japan from an Internet graveyard into one of the most wired nations in just half a decade. Whereas Japan's Internet penetration rate was merely 37.1 percent in 2000, it soared to 68.5 percent by 2006 (ITU, 2003; MIC 2007, p. 28). Thanks to *i-Mode*, by 2005 Japan had risen to become the first nation where more people access the Web through cell phones than through PCs (Williams, 2006).

In spite of many efforts to export its home-grown innovations, however, Japan's mobile technologies have not taken off in foreign markets. As the *New York Times* reported recently, using the phrase "Galápagos syndrome," "Japan's handset makers have little presence beyond the country's shores" (Tabuchi, 2009a, p. B1). Even the legend-

DOI: 10.4018/978-1-60960-037-2.ch013

Copyright © 2011, IGI Global. Copying or distributing in print or electronic forms without written permission of IGI Global is prohibited.

ary *i-Mode* service has failed to take root outside the country, including in Europe, North America, and Asia, and its limited popularity overseas is fading rapidly. As an explanation for this, critics have pointed at the overly complicated hardware and the "walled-garden" approach of Japanese mobile Internet services.

Although technological Darwinism has become a popular framework for understanding the evolution and divergence of the information and communication technologies (ICTs) in Japan, specific sociocultural factors in this unique development have rarely been explored. There are several studies that investigate the Internet penetration in Japan, but they tend to explain it predominantly from an economic or business perspective. For example, economic wealth is considered as one of the most important predictors of Internet use (Beilock & Dimitrova, 2003; Kiiski & Pohjola, 2002; Norris, 2001; Wareham, Levy, & Shi, 2004). The problem is, however, that Japan never fits into this explanation. In spite of being one of the world's wealthiest countries, Japan was very slow to adopt the Internet until the mobile Internet gained popularity.

In an attempt to account for the exceptional case of Japan, Norris (2001, p. 55) lumped it in with countries that are "relatively affluent yet display lower than average Internet penetration rates" like "Bahrain, Kuwait, the United Arab Emirates, and Saudi Arabia, as well as South East Asian states such as Singapore, Japan, and Brunei." She went on to argue that "they can perhaps best be explained by the extreme inequalities of personal income and wealth characteristic of these states, possibly combined with some cultural restrictions on use through restricting Internet Service Providers" (Norris, 2001, p. 56). In reality, Japan is far from being a country with "extreme inequalities of personal income and wealth." Although its income gap has widened with its long-term recession, Japan is still one of the countries with the least earning inequalities in the world. The country's income inequality has consistently been lower than that of many other successful Internet adopters like the U.S. and Canada (Jones, 2007, p. 9).

Although potential cultural factors are touched upon in Norris's analysis, the approach is problematic, not only because it is unclear what cultural elements those countries have in common but also because cultural forces are seen as "restrictions" on technological diffusion. In this formation, culture hardly plays a central role, understood either as barriers to or as facilitators of the adoption of technology. To be explored later, however, the "cultural restrictions" on the PC-based Internet during the 1990s are the very vector that helped morph Japan into one of the most connected nations in the world in the following decade. Interestingly, the same cultural forces shaping "mobile Japan" that has been envied until recently for being "blessed with some of the best-looking technology in the world" are now coming under attack for turning the technological powerhouse into an isolated island (Fitzpatrick, 2007).

This chapter explores this irony, focusing on the sociocultural revolution of Japan's mobile Internet, which skipped the PC and moved right into the cell phone. Against econo-centric views, this essay argues for a central role of culture in technological dissemination, shedding light on multifarious social, cultural, and ideological forces that cannot be reduced easily to quantifiable variables. In doing so, it also takes into consideration that the diffusion and use of the ICTs are often exposed to the random nature of "historical accidents" that "can neither be ignored nor neatly quarantined" in the analysis of technological diffusion (David, 1985, p. 332).

The Japanese wireless Internet is analyzed in comparative contexts to see how the shaping of it was influenced by earlier communication technologies such as the landline phone, pager, and word processor. Special attention is paid to the impact of the hardware-type word processor, which fascinated many Japanese during the 1990s by allowing them their first keyboard experience, and which, ironically, ended up blocking PCs from infiltrating into homes and from becoming the major platform used by Japanese to access the Web. Some popular assumptions will also be debunked such as the beliefs that the complexity of the Japanese script slowed down PC and Internet adoption, and that mobility is the key factor behind the popularity of the mobile Internet in Japan.

DIFFERENT "INTERNETS"

When exploring the Internet, many scholars tend to tear it away from society by focusing on the internal characteristics of the technology. For example, the Internet is widely defined by universal features like "anonymity," "decentralization," and "lack of regulation" (Connery, 1997, p. 170; Jazwinski, 2001, p. 185; Wolfe, 2001, pp. 12-13). Unlike what the singular proper noun suggests, however, "the Internet" is never a homogenous entity shared around the world. For instance, the Internet in Korea is far from being anonymous. Korea's real-name system requires by law every Internet user to provide his or her name and resident registration number to leave postings and comments on Web sites accessed by more than 100,000 visitors a day. Many government and commercial portals allow people to write opinions under their real names only after collecting extensive information about the registered users including their home address, email address, and land-line and cell phone number.

If the Internet lacks anonymity, it is not difficult to regulate the space as shown by the arrest of well-known blogger "Minerva" in 2009 after making unfavorable commentaries on the South Korean government's economic policy. In addition, most online services in Korea have built-in mechanisms of censorship such as *geum-chik-eo* (금칙어) or "banned words." A post can be rejected if it includes specific words or expressions considered crude or offensive. Some users skillfully get around the barrier and make their original points by slightly modifying the taboo words. Such efforts are not always rewarded, however, under the watchful eyes of the Web administrators.

Hardly decentralized is the Internet both in Korea and Japan. A common pattern found in those two countries is heavy reliance on commercial portals that provide one-stop access to a wide-range of information including news, blog, video, dictionary, game, shopping, etc, not to mention search and email. The centralization of the Net by a few dominant portal services partly explains the relative invisibility of Google in the two East Asian neighbors. In Korea, Google has just 2 percent of the market while domestic portal Naver's market share is 64 percent (Moon, 2010, p. 53). Google's performance in Japan is not very impressive, either. It receives only 33.7 percent of all search queries compared with 56.5 percent held by Yahoo's Japanized counterpart (Tabuchi, 2009b, p. B1). The centralization of the Internet is reinforced in the public sector, too. For instance, the Japanese government plans to create a comprehensive "one-stop public services portal" by 2013 (Low, 2009).

Another element that sets the Japanese Internet apart is its predominant platform to go online. Since the Internet started being embraced en masse from the 2000s in Japan, the nation has nurtured an Internet culture of radical difference. As mentioned, Japan's Internet penetration rate skyrocketed to 68.5 percent in 6 years from a mere 37.1 percent in 2000 (ITU, 2003; MIC 2007, p. 28). During this period, ironically, Japanese 20-year-olds using PCs to access the Internet actually dropped from 23.6 percent to 11.9 percent, meaning "20-year-olds now make up the same proportion of the total as 50-year-olds" (Norrie, 2007).

Because the Internet Age was brought by cell phones, Japanese Internet services have naturally evolved around mobile experience with portable devices with small screens. Mobage Town, Japan's largest mobile-only social networking service, for example, launched its PC version belatedly in 2008, but it maintains the original proportion formatted for the vertically long cell phone display, which is an odd choice for those unfamiliar with the mobile site (Figure 1).

What makes Japan's Internet different is not just the user interface design. The mobile-oriented Internet environment has spawned curious results. A good example is *keitai shousetsu* (携┌ 小┌) or "cell phone novel," a new literary genre influenced by mobile emailing. Cell phone novels literally mean novels written and read on cell phones. The short novels without serious plotting or character development are extremely popular; Five out of ten best-selling novels in 2007 in Japan were originally cell phone novels (Onishi, 2008).

Although referred to in the same way, the Internet used in different countries is strikingly different. So it would be fair to say that the Japanese and foreign users use different "internets." A comparative study of the Internet, in this sense, should be not only a *quantitative* but also a *qualitative* exploration of the technology. According to a 2009 survey by Japan's Ministry of Internal Affairs and Communications (MIC), about 75.3 percent of Japanese use the Internet (MIC, 2009). Statistically, Japan's Internet penetration rate is one of the world's highest, but there is an "enormous discrepancy between official statistics and empirical observation" (Clark, 2002; ITU, 2007).

About two-thirds of the Japanese "Internet users" log in from mobile phones, and they tend to show different use patterns from their PC-based counterparts. (Tabuchi, 2009b, p. B1). For instance, Internet usage on computers is 18.9 hours, but mobile users are likely to spend much less time on the Internet, averaging 8.1 hours per month (comScore, 2007). The difference is not just the amount of time spent online. The mobile Internet is used differently from the PC Internet, serving unique social functions in Japan (Ishii, 2004, p. 57). In Japan, the Internet has evolved into a highly private medium by combining with a personal *keitai* (portable) handset instead of a PC usually shared among family members. The mobile Internet, as a personal medium, has two major functions in Japan: entertainment and selective sociality.

People tend to use the same terms—"Internet," "surfing," and "email"—to refer to mobile experiences, but "surfing" on mobile devices cannot be the same as "surfing" the World Wide Web on a computer. PC users visit more varied websites than mobile users, and they tend to use different kinds of information. According to Ishii (2004, pp. 49), in Japan, PC-based Internet users access an average of 8.64 categories of Internet sites, while their mobile counterparts visit only 3.58 categories. The differences between the two groups are not just in the diversity but also in the quality of the information they pursue. In general, mobile users prefer lifestyle information while PC users prefer business information. About 85.2 percent of Japanese PC users visit search sites, but only 45.3 percent of their mobile counterparts visit them. Only fortune telling (23.5%) and matchmaking (8.0%) sites are visited more frequently via mobile phones than PCs, 16.7% and 3.54%, respectively (Ishii, 2004, p. 49).

The different use patterns are found not only in Japan but also in the U.S. According to a joint research by Telephia and comScore, PC users are more likely to use search sites like Google (56.7%) and Yahoo! (43.8%) than are their mobile counterparts. Mobile users surveyed visit the same sites much less: Google (13%) and Yahoo! (11.1%). The difference in the use patterns between PC and mobile users can partly be explained in terms of hardware restrictions. The display and input capabilities of mobile devices are severely limited compared to those of computers (Halvey, Keane, & Smyth, 2006). The interface limitations of the mobile Internet are likely to restrict information-seeking behavior and encourage passive and entertainment-oriented use patterns. Such hardware limitations, however, cannot fully explain Japan's unique evolution of Internet technologies and unique use patterns. Since many Japanese users had not experienced PC-based

Figure 1. Mobage Town's PC site

Internet before turning to the *keitai*, it would be fair to say the scaled-down version of the Internet was good enough for the specific needs of the Japanese market.

This does not mean that Japanese mobile users enjoy a limited amount of information. Mobile Internet services are full of interesting and useful information, and the number of carrier-controlled official sites has increased rapidly. For example, official sites of NTT DoCoMo's *i-Mode* were over 65,000 in 2006, which was dramatic growth from merely a few hundred when the service was first launched (NTT, 2006, p. 1). There is no denying, however, that Japanese use different "internets." The unique aspect of the Japanese mobile Internet is also found in their email use. Unlike PC-based email, mobile "email" in Japan represents a closed world, exchanged exclusively among those with existing relationships (Clark & Kay, 2005, p. 109). It is a serious mistake to assume that the Internet in Japan is the same as the one called *the Internet* in other countries.

JAPANSESE LANGUAGE: A BLESSING IN DISGUISE?

One of the most significant differences between Japan and other phonetic alphabet-using nations in terms computer penetrations is the way the portable word processor was accepted by the general public. In Japan, it was the electronic word processor instead of the typewriter that gave the general public the first typing experience. Japan has a relatively complicated writing system composed of two kinds of phonograms called *kana* (Hiragana and Katakana) and Chinese characters *kanji* (Ohiwa, 2004, p. 147). What complicates keyboard input is the existence of *kanji* consisting of a vast number of characters.

Commonly used *kanji* characters are over 2,000, which makes it hard to type Japanese words by combining limited keys. Due to this complexity, Japan was unable to develop typewriters that were cheap and easy enough for average people to buy and learn. As a result, many Japanese had to wait until the mid-1980s, when the industry started mass-producing affordable stand-alone word processors. When people in other countries were busy experimenting with personal computers, Japanese society was fascinated with word processors, the major breakthrough coming after a decades-long struggle. In 1984, the *New York Times* wrote:

While increasing numbers of American office workers are adapting to the computer keyboard after years spent at the familiar typewriter, their Japanese counterparts face a more fundamental question: How do you use a keyboard? The answer, if it is a keyboard for the Japanese language, is not easily. Some versions, with hundreds of keys and thousands of characters, are enough to give a typist nightmares. Indeed, as the Japanese struggle to automate their offices and develop their computer industry, one of the most formidable obstacles has been their own language. Only recently have Japanese computer companies managed to develop practical word processors for their own language. (Pollack, 1984, p. D17)

Japan did invent its first mechanical typewriter as early as 1915. However, it was more of a mini-printing machine than a typical desktop typewriter. For example, an "innovative" typewriter patented in 1929 by Kyota Sugmino had a character carriage composed of more than 2,000 *kanji* characters. A professionally trained typist had to pick every single character manually from the carriage, which was raised by a type bar and moved to a proper position before hitting the paper (Japanese Patent Office, 2002). Even expert typists could only achieve a speed of about 53 characters per minute on such *kanji* typewriters. Even simpler *kana* typewriters required lengthy training of between six months and a year (Gottlieb, 2000, p. 25). Until Toshiba made the prototype of the JW-10 in 1978 based on its own *kana-kanji* conversion system, most formal government and industry documents were produced on similar all-character typewriters by specially trained typists.

Difficulties involved in typing Japanese are not limited to the number of characters keyboards have to deal with. Since Japanese writing does not have word breaks, the *kana-kanji* conversion system must be made to morphologically analyze each phrase. The automatic system is not perfect, so the user has to keep intervening in the conversion process so that the system properly reflects the typist's intentions. In addition, Japanese has a lot of homonyms, making typing more brain-intensive than the usual keyboard input in other languages.

So just how does one type 6,000 characters with one finger? First, as with PC input, a Japanese writer must input characters from a syllabary of about 50 phonetic characters called Kana that represent the sounds that make up all Japanese words. [...] With Japanese, inputting the phonetic sounds is just the first input step, and in some ways it is the easier one. Next, the writer must convert some of these syllables into characters chosen from about 6,000 Chinese-derived Kanji characters. This task is complicated by the absence of word breaks in Japanese text. The conversion software, not infrequently, will guess wrong about where you intend word breaks and will offer a string of

Figure 2. Japanese typewriter

Japanese words—all perfectly correct word by word, but which are utter nonsense when strung together. A simple example, if English had no word breaks, would be if you typed hisid, intending to write hi Sid, and the software displayed from its dictionary of valid words the expression his id (Kay, 2004).

Although it is true that the Japanese language makes typing tougher than phonetic alphabets, it is overly simplistic to argue that the Japanese script itself has hindered people from adopting computers and the Internet. Typing Japanese on a PC keyboard may be cumbersome, but it would be even more so on a mobile phone with fewer keys. If so, how can we explain the emergence of *oyayubi-zoku* (親指族), or "thumb tribe" in Japan who masterfully tap out countless emails with one or two fingers? If the input system had been a major factor behind the slow adoption of the PC-based Internet in Japan, the dissemination of the *keitai* Internet should have been slower. On the contrary, Japan's Internet boom started with the introduction of mobile Internet services. "Japan was extremely slow in Internet takeup," but the *keitai* Internet brought a drastic change to the people who "weren't using the Net at all" (Beck & Wade, 2003, p. 123; 124; Clark & Kay, 2005, p. 107).

Difficulty is in the eye of the beholder, and it is truer when it comes to a native language. Those already familiar with other language input systems may find the Japanese version annoying, but it is not the case for the majority of Japanese users. What mattered more was the unique typing experience rather than the difficulty it involved. Japanese word processing on a regular keyboard became possible with advanced memory technologies. Japan was a world leader in that field, giving birth to the electronic word processor, but the innovative machine ended up being a distraction in the upcoming PC market. The hardware-type word processors satisfied the potential need that could have driven PC sales, but the people who had already invested in the writing machines were understandably hesitant to buy similar but still more expensive machines. Furthermore, unlike PCs, electronic word processing machines came with built-in monitors and printers.

Contrary to common wisdom, language-related difficulties in typing have not discouraged Internet use in Japan. They rather sparked explosive growth in Internet usage—not on PCs but on hand-held devices. Japan had long lagged behind in Internet acceptance, but things started changing when NTT DoCoMo launched the innovative *i-Mode* service in February, 1999. If Japanese typing had been as easy as its Roman counterpart, thumb-typing on mobile handsets would have felt very painful for Japanese users. Many Japanese, however, did not find much of a difference between the desktop keyboard and the mobile pad. Typing Japanese words, even on a full-sized QWERTY keyboard, requires complex learned procedures such as using arrow keys to choose correct *kanji* words. The commonly used method is typing phonetically on a *kana* or English keyboard, and the computer monitor shows several options of *kanji* words corresponding to a specific sound. The user then picks the target word from the list, manually. Such complexity forces even professionally trained typists in Japan to type fewer than 60 words a minute, while advanced typists in other countries using Roman alphabets can do more than 100 words a minute (Gottlieb, 2000 p. 25; Pollack, 1984).

Any text-entry methods on mobile devices are felt restrictive by most phonetic alphabet users due to the size of the keypads and limited use of the fingers. The typing rates of mobile phone users are two to five times slower than those of desktop keyboard typists (Starner, 2004, p. 97). The problems are not limited to speed; mobile users tend to make more mistakes in typing than their full-sized keyboard counterparts. Those using mobile input devices "tend to make the same or similar mistakes made by physically handicapped computer users" such as "[p]ressing the wrong keys, pressing the same key repeatedly by mistake, clicking the screen multiple times in error" (Barak, 2008). Many hand-held devices are now equipped with scaled-down QWERTY keyboards, but the issues of tiny button size and spacing have not been resolved. Since only two thumbs are allowed while typing on a mobile phone, the speed is similar to "hunt-and-peck" typing averaging around 23 words per minute (Starner, 2004, p. 100).

Hunt-and-peck typing has been used widely in Japan, even among professional typists, since the non-mobile era (Ohiwa, 2004, p. 147). In a sense, Japan has long been ready for mobile Internet technology. As a result, Japanese users have a definite advantage when it comes to text input on mobile devices. According to a quantitative study, Japanese college students can type 17 Japanese words per minute on a mobile pad, which is not much different from the 23 words per minute they can type on a desktop keyboard. Their English typing is much slower, however, both on mobile devices and on full-sized PC keyboards, and the gap between the two methods is much more significant: 5 and 14 English words per minute, respectively (Houser & Thornton, 2004). The Japanese typing rate of 17 words per minute on the mobile phone is quite impressive because the world's fastest mobile texter in 2003 typed 29 words per minute, which is six times slower than the desktop keyboard typing record of 192 words per minute (Starner, 2004, p. 97).

Many scholars have focused on the negative impact of the Japanese language on Internet penetration in Japan. It is commonly believed that difficult and slow Japanese input has had deterring effects on Internet use. More emphasis should be placed on how Japan's relative lack of keyboard input has paved the way for mobile text entry. The "two-stroke method" or "hunt-and-peck" typing had been used for a long time in Japan even before mobile phones became commonplace. Such experiences enabled a soft landing for the *keitai* Internet, making cell phone input "the de facto standard" in the early 2000s (Kay, 2004). The claim that Japanese scripts are responsible for the slow PC take-up needs to be verified, due not only to the successful launch of the mobile Internet, but also to the phenomenal popularity of word processors in Japan. If typing had been an issue, word processors, as well as computers, should have had difficulties in attracting users.

By 1993, 37.8 percent of Japanese households owned stand-alone word processors while only 11.9 percent had PCs (OECD, 1997, p. 87; Gottlieb, 2000, p. 3). In the same year, 47.3 percent of primary-school students and 46.3 percent of middle- and high-school students had word processors, but PC penetration rates were merely 20.5 percent and 22 percent, respectively, among primary-school students and middle/high school-students (Gottlieb, 2000, p. 4). PC ownership grew gradually, but more households had word processors until the late 1990s. For example, while over 40 percent of Japanese homes had a word processor by 1997, the percentage of PC-owning households was slightly over 20 (Gottlieb, 2000, p. 5; OECD, 2002, p. 191). When the *i-Mode* mobile Internet service was launched in 1999, attracting more than 10 million users in one and a half years, the PC was unfortunately sandwiched between the word processor and the cell phone (NTT, 2000).

So it would be more proper to say the potential demand for PCs has been absorbed by word processors and mobile phones than to argue that PC take-up was slowed down by language input issues. In Japan, PCs have competed with word processors and mobile phones—unlikely rivals in many countries. Two of the major functions of the computer, documentation and communication, have been performed successfully by the less expensive and more personal machines. Another major use of PCs is gaming, but it also failed to drive sales in Japan. Gaming consoles had already been popular since the late 1980s, when average people were not interested in computers. Nintendo released its 8-bit *Nintendo Entertainment System* (NES) in 1983 and the world's first hand-held video game console *Game Boy* in 1988, which became enormous hits. Since gaming, word processing, and communication had been done by separate machines, it is not surprising that PCs had difficulty appealing to Japanese customers.

PERSONALIZATION OF KEITAI

What has made the mobile Internet such a phenomenal success in Japan? Most importantly, the technology has effectively turned the Internet from a public medium into a personal one. In Japan, like a television set, a PC usually belongs to a family instead of to a specific individual. The popularity of mobile phones is indebted to privacy issues within homes (Clark, 2004; Ito, 2005). In other words, cell phones allow the Japanese to lead their "own lives" by giving them "personal space," by overcoming the problems of "cramped housing, lack of open spaces, and more surveillance at the workplace and in educational sites" (McVeigh, 2003, p. 27).

Many discussions of Japanese mobile Internet emphasize the Japanese lifestyle, especially the long hours spent commuting. This may have had an impact, to a certain degree, but it fails to account for the widespread failure of mobile Internet services in countries where long-distance commuting is not uncommon. In Korea, Japan's closest neighbor, for example, many people go to work or school using public transportation systems such as the bus, subway, and high-speed train, but the country has been significantly slower than Japan in adopting the mobile Internet.. It is a mistake to assume that mobility is the key factor behind the popularity of the mobile Internet. A quantitative study shows that the mobile Internet is used predominantly at home (76.8%), rather than in a car (24.0%) or on the train or bus (4.1%) (Habuchi, Dobashi, Tsuji, & Iwata, 2005, p. 102).

What matters more is the personalization, rather than the mobility, of the Internet platform. This point is proven by the fact that the number one reason people visit mobile Internet sites is to personalize their handsets. The most frequently visited sites are ringtone download sites (68.5%) and wall paper download sites (32.5%), which are basically to decorate and customize *keitai* terminals (Habuchi, Dobashi, Tsuji, & Iwata, 2005, p. 101; Matsuda, 2005, p. 124). It is understandable since mobile devices, especially the *keitai*, have been attached to individuals rather than to groups or institutions (Ito, 2004). A mobile phone not only belongs to a specific individual, but also enables the user to screen calls, increasing individual selectivity (Matsuda, 2005, p. 123). In Japan, the *keitai* has become the symbol of personal communication.

Users stated that they would never answer the phone of another individual, even of a spouse, and looking at another person's handset uninvited is socially unacceptable.

Phones are now almost always attached to individuals, and are valued as a guaranteed mechanism for connecting with a particular person, in sharp contrast to the home or office phone which presents the risk of someone else answering. (Ito, 2004)

It is important to remember, however, that the meaning of the *keitai* as a personal medium in Japan has never resulted from the nature of the mobile technology. The discourse has emerged in a complex socioeconomic setting of diverging interests and ideological underpinnings. As Okada (2005, p. 42) convincingly suggests, the *keitai* was originally developed as a "shared telephone" for organizational purposes. A *keitai* used to be leased to a representative individual within a group of people, such as a manager at a construction site with no fixed telephone line. It was very common for someone else to pick up the *keitai* when the owner could not answer it (Okada, 2005, p. 43).

The meaning of *keitai* has changed gradually, influenced most significantly by "the shift from a lease to a purchase model, which began in April 1994 and became standard in 1996" (Kohiyama, 2005, p. 64). This transition was driven partly by the elimination of deposit requirements, the reduction or waiving of activation fees, the introduction of affordable pricing policies, and the increased diversity of mobile phones in the market. Ownership through purchase has had a considerable impact on the change in the implications of the mobile phone from a public medium to a personal one.

In Japanese, the mobile phone is called a keitai, which might be roughly translated as "a portable," or "something you carry with you." In contrast to "the cellular phone" or "the mobile" which stress technology and function, the Japanese term stresses the relation between user and device. A keitai is not so much about a new technical capability or freedom of motion, but about a snug and intimate technosocial tethering, a personal device supporting communications that are a constant, lightweight, and mundane presence in everyday life. (Ito, 2004)

Before mobile phones became prevalent, pagers went through similar personalization from a business tool to a personal communication device. Electronic beepers maintained popularity among young users long after mobile phones started gaining popularity (Okada, 2005, p. 45). Some pager lovers even called mobile phones "tasteless" (Matsuda, 2005, p. 23). One of the major reasons was the pager was still considered *the* personal communication tool even though a better technology had come along. Pagers, commonly known by the brand name "Pocket Bell" (*poke beru*) in Japan, could be easily bought even by high-school students, unlike mobile phones were leased for business or personal purposes. A loaned device should be handled with caution. It is hard for users to personalize leased products by putting on stickers, inscribing names, or scratching them in a creative way.

The smart naming of the device also helped many Japanese to accept the business tool as a personal device. The name "Pocket Bell" may not sound impressive to native English speakers, but it effectively incorporated the beeper into the discourses of personal intimacy. Such individualization has happened to many other media such as radio, stereo, and television (Okada, 2005, p. 46).

Pocket Bell successfully followed the footprints of the *Walkman* and *Watchman* by establishing an association between a new technology and the familiar discourses of personality and individuality. Such individuality is represented often by the English word "my."

There is an entire genre of words like "my home" and "my car," for example, meaning individually owned cars and homes. A Japanese might ask a colleague if he is a my car commuter or takes the subway. Taking this a step further, one grocery store chain offers strawberry My Jam. (Kristof, 2005)

Mobile phones have replaced beepers—and eventually PCs—as the symbol of personal communication since the mid 1990s. The meaning of personal that the *keitai* came to bear has had a sizable impact on the use of the mobile devices. For example, the primary use of the *keitai* Internet is sending and receiving email (Clark & Kay, 2005; Ishii, 2004; Matsuda, 2005; Miyata, 2006). According to a survey, 57.7 percent of *keitai* users use the email function. The younger the users, the greater their reliance on mobile email: 89.2 percent for teens (12-19), 82.7 for those in the twenties, and 66.8 for those in their thirties (Matsuda, 2005, p. 124). In particular, female users tend to use mobile phones more as an Internet terminal (69.5%) than as a telephone (Habuchi, Dobashi, Tsuji, & Iwata, 2005, p. 100).

The *keitai* in Japan works on two different levels; one is a way to express personal statement, and the other is a means of personal communication. As discussed previously, Japanese Internet users access mobile Internet sites mainly to download ringtones and wall papers to personalize their machines. They function similarly to the mobile phone stickers and fashion straps that have grown into a huge industry.

The keitai is a locus that integrates Japanese subcultures. Young people use the latest J-Pop (Japanese pop music) for ringtones and Japanimation (Japanese anime and anime characters) like Pokémon, Doraemon, and Sen/Chihiro from the movie Spirited Away as mofits for wallpaper and keitai straps. A keitai is for them more than just a tool—it is something they are highly motivated to animate and to customize as a dreamcatcher, a good luck charm, an alter ego, or a pet. I might even call it an idol or a fetish and regard it as an animistic handy object (tokko) stretching a spiritual barricade (kekkai) around the body. (Fujimoto, 2005, p. 87)

The discourse of the *keitai* as a fashion statement has been reinforced as mobile manufacturers have introduced items in various styles and colors. The industry's effort to tailor products to personal taste is well reflected by advertising campaigns such as "link your style" (Tu-Ka) or "find your style" (KDDI) (Hill, 2003, p.192). The personalization of mobile devices can be found in any society, to some degree, but such practices are limited mostly to young female users.

In many countries, PCs, especially laptops and netbooks, have been associated with an individual by enabling personal access. In Japan, however, the mobile Internet has thrived at the cost not only of the PC-based Internet but also of computer use in general as mentioned previously. As a result, Japan's PC market is shrinking, making some manufacturers such as Hitachi stop producing computers for individual customers. Many Japanese PC makers, in an attempt to maintain their presence in the Japanese market, have focused on non-functional aspects such as design and color by making "PCs look more like audiovisual equipment—or even colorful objects—than computers." (Tabuchi, 2007). Sony's launch of "Lifestyle PC" laptops, which have received a lukewarm response in domestic and foreign markets, shows the dilemma of going online in the PC graveyard.

LEADING PRIVATE LIVES IN PUBLIC SPACE

In Japan, cell phones not only have a stronger connotation of individuality, but also are under tight social restrictions. Calling and receiving calls on cell phones in public is generally tolerated, if not welcomed, in many countries. On the other hand, many Japanese find it offensive as Okabe and Ito (2005, p. 105) describe:

Suppose that a ring tone breaks this silence, or somebody sitting in a subway car starts a keitai conversation. Most likely, people nearby will glance quickly at the source of the noise. If the offender speaks particularly loudly, she may get a glare or an expression of disapproval (even if there are people chatting more loudly in the next seat). This kind of scene is a familiar one in everyday life in urban Japan.

Calling someone on his or her *keitai* runs the risk of making the receiver a social offender in Japan. Many Japanese understandably see text messaging as a way to "converse on a train without bothering others" (McVeigh, 2003, p. 24). "Although many passengers type into *keitai* keypads or scroll through pages on tiny screens, nobody talks on *keitai*" (Okabe & Ito, 2005, p. 105). Mobile email was a "killer application" that has made cell phones an indispensable part of daily life by enabling Japanese to lead private lives in public space (Clark & Kay, 2005, p. 108). Commuters are frequently reminded that "Passengers are requested to refrain from using cell phones in trains" and "Thank you for not using your mobile phone" through signs and announcements in stations and trains (McVeigh, 2003, p. 29).

It is not surprising there is such a strong resistance in Japan to using mobile phones in crowded places. This is partly due to the well-established social manners of Japan where "[e]ven the sounds leaking from a young person's Walkman are considered a violation of this norm of silence" in public space (Okabe & Ito, 2005, p. 105). It is also involved in the widespread belief among Japanese that radio signals from mobile devices affect pacemakers worn by nearby passengers. In March 2000, the Tokyo metropolitan government decided to announce a much stronger message on subway trains: "Please turn off all cell phones during crowded hours" ('Phone problems', 2000).

Figure 3. Tokyo Metro's "Please do it at home" campaign posters

Nagoya soon followed by shutting down cell phone base stations on its subway platforms, fearing that hand-held phones may disrupt cardiac pacemakers (Edwards, 2004).

These fears are not completely baseless, since some medical studies show that "cellular phones that use digital signals can interfere with the rate of pacemakers" (Landler, 1995). Many Japanese, however, simply have their cell phones in silent mode on trains or buses instead of switching off so that they keep sending and receiving emails (Okabe & Ito, 2005, p. 215). Cell phone restrictions apply solely to making sounds such as voice calls and ring tones. So the negative attitudes towards public cell phone use should be understood as having a social rather than scientific origin.

MOBILE INTERNET AS AN EASIER AND SAFER ALTERNATIVE

The widespread diffusion of the mobile Internet in Japan has been facilitated by various ideological factors. One of these is the negative discourse of the Internet as a "lawless area" full of "libels, slanders, and even terror announcements" (Takeuchi, 2005). According to a survey, only 25.3 percent said that "most of or all of the information on the Internet is reliable and accurate." On the other hand, 69.7 percent of Koreans surveyed showed the same response ('World Internet Project', 2004). Such negative perceptions of the Internet in Japan have contributed to the general sense that the Internet is a "dangerous" place (Ohta, 1999). A 2006 survey shows that 78.6 percent of Japanese Internet users feel reluctant to enter their credit card number on websites (japan.internet.com, 2006).

The mobile Internet has thrived on such negative images of the PC-based Internet by providing a safer alternative. Unlike the PC-based Internet, the Japanese mobile Internet constitutes its own "walled garden" (Baker & Megler, 2001, p. 5; Clark, 2001). Mobile sites accessible via *keitai* phones are composed of "official sites" validated and approved by carriers, and "unofficial sites" the carriers do not take responsibility for. Official sites can be easily reached by clicking on hierarchically structured menus in a carrier's portal, but unofficial sites can be accessed only by entering the URL at the user's own risk. Each carrier controls its own official sites, blocking subscribers of the other mobile services from using its content (Baker & Megler, 2001, p. 4).

Furthermore, many of the mobile sites are inaccessible through the PC-based Internet, and vice versa. As Clark and Kay (2005, p. 107) point out, "the carriers have created separate 'Internets' that indeed offer e-mail communication through the Internet but are built of Web sites closed to non-subscribers." The "walled garden" approach, however, has given many users a sense of security.

NTT DoCoMo designed content to ensure that it would have features young women like—shopping sites, ring-tone downloads, and easy email capability. In addition, the "semi-walled garden" made young women feel safe: they would not be exposed to inappropriate content, yet they would not be prevented from searching further afield if they chose. (Baker & Megler, 2001, p. 5)

For many Japanese, mobile phones were an ideal alternative to PCs considering price, ease of use, privacy, portability, and intimacy with users. In addition, as stressed previously, the Japanese language did not pose additional difficulty in typing on a mobile keypad. With mobile phones replacing PCs to access the Internet, many users started forsaking the old machines. In Japan, the Internet-enabled *keitai* is a continuation of interpersonal communication devices like the pager and the cell phones rather than the PC. It should be noted here that the most popular application of the mobile Internet is to email instead of Web search. This fact shows that the needs of Japanese Internet users are more relationship-oriented than information-oriented. This explains why the demand for the Internet, which had been sluggish on PCs, exploded suddenly with the launch of the *i-Mode* service. Many Japanese use the mobile Internet as an always-on interpersonal communication tool, finding its core value in email, the unlikely "killer application."

Despite the striking differences between mobile handset use and Internet access from desktop PCs, the real killer application for Internet-enabled mobile telephones is exactly the same as the most popular application for PCs connected to the Internet: electronic mail. Japanese young people, in particular, adopted cellular e-mail communications with even more fervor than their European and U.S. counterparts embraced Internet-based instant messaging. It is not uncommon for 13-year-old girl in Japan to send and receive 100 e-mail messages per day via mobile handsets, mostly in the form of protracted, line-by-line conversations about school, meals, boys and extracurricular activities. Older teenagers, college students and even working adults are nearly as enthusiastic about mobile e-mail, and while the content of their communications may be less juvenile, it is still primarily personal rather than professional. (Clark & Kay, 2005, pp. 108-109)

Mobile Internet users, in Japan, tend to send more emails to fewer addressees than their PC counterparts. According to Ishii (2004), mobile email users in Japan send 25.9 emails a week while PC-based Internet users send just 15.2 emails. Interestingly, mobile email is sent to a smaller number of people (5.5) than PC-based email (8.6).

Qualitative differences are also found between PC and mobile emails. While mobile email is sent predominantly to friends people meet on a regular basis (58.4%), PC email is sent mostly to friends hard to meet up with regularly (49.6%). Mobile email is used less frequently for business purposes than PC-based email. While only 10.8 percent of mobile email is used for business contacts, almost three times as many emails are sent via PCs (31.6%) for the same purpose. These results support that, in Japan, mobile email works mainly as a social cement to reinforce existing relationships.

CONCLUSION: "JAPANESE-STYLE" IT REVOLUTION

As discussed, the Internet flourished on *keitai* devices as an interpersonal medium that succeeds the beeper and the cell phone, and as a personal statement that succeeds the *Walkman*. Behind the massive popularity of the *keitai*-based Internet were also patriotic discourses such as "*Keitai* are saviors of the Japanese economy" and "*Keitai* represent a Japanese-style IT revolution" (Matsuda, 2005, p. 32).

In the background of these discourses, there is a common understanding that Japan has fallen behind the United States, Scandinavian countries, and Korea in the areas of personal computers and the Internet. Hence, with the prolonged recession since the 1990s as another contextual aspect, the mobile Internet, has been referred to as an opportunity. [...] [T]he mobile Internet is quite often discussed in the context of the ideology of Japanese uniqueness. Such discourse naturally remind us of "the glory of the past" when Japanese small cars and electronics dominated the world. (Habuchi, Dobashi, Tsuji, & Iwata, 2005, p. 96)

Ideologies of *keitai* technonationalism have been articulated through "Japaneseness." While Western rivals are good at dealing with large devices such as computers, the Japanese, who are "naturally more concerned with precision," excel at manufacturing tiny, high-tech devices (Habuchi, Dobashi, Tsuji, & Iwata, 2005, p. 96; Morita, 1986 in Mastuda, 2005, p. 34). The discourse of "Japanese uniqueness" has been associated with the mobile-based wireless Internet since it reminded them of "the glory of the past" made possible by precision engineering (Habuchi, Dobashi, Tsuji, & Iwata, 2005).

The *keitai* Internet has also been described as a solution to the digital divide plaguing many countries using the PC-based Internet. *Keitai* phones have been offered as cheaper and easier terminals to access the Internet than PCs. Although the *keitai* Internet is not quite the Internet in the traditional sense, the explosive increase in "Internet users" shows that it has been accepted favorably by many Japanese.

Traditional socioeconomic analyses of the Internet fail to take into consideration these semiotic aspects. Furthermore, various historical accidents make it even more difficult to apply a one-size-fits-all model to technological evolution. As discussed, when the PC was introduced to customers in Japan, its people were already fascinated by the hardware word processor that allowed them their first keyboard experience. PC ownership grew gradually, but until the late 1990s more people used their beloved word processors. The Internet had to wait until 1999 when the *i-Mode* mobile Internet service was launched, attracting more than ten million users in one and a half years (NTT, 2000). The PC came to Japan unfortunately sandwiched between the word processor and the cell phone and was destined never to see its heyday in the country.

Such unexpectedness and randomness of history makes it virtually impossible to make a generalized prediction of the fate of a technology. For this reason, David (1985, p. 332) concluded that "it is sometimes not possible to uncover the logic (or illogic) of the world around us except by understanding how it got that way" because "[a] *path-dependent* sequence of economic changes is one of which important influences upon the eventual outcome can be exerted by temporarily remote events, including happenings dominated by chance elements rather than systematic forces." Such difficulty is complicated even more by the symbolic and ideological values certain technology comes to bear for members of a society.

The different trajectories of ICTs in Japan prove that technology does not exist as a discrete entity but as a discursive formation that encompasses multifarious social, cultural, economic, and most importantly, ideological forces. Any technology involves meaning-making processes in which various social agents participate with different—sometimes harmonious, sometimes conflicting—intentions. As Baudrillard (1998) suggested, the consumption of any object entails signs and values, and technology cannot be an exception. It would be a serious omission to turn a blind eye to how a technology is made to have a meaning within a society. As explored in the case study of Japan, different discourses of the Internet have led to different adoption rates, and eventually to different technologies of "internets." This fact shows that the adoption and use of a technology is a subject properly dealt with not only by sociologists but also by semioticians.

REFERENCES

Baker, G., & Megler, V. (2001). *The semi-walled garden: Japan's "i-mode phenomenon*. New York: IBM.

Barak, S. (2008, July 2). Mobile users are all fingers and thumbs. *Inquirer*. Retrieved Aug. 9, 2008, from http://www.theinquirer.net/gb/inquirer/news/2008/07/02/mobile-users-similar

Baudrillard, J. (2003). *The consumer society: Myths and structures*. London: Sage. (Original work published 1998)

Beck, J., & Wade, M. (2003). *DoCoMo—Japan's wireless tsunami: How one mobile telecom created a new market and became a global force*. New York: AMACOM.

Beilock, R., & Dimitrova, D. V. (2003). An exploratory model of inter-country Internet diffusion. *Telecommunications Policy*, *27*(3-4), 237–252. doi:10.1016/S0308-5961(02)00100-3

Clark, T. (2001). Baghdad by the bay. *Japan Internet Report, 59*. Retrieved August 1, 2008, from http://www.jir.net/jir7_01.html

Clark, T. (2002). How to evaluate the number of "Internet users" in Japan. *Japan Internet Report, 65*. Retrieved May 2, 2009, from http://www.jir.net/jir7_02.html#4

Clark, T. (2004, October). Mobile communications and the wireless internet—the Japanese experience. *Receiver, 11*, 1-4. Retrieved May 2, 2009, from http://www.vodafone.com/flash/receiver/11/articles/pdf/11_02.pdf

Clark, T., & Kay, C. (2005). *Saying yes to Japan*. New York: Vertical. comScore. (2007, July 9). *comScore publishes the first comprehensive review of Asian-Pacific usage*. Retrieved April 4, 2009, from http://ww.comscore.com/press/release.asp?press=1520

Connery, B. A. (1997). IMHO: Authority and egalitarian rhetoric in the virtual coffeehouse. In Porter, D. (Ed.), *Internet culture* (pp. 161–179). London, New York: Routledge.

David, P. A. (1985). Clio and the economics of QWERTY. *The American Economic Review, 75*(2), 332–337.

Edwards, G. (2004). Nagoya shuts down subway cell phone usage. *Engadget*. Retrieved October, 15, 2006, from http://www.engadget.com/2004/08012/ nagoya-shuts-down-subway-cellphone-usage

Fitzpatrick, M. (2007, September 27). Why mobile Japan leads the world. *Guardian*. Retrieved November 1, 2009, from http://www.guardian.co.uk/technology/2007/sep/2 7/guardianweeklytechnologysection.mobilephones

Fujimoto, K. (2005). The third-stage paradigm: Territory machines from the girls' pager revolution to mobile aesthetic. In Ito, M., Okabe, D., & Matsuda, M. (Eds.), *Personal, portable, pedestrian* (pp. 77–101). Cambridge, MA: MIT Press.

Gottlieb, N. (2000). *World-processing technology in Japan: Kanji and the keyboard*. Richmond, VA: Curzon Press.

Habuchi, I. (2005). Accelerating flexibility. In Ito, M., Okabe, D., & Matsuda, M. (Eds.), *Personal, portable, pedestrian: Mobile phones in Japanese life* (pp. 165–182). Cambridge, MA: MIT Press.

Habuchi, I., Dobashi, S., Tsuji, I., & Iwata, K. (2005). Ordinary usage of new media: Internet usage via mobile phone in Japan. *International Journal of Japanese Sociology, 14*, 94–108. doi:10.1111/j.1475-6781.2005.00071.x

Halvey, M., Keane, M., & Smyth, B. (2006, April). *Time-based patterns in mobile-Internet surfing*. Paper presented at the meeting of the CHI conference, Montréal, Canada.

Hill, R. (2003). A mobile phone of one's own: Japan's "Generation M". *New Zealand Journal of Asian Studies, 5*(1), 178–194.

Houser, C., & Thornton, P. (2004, March). *Japanese college students' typing speed on mobile devices*. Paper presented at the meeting of the IEEE International Workshop on Wireless and Mobile Technologies in Education, Taipei, Taiwan.

http://www.nytimes.com/1 995/04/29/business/cellular-phones-may-affect-pacemakers.html

Ishii, K. (2004). Internet use via mobile phone in Japan. *Telecommunications Policy, 28*(1), 43–58. doi:10.1016/j.telpol.2003.07.001

Ito, M. (2004, October). *Personal portable pedestrian: Lessons from Japanese mobile phone use*. Paper presented at the meeting of the 2004 International Conference on Mobile Communication, Seoul, Korea.

Ito, M. (2005). Mobile phones, Japanese youth, and the replacement of social contact. In Ling, R., & Pedersen, P. (Eds.), *Mobile communication and the Re-negotiation of the public sphere* (pp. 131–148). London: Springer-Verlag.

ITU. (2003). *Promoting broadband: The case of Japan*. Geneva: International Telecommunication Union.

ITU. (2007). *ICT statistics 2007*. Retrieved July 14, 2008, from http://www.itu.int/ITU-D/icteye/DisplayC ountry.aspx?countryId=120

Japanese cybercultures (pp. 19-33). London & New York: Routledge.

Japanese Patent Office. (2002). *Kyota Sugimoto (Japanese typewriter)*. Retrieved July 26, 2008, from http://www.jpo.go.jp/seido_e/rekishi_e/kyota_sugimoto.htm

japan.internet.com. (2006, June 28). 利川しやすいが抵抗も感じるクレジットカ「ド~オンライン決「調杳 (Easy to use, but still reluctant to use credit cards online). Retrieved August 1, 2008, from http://japan.internet.com/research/20060628/1.html

Jazwinski, C. H. (2001). Gender identities on the World Wide Web. In Wolfe, C. R. (Ed.), *Learning and teaching on the World Wide Web* (pp. 171–189). London: Academic Press. doi:10.1016/B978-012761891-3/50012-6

Jones, R. S. (2007). *Income inequality, poverty and social spending in Japan: Economics Department working papers No. 556*. Paris: OECD.

Kay, C. (2004, March 19). One finger, 10 keys and 6,000 characters. *Globalization Insider*. Retrieved July 17, 2008, from http://www.lisa.org/archive_doma in/newsletters/2004/1.3/kay.html

Kiiski, S., & Pohjola, M. (2002). Cross-country diffusion of the Internet. *Information Economics and Policy*, *14*(2), 293–310. doi:10.1016/S0167-6245(01)00071-3

Kohiyama, K. (2005). The development of mobile communications in Japan. In Ito, M., Okabe, D., & Matsuda, M. (Eds.), *Personal, portable, pedestrian* (pp. 61–74). Cambridge, MA: MIT Press.

Kristof, N. (1995, February 21). Japan's favorite import from American: English. *New York Times*. Retrieved July 30, 2008, from http://query.nytimes.com/gst/fullpage.html?res=990CE0DC-1430F932A15751C0A963958260

Landler, M. (1995, April 29). Cellular phones may affect pacemakers. *New York Times*. Retrieved March 31, 2010, from

Low, C. (2009, July 13). Japan plots one-stop public services portal. *FutureGov*. Retrieved March 27, 2010, from http://www.futuregov.net/articles/2009/j ul/13/japan-plots-one-stop-online-public-services-portal

Matsuda, M. (2005). Mobile communication and selective sociality. In Ito, M., Okabe, D., & Matsuda, M. (Eds.), *Personal, portable, pedestrian* (pp. 123–142). Cambridge, MA: MIT Press.

McVeigh, B. (2003). *Individualization, individuality, interiority, and the Internet: Japanese university students and e-mail* (Gottlieb, N., & McLelland, M., Eds.).

MIC. (2007). *2007 Information and Communications in Japan White Paper*. Tokyo: Ministry of Internal Affairs and Communications.

MIC. (2009, April 7). 平成*20*年「通信利用動向調杳」の結果. Retrieved November 7, 2009, from http://www.soumu.go.jp/menu_new s/s-news/02tsushin02_000001.html

Miyata, K. (2006, June). *Longitudinal effects of mobile Internet use on social network in Japan*. Paper presented at the meeting of the International Communication Association, Dresden, Germany.

Moon, I. (2010, March 1). Korean tech is losing its cool. *Bloomberg BusinessWeek*, 52-53.

Norrie, J. (2007, April 14). A nation's youth all thumbs with PCs. *The Age*. Retrived July 20, 2008, from http://www.theage.com.au/articles/2007/04/13/1175971346787.html

Norris, P. (2001). *Digital divide: civic engagement, information poverty and the Internet in democratic societies*. New York: Cambridge University Press.

NTT. (2000, August 7). *"i-mode" subscriptions top 10 million*. Retrieved August 1, 2008, from http://www.nttdocomo.com/ pr/2000/001031.html

NTT. (2006). Keyword search service expands mobile convenience. In *Mobility*. Tokyo: Nippon Telegraph & Telephone.

OECD. (1997). *Information technology outlook 1997*. Paris: OECD.

OECD. (2002). *Information technology outlook 2002*. Paris: OECD.

Ohiwa, H. (2004). Keyboard training in an information-oriented society. *IFIP International Federation for Information Processing, 14*, 147–152. doi:10.1007/1-4020-8136-7_15

Ohta, S. (1999). The shift to Internet EDI and Web EDI. *JEDIC Newsletter, 39*. Retrieved August 1, 2008, from http://www.ecom.jp/e_jedic/activity/news44-35/n39/koen1_e.htm

Okabe, D., & Ito, M. (2005). *Keitai* in public transportation. In Ito, M., Okabe, D., & Matsuda, M. (Eds.), *Personal, portable, pedestrian* (pp. 205–217). Cambridge, MA: MIT Press.

Okada, T. (2005). Youth culture and the shaping of Japanese mobile media: Personalization and the *keitai* Internet as multimedia. In Ito, M., Okabe, D., & Matsuda, M. (Eds.), *Personal, portable, pedestrian* (pp. 41–60). Cambridge, MA: MIT Press.

Onishi, N. (2008, January 20). Thumbs race as Japan's best sellers go cellular. *New York Times*. Retrieved March 10, 2010, from http://www.nytimes.com/2008/0 1/20/world/asia/20japan.html

Phone problems: Cellular phone etiquette becomes a major issue. (2000, July 5). Retrieved October 15, 2006, from http://web-japan.org/trends 00/honbun/tj000704.html

Pollack, A. (1984, June 7). The keyboard stymies Japan. *New York Times,* D17.

Starner, T. (2004). Keyboards redux: Fast mobile text entry. *IEEE Pervasive Computing / IEEE Computer Society [and] IEEE Communications Society, 3*(3), 97–101. doi:10.1109/MPRV.2004.1321035

Tabuchi, H. (2007, November 5). Japanese bypassing PCs in favor of an array of electronic gadgets. *Seattle Times*. Retrieved March 27, 2010, from http://seattletimes.nwsource.com/html/businesstechnology/2003994240_btjapanpcs05.html

Tabuchi, H. (2009a, July 19). Why Japan's cellphones haven't gone global. *New York Times*, B1.

Tabuchi, H. (2009b, November 29). In Japan, an odd perch for Google: Looking up at the leader. *New York Times*, B1.

Takeuchi, K. (2005). The challenge of Jan-Jan in participatory news. *OhmyNews*. Retrieved April 15, 2009, from http://english.ohmynews.com/articleview/arti cle_view.asp?menu=c10400&no=234393&rel_no=1

Wareham, J., Levy, A., & Shi, W. (2004). Wireless diffusion and mobile computing: Implications for the digital divide. *Telecommunications Policy, 28*(5-6), 439–457. doi:10.1016/j.telpol.2003.11.005

Williams, M. (2006, July 5). More mobile users than wired in Japan. *InfoWorld*. Retrieved May 2, 2009, from http://www.infoworld.com/d/networking/ more-mobile-internet-users-wired-in-japan-259

Wolfe, C. R. (2001). Learning and teaching on the World Wide Web. In Wolfe, C. R. (Ed.), *Learning and teaching on the World Wide Web* (pp. 11–22). London: Academic Press.

World Internet Project. (2004, January 14). Retrieved January, 9, 2007, from http://www.anderson.ucla.edu/x965.xml

ADDITIONAL READING

Abbate, J. (1999). *Inventing the Internet: Inside technology*. Cambridge, MA: MIT Press.

Aizu, I. (2002, September). *A comparative study of broadband in Asia: Deployment and policy*. Paper presented at the meeting of the "Asian Economic Integration: Current Status and Prospects" symposium, Tokyo, Japan.

Akiyoshi, M. (2004). *Unmediating community: The non-diffusion of the Internet in Japan, 1985-2002*. Unpublished doctoral dissertation, University of Chicago.

Akiyoshi, M., & Ono, H. (2005, August). *Digital inequality and the social shaping of the Internet in Japan*. Paper presented at the meeting of the American Sociological Association, Philadelphia, PA.

Bell, D. (2001). *An introduction to cybercultures*. London, New York: Routledge.

Coates, K., & Holroyd, C. (2003). *Japan and the Internet revolution*. New York: Palgrave Macmillan. doi:10.1057/9781403990075

Ducke, I. (2007). *Civil society and the Internet in Japan*. New York: Routledge.

Gottlieb, N., & McLelland, M. (2003). (Eds.). *Japanese cybercultures*, London: Routledge.

Hirata, K. (2002). *Civil society in Japan*. New York: Palgrave Macmillan. doi:10.1057/9780230109162

Ito, M., & Matsuda, M. (2005). (Eds.). *Personal, portable, pedestrian: Mobile phones in Japanese life*. Cambridge, MA: MIT Press.

Kingston, J. (2004). *Japan's quite transformation: Social change and civil society in 21st century Japan*. New York: RoutledgeCurzon.

MacKenzie, D., & Wajcman, J. (Eds.). (1999). *The social shaping of technology*. Buckingham, Philadelphia: Open University Press.

Nakayama, S. (2002). From PC to mobile Internet: Overcoming the digital divide in Japan. *Asian Journal of Social Science, 30*(2), 239–247. doi:10.1163/156853102320405834

Natsuno, T. (2003). *The i-mode wireless ecosystem*. West Sussex, UK: John Wiley & Sons. doi:10.1002/0470859199

Naughton, J. (1999). *A brief history of the future: The origins of the Internet*. London: Phoenix.

NTT. (2007). *NTT DoCoMo to offer new flat-rate data communications billing plans*. Retrieved August 3, 2008, from http://www.nttdocomo.com/pr /2007/001316.html

Pinch, T., & Bijker, W. (1984). The social construction of facts and artifacts. *Social Studies of Science, 14*, 399–441. doi:10.1177/030631284014003004

Smith, N. P. (1997). Computing in Japan: From cocoon to competition. *Computer, 30*(3), 26–33. doi:10.1109/2.573649

Sterne, J. (1999). Thinking the Internet: Cultural studies versus the millennium. In Jones, S. (Ed.), *Doing Internet research: Critical issues and methods for exploring the Net* (pp. 257–287). Thousand Oaks, CA: Sage Publications.

Sugiyama, M. (2000). Media and power in Japan. In Curran, J., & Park, M. (Eds.), *De-Westernizing media studies* (pp. 191–201). London, New York: Routledge.

Taira, K. (1988). Economic development, labor markets, and industrial relations in Japan, 1905-1955. In Duus, P. (Ed.), *The Cambridge history of Japan: The twentieth century* (pp. 606–653). Cambridge: Cambridge University Press.

KEY TERMS AND DEFINITIONS

Keitai: A Japanese word for "cell phone." Keitai literally means "portable" or "something you carry with you."

i-Mode: NTT DoCoMo's revolutionary mobile Internet service launched in February 1999. This Japanese wireless Internet service has increased Internet users drastically, creating unique Internet cultures.

Galápagos Syndrome: Evolution and divergence of a product isolated from world markets.

Technological Darwinism: A way to see technological development in terms of evolution theory.

(Hardware) Word Processor: A word processing machine that appeared between the typewriter and word processing software. Portable word processors were more popular than PCs in Japan during the 1990s.

Historical Accidents: Radom historical factors that influence technological diffusion and development but are too unpredictable to be neatly quarantined for analysis.

"Internets": A concept to argue that the Internet is not one but many.

Chapter 14
Exploring the Diffusion of Mobile Phone among Zimbabwean Migrants in Botswana

Do Kyun Kim
University of Louisiana at Lafayette, USA

ABSTRACT

From the early 2000s, the population mobility from Zimbabwe has drastically increased due to the collapse of the national economy and political instability. While the Zimbabwean migrants in Botswana have experienced horrific social, economic, and political difficulties, the mobile phone adoption rate among them has skyrocketed. Based on the theory of diffusion of innovations, this study investigates the influence of mobile phone diffusion among Zimbabwean migrants at the individual, community, and governmental levels.

EXPLORING THE DIFFUSION OF MOBILE PHONE AMONG ZIMBABWEAN MIGRANTS IN BOTSWANA

International population mobility is a common feature in the contemporary history of the southern African countries (International Organization for Migration, 2005). Some countries in the region even allow people from neighboring countries to stay legally for a few months without a VISA or a type of permission to enter and work in the host countries. The key factors driving people to migrate vary, including job opportunity, poverty, disease, human rights abuse, as well as political instability. As a result, separation from family members and disconnection with previous communication networks are easily observable in most southern African countries (Peberdy, 2000).

DOI: 10.4018/978-1-60960-037-2.ch014

Botswana is a southern African country, often called a sub-Saharan country. The population of Botswana was approximately two million people in 2009 (USA CIA, 2009). Although the country has a comparably small population, its economy is one of the strongest in the entire African continent: GDP per capita in 2008 was in excess of USD 13,900 and the unemployment rate was less than 7.5%. In contrast to the strong

Copyright © 2011, IGI Global. Copying or distributing in print or electronic forms without written permission of IGI Global is prohibited.

economy, Botswana's HIV/AIDS infection rate reached 25% among its total population in 2009 (UNAIDS, 2009).

Botswana is now experiencing a new but notable social phenomenon related to international migration. From the early 2000s, the population mobility from Zimbabwe to Botswana has drastically increased due to the collapse of the national economy and political instability in Zimbabwe (Africa Monitor, 2007). Prior to the massive migration, the Botswana government allowed Zimbabweans to stay and work legally for three months per year without a government permit. This open border policy motivated Zimbabweans to move into the rich country of Botswana (Moroka, 2005). More recently, however, responding to the massive international migration, the Botswana government is now concerned about the existing and predictable consequences of the Zimbabwean migration into Botswana.

Although Zimbabweans are the major migrant group in Botswana, most of them experience severe hardships in their lives, which are unique as migrants and different from Batswana – people of Botswana nationality. Most of them live apart from their family members, who have stayed in Zimbabwe, have unstable income sources, and live in very poor housing conditions, often without indoor sanitizing facilities. For instance, many Zimbabwean households in Botswana share one water conduit together and use one toilet, which is located in a corner of the residential block and does not have a flushing device. If a migrant is single or lives apart from his/her family members, in most cases, the person shares a room with many other people, because the cost of housing in an urban area is very expensive in spite of the poor quality of the place.

Even in such dire circumstances, a large portion of Zimbabwean migrants have mobile phones. Although no statistical data is available for the rate of mobile phone adoption among Zimbabweans, a high rate of adoption can be surmised by the abundant possession and use of mobile phones that was observed. It was easily noticed that many people on the street frequently checked their mobile phones to see if anyone had called.

Capturing these interesting scenes, this study explores why and how the diffusion of the mobile phone occurs among Zimbabwean migrants in unique situations as foreigners in Botswana and analyzes how the diffusion of the mobile phone affects Zimbabwean migrants' individual lives, their migrant community, and the governmental policies towards international migrants. The underpinning theoretical frame for this study is the diffusion of innovations (DOI). Thus, the literature review covers factors affecting the diffusion of innovations in general and mobile phones. For data collection, this study employed ethnography with other qualitative methods including interviews and observation.

DIFFUSION THEORY AND MOBILE PHONE DIFFUSION

Diffusion research in the field of communication studies investigates how an individual in a social system adopts an innovation introduced through certain communication channels (Rogers, 2003). In this definition, three elements of diffusion constitute the key to understanding how the diffusion of mobile phones occurs among the Zimbabwean migrants in Botswana: innovations, social systems, and communication channels. First, an innovation can be anything, such as an object, idea, information, practice, program, etc., if an individual perceives it as new, even though the perceived innovation is already prevalent among others. Second, a social system refers to any social variable that may affect a potential adopter's decision making process. It subsumes culture, history, tradition, politics, economic situation, and rules established by institutionalized human interaction. Third, a communication channel is any means that can deliver messages and actions, including individuals and mass media, such as television, radio, Internet, mobile phone, interpersonal communication, etc. These three elements always interplay in the process of diffusion.

In the theory of diffusion of innovations, some clusters of diffusion fertilizers stimulate the innovation diffusion: perceived attributes of innovations, diffusion strategies by change agents, and social systems. First, the diffusion of innovations is influenced by how an innovation is perceived by a potential adopter. In general, five perceived attributes of innovations determine innovation diffusion. The five perceived attributes consist of relative advantage, compatibility, complexity, trialability, and observability (Rogers, 2003). If an innovation is perceived as advantageous and is consistent with the existing values, past experiences, and needs of potential adopters, the innovation is more likely to diffuse. In addition, if an innovation is easy to use, accessible, and frequently observable, the innovation is also more likely to diffuse.

The second cluster is purposively designed diffusion strategies. Currently, diffusion scholars not only analyze already existing patterns of diffusion, but also aggressively design strategies for the future diffusion of innovations. It is called "purposive diffusion" (Singhal & Dearing, 2006). Interestingly, it is very hard to confidently say that the most widely diffused innovations are the best. For instance, is Coca-cola the best beverage in the world, even though it is the best-selling beverage? Well, maybe or maybe not. In addition, many people fail to follow governmentally regulated laws, for example, as seen in the increasing number of illegal drug users. In this sense, innovation diffusion often depends on diffusion strategies rather than the quality of innovations (Kim et al, 2007).

The third cluster is characteristics of a social system. A social system in diffusion research refers to a socially constructed variable that affects a potential adopter's decision making process. It includes organizational and institutional structures, such as culture, history, tradition, governmental policy, economic structure, and technology. From the diffusion perspective, the relationship between individuals and social systems can be viewed in two reciprocal ways. First, a social system shapes or enforces an individual's adoption of an innovation; second, the individual's adoption of an innovation can change a social system. Either way, the diffusion of innovation reflects the process of social change (Dearing & Kim, 2008). While the perceived attributes of innovations are more an individual variable that affects personal decision-making, a social system is an external force to drive people to adopt a certain innovation. Since communication refers to interactions with others, the adoption of a communication medium, such as the Internet and mobile phone, is certainly related to the characteristics of a social system where an individual is situated.

For the diffusion of the mobile phone, a growing number of research has dealt with different social systems influencing innovation adoption in different places in the world, e.g., Chile (Ureta, 2008), China (Ngan & Ma, 2008), Chana (Burrell & Anderson, 2008), Italy (Fortunati, 2002a), Israel (Lemish & Cohen, 2005), and the United States (Campbell & Russo, 2003). Coupled with the empirical evidence of influence of a social system on the adoption and use of mobile phone, the diffusion of the mobile phone has been analyzed with several other theories that share very similar or the same idea as diffusion theory in dealing with diverse social influences.

For example, Social Influence (SI) theory shares some assumptions with the theory of diffusion of innovations, such as the influence of social system and other people in close contact. Specifically, the SI model explains that the attitude, statement, and behavior of others in close contact shape an individual's evaluation of communication technologies and affect the adoption of them (Fulk, 1993; Fulk, Schmitz, & Ryu, 1995; Fulk, Schmitz, & Steinfield, 1990). In other words, contextual social factors encourage an individual to use communication media and provide an answer for why an innovation diffuses as well. From this perspective, both the SI model and diffusion theory focus on communicative interpersonal relationship as one of the most significant determinants in shaping an individual's attitude and innovation adoption.

In addition to the SI model, the communication network analysis, embedded in the theory of diffusion of innovations, also provides a detailed theoretical explanation about an individual's adoption of the mobile phone. Diffusion theory explains two types of communication networks which a society system may have: interlocking personal network – a network where individual members of a network interact with one another, and radial personal networks – a network that links individuals to a focal person but without interaction between one another (Rogers, 2003). Both types of communication networks can contribute to the diffusion of an innovation, depending on the characteristics of a social system.

Social learning theory (Bandura, 1977, 1986) highlights the influence of observation, whereby a person is influenced by observed behaviors, motivating the viewer to alter their behavior. Through routine communication with others, an individual naturally observes and imitates what significant others, often informal opinion leaders, do, adopting their behavior. This mimicking behavior becomes a crucial determinant in an individual's decision making when adopting an innovation. Along with the social influence model and social learning theory, the theory of diffusion of innovations also recognizes the importance of an individual's acquaintances and how frequently communication occurs between them in accelerating the diffusion of an innovation (Dearing 2009, Dearing & Kim, 2008, Kim et al, 2007, & Rogers & Kincaid, 1981).

Empirically, several studies indicate other social influences in the diffusion process as well. For instance, some scholars point out that the diffusion of communication technologies (ICT) correlates with income and government trade policies (Baliamoune-Lutz, 2003; Leung & Wei, 1999; Rice & Katz, 2003). However, the relationship between ICT adoption and the level of education has been controversial among scholars with different research subjects. As opposed to this type of study, which tries to find the socioeconomic and demographic factors influencing the adoption of an innovation, the investigation on communicative variables influencing an individual's innovation adoption constitutes another stream of the diffusion research.

Historically, human society has invented more and more communication channels primarily due to the development of new technologies. Communication channels have evolved from human languages (interpersonal communication) and letters to technology mediated communication tools, including landline phones, mobile phones, and Internet mediated communication tools. With all these communication channels, which communication channel would be the most influential in the diffusion of innovation? Perhaps, broadly speaking, mass media and interpersonal communication would be the most common answer. Both types of communication channel certainly have their own unique strengths and weaknesses, depending on the case and purpose of diffusion. However, when considering the adoption and use of the mobile phone among Zimbabwean migrants, the question can be answered through the analysis of early adopters, those who try an innovation earlier than others and, at some point of time, motivate and encourage others to adopt the innovation, since few Zimbabweans possess other electronic mass media tools, such as TV and Internet.

Based on decades of research, diffusion scholars were able to generalize some features of early adopters in the diffusion process. For instance, an early adopter has a higher socioeconomic situation, more cosmopolite character, higher education, and more interpersonal channels and social participation (Rogers, 2003). However, the diffusion of a new communication technology often shows many unique features that are different from the traditional generalizations. For instance, Baliamoune-Lutz (2003) found that the factor of higher education was not necessarily an indicator of early mobile phone adoption. In addition, a large number of studies, e.g., Akiyoshi and Ono (2008), Leung and Wei (1999), & Srivastava (2005), show that most early adopters of the mobile phone are young adults in high school or college or still in an early stage of their careers. Therefore, they are more likely to have lower positions in their work places with less income than older people with higher income and are not necessarily a more educated group of people. However, the younger are more willing to try an innovative communication tool. Although some controversy still exists regarding mobile phone diffusion, the theory of diffusion of innovations emphasizes that the interpersonal communication channel often functions as a more important determinant for behavioral decision making on purchasing and using a mobile phone (Kim et al., 2007).

DATA COLLECTION

The data collection was conducted during the summer of 2009 in Gaborone, Botswana. The sites of data collection included major shopping malls in the city, which were popular places for Zimbabwean migrants to seek part-time employment, and residential areas where most Zimbabwean migrants lived collectively. Groups of migrants often sit in the corners of parking lots at the shopping malls and waited for someone to offer them daily work. In residential areas, people waited near the main streets of the town in the hope of getting some work for the day. Many employers who needed part-time or temporary workers also frequently visited those places to hire Zimbabwean migrants, mostly for daily work.

This research employed ethnography as a method of data collection, including observation and interviews. A data collection team was formed with a professional researcher and three assistants, two males and one female. The data-collection assistants were legal Zimbabwean migrants who had worked in Botswana for several years and also had connections with illegal immigrants as well. The data collection team regularly visited the data collection sites, including the parking lots of malls, residential areas, and stores where research subjects were working, to collect data. Interviews were conducted in English, because many Zimbabweans in Botswana were able to speak English as well as their ethnic languages.

The demographic background of the migrants varied. Both males and females were seeking job opportunities together at a site, and their ages ranged from the late teens to the early forties. They also had a variety of specialties, such as hair designer, plumber, construction worker, etc. However, they were more than willing to try any sort of work, if it would make money. Most of them were alone in Botswana, separated from their family members living in Zimbabwe. In terms of educational level, older people showed much higher educational levels than younger people. This is because when the older were students, Zimbabwe was politically and economically stable and, thus, able to offer better educational opportunities to their students.

Findings: Mobile Phone in a New Life

Mobile phone diffusion has been remarkable in Botswana. In 2009, Botswana's total population was about 2 million people. At the same time, the number of mobile phone users had reached approximately 1.8 million, while the use of landline phones was less than 0.15 million (USA CIA, 2009). Graph 1 shows the dramatic increase of mobile phone adoption in Botswana in recent years (World Bank, 2006). In contrast, the diffusion rate of other communication or mass media tools remained much lower than the mobile phone. Showing this trend, a mobile phone has become a necessary item in an individual's lives in Botswana. Based on the proliferation of mobile phones, this study first explored why and how Zimbabwean migrants adopted mobile phones in their unique situations as economically poor international migrants in Botswana.

Individual Needs

According to Campbell and Russo (2003), United States college students have three main reasons for possessing a mobile phone: 1) maintaining communication with significant others, such as family members and close friends; 2) communicating long-distance at no additional cost; and 3) receiving a recycled mobile phone from a significant other when the other changes to a new mobile phone. Many other studies have shown similar results as well. However, the mobile phone adoption among Zimbabwean migrants in Botswana demonstrates a special and/or stronger need for a key communication tool—personal survival in a foreign country.

When a Zimbabwean comes across the border into Botswana, the person enters a new life in a new country. Even though Botswana has countless Zimbabweans already there, the new country is not as familiar as their own country. All the differences, such as culture, social surroundings, and people, are surely very exotic. Considering their new situations, what motivated a migrant to have a mobile phone, even someone with a very low income and poor housing situations? In addition, how does the use of a mobile phone affect the individual's life as a foreigner in Botswana?

It was found that, a migrant, often isolated from the main stream of Botswana society, had to have a communication channel to access possible sources of information. Because the purpose of migration for most Zimbabwean migrants was usually to find a source of income, they needed and wanted to have as many sources of information as possible immediately after they moved to Botswana. An interviewee who was in his early 30s told this story:

I came here this spring with my friend from my town (in Zimbabwe). He was here before and he said we can find jobs here. But, he got a job in other city...so I had to find my job myself...I didn't know much about Gaborone, but I have to make money. So, I bought this phone to talk to my friends here. (he showed an old type of mobile phone, a

Figure 1. Source: World Development Indicators (WDI) database, World Bank, 2006)

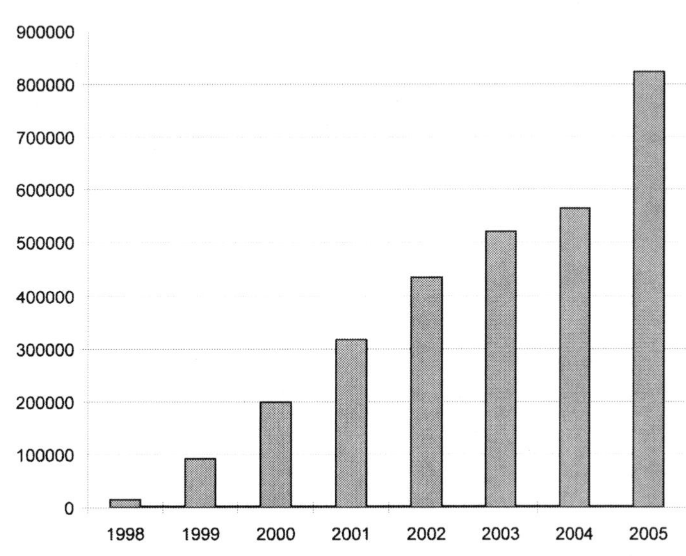

European brand.) ... we share job information and they often let me know jobs I can take.

His story shows that migrants purchase mobile phones to contact their information sources and find job opportunities in their host country as most of the people in their communication networks share the same situation. Being able to communicate with people in the same situation was paramount to survival, enabling a person to get helpful and often necessary information to a migrant in a foreign country, such as housing, job information, and medical advice. Highlighting the advantages of a mobile phone, Markus (1987) notes that the more people are in a communication network the greater the incentives are to join a network. Zimbabwean migrants have found that the mobile phone enabled them to reach the greatest number of potential job opportunities and other necessary information in a fast and efficient way.

Social Influence: Communication Patterns

Another reason for the rapid diffusion of mobile phones was that Zimbabwean migrants in Botswana relied heavily on interpersonal communication. Although ample information was available for free through other mediums of information delivery, such as village newspapers and information brochures, the migrants depended more heavily on interpersonal communication for their decisions as a male worker explained:

I got a job that gave me this thing (he had a large scar on his arm). It was last year. I got that job from that paper (pointing to a stack of free newspapers in a corner of the parking lot). I was hired and worked in a construction site. But, it was very very dangerous work.... There was no safety tool and they operated big machines. I was scared, but, I needed money. So, I worked there and got a big hammer falling on my arm. So, now, I don't want to work in the dangerous place, never, never... now, usually, my friends let me know some jobs that I can take... if I find a job, I always ask my friends the job is safe.

As his experience shows, migrants often need more descriptions of the work that most employers did not provide when contracting the workers for an unsafe job. In other words, information obtained from printed information sources were viewed as less credible than information obtained from friends and/or peers, especially in a new country. Therefore, a migrant relies on interpersonal communication, and this, in turn, encouraged the migrants' adoption of the mobile phone.

Many employers showed a similar reliance on interpersonal communication. When interviewing store owners who hired Zimbabwean workers, an owner told the research team the following:

I hired these Zimbabweans because my friends told me they work hard. I first hired one person, and the other through the one who I first hired. Now, I have five Zimbabwe people working in my store. They have a good connection. So, if I need more, it is not too difficult to get another one because they get one for me ... I cannot believe one just stopping my store and ask hiring... There are many thieves around here...

As shown in the interviews, both potential employers and migrant workers considered interpersonal communication as a more credible source of information.

Symbol of New Life

The possession of a mobile phone was also seen as a symbol of a new lifestyle in new social surroundings. In fact, many people mentioned the symbolic function of the mobile phone, e.g., "*you know?... if you want to live here, you need this.*" Although this may sound more like a need-based statement, the nonverbal cues clearly signified his exaggerated pride as a mobile phone user.

When the Zimbabwean migrants enter Botswana, most are very poor with little or no money to purchase anything. Their first concern is housing, a place to stay. In contrast to neighboring countries, the extremely high cost of housing in Botswana becomes an initial obstacle for most of Zimbabwean migrants. However, after finding housing, the migrant's next concern is about job information. At this point, a migrant may feel the need of a mobile phone and, if possible, purchase it. Eventually, the acquisition and use of a mobile phone construct a new and evolving lifestyle for the migrant.

Furthermore, the lifestyle of most Zimbabwean migrants is not very stable. They have to move frequently from place to place, depending on the cost of housing and job opportunities, and their level of income also varies from day to day. At the same time, they need to maintain constant communication with the members of their communication networks which are often their only source of job information and medium of maintaining friendships.

Finally, the possession of a mobile phone often symbolizes membership in a migrant society. Since the mobile phone is a major communication tool among the migrants in a new environment, the possession of a mobile phone symbolizes an accumulation of experience, as well as an information center, among Zimbabwean migrants. From the diffusion perspective, mobile phone users are more likely to be informal opinion leaders who are easily accessible and provide advice to their followers. Much like informal opinion leaders, the characteristics of early adopters of mobile phones are such that they are more likely to have necessary information, higher incomes, and more contacts. Theoretically and empirically, informal opinion leaders are also known to be influential in shaping people's attitude and changing their behaviors and people often mimic them and follow their opinions.

Inexpensive and Convenient Mobile Phone Services and Population Mobility

In addition to a culture that appreciates interpersonal communication, a systemic factor contributed to accelerating the diffusion of the mobile phone. Most Zimbabwean migrants use prepaid mobile phones, because they did neither want to reveal their identification, which is required for a phone contract, nor have regular income to pay a monthly service charge to a mobile phone service. In addition, unlike the mobile phone service system in the United States, incoming calls are free of charge in Botswana. This inexpensive and convenient system of mobile phone service in Botswana certainly motivated potential adopters and encouraged even poor migrants to get mobile phones. Reflecting this pattern, it was easy to observe that many phone card sellers approached people on the street to sell prepaid mobile phone cards with pin numbers for various minutes of usage.

Since mobile phones have been diffused by inexpensive and convenient services, enabling people to maintain not only their communication networks in a foreign country but also their distant networks, mobile phone diffusion has induced a higher level of domestic and international population mobility. A migrant said:

I was the first person in my family who came to Botswana... After I got this phone, I could talk with my family in Zimbabwe... I said my brother came here, so he was with me for some months... Of course he has a mobile phone because he is working in Francistown now. So, I think I move to Francistown too.

If it was hard to obtain and maintain a mobile phone, communication between this person and his family or other migrants would be more difficult than it is with his mobile phone. In addition, as he stated, a large number of migrants in the interview groups first considered moving to Botswana because of their communication with people who already worked in Botswana. This observation also supports the importance of interpersonal communication in the migrants' decision to move and communication with others from a distance.

ANALYSIS AND DISCUSSION: DIFFUSION OF MOBILE PHONE AND TRANSFORMATION

The contemporary history of southern Africa, including Zimbabwe, Botswana, Namibia, Zambia, and South Africa, demonstrates a high level of population mobility: people in the region have been moving in and out of their countries to seek job opportunities mainly due to the domestic political and/or economic instability in their home countries. While the high mobility rate of people across the countries implies a fast adoption of convenient communication methods (e.g., mobile phones) to communicate with people from a distance, poverty can be seen as a strong obstacle to the diffusion of mobile phones among people from other countries. However, the adoption of the mobile phone has skyrocketed among people in the region. As findings of this study revealed, the migrant's need for information, their reliance on information from interpersonal communication, cheap and convenient mobile phone service, and the symbolic function of the mobile phone have encouraged Zimbabwean migrants to readily adopt mobile phones. Based on the findings, the salient mobile phone adoption among Zimbabwean encouraged researchers' further analyses on implications of the new communication phenomenon not only among Zimbabwean migrants in Botswana, but also international migrants in southern African countries in general.

Coupled with the historical population movements, the diffusion of the mobile phone has transformed cultural, political, and economic structures in the southern Africa region. First, the spread of the mobile phone can explain two phenomena — frequent communication and a larger boundary of communication. Frequent communication among people using mobile phones increased not only the efficacy of information sharing, but also individual dependency on others. Although the role of mobile phone communication among Zimbabwean migrants is primarily information sharing, the diffusion of the mobile phone apparently provides more chances for the owners to communicate with others. As a result, the frequency of mobile phone communication often induces dependency on other people in a foreign country. An interviewee stated that "*whenever I have free time, I used to call my friends. I don't know why... but, my hands go to my phone all the time.*" Additionally, frequent communication through mobile phones strengthens a sense of belonging to a communication network (Fortunati, 2002b), and, accordingly, the group coherence among migrants is more likely to become rigid. In other words, an individual seeks more communication opportunities to be closely affiliated with a larger group that serves as a main source of information and friendship in a foreign country.

In terms of the communication boundary, whereby a person can communicate with significant others at any time and in any place, the use of mobile phone overcomes the limitations of time and space in communication and, consequently, extends the boundary of communication (García-Montes, Caballero-Munoz, & Perez-Alvarez, 2006). This also reinforces reconstruction of the concept of "space" from a static concept of space to a dynamic space that is not limited by addresses and precise spatiality of geography (Fortunate, 2002b). This connectivity across time and place has diluted the traditional concept of collective identity, which emphasizes co-existence at the same time and place in constructing an identity (Srivastava, 2005).

Second, information immediacy and communication immediacy featured in the mobile phone communication (Fortunati, 2002b) created the culture of present in the society. According to García-Montes & Carretero (2000), the use of the mobile phone affects the construction of self with two main consequences — fragmentation, which is the loss of the past due to a context of current behavior, and irresponsibility, which disregards the consequences for actions. Due to these two features of mobile phone use, a migrant's identity only reflects the present self which is independent from the past and the future. In this situation, everything may be spontaneous and fragile.

I don't think my future seriously now because... I don't know... I have a job and working hard. That's important. Most of people here do different things from their jobs in Zimbabwe. They don't care. I don't care either. And, I have no time to think about my future... The important thing is where is job now... I do everything I can now, and people think also the same.

It is interesting to ponder the tension between a heavy reliance on interpersonal communication, a finding of this study, and spontaneous personal identity in a foreign society that leads Zimbabwean migrants to adopt mobile phone usage. In a less stable situation that requires frequent moves, a person may feel isolation or disconnection from existing communication networks, and the perceptional isolation causes the person to redefine the self with the present self. However, under the assumption of a human as a communicative being, the present self has a limitation in its influence on an individual. As a result, people actively seek a way to maintain the relationships with their peer groups. This effort of creating and maintaining relationship encourages the migrants to adopt mobile phones. In this sense, a mobile phone becomes a new symbolic device that connects spontaneous beings in new social surroundings to previous communication networks and even enlarges communication networks in a new place.

Third, the emergence of strong communication networks among migrants, resulting from the widespread usage of mobile phones, has triggered governmental policy makers to pay more attention to the migrants. Due to their frequent mobility and ability to rapidly share information, the government has become aware of the potential for serious political problems, if the government is unable to control the migrants. However, from the migrants' perspective, the possession of a mobile phone can increase an individual's security, as well as the ability to gather more and more recent information. The studies by De Miguel and Barbeito (1997) in Spain and Palen et al. (2001) in the United States also support that an outcome of mobile phone acquisition is an increased sense of personal security. However, on a larger level, the rapid movement of migrant groups within a country, who try to escape government surveillance and control, may threaten the host country's national security. The political riot in Seattle, Washington, during the 1999 WTO convention, exemplifies this political concern. Rapid communication with new communication technology enabled protesters to more effectively organize mass political strikes against the WTO meeting and allowed them to move quickly to escape the police (Srivastava, 2005). Applying this concern to the present study, fast mobile phone-based information sharing would also create a serious political problem in controlling migrant populations in host countries.

As another governmental level of concern, the economic influence of migration has been a long-lasting issue of discussion in most host countries of Africa. One of the major economic concerns of a host country is the leakage of national economic capital to the migrants' home country. Because the reason for international migration is mostly related to economic reasons, e.g., making money and sending money and commodities to the family members in migrants' home country, the more migrants come to a host country, the more likely it is that the host country's national economic capital will be leaked to other countries. Related to this economic prediction, the higher mobile phone adoption among Zimbabwean migrants, enabling them to communicate with people in their home country, can increase the rate of international migrations from Zimbabwe to Botswana and, as a result, it is very likely that more of Botswana's national economic capital would be leaked to Zimbabwe. In addition, since the Zimbabwean migrants are mostly illegal, the Botswana government hardly imposes tax on the migrants' income. Furthermore, most of the international money and commodity transactions are neither monitored nor regulated by host countries, and most migrants use untraceable means to deliver money or commodities to family members in their home countries. One interviewee noted:

We cannot open a bank account here, and the bank charges a lot for sending money. My family in my country also does not have a bank account…most people here send money and presents by friends who go back to my country. They deliver things to my family too.

Researchers observed many bus terminal employees in Botswana load numerous bags and boxes on top of buses destined for long-distance and international destinations. Passengers were often asked to deliver money and economic commodities from Botswana to friends and family in foreign countries. With some payment, many bus drivers also delivered boxes and envelopes containing money and other miscellaneous things as well. When Zimbabwe suffered a serious drought a few years ago, buses delivered an unaccountable number of bottled water. Again, the items were tax-free and delivered without a governmental customs inspection.

CONCLUSION AND SUGGESTION

The diffusion of mobile phones in southern African countries has proliferated within a short period of time. This new communication phenomenon has engendered many affects and consequences on society both expected and unexpected. Based on this background, this article analyzed the diffusion and influence of mobile phones among Zimbabwean migrants in Botswana. The analysis of ethnographic data revealed that the diffusion of the mobile phone among the Zimbabwean migrants has affected a generation with a new communicative identity and the development of strong interdependent relationships among the migrant community. It has also transformed cultural, political, and economic landscapes.

Considering the rapid diffusion of the mobile phone and its profound and diverse influences in the region, researchers should call for more interdisciplinary studies related to the new communication phenomenon using mobile phones and examine the effects of mobile phone diffusion in southern African countries. In addition to an academic effort, governments in the region should also establish an intergovernmental collaboration to respond to changes led by the diffusion of mobile phone and international population mobility, expecting both the positive and negative consequences in political, economic, and public health relationships in the region.

As analyzed through this study, the hand-size communication device can generate as much power and energy as a nuclear weapon and transform society in unimaginable directions. Hopefully, this inevitable power is directed toward a desirable way that creates a better future in the region.

REFERENCES

Africa Monitor. (2007, May). Inflation subsides. *Africa Monitor: Southern Africa, 12.*

Akiyoshi, M., & Ono, H. (2008). The diffusion of mobile internet in Japan. *The Information Society*, *24*, 292–303. doi:10.1080/01972240802356067

Baliamoune-Lutz, M. (2003). An analysis of the determinants and effects of ICT diffusion in developing countries. *Information Technology for Development, 10*, 151–169. doi:10.1002/itdj.1590100303

Bandura, A. (1977). *Social learning theory*. Englewood Cliffs, NJ: Prentice-Hall.

Bandura, A. (1986). *Social foundations of thought and action*. Englewood Cliffs, NJ: Prentice-Hall.

Burrell, J., & Anderson, K. (2008). 'I have great desires to look beyond my world': Trajectories of information and communication technology use among Ghanaians living abroad. *New Media & Society, 10*, 203–224. doi:10.1177/1461444807086472

Campbell, S. W., & Russo, T. C. (2003). The social construction of mobile telephony: An application of the social influence model to perceptions and uses of mobile phones within personal communication networks. *Communication Monographs, 70*, 317–334. doi:10.1080/0363775032000179124

De Muguel, A., & Barbeito, R. L. (1997). *El impacto de la telefonía móbil en la sociedad española* [The impact of mobile telephony on Spanish society]. Madrid: Ericsson.

Dearing, J. W. (2009). Applying diffusion of innovation theory to intervention development. *Research on Social Work Practice, 19*, 503–518. doi:10.1177/1049731509335569

Dearing, J. W., & Kim, D.-K. (2008). *Diffusion of information and innovation. International Encyclopedia of Communication*. London: Blackwell Publishing.

Fortunati, L. (2002a). Italy. Streotypes: true or false. In Katz, J., & Askhus, M. (Eds.), *Perpetual contact: Mobile communication, Private talk, Public performance*. Cambridge, UK: Cambridge University Press. doi:10.1017/CBO9780511489471.006

Fortunati, L. (2002b). The mobile phone: Towards new categories and social relations. *Information Communication and Society, 5*, 513–528. doi:10.1080/13691180208538803

Fulk, J. (1993). Social construction of communication technology. *Academy of Management Journal, 36*, 921–950. doi:10.2307/256641

Fulk, J., Schmitz, J., & Ryu, D. (1995). Cognitive elements in the social construction of technology. *Management Communication Quarterly, 8*, 259–288. doi:10.1177/0893318995008003001

Fulk, J., Schmitz, J., & Steinfield, C. W. (1990). A social influence model of technology use. In Fulk, J., & Steinfield, C. (Eds.), *Organizations and communication technology* (pp. 117–139). Newbury Park, CA: Sage.

García-Montes, J. M., Caballero-Muñoz, D., & Pérez-Álvarez, M. (2006). Changes in the self resulting from the use of mobile phones. *Media Culture & Society, 28*, 67–82. doi:10.1177/0163443706059287

García-Montes, J. M., & Carretero, M. J. (2000). La psicologa social de la telefonía móvil. [Social Psychology of Mobile Telephones] In Caballero, B., Méndez, M. T., & Pastor, J. (Eds.), *La mirada psicosocial: Grupos, procesos, lenguajes y culturas*. Madrid: Biblioteca Nueva.

International Organization for Migration (IOM). (2005). *HIV/AIDS, population mobility and migration in Southern Africa: Defining a research and policy agenda*. Geneva: International Organization for Migration.

Kim, D.-K., Chitnis, K., Vasanti, P. N., & Singhal, A. (2007). Opinion leadership in Indian Villages and Diffusion of e-Choupal. *Journal of Creative Communication, 2*(3), 335–350.

Lemish, D., & Cohen, A. A. (2005). On the Gendered nature of mobile phone culture in Israel. *Sex Roles*, *52*, 511–521. doi:10.1007/s11199-005-3717-7

Leung, L., & Wei, R. (1999). Who are the mobile phone have-nots? *New Media & Society*, *1*, 209–226.

Markus, M. L. (1987). Toward a "Critical Mass" theory of interactive media. *Communication Research*, *14*, 491–511. doi:10.1177/009365087014005003

Moroka, T. M. (2005). Family economic pressure and health outcomes in low-income Francistown, Botswana: The reverse buffering role of social support. *Journal of Social Development in Africa*, *20*, 106–127.

Ngan, R., & Ma, S. (2008). The relationship of mobile telephony to job mobility in China's Pearl River Delta. *Know Techn Pol*, *21*, 55–63. doi:10.1007/s12130-008-9046-8

November 21, 2009, from https://www.cia.gov/library/public ations/the-world-factbook/geos/bc.html

Palen, L., Salzman, M., & Youngs, E. (2001). Discovery and integration of mobile communications in everyday life. *Personal and Ubiquitous Computing*, *5*, 109–122. doi:10.1007/s007790170014

Peberdy, S. (2000). Mobile entrepreneurship: Informal sector cross-border trade and street trade in South Africa. *Development Southern Africa*, *17*, 201–219. doi:10.1080/713661400

Rice, R. E., & Katz, J. E. (2003). Comparing internet and mobile phone usage: Digital divides of usage, adoption and dropouts. *Telecommunications Policy*, *27*, 597–623. doi:10.1016/S0308-5961(03)00068-5

Rogers, E. M. (2003). *Diffusion of innovations* (5th ed.). New York: Free Press.

Rogers, E. M., & Kincaid, D. L. (1981). *Communication networks: Toward a new paradigm for research*. New York: Free Press.

Singhal, A., & Dearing, J. W. (2006). *Communication of innovations: A journey with Ev Rogers*. Thousand Oaks, CA: Sage Publications.

Srivastava, L. (2005). Bobile phone and the evolution of social behaviour. *Behaviour & Information Technology*, *24*, 111–129. doi:10.1080/01449290512331321910

UNAIDS. (2009). *09 AIDS epidemic update*. Retrieved November 15, 2010, from http://data.unaids.org/pub/Report/20 09/JC1700_Epi_Update_2009_en.pdf

Ureta, S. (2008). Mobilising poverty?: Mobile phone use and everyday spatial mobility among low-income families in Santiago, Chile. *The Information Society*, *24*, 83–92. doi:10.1080/01972240701883930

USA, Central Intelligent Agency (CIA). (2009). *The world factbook*. Botswana: Retrieved.

World Bank. (2006). *Facts and Figures*. Retrieved November 21, 2009, from World Development Indicators 2006.

Chapter 15
The Triumph of *Shanzhai*: No Name Brand Mobile Phones and Youth Identity in China

Mei Wu
University of Macau, China

Hongye Li
University of Macau, China

ABSTRACT

This chapter attempts to explore the connection between the popularity of shanzhai (no-name brand) mobile phones among urban youth and their social characteristics in contemporary China. Through qualitative analyses of in-depth interviews both shanzhai and non-shanzhai users and selected online group discussions of shanzhai workers, this connection is examined in five realms: political attitudes, practical rationale, social values, interest-seeking and self identification. The findings indicate that the youth's attitudes towards shanzhai phones are prominently based on practical rationale, followed by their craving for entertainment and fashion. The political orientation of anti-mainstream, anti-authority and anti-brand is not prominent. To use a brand imitation is not so much a vanity issue for social status elevation, as a need of self experience within an affordable means. Finally, young people who are engaged in the shanzhai industry do form their own cycle and build their identity strongly associated with the "shanzhai" both as a business career and political mission.

INTRODUCTION

One of the latest buzz words in China's popular culture of 2008 is "*shanzhai*" (literal translation: mountain village), which originally meant a fortress in a mountain occupied by outlaws. However, in its post-modern connotation, it refers to a wide range of imitations of brand names. It is associated closely with a phenomenon of producing and consuming low-cost products, which imitate brand names, mostly mobile phones and digital products.

DOI: 10.4018/978-1-60960-037-2.ch015

The "*shanzhai*" mobile phone, as the name indicates, carries a connotation of rebellion, resistance, periphery, and outlawing. Foremost, it is an antonym of the "brand-name" phone. The "brand-name" phone signifies the celebrated name, good quality, high price, recognized prestige and an

Copyright © 2011, IGI Global. Copying or distributing in print or electronic forms without written permission of IGI Global is prohibited.

enviable possession of social status. Whereas the "*shanzhai*" phone is associated with the no-name, shoddy quality, low price, dubious reputation and a distasteful piece of low class production. The popularity of *shanzhai* phones poses a series of intriguing questions: what does it signify for the contemporary urban youth in China? Does it indicate a resistance against the authority of the "brand-name"? What other meanings does the *shanzhai* phone also embody? To what extent do these meanings connect to the characteristics of urban youth in China today? In other words, if the mobile phone is a central artifact of the youth's self-image (Fortunati 2001) and personal style (Katz & Aakhus, 2002), how does the *shanzhai* version symbolize the construction of self identity for the young generation in contemporary China?

This paper attempts to identify the connections between the popularity of *shanzhai* mobile phones among young people and the characteristics of youth described in the public discourse in contemporary China. It explores how young people's choice and use of mobile phones, especially their attitudes towards *shanzhai* mobile phones, express their styles and characteristics, particularly their pursuit of a distinct personality. The inquiry centers on deciphering the *shanzhai* phenomenon in relation to the youth culture discourse which emphasizes the defiant and rebellious attitude of youth towards the mainstream culture and their construction of a sort of anti-mainstream or anti-brand identity. The central research question is: How does the *shanzhai* mobile phone connect to the characteristics of urban youth in China today? This central issue is explored by two sub-issues:

a) What are the meanings of the *shanzhai* mobile phone constructed by phone users in relation to the characteristics of youth?
b) What are the meanings of the *shanzhai* mobile phone constructed by *shanzhai* workers in relation to the characteristics of youth?

SHANZHAI MOBILE PHONES

"*Shanzhai*" is not a brand name of mobile phones, but a generic term for all the small brands, non-brands and even pirated handsets. Initially, it referred to a low technological production which copied brand-name mobile phones and other digital products. Gradually, it was used to describe a type of industrial phenomenon in which small and unknown manufacturers attempted to imitate a variety of brand-name products. The word "*shanzhai*" can be used both as an adjective ("*shanzhai* phones") and a verb (e.g. to "*shanzhai*" a phone). The "*shanzhai* production" is mainly characterized as imitation, primitive technology, small operation, low cost, swift market adaptation, customer-tailored design and so on. "*Shanzhai* products" therefore are associated with brand-name imitation, low prices, shoddy quality, practical functionality, novel designs, alternative styles, etc.[1]

According to media reports, *shanzhai* mobile phones have occupied nearly 30% share of the domestic market in China. Globally, Chinese *shanzhai* phones sold about 150 million sets in 2007 when the "*shanzhai* phones" suddenly rose to prominence. This figure accounts for 1/10 of the total global sales (Yin 2009). Traditionally, *shanzhai* phones were mostly bought by migrant workers and low-income consumers. However, as *shanzhai* phones outperformed brand-name phones in price, functions, durability and designs, they became a trendy artifact for teenagers and urban youth in such metropolitan cities like Beijing, Shanghai, Guangzhou and Shenzhen. The most famous centre of *shanzhai* mobile phone manufacturing and trade is in the Huangqiangbei district in Shenzhen. This area of 1.45 square kilometers is the centre of 27 electronic markets and nearly 30 companies which generate RMB37 billion a year.

The MTK chip was central in bringing the *shanzhai* mobile phone to market. It revolutionized the manufacturing process, created a consistent quality in every handset and was at par with the major Western mobile phone manufacturers' specifications. In addition, the handsets could be custom made to clients' needs. For example, the colour and shape of the case could be tailored to a particular client's specifications, the ringtones tuned to any desired style, the monitor adjusted to size, pixels and colour combination, etc. In sum, the *shanzhai* handset became very competitive with most world brands and was offered at a much lower cost than the major brands from Japan, the USA and Europe. The *shanzhai* model of mobile phone manufacturing is now similar to the assembly of a computer – the MTK chipset is like CPU, memory ram, hard disk, and a video card. Once the core parts are put into place, all a manufacturer needs to do is to adjust the handset according to the customer's demands to choose different cases, ringtone styles, screens and other accessories. Thus a variety of brand new models of handsets are ready for the market. Not only is the quality of these phones quite acceptable, they offer fresher models and trendy designs. For example, during the 2008 Beijing Summer Olympics, there were *shanzhai* phones which looked like the "Bird's Nest" stadium and "Fuwa" mascots. During the 2008 U.S. presidential election, "Obama" *shanzhai* phones were sold for only US$30 in Kenya, and they had Obama's name and his slogan "Yes We Can" on the phone (Shanzhaiji 2009).

BRANDS AND *SHANZHAI* PHONES

The brand is explicit of a symbolic power. It exemplifies wealth, success and celebrity. It commands public prestige and consumerist authority. Above all, it is a manifestation of materialistic hegemony. It is commonly recognized that brand names play an important role in the construction of youth identity. The brand affinity is associated with three components: Authority, Identification and Approval (Morgan 1998). The authority refers to the heritage and trust embodied in the brand; the identification is related to a private construct of meanings and self-image; and the approval connected to the public ratification of the image and the authority of brand prestige. The consumption of a brand is in part expressive of the consumer's self definition (Morgan n.d.). As adolescents and young adults are in constant need of reassurance of their peer groups' approval for identity enhancement, the brand possession is a way to transfer the symbolic meanings from the group to the self who reflexively evaluate the responses (Dittmar 1992; Auty 2001).

Earlier studies found that Chinese consumers place more importance on the social value of brands than Westerners (Tse 1996). They are more likely to perceive the brand possession as a social status symbol, and use it for differentiating in-groups and out-groups. The former exert greater influence on brand-name preferences (Tse 1996). Chinese teens have also been discovered to be more positive towards foreign (Western) brands than their Japanese counterparts (Eckhardt and Houston 2002; Parker, Schaefer and Hermans 2006). They tend to perceive foreign brands as being more prestigious and therefore an impressive symbol for image enhancement.

However, the *shanzhai* phone is the antithesis to the brand-name. It is a no-name, cheap, imitation, low-tech, poorly crafted, of shoddy quality, and suspicious reputation which is often akin to piracy, counterfeits and illegal knock-offs. Until 2008 a typical *shanzhai* consumer was mainly associated with poorly dressed migrant workers from the countryside wandering to a city in search of a job. It is beyond anyone's imagination that such an obnoxious object would become a symbol of "cool culture" for urban youth. The fad of *shanzhai* has quickly expanded to a wider social and cultural landscape. Any famous and prestigious cultural product and event, be it a box-office movie, a CCTV gala, the Olympic Torch Relay, the Nobel Prize, could all be made into a *shanzhai* version in which ordinary youngsters thrive on their own parodying talent, grass-roots creativity, media-tech proficiency and juvenile contestation.

Research on the user culture of *shanzhai* phones is very rare. Naomi Klein (2002) in her best-selling book *No Logo* points out that there is a rising grassroots resistance in the age of global capitalism in which people's lifestyle is dictated by the hegemony of brands. This oppositional approach to the no-name brand phenomenon is one of the underlying constituents for this study. In addition, the literature which examines the consumption pattern of counterfeit goods may serve, to a certain extent, as a relevant reference point. Rutter and Bryce (2005) identify three discursive constructs in framing the issue: the deviant perspective which conceives the consumer of counterfeit goods as the "other". The act is unethical, isolated and marginal in a subversive subculture. The second is the oppositional perspective (e.g. cultural studies) which frames the consumption of brand-name imitations as political opposition to the corporate profiteering (Blankenship, 1986; Kwong et al 2003). The third is the perspective emphasizing the routine consumption practice. It argues that the consumption should be recognized as everyday leisure and routine. It shares more with an established consumption practice than either illegal or oppositional performance of subcultures (Miller 1998, Gronow and Warde 2001; Rutter and Bryce 2005). In their study of counterfeit consumption in the UK, Rutter *et al* discovered that engagement with counterfeit goods is broadly distributed across the country and cost was the top motivation. This pattern echoes that of consumption of legal goods.

In addition, there are other non-price determinants in consumers' decisions towards a counterfeit purchase. Wee, Tan and Cheek (1995) identify three categories: psychological factors, product factors and demographic factors. Each category contains several elements:

- Psychological attributes: Attitude, brand status, materialistic outlook, novelty seeking, and risk taking;
- Product attributes:
    - Functional: durability, physical appearance, quality, purpose, image;
    - Perceived fashion content
- Demographic attributes include age, education, income and so on.

CHARACTERISTICS OF CHINESE URBAN YOUTH AND *SHANZHAI* PHONES

Although academic research in *shanzhai* studies is still limited, popular discourses in the Chinese mass media place varied emphasis on it. One is the economic discourse which examines the business model of the shanzhai production and distribution (Zhong, 2009); the second is the technological discourse focusing on the shanzhai's implications for technological innovations. The third is the youth culture discourse which views the shanzhai phenomenon as representative of the characteristics of contemporary youth.

It is important to note that the current young generation of China's "mobile culture" is very unique in Chinese history. They are the first generation of the "single children" who were born after 1979 when China implemented the most strict family planning scheme, "the one child policy" in its urban families. There are over 100 million single children in cities (Ren 2006). About 70 percent of urban families are "single child" families. Many studies and popular writing describe this generation as the "Me Generation" who are much more pampered and thus more self-centered than the previous generations. They are also believed to be more materialistic, consumption-oriented (Zhong 2007; Ren 2003), brand-conscious (Zhao, Yin & Han 2006) and vulnerable to the addictive use of such technological artifacts such as the Internet, electronic/online games, MP3, even mobile phones. How would such a generation be attracted to *shanzhai* phones? Low prices for them may not be the decisive factor in their purchase. Some argue that the appeal of the *shanzhai* phone to the young people of the 80's and 90's lies more in their user-friendly designs and multi-functional performance, which are in keeping with the social characteristics of the new generation who seek change and novelty (Zhao n.d.). Other reports paint the *shanzhai* mobile phone as a signifier of a rebellion for the white-collar urban youth who toil at their tasks in front of office computers. They often feel suffocated in their environment of files, meetings and tedious work that they are in need of a toy to release their work-related stress, and the "*shanzhai* phone" falls into this role of a "toy" with a rebellious spirit (*Southern Weekly*, 2008).

Based on various references, we summarize the following 11 characteristics of contemporary Chinese urban youth which are raised in the popular discourse as connected to the symbolic meanings of the *shanzhai* phone. These attributes fall into five realms: political attitudes, practical rationale, social values, interest-seeking and self identification:

Political attitudes:

1) Anti-mainstream, anti-authority and anti-brand

Youth are in defiance of authorities. They express themselves through various ways to attempt to create new rules so as to realize the awakening of the self-awareness (Zhang & Wu, 2008). *Shanzhai* phones are believed to represent their attitude of defiance against the authority of mainstreams and big brands.

2) Patriotic

The youth of the 80's and the 90's tend to hold a more affirmative attitude towards China's current development status than the previous generation. They display a high recognition for the country, and identify strongly with the Chinese culture, and Chinese nation. Their patriotic enthusiasm and unity of edifying the spirit for the country were explicitly demonstrated during the Beijing Summer Olympics and the Wenchuan earthquake catastrophe in 2008 (Fang, 2009). This patriotic sentiment may influence their attitude towards the consumption of Western brands and turn their attention to local and no-name brand products like *shanzhai* phones.

Practical rationale:

3) Rational and pragmatic

Rationality is considered as a significant consumption pattern of the current youth. After an impetuous consumption period, the youth will gradually calm down and enter the rational stage. Making both ends meet, the youth tend to accept the pragmatic consuming model (Ye, 2005). They appear to be thrifty consumers (Gong 2002), which also applies to their consumption of mobile phones.

Social values:

4) Competition-oriented

The 80's and 90's generations who were born after China's reform and, opening up policy was implemented, grew up in an environment full of fierce competition. They have to strive their best to be recognized and to achieve their goals. The *shanzhai* phone symbolizes to some degree a competitive spirit and a local underdog competing with, and against, multinational corporate giants.

5) Upholding equality

The young generation pays more attention to the fairness and equality of society. This consciousness of equality goes beyond a political right, and extends as well to a consumption domain in a sense that they believe the general public should enjoy a high-tech consumption experience. The ownership of a mobile phone brings the freedom of acquiring information and the feeling of equality. The *shanzhai* mobile phone enables ordinary people to also enjoy the feeling of owning and using luxury mobile phones. A high-end handset is no longer a symbol of the rich, and it is available to the general public to use.

6) Participatory and do-it-yourself (DIY)

Young people are likely to participate and get involved in a DIY process. The *shanzhai* mobile phone is considered to provide the youth with a platform to participate directly and immediately in the seeking of novelty and luxury. Their obsession with the *shanzhai* phone is similar to their fascination with the computer which can be played with their own hands. The *shanzhai* mobile phone is also a combination of primary accessories and involves the DIY culture (Wang, 2008). The user-friendly attribute of *shanzhai* mobile phones corresponds to young people's favour for participation. The distinctive-appearance attribute of *shanzhai* mobile phones meets young people's favour for participation.

A *shanzhai* mobile phone exceeds the brand-name counterparts in offering a participatory DIY process with its ultra multi-function and user-friendly attributes -- its large screen, intelligence, TV functions, double cameras, the double caliper double treats, multimedia broadcasting and a super wait function (Yu & Zhang, 2005).

7) Tolerance of diversity

The young generation, with active thoughts and talent in imitating, tend to get in touch with and tolerate diverse new things, either in opinions or lifestyles. They are likely to accept and distribute different popular cultures. They are much more acceptable of a diverse range of new products, styles and consumption patterns (Yu, 2007).

Interest-seeking:

8) Technological novelty-seeking

Young people tend to follow technological trends and go after products perceived as innovative in technology and imagination. Novel things and simulative activities are adopted among the youth community at the fastest speed. The *shanzhai* phone embodies some sort of technological novelty that appeals to youngsters.

9) Entertainment seeking

Yong people tend to be attracted by the entertainment factor in the mobile phone designs. The new entertaining services are always needed to provide the youth an access to the up-to-date environment (Zhang, 2004). To some extent, mobile phones are considered as a "fashionable toy". Among the young people, the pattern of using the mobile phones is always called "play". Some of the master users are called "mobile players". These terms just indicate the entertaining attribute of the *shanzhai* mobile phone use. Under the support of new technology, an entertaining and freer environment is brought to the youth (Zhang&Wu, 2008).

10) Fashion seeking

Following a fashionable trend is the most significant manifestation of the youth's culture and lifestyle. Yong people with an open mind and strong curiosity pay more attention to personality and pursue a trendy expression of one self. They tend to accept more readily a new object, concept and action if it is considered "cool" and trendy. Young people play a leading role in society in setting new consumption styles (Yu, 2007).

Self identification:

11) Experience-centered and identity-searching

The *shanzhai* handset provides a luxurious platform for the youth to have a brand new personalized experience. A mobile phone is not a luxury, but a piece of consumer good for commoners to experience (Ye, 2006). The youth have come to the period of the rising experientialism. Experiencing another status, an exotic life, and the creativity -- this has become one essential part for the youth life (Ye, 2006). The consumption is also a process of experience. As the social consumption trend becomes diverse, young people have more choices. Unrestricted and personalized consumption becomes possible for them (Ye, 2006). Consuming mobile phones is no longer for showing off, but a consuming experience which can be enjoyed by everybody.

As mobile phones with high-quality and high-tech features are no longer the most popular, people are more likely to choose mobile phones with a personality (Ye, 2006). Thus a phone becomes an expression of self-identity. The *shangzhai* phone offers a variety of ways for a user to demonstrate his/her individuality through its distinctive design corresponding to his/her own preferences. (Yu&Zhang,2005).

This paper is to attest the verification of the above-stated attributes through a qualitative analysis of in-depth interviews and selected online group discussions. In line with the youth culture discourse, this paper conceptualizes that the use of *shanzhai* phones by young people signifies some sort of political statement against the authority of the mainstream and the brand culture or of patriotic sentiment supporting China's brands. For analysis, youth's consumption and usage pattern of mobile phones, especially these young people's attitudes towards *shanzhai* mobile phones, are regarded as representative of their social characteristics summarized in the above-stated section. We expect to substantiate these arguments in the youth culture discourse through the methods of interviews and online discussion analysis to uncover whether the resistant emotion or anti-mainstream attributes of the youth are articulated in their attitudes toward *shanzhai* mobile phones, and whether other attributes are also shown in their mobile phones' consumption and usage patterns. This study explores this relation primarily from two perspectives: the mobile phone users (including users and non-users of *shanzhai* phones) and those who work in the *shanzhai* industry (*shanzhai* workers). Due to these two groups' differences in knowledge of, and orientations towards, *shanzhai* mobile phones, they are separately analyzed in respect to their discourse of, and attitudes towards, the *shanzhai* phone.

METHODOLOGY

This study adopts two methods: a qualitative analysis of in-depth interviews and a textual analysis of QQ group discussions.

The interview analysis involved in-depth interviews of 16 young people, including 9 males and 7 females, aged between 17 and 38. Among them, 2 were 18, 12 between the age group of 19-30, 2 in the age group of 31-38. The interviewees came from eight Chinese cities including big cities like Beijing, Shanghai, Guangzhou, Shenzhen, mid-sized cities like Nanchang (Jiangxi Province), Quanzhou (Fujian Province), and small cities like Huludao (Liaoning Province) and Maoming (Guangzhou Province). Among them 9 were *shanzhai* phone users and 7 were non-users of *shanzhai* phones. 13 were selected through a snowball method of recommendations by friends, acquaintances or interviewees to conduct the face-to-face interviews. The remaining 3 were selected for online interviews from the *shanzhai* QQ groups observed in this study. They were selected purposely based on their high level of participation in the QQ groups analyzed. The average interview lasted about 30 minutes. The face-to-face interviews were recorded and later transcribed word-by-word into digital document files, while the data of on-line interviews were downloaded and then sorted into the files. (See Figure 1 for demographic information of the interviewees.)

The 11 characteristics of urban youth stated above were used as the principal classification of

Figure 1. Demographic information of interviewees

No	Name*	City	Age	Gender	Education	Occupation	Use Type*
1	Alex Guo	Guangzhou	19-25	M	High School	Office Worker	U/W
2	Amy Zhou	Maoming (Guangdong)	-18	F	High School	Student	U
3	David Hu	Huludao (Liaoning)	31-38	M	Undergraduate	Businessman	N/W
4	Grata Gan	Shanghai	19-25	F	Undergraduate	Office Worker	N
5	Hannah Hu	Shanghai	19-25	F	Undergraduate	Office Worker	U
6	Jerry Xia	Shanghai	19-25	M	Graduate	Student	N
7	Kevin Guo	Shanghai	26-30	M	Graduate	Student	U
8	Kitty Yang	Maoming (Guangdong)	-18	F	High School	Student	U
9	Liz Yang	Guangzhou	26-30	F	Graduate	Student	U
10	Mr. An	Shenzhen	19-25	M	High School	Office Worker	U/W
11	Mr. Chao	Nanchang (Jiangxi)	19-25	M	Undergraduate	Office Worker	N
12	Mr. Dong	Beijing	31-38	M	Undergraduate	Businessman	U
13	Mr. Li	Shanghai	19-25	M	Undergraduate	Office Worker	N
14	Mr. Liu	Shenzhen	26-30	M	Undergraduate	Office Worker	U
15	Ms. Yang	Maoming (Guangdong)	26-30	F	High School	Office Worker	N/W
16	Sally Yang	Quanzhou (Fujian)	19-25	F	Undergraduate	Student	N

Note

Name: To protect the interviewees' privacy, their first names are changed to English names. 6 of them prefer not to give their first names.

Use Type: U: Shanzhai users; N: Non shanzhai users; W: Shanzhai workers

categories. Under each category, several subcategories were created during the qualitative analysis.[2]

The QQ group discussion analysis was conducted based on the four QQ group discussion content. Through our online search and observation, the four QQ groups, which were highly active in the *shanzhai* business, were purposely selected. Most of the members in the QQ groups were in the business of *shanzhai* manufacturing, promotion and distribution. The content of the four QQ group discussions was downloaded between Sept.19, 2008 and Feb. 29, 2009.

The four QQ group are:

- The Home of *Shanzhai* Mobile Phone Friends (hereafter referred to as *Shanzhai* Friends, and the corresponding bbs address: http://qun.qq.com/air/#53125564/bbs
- Directive Groups of *Shanzhai* Kings (hereafter referred to as *Shanzhai* Kings), and the corresponding bbs address: http://qun.qq.com/air/#68156093/bbs
- China *Shanzhai* Mobile Phone Communication Platform (hereafter referred to as *Shanzhai* Platform), and the corresponding bbs address: http://qun.qq.com/air/#38505366/bbs
- *Shanzhai* Base (hereafter referred to as *Shanzhai* Base), the corresponding bbs address: http://qun.qq.com/air/#49955640/bbs

Based on a preliminary summary of various discussion points in the QQ groups' contents, we then selected the *Shanzhai* Friends as the focus of analysis because it offered more information, opinions or suggestions regarding members' choice, usage and attitudes towards *shanzhai* mobile phones. The other three groups were more business-oriented information centers. This part of the analysis primarily was concerned with how young *shanzhai* practitioners perceived and promoted the *shanzhai* mobile phone in particular and *shanzhai* as a business in general. These issues were examined based on the following five categories: a) Definition of *shanzhai* mobile phones; b) Identification with the *shanzhai* business; c) Attitudes towards *shanzhai*; d) Promotional strategies; and, e) Knowledge of *shanzhai* phones and the *shanzhai* business.

The results were then further analyzed with the 11 attributes in the interview analysis. Both analyses of interviews and QQ discussion content were conducted through a qualitative research application tool.

RESEARCH FINDINGS

Combined findings from the two analyses indicate the following the orientations in constructing the meaning of *shanzhai* phones which connects to the characteristics of contemporary Chinese urban youth:

Practical Rationale

The interviewee's attitudes towards *shanzhai* mobile phones are prominently featured in connection with their characteristic of practical rationale. They consume *shanzhai* phones predominantly out of economic consideration of the handset's cost-performance ratio as an intelligent and thrifty consumer.

All the 16 interviewees agreed that the dominant advantage of *shanzhai* mobile phones is their price, in particular in their consideration of the cost-performance ratio. Although most of them considered the quality is the first consideration, 15 of the 16 interviewees stated that the low price is one of the most attractive factors for their purchase intention. Ten out of the 16 weighed more on the cost-performance ratio. The focus in their discussion was often around whether the mobile phone was worth the asking price. Mr. Dong, a businessman said,

"I pay most attention to the cost-performance ratio. Before the final decision, I often shop around. And the reason why I support the shanzhai phone is mainly for its high cost-performance ratio.... I have chosen the shanzhai phone because I feel it is user-friendly, solid and durable. I pay more attention to signals. The shanzhai phone of mine is better in signals. In many areas where signals are weak, other people's phones can not receive any signal, but my phone can. And my phone is good with battery. Because I have many phone calls. The battery of brand phones will only last for one day, but my shanzhai phone can last for three days. Furthermore, the volume must be loud enough because I am often outside of the office." (Mr. Dong, 38)

For *shanzhai* business people in the QQ group, they emphasized that low prices are essential and the cost/performance ratio is most important for the *shanzhai* phone. The words they repeatedly used to describe the *shanzhai* phone included: cheap, bargain price, good performance/price ratio. An interesting metaphoric description is: the price you pay for a mobile phone is equivalent to that for a Chinese cabbage;

Entertainment and Fashion Seeking

The second prominent point is the entertainment and trendy orientation demonstrated in the youth's attitudes towards the *shanzhai* phone. Young people constantly seek technological novelty, entertainment and fashion. Our findings indicate that young people felt attracted to the *shanzhai* phone because they discovered a lot of entertainment and fashion in it.

The entertainment attribute of the *shanzhai* phone was evidenced in the interviews. From most interviewees' discourse, mobile phones tend to be described as a "toy". To use a phone was referred to as to "play"; master users have then become "players." The *shanzhai* phone has many entertainment functions including polyphonic+mp3/mp4, the appearance of the design, radio and e-books, playing QQ, watching TV programs and films, taking photos, etc. From the descriptions of the interviewees, the *shanzhai* mobile phone is more like a toy both in the appearance and in the functions. Liz Yang's words can summarize their common opinions:

"I have played with the shanzhai mobile phone, and felt it pretty good. It was just like fashionable gadgetry. I killed time by playing games with it, listening to the radio and reading e-books. All kinds of entertainment. Many people like to watch films or TV on their phones. This was especially popular during the Beijing Olympics. There are also people who like reading e-books or chatting on QQ. [The shanzhai phone] meets this trend of entertainment. Its appearance designs have also become more entertaining. The shanzhai phones are quite impressive in their decorative, cartooning, toy-like and colorful designs. In sum, they are very entertaining and attract the eyeballs of youngsters. Some shanzhai phones look very delicate and cute. I have seen a mini iPhone, with Hello-Kitty in the back, and it also has a cover which turns into a TV screen. The cover can also rotate. It is quite lovable. I'm fond of cute ones. The Hello-Kitty ones in particular meet the desire of young girls for fashion-seeking. They are also in clear and brisk pink color which is my favorite. (Liz Yang, 27)

Seeking a fashion in the appearance, function, pattern/style of mobile phones is demonstrated in youths' choice of phones, especially for female users. In their rating of various attributes of phones, appearance ranks second, followed by the quality, but before the price. The youth tend to go after a fashionable trend. Owning one fashionable mobile phone is one important channel to receive acknowledgement from the surroundings.

As people are seeking for the fashionable appearance, "*niu*" (pinyin, this word expressing one's admiration emotion) is used for describing the freshest mobile phones, which expresses youth's enthusiasm and identification for the fashion.

2008-11-3 14:02:00 *Shanzhai* Orange

The purpose of buying shanzhai mobile phones is just for its good-looking appearance. Among the shanzhai mobile phones, the fashionable ones attract people more.

2008-10-7 17:40:51 Xiaoyujie

I just want to get a mobile phone with a "niu" model. I do not care about the price.

2008-10-7 17:42:26 Crazy for *Shanzhai*

What is the meaning of "niu"?

2008-10-7 17:42:37 Crazy for *Shanzhai*

How to measure it?

2008-10-7 17:42:58 Xiaoyujie

I hope I will get more attraction if I use it when I am walking on the road.

2008-10-7 17:43:22 Kural

I think if you just wear one shoe. You can get it.

Moreover, in view of catching up with swiftly changing fashions of the mobile phones, one view was repeatedly emphasized in QQ discussions that paying too much for brand-name mobile phones was a deplorable way of seeking fashion.

2008-10-5 16:04:31 *Shanzhai* Orange

Shanzhai mobile phone which just is sold for several hundreds for me is very good.

2008-10-5 16:05:13 Crazy for *Shanzhai*

If you buy a mobile phone which is 4,000-5,000, you will certainly feel regretful after 3 months!

2008-10-5 16:05:43 *Shanzhai* Orange

Anyway, even if it is the best mobile phone, you will feel tired of it after holding it for 10 minutes in your hand.

2008-10-5 16:06:19 Xiao Maodou

I agree with you. I still think my C119 is good enough. I will not care even if it is lost or is stolen. It is just 100 RMB.

2008-10-5 16:07:05 Xiao Maodou

Mobile phone is just like a toy with a rather high updating trend. It is not necessary to seek the fashion trend by buying expensive brand-name mobile phones. If you want to make yourself look very respectful, you may buy various mobile phones with multi-level functions. Thus you can change it daily.

2008-10-5 16:07:30 Xiao Maodou

I am doing like this. I have got various mobile phones. Ha-ha.

Hence, *shanzhai* mobile phones were regarded favorably by young fashion seekers since they made it possible for frequent phone replacement at a low cost. With a very short life cycle of a handset and low prices, it is easier to follow a trend by obtaining a *shanzhai* phone.

2008-11-26 20:29:08 Laobai

I still like buying shanzhai mobile phones.

2008-11-26 20:29:13 Laobai

You can buy a new one if it can not be used anymore.

2008-11-26 20:29:15 Laobai

Like this, you can try a fresh model continuously.

Although the *shanzhai* phone is known as being imitative rather than innovative, some interviewees still found novelty in it. One interesting example was about a *shanzhai* phone which was quite popular with secondary school students in Maoning, Guangdong. It had a voice-changing function so that students could use that phone to call their teacher for a sick leave pretending to be their parents or other adults.

But a consensus among the interviewees and QQ groups was that most of the *shanzhai* phones were imitative and the technological innovation was rather rare.

2008-11-16 22:02:03 Qing Ping

There is no innovation. They use the raw and low-cost materials to produce imitations of brand models. There is no after-sale service, no quality insurance. Thus the cost will decrease and the mobile phones which the consumers get will be cheap.

2008-10-7 17:45:36 Xiaoyujie

A shanzhai mobile phone also has an original version.

2008-10-7 17:45:54 Crazy for *Shanzhai*

The original one?

2008-10-7 17:45:55 Cool Digital

Some of the appearances are original.

2008-10-7 17:46:19 Crazy for *Shanzhai*

OMG! Even if the appearance is rather unique, it is also imitative. You might as well buy an original one! All of the shanzhai are anti-mainstream.

Political Attitudes

Our findings indeed demonstrated the attitude of defiance against the monopoly of brand-name mobile phones, specially the foreign brands, but this oppositional orientation is definitely not prominent. 13 out of 16 interviewees expressed the view that *shanzhai* phones contributed to the breaking up of the monopoly of brand-name mobile phone companies.

Sally Yang, a university student, said,

"I do hope shanzhai phones will grow and surpass the brand phones. But I also hope they should be kept at a low price, better quality, and being trendy."(Sally Yang, 20)

Eight out of these 13 people expressed their support for Chinese brands; however their support was mainly based on the condition that Chinese brands have the similar level of quality as foreign brands. Hannah said,

"I wanted to buy a shanzhai mobile phone just based on the thought of supporting Chinese own brands. Some Chinese brand goods are both good quality and low price; they are just as good as foreign brands. They have a better price-performance ratio." (Hannah Hu, 24)

Jerry Xia stated:

"If the quality of Chinese goods is good enough, I prefer to buy them.... But the current situation is that the quality of domestic brand phones is not so good. I hope the market system will be improved and Chinese will establish their own brands." (Jerry Xia, 24)

However, for young *shanzhai* workers in QQ groups, although they had different opinions about *shanzhai* phones and some of them never used one, they demonstrated a strong political attitude towards supporting and developing Chinese brands. The dominant rhetoric is support for China's own production, which reflects in the recognition and support for such Chinese brands like Lenovo, Tianyu, Jinli, and so on. They also viewed the *shanzhai* phone as part of Chinese production and hoped that Chinese brand productions would occupy a space in the world market.

2008-10-14 21:18:51 No.9(CUOTO)

Support China's brands.

2008-11-3 14:00:00 Zhanbaini

Tianyu is the standby of China's brands. I support it.

2008-9-20 12:38:48 Zhongzhengkeji

Shanzhai mobile phones are suitable for Chinese or even the whole world. GOOD.

In regard to their attitude towards brand mobile phones, they recognize that brand phones may be slightly better in quality and after-sales service, but they are too expensive.

2008-11-16 22:23:17 ZhongzhengkejiF488E

Do you want the brand mobile phones with sky-high price occupy the Chinese market?

One of the prominent points raised in their discussion is that *shanzhai* phones are more suitable for Chinese people and the Chinese market than some Western brands. One example is when they talked about the Blackberry:

2008-9-27 16:57:36 Xiaohan

Blackberry does not fit for Chinese.

2008-9-27 16:58:32 Xiaohan

It is made for America for its e-mail function.

2008-9-27 16:58:41 Xiaohan

Multifunction is not enough.

2008-9-27 16:59:23 Heshen888

I have played it, but I did not like it.

2008-9-27 17:05:00 Xiao Laoda

Chinese are not interested in the push mail technology.

Social Values

Among the four characteristics in this category, two attributes are repeatedly exhibited: the positive recognition of the competitive spirit and endorsement of equality in consumption.

In the interviews, the competitive spirit is positively recognized as playing a key role in the success of *shanzhai* products. If the *shanzhai* phone can succeed in the marketplace, it means it is more competitive.

Several interviewees held the attitude of "survival of the fittest" towards the existence and development of the *shanzhai* mobile phones in the marketplace. Kevin Guo and Mr. Li both stressed the point that "being is rational." Kevin Guo said:

"I support shanzhai mobile phones. Being is rational and valuable. And the shanzhai phenomenon is an outcome of social development at a certain stage. Just like our country's economic system, public ownership as the main body, other economic ownerships coexist. And the market economy is better than the planned economy. To some extent, the shanzhai mobile phone is like a private economy with low cost and small scale, but the role it plays cannot be ignored. It applies to the necessary demand. One day, if it is out of the date, it will disappear automatically without even our opposition." (Kevin Guo, 26)

Another recurring theme expressed by the interviewees and in the QQ groups is the freedom of choice in commodity consumption which is made possible by *shanzhai* production. This is resonant with the young generation's characteristic of upholding equality. It is interesting to note that this equality attitude is extended to the sphere of consumption. The technology-sharing attribute embodied in the *shanzhai* phone is emphasized.

14 people regarded mobile phones as technology sharing goods rather than a status symbol. Mr. An, who is a salesman for *shanzhai* mobile phones, often stated that selling *shanzhai* mobile phones was not just for money, but for sharing benefits with Chinese people. "I think *shanzhai* mobile phones are good stuff suitable for the masses." (Mr. An, 23) His point was also echoed by Amy Zhou who said: "*Shanzhai* phones are products for the masses. They serve the interests of consumers. To become a luxurious brand is not the only way to make it. (Amy Zhou, 18)

Kevin Kuo said:

"The mobile phone is a kind of technology-sharing product. Now, I can pay a lot to buy a brand-name iPhone, and also I can buy a shanzhai version, such as a shanzhai Orange. It is also pretty good. Many people around me bought it just for experiencing the iPhone." (Kevin Guo, 26)

Self-Identification

It is explicit from the interviews that the mobile phone involves a very distinctive and individualistic experience. When the interviewees described what mobile phones they would like to own, a distinctive artifact with one's individualistic characters is often mentioned, which reflects their desire to seek individualism and also to be recognized by one's group to have a distinctive identity. When they were asked whether they would feel 'losing face' if they buy or use *shanzhai* products, Sally Yang said:

I have never thought that way, it is only a handset, if it has a good quality and is fashionable, I don't feel 'losing face.' On the contrary, [using shanzhai] demonstrates a very individualistic behavior, and if possible, I want to have a unique one, which can show my distinctiveness from others. (Sally Yang, 20)

Amy Zhou also liked to have a unique phone, but she considered using an unpleasant-looking phone "face-losing." She said:

"If a phone is too ugly, I would rather not to use it. I care a little more about "my face". I hope to use more respectful phones. When I go to university, I will buy a brand-name phone.... I don't like to own the same phone as others, but if these people are my friends, I will feel OK to be the same with them. Of course, it is best to have a unique one. I have a thought to have a phone tailor-made according to my own design." (Amy Zhou, 18)

However, David Hu said that he did not go for the self-representation of the mobile phone, but rather for the phones which were conceived by people of his social circle as a representation of social status who identified more with brand-name phones. He usually changed his handset every two or three months. He said:

I will spend RMB20,000-30,000 a year for the replacement of mobile phones. I will lose interest in the handset if too many people around me use the same one. The phones I use do not represent my personal character, but it is the thing you hold in your hand and it has to appear to have a social status. In addition, all my friends use the mobile phone over RMB3000 apiece. People from the Northeast pay most attention to "face." I have been suffering from it. But the mobile phone must not be lower-class.... That's the social circle of my life, people care about "face", and go after the fashion; people in this class are just like that. (David Hu 31)

However, for discussants in the *shanzhai* QQ groups, our findings indicated that they identified quite strongly with the *shanzhai* phone and *shanzhai* business. One indicator is their QQ names. Many of them tended to name themselves with "*shanzhai*" or "*zhai*". Nearly 30 percent of QQ names were directly or indirectly related to *shanzhai* in the two groups we conducted a name count. In the *Shanzhai* Friends, 12 out of the 59 members named themselves related with "*shanzhai*" or "*zhai*", such as:

- Crazy for *Shanzhai* (我为山寨狂),
- *Shanzhai* Captain (山寨城管队长),
- King of *Shanzhai* Gang(山寨帮派老大),
- Supporting Chinese *Shanzhai* Mobile Phones(支持中国山寨机),
- Carrying *Shanzhai* Through to the End(将山寨进行到底) and so on.

In *Shanzhai* Platform, 36 out of the 114 participants called themselves "*shanzhai*" or "*zhai*" and associated them also with their location. For example,

- General King of *Zhai*- Zhang (总寨主-张),
- Shenzhen Contemporary *Zhai*-Li (深圳现代寨-李),
- Chongqing *Zhai*-No.9(重庆寨-老九),
- Ningbo *Zhai*-Purchasing Agent(宁波寨-机采购).

Therefore we can see that many members, although mainly coming from Shenzhen, were from different places. They gathered in the QQ group exchanging information. This phenomenon is also evidenced in the QQ group *Shanzhai* Base, and the group master even required the members to put the word "*zhai*" before their names. And over 198 among 408 followed this suggestion.

The group sites also carried slogans for the promotion of *shanzhai* mobile phones or the so-called *shanzhai* careers.

Shanzhai Kings:

"Sharing how to operate shanzhai mobile phones in the future development";

Shanzhai Friends:

"Shanzhai spirit: nothing can defeat us!! This group is for the communication of shanzhai enthusiasts and we are not welcoming the advertisements!!"

And also in the group members' discourse, you can easily find promotional words like, "China needs to open to the world and follow the steps with the developing countries, and *shanzhai* should be promoted to the outside word." Others include: "*Shanzhai* mobile phones are good enough and fit for everybody"; "Defeat Nokia and develop *shanzhai*"; "The powerful *shanzhai* mobile phones," etc.

In sum, although the members have not clearly discussed their aims for undertaking the business with *shanzhai* mobile phones, through an analysis of the QQ names and their discourse, roughly four levels of identification are discovered in the process:

For a living→ interest in *shanzhai* →enthusiasm for *shanzhai* →*shanzhai* career.

Mr. An and Alex Guo were two *shanzhai* workers selected from the QQ groups for in-depth interviews. Asked about the purpose of doing business with *shanzhai* mobile phones, Mr. An, the seller of mobile phones, interpreted,

"My original objective was for making money and for a living. With gradually carrying out this business and communicating with fans of shanzhai phones, I also hoped the Chinese own mobile phones will get ahead in the mobile phone market."(An, 27)

Alex Guo, the other interviewee, said, "I am engaged in *shanzhai* mobile phone design for my part-time job. I am not just interested in it, but I also hope one day it will become my career."

DISCUSSION AND CONCLUSION

Several elements could be drawn from our findings:

The youth's attitudes towards *shanzhai* mobile phones prominently reflect their practical rationale. They consume *shanzhai* phones predominantly out of economic consideration of the handset's cost-performance ratio as an intelligent and thrifty consumer. In the previous studies and the popular media discourses, there are many negative comments towards Chinese youths' consumption patterns in the present time. They are often depicted as much more pampered with an extravagant or wasteful consumption style. In this study, we have found that they are indeed consumption-oriented, brand-conscious and vulnerable to the addictive use of the mobile phone, but they are also very rational and pragmatic. The *shanzhai* mobile phone, combined with features of low prices, practical functionality, fairly good quality, and so on, is just the right commodity. The popularity of the *shanzhai* phone among the young people is foremost a practical economic choice.

The second eminent point is the entertainment and trendy orientation demonstrated in the youth's attitudes towards the *shanzhai* phone. This obviously relates to the subculture of current Chinese youth, congruent with what Yu summarizes by stating that the current youth are likely to be participants in entertainment-seeking and enjoyment processes (2005). And this subculture is most often associated with an entertainment-focus. It is even called the youth entertaining subculture (Zhong, 2007). Although the mobile phone is mostly instrumental, the youth also regard their mobile phones as a "gadget" for killing time and they are willing to buy new and posh phones to "play" with and to look "fashionable".

Our findings show that entertainment is the most prominent in our interviewees' meaning of the *shanzhai* phone. Compared with the brand-name mobile phone, the *shanzhai* phone is always regarded as a funny "toy". Many people buy it just for its "playful" attribute and can be used for killing time. It can be said that the *shanzhai* phone's entertaining nature just satisfies the current youth's willingness to indulge in entertainment. Similar to the cool style of fashion representation, the *shanzhai* phone represents a life style of involving both entertaining and trendy elements that appeal greatly to young people. Some Chinese scholars have referred to this entertaining subculture as a kind of postmodern consumption culture, which aims to deconstruct tradition and authority (Zhong 2007).

In this study, our initial thought was that the *shanzhai* phone use may reflect mainly the political orientation of the young generation's resistance towards the traditional and mainstream culture in metropolitan cities. Obviously, we have identified some elements of anti-mainstream, anti-authority and anti-brand attitudes, as well as patriotic sentiments to own and produce China's brands, but this political orientation is definitely not prominent. And we also noticed that the youngest generation of the 90s are most resistant towards the mainstream, but pay least attention to the politics of brand than the other two generations of the 70s and 80s. This is consistent with some of the known characteristics of the present-day youth subculture that they are apparently rebellious and look for alternatives, but yet they are less politically emotion-charged. *Shanzhai* mobile phones represent a symbol of rebellion, resistance, fringe, and outlawing to some extent, and these attributes echo the current youth's yearning for rebellious expression to the outside world. The resistant element does exist in the use of *shanzhai* phones by the young Chinese, but it is not a compelling one.

In the social value orientation, one recurring theme is the freedom of choice in commodity consumption which is made possible by *shanzhai* production. This is resonant with the young generation's characteristic of upholding equality. Another theme that warrants attention is young people's identification of the competitive spirit with the *shanzhai* mobile phone. In their discourse, the competitive spirit is recognized and emphasized as one of the principles in evaluating the popularity of *shanzhai* products. If the *shanzhai* phone can succeed in the marketplace, it is because this indicates the *shanzhai*'s competitiveness in the market. This attitude may reflect the highly competitive environment young people face today from school life to the job market.

Furthermore, it is the issue of brand imitation which has been closely associated with the *shanzhai* phone from its outset. To what extent, do the brand-names of mobile phones and the social status they carry play a part in the use of *shanzhai* phones?

In Martineau (1968)'s research, a person's consumption pattern is a symbol of his/her social status. Image attributes reflect how consumers feel about product use which associates them with a desired group, role, or self-image (Lefkoff-Hagius and Mason, 1993). While, in this study, 14 out of 16 interviewees consider that the mobile phone consumption is not related to the social status; they also state that they will not feel lack of respect if they use *shanzhai* mobile phones. They tend to regard mobile phones as high technology commodities which should be shared by common people. Only one interviewee admitted that he has to pay much for mobile phones in order to gain more respect and promote his image among his peers. The QQ group discussion even ridiculed those who still considered the mobile phone a social status by citing an example that employees usually used more expensive phones than their boss.

The next question is what motivates the consumption and production of brand-name imitations? As it was pointed out, some of the incentives may be just to seek novelty and follow a trend. Consuming brands is a low cost means of satisfying curiosity and need for experimentation (Wee, Tan and Cheok, 1995). This pattern is also evident in the *shanzhai* phones' consumption for the urban youth. First, the youth are the prominent group for seeking novelty and fashion. In this study, we've found the youth have a high recognition for technology innovation and fashion. They are very keen on getting the newest "gadget" in their hands, to feel it, to test it and to play with it. Jut to have an immediate and intimate experience with the artifact to satisfy their curiosity for the latest "toy." And in view of the short life cycle of mobile phone models and the brand-name phones' high price, it is not necessary to pay much more for brand-name products which may be in vogue for only a limited period of time. But *shanzhai* phones provide a possibility for them to meet their curiosity and need for experiencing yet another handset with a low purchase cost. Once again, it is out of a practical rationale as a thrifty consumer—to obtain an experience of "brand-name gadgets" through an affordable cost in satisfying their curiosity for technological novelty, the need for self experience and do-it-yourself, and most importantly, the urge for expression of self-identity and self-uniqueness. So it is not so much a vanity concern for social status elevation, but rather as a need of self experience and self expression within an economic means of affordability.

Finally, this study finds that young people who are engaged in the *shanzhai* business of production and promotion do form their own cycle and build their identity strongly associated with the "*shanzhai*" both as a business career and political mission. No matter what their own attitudes towards the *shanzhai* mobile phone itself or its business development, they are purposeful in developing China's own brands and producing mobile phones deemed suitable for the Chinese market. They build QQ groups on the Internet as collective platforms, name themselves with *shanzhai* related words, and promote the *shanzhai* business with catchy slogans and political enthusiasm. And they constantly communicate and share information about *shanzhai* mobile phones, not only for sales promotion and market development, but also for having a grand mission in mind to improve and expand the *shanzhai* business. As business people, their first and foremost concern is to make money. However, they are also concerned about developing Chinese brands. From the attribute analysis of differences between ordinary users and the practitioners in the interviews, it is clear that *shanzhai* businessmen tend to be more anti-mainstream, anti-authority and anti-brand oriented than ordinary users. They also demonstrate more patriotic sentiment. The *shanzhai* mobile phone is no longer just a business product. And it is also an artifact reflecting their passion to seek more technological innovation and attaching great importance to competition and equality. In a word, for these people, their connection with *shanzhai* mobile phones continuously constructs their identity both as businessmen and also as promoters of Chinese "brands".

We recognize limitations of this study, specifically, that the sample size of 16 interviewees may not be representative enough for the understanding of the urban youth in contemporary China. As a complicated social and cultural issue, the *shanzhai* phenomenon closely connects to the process of globalization. It is rooted in the world's economic structure whereby brands are separate from their manufacturers, and Western brands reign over the global consumer market. The relationship between *shanzhai* and globalization may deserve future investigations.

REFERENCES

Blankenship, L. (1986). The conscience of a hacker. *Phrack, 1*(7), phile 3 of 10, from http://www.phrack.org/archives/7/P07-03

Dittmar, H. (1992). *The Social Psychology of Material Possessions*. Hemel Hempstead: Harvester Wheatsheaf.

Eckhardt, G. M., & Houston, M. J. (2002). Cultural paradoxes reflected in broad meaning: McDonald's in Shanghai, China. *Journal of International Marketing, 10*(2), 68–82. doi:10.1509/jimk.10.2.68.19532

Fang, N. (2009). *Shanzhai shouji chanye yanghuo zhusanjiao 300wanren* [*Shanzhai* phone industry supports livelihood of 3 million in the Pearl River Delta]. *Southern Net*, 02-17. Retrieved February 17, 2009, from http://tech.163.com/09/0217/03/52AV0FNJ000915BE.html

Fortunati, L. (2001). The mobile phone: An identity on the move. *Personal and Ubiquitous Computing*, *5*(2), 85–98. doi:10.1007/PL00000017

Gong, C. (2002). Kuwenhua, qingnian jiazhiguan, shehui zhuanxing [Cool Culture, youth values and social transformation]. *Youth Research*, *2*, 30–35.

Gronow, J., & Warde, A. (Eds.). (2001). *Ordinary Consumption*. London: Routledge.

Katz, J. E., & Aakhus, M. (Eds.). (2002). *Perpetual Contact*. Cambridge, UK: Cambridge University Press. doi:10.1017/CBO9780511489471

Klein, N. (2002). *No Space, No Choice, No Jobs, No logo*. New York: Picador.

Kwong, K. K., Yau, O. H. M., Lee, J. S. Y., Sin, L. Y. M., & Tse, A. C. B. (2003). The effects of attitudinal and demographic factors on intention to buy pirated CDs: The case of Chinese consumers. *Journal of Business Ethics*, *47*, 223–235. doi:10.1023/A:1026269003472

Lefkoff-Hagius, R., & Mason, C. H. (1993). Characteristics, beneficial, and image attributes in consumer judgments of similarity and preference. *The Journal of Consumer Research*, *20*, 100–110. doi:10.1086/209336

Martineau, P. (1968). Social class and spending behavior. *Journal of Marketing*, *23*, 274–278.

Miller, D. (1998). *A Theory of Shopping*. Cambridge, UK: Polity Press.

Morgan, R. (1998). Linking brand imagery with preference. In Greig, I. (ed), *Brand Choice Modeling*. ESOMAR, 63-84.

Parker, R. S., Schaefer, A. D., & Hermans, C. M. (2006). An investigation into teens' attitudes towards fast-food brands in general: A cross-cultural analysis. *Journal of Foodservice Business Research*, *9*(4), 25–39. doi:10.1300/J369v09n04_03

Ren, M. (2003). Qingshaonian wenhua yihua xianxiang de shehuixue fenxi [Sociological analysis of the cultural alienation of youth and juveniles]. *Contemporary Young Research*, *1*, 21–24.

Ren, X. (2006). Chengshi qingnian xiaofei wenhua jiedu [Interpretation of urban youth consuming culture]. *Journal of Beijing Youth University for Political Science*, *1*, 23–26.

Rutter, J., & Bryce, J. (2005). *Consumer Engagement with Counterfeit Goods*. Report for the Organised Crime Task Force, Northern Ireland Office. Retrieved from http://www.allianceagainstiptheft.co. uk/downloads/pdf/Fake-Nation.pdf

Shanzhaiji. (2009). *Daihao MI Obama* [Coded MI Obama]. Retrieved January 16, 2009, from http://www.shanzhaiji.cn/info /20090116/7035.html

Southern Weekly. (2008). To defend the spirit of *shanzhai* mobile phones. Retrieved August 4, 2008, from http://news.qq.com/a/20080804/000440.htm

Tse, D. (1996). Understanding Chinese people as consumers: Past findings and future propositions. In Bond, M. (Ed.), *The Handbook of Chinese Psychology* (pp. 352–363). Hong Kong: Oxford University Press.

Wang, Y. (2008). Shanzhaiji yu guochanshouji de fazhan [*Shanzhai* mobile phones and the development of mobile phones made in China] [Humanities and Social Sciences]. *Journal of Hubei University of Economics*, *12*, 71–72.

Wee, C. H., Tan, S. J., & Cheek, K. H. (1995). Non-price determinants of intention to purchase counterfeit goods: An exploratory study. *International Marketing Review*, *12*(6), 19–46. doi:10.1108/02651339510102949

Xi, G. (2008). Woguo shouji yonghushu 6.08yi [The number of the mobile phone users has reached 608 million]. *People.com.* Retrieved February 43, 2009, from http://www.chinaz.com/News / IT/09243XT2008.html

Ye, D. (2006). Shouji waiguansheji de yulehua qushi [Entertainment tendency of mobile phone contour design]. *Packing Engineering, 27*(3), 178–180.

Ye, Q. (2005). Shehuixueshiyexia dangdai chengshi qingnian de xiaofeiguan [Contemporary urban youth's consumption value under the sociology perspective]. *Journal of Hefei University of Technology, 19*(5), 144–148.

Yin, S. (2009). Shanzhai shi dianfu ["*Shanzhai*" subversion], *21Century. Business Review (Federal Reserve Bank of Philadelphia), 04*, 20.

Yu, J. (2005). Shouji duanxin shishang dui daxuesheng jingying wenhua de chongji [Shocks to college students elite culture from mobile phone short messages]. *Journal of Yuxi Teachers'. College, 21*(10), 88–90.

Yu, S., & Zhang, S. (2005). Chuangzao renxinghua de shouji shijie [Creating human-based mobile phone world]. *Chinese Journal of Ergonomics, 11*(4).

Yu, Y. (2007). Qingnian wenhua shiye kan qingnian shishang [Youth fashion from youth culture view]. *Journal of Beijing Youth University for Political Science, 16*(2), 12–15.

Zhang, L., & Wu, W. (2008). Dangdai chengshi qingnian tezheng yuqi yingxiang yinsu de fenxi [Characteristics of contemporary urban youth and the analysis of influencing factors]. *Journal of Shanxi College for Youth Administrators, 21*(4), 9–11.

Zhang, Z. (2004). "Yule" shouji fazhan de xinqushi ["Entertainment" as the new trend of mobile phone development]. *Computer Knowledge and Technology, 6*, 26–28.

Zhao, B., Yin, Q., & Han, K. (2006). Daxuesheng shouji shiyong qingkuang diaocha [Survey of the usage of mobile phones among university students]. *Market Research, 10*, 15–19.

Zhao, X. (2007). Qianxi dangdai qingnian shenghuo fangshi de tezheng ji fazhan qushi [The analysis of characteristics and developing trend of young people's life style]. *Youth Study, 2*, 18–20.

Zhong, D. (2009). *Youyizhong moshi jiao shanzhai* [There is one model called *Shanzhai*]. Wuhan, China: Wuhan University Press.

Zhong, Y. (2007). Houxiandai xiaofei: qingshaonian yule yawenhua jiqi fasheng [Postmodern consumption: Youths' entertaining subculture and their occurrence]. *Current Youth Study, 1*, 31–34.

ADDITIONAL READING

Grant, I. (2007). Why young consumers are not open to mobile marketing communication. *International Journal of Advertising, 26*(2), 223–246.

Guo, Z., & Wu, M. (2009). Dancing thumbs: Mobile telephony in contemporary China. In Zhang, X., & Zheng, Y. (Eds.), *China's Information and Communications Technology Revolution* (pp. 34–51). London: Routledge.

Hu, K. (2008). Made in China: The cultural logic of OEMs and the manufacture of low-cost technology. *Inter-Asia Cultural Studies, 9*(1), 27–46. doi:10.1080/14649370701789617

Instituto de la Juventud. (2002). *Revista de Estudios de Juventud (Instituto de la Juventud).* Madrid: Author.

Ishii, K., & Wu, C. I. (2006). A comparative study of media cultures among Taiwanese and Japanese youth. *Telematics and Informatics, 23*, 95–116. doi:10.1016/j.tele.2005.05.002

Ling, R. (2007). Children, youth, and mobile communication. *Journal of Children and Media*, *1*(1), 60–67. doi:10.1080/17482790601005173

Pang, L. (2004). Piracy/Privacy: The despair of cinema and collectivity in China. *Boundary 2*, *31*(3), 101–124. doi:10.1215/01903659-31-3-101

Wang, J. (2005). Youth culture, music, and cell phone branding in China. *Global Media and Communication*, *1*(2), 185–201. doi:10.1177/1742766505054633

Wang, J. (2008). *Brand New China: Advertising, Media, and Commercial Culture*. Boston: Harvard University Press.

Wu, M., & Lin, H. (2009). Magic wand: Mobile phones and Fujian entrepreneurs in China. *International Journal of Sociotechnology and Knowledge Development*, *1*(4), 37–51.

Yoon, K. (2006). The making of neo-Confucian cyberkids: Representations of young mobile phone users in South Korea. *New Media & Society*, *8*(5), 753–770. doi:10.1177/1461444806067587

KEY TERMS AND DEFINITIONS

Mobile Phone: A portable telephone connected to the telecommunication system by radio instead of by a wire. It can be used anywhere where signals can be received.

***Shanzhai*:** A Chinese word literal translated as a "mountain village," which originally meant a fortress in a mountain occupied by outlaws. However, in its post-modern connotation, it refers to a wide range of imitations of brand names. It is associated closely with a phenomenon of producing and consuming low-cost products, which imitate brand names, mostly mobile phones and digital products.

Chinese Urban Youth: Young people living and working in cities in China.

Characteristics of Youth: Noticeable attributes of young persons as a collective group.

Brand: A generic term referring to the distinctive combination of a name, logo and words that corporations use to distinguish their product from others in the market.

ENDNOTES

1. Reports in Western media are often misleading in translating *shanzhai* products as "counterfeits". The *shanzhai* mobile phone is a very complicated phenomenon which can be roughly divided into three categories (Kan Kaili): a) The completely illegal production of pirated mobile phones; b) Operations combined with brand-name imitations with some degree of their own designs and innovations; c) The production of mobile phones of less well-known brands, particularly those manufactured through the use of the cell phone chipset made by MediaTek, a Taiwan-based company specializing in developing components for wireless communication. Some of these phones have their own brands which have become increasingly known on the market as Tianyu, Jinli, CECT, etc. This study defines the *shanzhai* phone in line with the last category of *shanzhai* phones. The production of counterfeits should not be considered as a *shanzhai* model, but an illegal operation.
2. For detailed categorization, please refer to Hongye Li's MA thesis "Challenging the Mainstream:
   Youth Identity and the Popularity of Shanzhai Mobile Phones in China". University of Macau. 2009.

Chapter 16
Convergence of Mobile TV with Everyday Life[1] and Culture

Seung-Hyun Lee
University of North Carolina at Greensboro, USA

ABSTRACT

Mobile communication technology has become one of the most popular technologies, profoundly associated with many users' everyday life. Mobile communication technology has evolved from a simple communication tool to mobile multimedia. This chapter concentrates on 'mobile TV,' the convergence of digital broadcasting and telecommunications. The main purpose of this chapter is to explore the diffusion and adoption of mobile TV-DMB (Digital Multimedia Broadcasting) in its early stages and examine the use of mobile TV-DMB in a wide range of everyday practices. This chapter also focuses on the factors that influence the adoption and use of mobile TV-DMB, and a contextual understanding of DMB that may affect media use patterns in a wide range of social and cultural contexts. The results of this study underline the potential of the broad integration of mobile TV-DMB into people's everyday life.

INTRODUCTION

Media use is profoundly associated with culture and individuals' everyday lives and practices of those who have grown up with digital technologies such as the computer, the Internet, and the mobile phone. For the young generation, the use of these media has become routine. Particularly, mobile media, usable anytime and anywhere, is now becoming more widely connected to people's daily life and culture. People carry and use their own mobile media everywhere and they are engaged with it in a wide range of social and cultural contexts. As Fujimoto (2006, p.87) points out that "the *keitai*[2] is a locus that integrates Japanese subcultures," mobile media culture integrates with many subcultures, such as the choice of ringtones and straps, e-mail, texting, picture messaging, and social networking. In this regard, the use of mobile media might be incorporated into the public's everyday life, affecting the patterns of media use and creating new trends in a specific

DOI: 10.4018/978-1-60960-037-2.ch016

Copyright © 2011, IGI Global. Copying or distributing in print or electronic forms without written permission of IGI Global is prohibited.

culture (Lee, 2008). The adoption of mobile media will likely have profound social and cultural effects. In their study of mobile TV, Constantiou, Papazafeiropoulou, and Vendelø (2009) underline the importance of culture in the adoption of mobile TV and in differentiating the preferences for mobile TV content services.

Because mobile communication technology has become one of the most popular technologies, closely associated with everyday routines for many users, mobile communication technology -- such as the mobile phone -- has expanded into mobile multimedia, which can be used not only for communication, but also for mobile Internet, music, games, banking, shopping, and even television. In other words, mobile technology has evolved from a simple communication tool to mobile multimedia, combining the capabilities of the Internet, camera, video, and TV. As the convergence of text, audio, video, and visual images (May & Hearn, 2005), mobile multimedia play an increasingly important role of transmitting and disseminating information and entertainment, in addition to voice communication and text messaging. Mobile multimedia represents four distinctive attributes - convergence, immediate accessibility, individuality, and portability, allowing users to do many things with one, multi-purpose device. These characteristics indicate vital implications for mobile multimedia use in that they can be used regardless of time and place in everyday life.

The latest evolution of mobile technology includes the launch of mobile TV. The convergence of digital broadcasting and telecommunications, so-called "Mobile TV," is an emerging mobile multimedia technology (Lee, 2007). The idea of delivering live TV and audio and video content to mobile phones, personal digital assistants (PDA) and laptops has resulted in mobile TV. Mobile TV refers to the integration of live TV, personalized time- and place-shifted TV, and on-demand audiovisual content delivered to mobile devices. It is called 'TV on the palm or on my hand' (Burns, 2007; Europa, 2007; Reardon, 2005; Suh, 2005). Mobile TV is the convergence multimedia of telecommunications as an extension of mobile phones, broadcasting as an extension of television, and the information technology as an extension of the Internet.

This chapter explores the diffusion, adoption and early use of mobile TV-DMB (Digital Multimedia Broadcasting) in its initial stages of adoption and characterizes the use of mobile TV-DMB in a wide range of everyday practices. It also focuses on examining the factors that influence the adoption and use of mobile TV-DMB, based on the diffusion of innovations and adoption approach (Rogers, 1995). It aims to depict a contextual understanding of mobile TV-DMB that may affect media use patterns in a wide range of social and cultural contexts. It proposes two main research questions, "What factors influenced the adoption and use of mobile TV?" and "How was mobile TV used in public places?"

This chapter draws from a combined online survey, in-depth interviews and observation, which were conducted in Korea during April and May of 2006 to collect data on the adoption and use of mobile TV-DMB in its infancy. The researcher recruited participants from among DMB users nationwide through several DMB-related online community websites. Because the survey was conducted only one year after the S-DMB service launch and six months after the T-DMB service launch, and because mobile TV-DMB with the low penetration rate was in its infancy at the time of the survey, a random sample of the population through random-digit-dialing (RDD) would generate a very small sample of DMB users.[3] Although the sample might not represent the overall Korean population, I believe that the findings of this study will provide insights into the nature of mobile TV-DMB use in its early stages in Korea. A total of 205 valid responses were used for this study and the response rate of the survey was approximately 41%. Among the respondents, 94% were male and 6% were female.

The average age of the respondents was 28 years old (M= 27.57, SD=7.17).

This chapter also includes the interviews and observations of fifteen mobile TV-DMB users, consisting of fourteen males and one female. They consist of two graduate students, six undergraduate students, one self-employed, and six professionals. Nine participants were between 20 and 30 years old, and five were between 31 and 40 years old. There was one participant who was older than 40 years old. Half the participants estimated that their commute took between 30 minutes and one hour, and half estimated between one and two hours. Nine participants used S-DMB, while six used T-DMB. Only one participant used a DMB set-top box with a monitor in his car, and two used a portable multimedia player (PMP) for DMB use. The rest of them used a combination of mobile phones with DMB. I observed these participants' actual DMB use in their everyday lives during their morning and evening commute for a week in May 2006. I observed how they used DMB while they were taking public transportation (bus or subway) or waiting for it, while driving a car, while walking outside, or in public places such as a restaurant. At the same time, I conducted an interpersonal in-depth interview with each participant asking about their use of mobile TV-DMB in everyday life (Lee, 2008).

For this research, several variables were measured: factors influencing the adoption and use of mobile TV-DMB; DMB use behavior, including frequency, time, and location; adoption time of DMB; and demographic variables.[4] For data analysis, exploratory factor analysis, Pearson correlation analysis, and content analysis, were performed. After factor analysis,[5] four factors were categorized into three groups: *'entertainment'* into intrinsic factors; *'innovativeness'* into personal factors; and *'immediate accessibility'* and *'usefulness'* into instrumental factors.

This chapter begins with a brief overview of the emergence and development of mobile TV-DMB in Korea as structured by particular social, cultural, and historical contexts. It describes how and why mobile broadcasting is embedded within Korean society, using a social and cultural perspective to understand the phenomena surrounding mobile TV-DMB in Korean culture. After presenting the factors influencing the adoption and use of mobile TV-DMB and its use in the boundary between private and public places, the chapter concludes by discussing the implications of mobile TV-DMB use.

MOBILE TV

Emergence and Development of Mobile TV-DMB in Korea

In the history of mobile TV, Korea became the first country in the world to launch mobile TV broadcasting for mobile phone users via both satellite and terrestrial networks.[6] Korea developed a novel standard, DMB (Digital Multimedia Broadcasting), for mobile broadcast TV services. To expand the capability of the limited spectrum resources in Korea, Korean telecommunications companies have focused on developing the digitalization of radio broadcasting. In this process, the compression and transmission technology used for DAB (Digital Audio Broadcasting) has been developed to provide not only digital radio broadcast but also video and data services. The service was named DMB in Korea in late 2002 (Teng, 2005). SK Telecom (SKT) of Korea founded TU Media Corporation in 2003 as a DMB consortium of 155 enterprises including terrestrial broadcasting, program providers, devices and gap filler[7] makers, and agencies in Korea (Suh, 2005; TU Media, 2004).

TU Media launched the world's first commercial satellite DMB (S-DMB) service to mobile phone subscribers on May 1, 2005, and the free-to-air terrestrial DMB (T-DMB) was launched on December 1, 2005 (see Figure 1). T-DMB is an advanced version of European DAB that incorporates MPEG-4 multimedia technology with the Eureka-147 DAB standard (Teng, 2005). Advertiser-sponsored T-DMB service is run by six regional operators–KBS, MBC, SBS, YTN DMB, Korea DMB and U1 Media–and retransmits traditional television programs to handheld devices via terrestrial networks. In contrast to it, S-DMB service based on subscription fees provides more diversified content such as dramas, sports, movies, animations, audio and data broadcasts, and games (Kim, 2004; NIA, 2007a; Suh, 2005; TU Media, 2004; also, see Table 1).

Unlike downloading video clips, mobile TV-DMB is like watching live TV, regular terrestrial, cable or satellite TV on mobile devices such as mobile phones. Beyond its entertainment function, DMB can also be used as an information and communication media as well, particularly if the DMB mobile phone is used as the combination of broadcasting TV and communication service. The DMB mobile phone is the most popular device for DMB use (Lee, 2007). By providing news, information, and audio and video entertainment, mobile TV-DMB with portability are gradually being diffused, particularly to those who commute by bus, subway, or car in urban areas.

As of April 2007, about 11 percent of the total Korean population used DMB. The number of S-DMB subscribers has reached 1.13 million, while a total of 4.04 million T-DMB devices have been sold. The free service and the expansion of DMB coverage have influenced the rapid increase of T-DMB service use (NIA, 2007a, 2007b). After two years, in June 2009, a total of 20 million T-DMB devices has been sold, while the number of S-DMB subscribers has reached 2 million in Korea (Kim, 2009). T-DMB shows the rapid growth, compared to S-DMB. The total 22 million accounts for approximately 45 percent of the Korean population. This implies that free service of DMB has resulted in the fast diffusion of DMB and attracted more consumers. Beyond the growth of mobile TV-DMB, Korea already showed the high penetration rate of mobile phone use. The next section describes the background that has led to the rapid growth of mobile technology in Korea.

Figure 1. Model of Satellite and Terrestrial Mobile TV-DMB

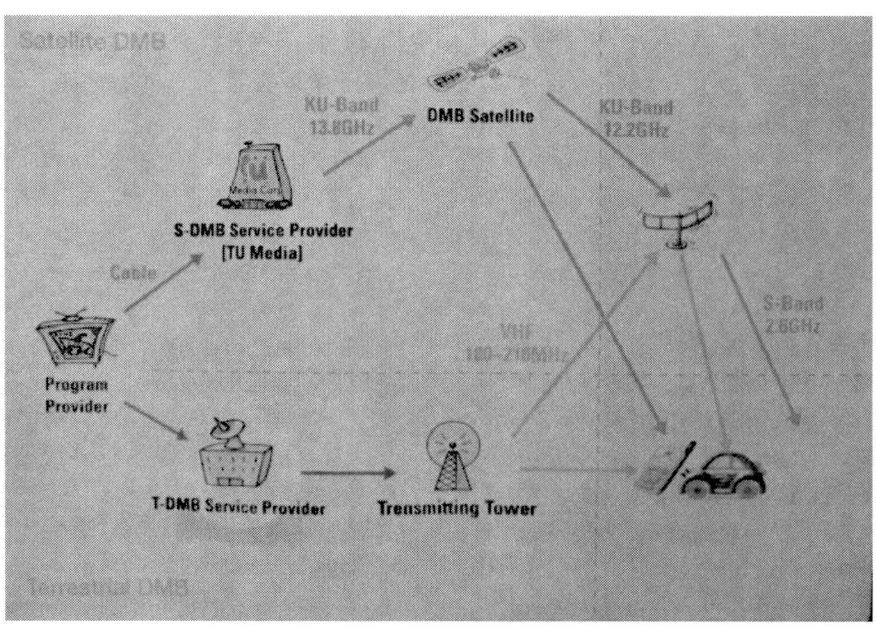

Table 1. Comparison between S-DMB and T-DMB of Mobile TV

	Satellite DMB (S-DMB)	Terrestrial DMB (T-DMB)
Characteristics	Commercial service	Public & universal service
Coverage	Nationwide	Seoul metropolitan area → Nationwide
Channels	12 Video, 26 Audio, Data	7 Video, 13 Audio, 1 Data
Business Profit model	Monthly fee ($10) Activation fee ($20)	Free, Advertisement
Start date	May 1, 2005	December 1, 2005
Subscribers / Sold devices	2 millions (June, 2009).	20 millions (June, 2009).
Operators	TU Media Corp. (1)	KBS, SBS, MBC, YTN-DMB, U1 Media, Korea DMB (6)
Program providers	Terrestrial and cable television broadcasting, TU Media, other media companies	KBS, SBS, MBC, YTN-DMB, U1 Media, Korea DMB
Regulatory	No retransmission of terrestrial TV programs →Now available	Retransmission of terrestrial TV Programs

* Sources: Kim (2009), NIA (2007a), Suh (2005), & TU Media (2004).

Background and Growth of Mobile Technology in Korea

Along with the growth of the IT (Information and Technology) industry, Korea shows the rapid growth of mobile technology, particularly the high penetration rate of mobile phone use. This fast growth of mobile technology in Korea results from several factors. First, the Korean government has played a central role in the mobile market. Korean government policies have promoted market liberalization and competition among mobile operators. Consequently, mobile operators have actively developed new mobile technologies and launched new mobile services, which has resulted in increasing the use of mobile technology and the strong consumer culture in Korea. Second, in the "U-Korea"[8] (or ubiquitous Korea) project, the Korean government has concentrated on the development of mobile telecommunications as a

Figure 2. DMB mobile phone

key strategic industry. It consequently led in the global information and technology market, and also in building "m-government" (or mobile government) that allows people to access administrative information and obtain public services through mobile devices since 2002 (NIA, 2007b). Third, the astonishing information demand and an active and mobile youth culture have influenced the increased mobile technology use. As an example of youth culture and the socio-political aspect in mobile phone use, mobile phones facilitated mobilization among young Korean voters during the presidential election of 2002 (Yoon, 2007). Miyata, Boase, Wellman, & Ikeda (2006) underline a hybrid Japanese culture of mobile phone with youth culture.

Mobile phones have become incorporated into youth culture to such an extent that text messages, air time, and..... binding youth closely together. Text message exchanges are often incorporated into face-to-face contact with peers during hang-out time. (p.144)

Overall, government policies, strong consumer culture, and information demand have combined to foster the development of mobile technologies in Korea. In addition to these factors, the urban ecologies of cities such as Seoul, characterized by high public transportation use, heavy traffic, and dense population, have likely contributed to the development of mobile technology, especially the advent of mobile TV-DMB in Korea.

The development of mobile technology provides a convergence service that combines telecommunications and informatics, that is, a convenient information-based environment. A variety of functions, such as Internet access, subway and bus tickets, dictionaries, alarm clocks, and even credit cards, are being integrated into one small device, the mobile phone. People carry their mobile phone everywhere and depend on it increasingly in their everyday lives. Mobile phones are becoming a daily necessity in Korean culture, creating profound impacts both on users' daily lives and on the nation's industry (Korea net, 2006; MIC, 2005 &2006; NIA, 2007a). The development of digital and mobile technology also made possible the transmission and dissemination of traditional television programs to the mobile phone and the Internet. Mobile multimedia technology expands its applications and transcends temporal and spatial limitations in media use. It enables users to experience mobile multimedia anytime and anywhere. In sum, the highest penetration of broadband and the high use of mobile phones and the Internet in Korea (see Figure 3) have fostered mobile multimedia culture and the emergence of mobile TV. Like Japan where shows the high penetration rate of Internet mobile phone use, is Korea being become a networked and "mobile-ized"(Miyata et al., 2006, p.144) society?

Diffusion and Adoption of Mobile TV-DMB

Many previous studies of new media have employed the diffusion of innovations and adoption approach (Rogers, 1995) to examine the diffusion and adoption of new media. This study of mobile TV-DMB also adapts the diffusion of innovations and adoption approach to see how mobile TV-DMB is introduced, diffused, and integrated into the people's everyday life. This approach focuses on the study of new technology diffusion, the aggregate of individual users as the level of analysis, motivations for adopting new media, and individuals' adoption decision and media choice. This approach describes the adoption process of an innovation at an individual level (Pedersen & Ling, 2003). Rogers (1995, pp.10-1) defines diffusion as "the process by which an innovation is communicated through certain channels over a period of time among the members of a social system," and innovation as "an idea, practice, or object that is perceived as new by an individual or other unit of adoption."

Figure 3. Development of Telecommunications in Korea (Source: Koo (2001), MIC of Korea (2003 & 2006), NIA (2007a, b), NCA (2003))

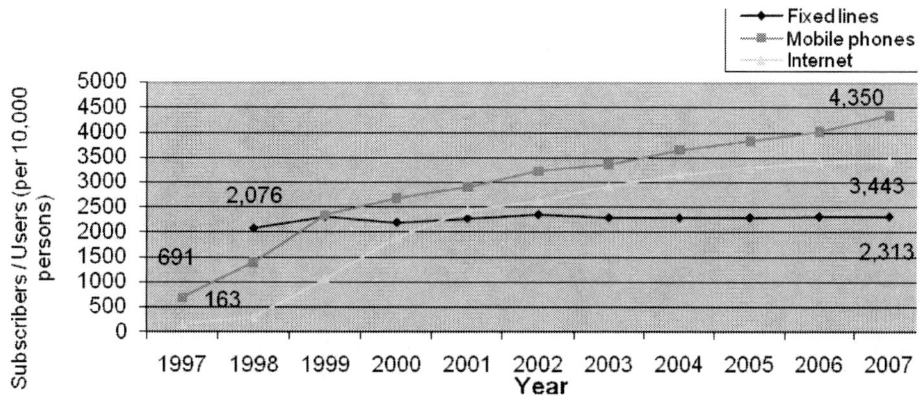

Based on the relative adoption time, Rogers (1995, pp.183-5) groups adopters into five categories that follow a standard S-curve[9]: innovators (2.5%), early adopters (13.5%), early majority (34%), late majority (34%), and laggards (16%), all of whom differ from each other in terms of demographics, socio-economic status (SES), media use, and personality. Particularly, innovators and early adopters are highly motivated by their own interest in an emerging new technology, so they adopt it earlier than the mass, and then integrate it into their environments and lives (Atkin, Jeffres, & Neuendorf, 1998; Lin, 1994; Trepte, Ranné, & Becker, 2003). This study of mobile TV-DMB employs the conceptual framework of diffusion of innovations and adoption approach (Rogers, 1995) to characterize early adopters of mobile TV-DMB. Some critics argue that Rogers' diffusion of innovations approach has not adequately explained the reason of the low diffusion rates that are observed in many countries and the reason why some innovations have not gone through their adoption curve (Constantiou et al., 2009). However, it is necessary to consider that there is a gap between the development of an innovation and the actual adoption and implementation of the innovation, thus researchers should consider other factors that influence the diffusion and actual adoption of an innovation in specific circumstances.

The diffusion of innovations and adoption approach has also been used to identify the factors that influence the innovation adoption and use. Previous studies found important determinants such as individuals' socioeconomic status and intention to use a specific technology or service, and the attributes of new media that affected the decision to adopt and use a new technology in the early stage. These factors were crucial elements of user analysis (Ruggiero, 2000). According to previous studies, the interest in trying out new technologies was the consistent predictor of new media adoption (Atkin et al., 1998; Lin, 1994; Trepte et al., 2003). Goldsmith and Hofacker (1991) called this strong interest in specific consumer goods "*consumer innovativeness*," which was described as a personal predisposition of interest and affinity to new technology. Rogers (1995, p.252) identifies innovativeness as "the degree to which an individual or other unit of adoption is relatively earlier in adopting new ideas than other members of a system."

The factors found from the previous studies of new media adoption and use can be categorized into three groups: intrinsic factors such as relaxation and entertainment; instrumental (task-oriented) ones such as information seeking and accessibility; and personal trait factors such as innovativeness (O'Keefe & Sulanowski, 1995; Leung & Wei, 2000; Stafford & Stafford, 2001). Intrinsic media use, as known ritualized media use, means a habitual media use to pass time, while instrumental media use reflects active media use, such as information seeking (Katz, Blumler, & Gurevitch, 1974; Rubin, 2002). Personal traits have influenced the adoption and use of new media (Lin, 2004).

Leung and Wei (2000) found that mobility and immediate access strongly influenced the mobile phone use on buses, cars, and trains, or in malls and restaurants. According to the mobile TV-DMB study by Shin (2006), informativeness, entertainment and interactivity were significantly related to the positive attitude toward DMB and influenced its usage and satisfaction. The results of his DMB study indicated that entertainment was the most important determinant for mobile TV use behavior and interactivity was also a key determinant. Depending on available features of new media, there are consistent and new factors that influence the adoption and use of new media. In addition, the convergence of new media with digital technology has changed the patterns of exposure to media (Ruggiero, 2000), affecting media experience and communication modes. This study focuses on examining whether these factors also influence the adoption and use of mobile TV-DMB and whether they are also important elements of user analysis. Understanding these factors is crucial for research on mobile TV-DMB in that it can provide insight into the adoption and use of mobile media and individual adoption patterns (Rubin, 2002).

Intrinsic Factors: Entertainment

As an intrinsic factor, '*entertainment,*' was identified as the strongest factor (α=0.85) in the adoption and use of mobile TV-DMB for relaxation, escape, fun, and passing time. Mobile TV-DMB is perceived as a new form of entertainment. Many respondents indicated that the main reasons to use DMB were to pass time and to entertain themselves while they were using public transportation during commute or waiting for someone or something. This factor is in line with the findings from many previous studies of new media and the study of mobile TV-DMB by Shin (2006). The respondents, whose commute took a long time and who spent a long time on public transportation or in a personal car, tended to use DMB to pass time. This current study confirmed that '*entertainment*' is the consistent and key element to predict and examine the adoption and use of new media.

Because it takes one and a half hours by public transportation from home to school, I watch DMB to pass time. (32-year-old male graduate student)

Personal Trait Factors: Innovativeness

'*Innovativeness*' (α = 0.78) was categorized as a personal trait factor. The innovativeness refers to an individual's tendency to seek novelty. The respondents tended to adopt and use DMB for its novelty and out of their curiosity about the idea that people can watch TV via mobile phones. The respondents also tended to consider themselves fairly innovative in the adoption of a new technology. They gather information on a new technology more actively and earlier than others, purchase and use it earlier than others, and actively convey their opinions about it to others. In addition, they expressed that they did not prefer to stick with old technology. The respondents, who had a greater interest in a new technology and preferred to experience it earlier than others, adopted DMB earlier than others. A male professional mentioned that "*I think the idea of having a TV on my hand is great in the information age.*"

The findings of this study present that '*show-off*', which was categorized into innovativeness factor, had an effect on the adoption and use of

DMB. Interestingly, the respondents tended to feel superior to others because of using a new technology and enjoy the sense of being far ahead of their times by using an exclusive DMB. From the open-ended responses of the DMB survey, many respondents expressed that "*It is my own fashion and style.*" and "*I wanted to experience a new broadcasting media earlier than others.*" The respondents indicated that they garnered attention from friends, passersby, or strangers, and became an object of the envy of others when they used DMB. They also pointed out that people in public spaces sometimes looked at DMB users with curiosity, giving them a feeling of superiority. A 47-year-old male (self-employed) respondent expressed that he heard from others, *"You've become younger since you started using DMB."* He underscored his personal satisfaction as an early adopter. Some previous studies found that younger people tended to boast about their special ringtones and the straps of mobile phones, particularly in Korean and Japanese cultures. However, not many new media studies pointed out the impact of 'show-off' on the adoption and use of new media.

Among demographic variables, age and the level of education were highly related to the adoption and use of DMB. The findings showed that the early adopters of mobile TV-DMB were predominantly younger users aged between 21 and 30 years (50.7%), with higher levels of education (43% with a bachelor's degree). Also, the adoption and use of mobile TV-DMB was significantly associated with commute time by public transportation or personal car. The commute time had an impact on the early adoption of DMB ($r=0.15$, $p<0.05$). The respondents whose commute time took longer tended to adopt DMB earlier than others. In addition, evening commute (36%) was the most popular time for the respondents to use DMB, particularly for entertainment programs, followed by morning commute (32%) for news programs. In terms of DMB use patterns, more than half (52%) of the respondents had used DMB everyday during the week before the survey. Also, around 35% of the respondents spent between 31 minutes and one hour for DMB use on an average week day. Unlike previous studies of mobile media adoption, this study underlines the impact of commute time on the adoption and use of DMB.

Another interesting and important finding is that the adoption and use of mobile TV-DMB was significantly related to individuals' everyday life, such as a busy lifestyle of school or work, circumstances, and private space. Some respondents pointed out that because they spent their most time outside of home, they had no time to watch TV at home.

Because of my busy life, my television viewing time decreased more and more, and I couldn't find time to watch my favorite TV programs. So I purchased a DMB mobile phone to watch it during my spare time outside of home. (34-year-old male professional)

Although I am a student, I spend most of my time outside of home and I can't watch TV at home. Now, I often use DMB at school. (21-year-old male college student)

Although mobile TV-DMB was originally designed to be used outside of home, especially on the move, it was also frequently used at home as a matter of personal preference and specific situation and environment. The respondents frequently used DMB at home every day ($r=0.54$, $p<0.01$). In contrast to the assumption that DMB might not be used at home because there is a bigger television screen at home (Lee, 2006), some respondents used it frequently at home for the specific reasons. Some respondents preferred to watch their favorite programs through DMB alone in their own private room at times when other family members occupied the television or when traditional television did not show their favorite programs. In addition, some younger respondents who live alone and do not have their own television tend to watch DMB at home as a substitution for traditional television. These findings imply that mobile TV-DMB has the potential of complementing traditional television at home in particular situations, and DMB use is related to personal environment and private space. Miyata et al. (2006) underlines that Internet use through mobile phones has already been integrated into daily life of the Japanese.

Instrumental Factors: Immediate Accessibility, Convenience and Usefulness

In the line with the results from the previous studies, '*usefulness*' (α=0.78) as an instrumental factor signifies that the respondents consider that the use of DMB can make their spare time useful. Most respondents mentioned that they used DMB because it was convenient, useful, and easy to carry, watch, and use it at anytime and anywhere as the convergence of multimedia. They used DMB for the convenience of an all-in-one device. The respondents indicate that DMB replaces MP3, radio, mobile phone games, or newspaper in part, specifically on public transportation during daily commute, waiting time, or spare time. A 21-year-old female worker emphasized that

I don't need to carry a mobile phone and a CD player or MP3 player separately. Through my DMB mobile phone, I can listen to radio and music as well as make or receive a call.

As another instrumental factor, '*immediate accessibility*' (α=0.62), included DMB's mobility, immediacy, and convenience. '*Immediate accessibility*' refers to the ease and convenience of use and the immediate availability of media. The respondents adopted and used mobile TV-DMB primarily to view real-time TV programs or their favorite programs anywhere and anytime, to substitute for other media, and to access news and information immediately. Many respondents indicated that they used DMB because of its immediate accessibility to the latest news and information. This study found the significance of immediate accessibility to news and information in the adoption and use of DMB. This finding implies that immediate accessibility is an essential factor in the adoption and use of mobile media. The previous study showed that immediate accessibility was a significant factor on the adoption and use of mobile phones (Leung & Wei, 2000). The majority of the respondents (91%) mentioned that they could gain diverse information, such as real-time news, traffic information, or weather, through DMB faster than through other media. Some feel as if they are on the cutting-edge.

In sum, immediate accessibility and entertainment are considered the most important factors in the overall use of mobile TV-DMB. In fact, the respondents who pursued immediate accessibility or entertainment from DMB use tended to use it more often and spend more time on it, especially in the evening. The results imply that mobile TV-DMB shows the potential of complementing traditional media, such as television, radio, and newspaper, on the move, particularly in the evening time, and the relations between DMB use patterns and individuals' daily life.

Mobile TV-DMB use in Public Spaces

One important issue in the use of mobile media is the manners of mobile media users in public spaces and people's perception of manners (Ling, 2004). One of the social issues in mobile phone use in public spaces is "the colonization of public spaces with private conversations" (Okabe & Ito, 2006, p.207). Particularly in Japan, mobile phone use in public spaces has been considered a social issue, disrupting the social norms (Matsuda, 2006). *Keitai* use in Japan is restricted on train and bus. Previous studies have pointed out an ethical issue. Poor etiquette, such as loud conversation over mobile phones, disrupts public spaces. The studies of mobile phones (Ling, 2004; Matsuda,

2006; Okabe & Ito, 2006) discussed the social issues of manners in mobile media use in public spaces, showing the development of social order for mobile phone use in public spaces. This study discusses whether DMB use in public spaces is considered a social issue and an invasion of public spaces, and whether this issue is considered a specific cultural phenomenon in particular countries.

The respondents used mobile TV-DMB more frequently on public transportation such as bus, subway or train, or in public places such as restaurants or schools. In contrast to the previous studies of the social issue of mobile phone conversation in public spaces (Ling, 2004; Matsuda, 2006; Okabe & Ito, 2006), this study found that the majority of the respondents (78%) did not think that DMB use on public transportation or in public places would cause any social problems or annoy others as long as users do not make any noise from their use. They also thought that because it was an early stage of DMB and its penetration rate was low, DMB use in public spaces did not bring about serious social issues at that point. However, they pointed out that any mobile TV users should watch or listen to it with a headset and a low volume setting, not with a speaker function. They also suggested that users should control themselves, take care not to trouble others, follow the rules of etiquette, and try not to attract attention or make noise when they use it in public spaces. A quote from the open-ended responses of the DMB survey indicates that "*The problem would be an intrusive sound, not DMB use itself in public places.*" Some respondents expressed that "*DMB use in public spaces would not be a problem at all because it is a personal media and a personal choice to enjoy a contemporary technology.*" They did not consider that it was an invasion of public spaces. Rather, they expect that DMB use might become popularized soon and be similar to mobile phone use in public spaces, so it would not be a problem and could even improve social situations because people can watch and enjoy it together and share information, especially important news. A female respondent indicated that *"Mobile TV might even become an information conveyer."* One interesting response was that *"I felt rather uncomfortable when I used DMB on public transportation because a person next to me tried to watch it."*

However, there are some opposing voices against DMB use in public spaces. Some respondents saw users who watched DMB without a headset and turned up the volume. They pointed out that those users did not follow etiquette, but pursued their own convenience only. They also argued that DMB would attract public attention, but might get rejected by others, so DMB users ought to restrict themselves not to use it in public spaces if possible. Some respondents pointed out that there were some DMB users who tried to show off their use and some others felt some form of generation gap. In addition, some respondents pointed out some potentially dangerous situations with DMB use while walking on the street or in a standing position on public transportation, especially in a sudden stop or start situation. They argue that these situations can cause an accident and harm others because there are a lot of commuters during commute, so DMB users standing or walking should not watch it.

I planned to watch it while walking on the street, but it seemed to be dangerous because I almost had an accident. (24-year-old male college student)

Moreover, some users indicated that they sometimes passed a bus stop where they had to get off while they were watching mobile TV. These findings imply whether the mobile-ized Korean society needs to develop social standards and regulations for DMB use in public spaces.

DISCUSSION

This study of mobile TV-DMB presents several significant points from the findings. First, adoption and diffusion of innovations approach emphasizes

Figure 4. T-DMB use in a bus

the significance of distinguishing adopter segments and shows patterns of adoption at each different stage. It distinguishes each adopter category in terms of communication behavior, personality and values, and socioeconomic status. However, this distinction of each adopter category is overly simplified without reflecting a complex reality and other important factors surrounding the adoption of innovation. Particularly, the distinction between innovators and early adopters is not clear enough. Rather, they overlap each other and show similar characteristics in the adoption of innovation. This adopter category criteria need to be modified and extended so that this approach can be applied to technology transition, in general, and mobile multimedia, specifically.

Second, many previous studies of the adoption and use of new media have centered on the relations between new media use and frequency, time, demographic variables, socioeconomic status, and the attributes of new media. However, different core elements should be considered and examined for the study of adoption and use of mobile multimedia. This current study shows that personal circumstances and lifestyle are also significant elements for the adoption and use of mobile-TV. In addition, 'entertainment' as the consistent and key element is used for most studies of the adop-

Figure 5. S-DMB use in a subway

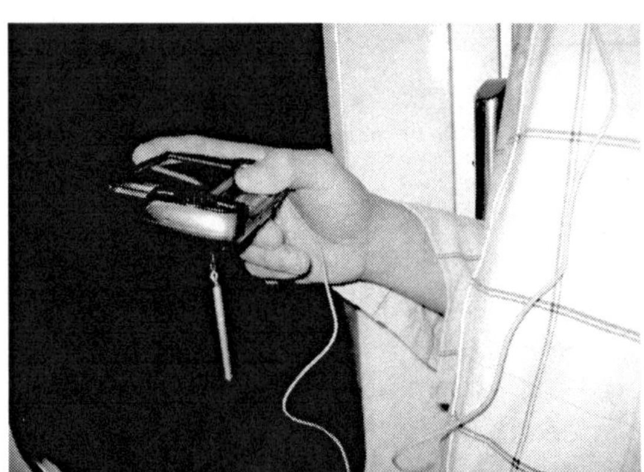

tion and use of new media, while innovativeness variable is limited to the study of the early stage of new media.

Another significant point is that culture should be considered and examined in the study of new media in that different cultural contexts show different adoption patterns and penetration rates of an innovation. Particularly, mobile multimedia with diverse functions integrate subcultures and create new mobile culture. This study also points out that mobile multimedia show the potential of complementing traditional media in certain situations. As mobile media are becoming the convergence of diverse media, it is significant to consider and examine the impacts of mobile multimedia on traditional media use. In relation to the convergence of mobile multimedia, it is also necessary to focus on how mobile multimedia erode the boundary between public and private spaces and how a mobile-ized society can develop social standards and regulations for mobile multimedia use in public spaces.

FUTURE RESEARCH DIRECTION

This study shows some limitations. One limitation is the sampling method that the study used. Because of the low penetration rate of mobile TV-DMB service at the time of the survey, this study used a convenient sample of subjects, rather than a random sample of the population, which would generate a very small sample of subjects. In the early stage of an innovation, it is difficult to find users of an innovation through a random sampling method such as random digit dialing (RDD). Although the sample might not be a representative of the whole population of Korea, I believe that the findings of this study are informative about the nature of mobile TV-DMB use in its early stage in Korea. Although this study does not compare with other countries, it describes the Korean case so that future researchers can make the comparisons. The results of this study would be helpful to understand the early stage of mobile TV-DMB, given that there was not abundant academic empirical research on its early stage phenomena.

The small sample size and the unbalanced sample in terms of gender are another limitation of this study. The sample in this study is overwhelmingly male, making this research essentially a study of younger male early adopters. At the time of the online survey in 2006, the penetration rate of DMB was approximately 10% of the total Korean population. Among them, male users of DMB represented 80%. Most previous new media studies also show a similar gender unbalance in the early adoption of new technologies. In many cases of new technology adoption in Korea, males have been dominant early adopters of new technologies such as mobile phones, the Internet, and electronic games.

In this regard, I suggest several future research directions. Future research should encompass a larger sample of mobile TV users in order to account for the diffusion, adoption, and use of mobile TV in the general population. Also, it is necessary to recruit more female users to see if there are other gender differences in diverse aspects and situations of mobile TV use. In addition, it is worthwhile to compare mobile TV users and non-users in future research.

Future research should focus on comparing diffusion and adoption process and use of mobile TV across countries with a wider perspective to explore cultural impacts on mobile TV use and individual users in particular political, geographical, economic and cultural context. Also, future research should test mobile TV adoption and use by applying alternative theoretical approaches, which can take into account the individual adoption and use. Future research needs to focus on exploring several points, including the relations between mobile multimedia technology development, culture, and society; the effects of mobile TV on traditional media use and everyday life; the relations between new features such as interactivity and data services of mobile TV and its use; and heavy versus light habitual mobile TV use in everyday life. In addition, it is crucial to address social relations in everyday life of mobile TV users and mobile TV manners and people's inconsiderate use of mobile TV in public spaces. It is necessary to further discuss and develop methodologies for research on mobile multimedia of the early stage.

CONCLUSION

Like the findings from the previous studies of new media, the dominant early adopters of mobile TV-DMB were also younger users with high education levels. As the early adopters, they were highly interested in a new technology and preferred to experience it earlier than others. The findings of this study underline that mobile TV-DMB as a convergence mobile multimedia technology can be broadly integrated into the public's everyday life for the purposes of entertainment and immediate access to news and information and for substituting or complementing other media, regardless of time and space. Mobile TV-DMB is widely being used on public transportation, particularly during commute, waiting time, or traffic congestion, in order to pass time, to be stimulated or to make their spare time between activities useful. It is also used highly by those who use a car everyday for a personal purpose or for business. These results satisfy the original goal of mobile TV-DMB. In addition, the results imply that mobile TV-DMB use on the move or outside of the home or even at home has the potential to be integrated into individual users' everyday lives and social and cultural contexts, particularly in an urban ecology. However, it is necessary to examine and discuss how mobile TV-DMB is incorporated into people's everyday life and affects it in different cultural and economic contexts, whether it can complement other media, and how well it satisfies people's needs.

The results of this study also highlight that immediate accessibility, entertainment, and innovativeness factors are significant predictors of mobile TV-DMB use behavior. Particularly, the instrumental factor, immediate accessibility of mobile TV-DMB regardless of time and place, is a significant factor for its use and the most inclusive predictor of overall DMB use behavior. However, it is significant to explore how mobile TV-DMB influences both personal and mass media consumption behavior by overcoming the spatial and temporal limitations of traditional media, and whether DMB use in public spaces causes some social problems. In addition, it is necessary to focus on the latent social issues, some risks to health and safety, or some other negative effects that it might bring about when its use is widespread in a society, particularly if some people use it without following social order or mobile etiquette.

Mobile multimedia is a trend in the development of contemporary mobile media. More and more people want to use mobile multimedia, which provide immediate accessibility and convenience with diverse functions, from the Internet access to entertainment. Users can experience different media through an all-in-one gadget. This research will make a major contribution in creating a baseline as we transition into greater pervasiveness of mobile, digital, entertainment, and information technologies. It will help us understanding early mobile TV user behavior and the changing role of mobile technology, from communication to information and entertainment media that might be integrated into people's life. This study will provide insights into the diffusion of convergent mobile multimedia and the technological integration in the digital media environment.

REFERENCES

Atkin, D. J., Jeffres, L. W., & Neuendorf, K. A. (1998). Understanding Internet adoption as telecommunications behavior. *Journal of Broadcasting & Electronic Media*, *42*, 475–490.

Burns, S. (2007). *Mobile Tech Today*. Retrieved November 7, 2007, from http://www.mobile-tech-today.com /story.xhtml?story_id=56350

Constantiou, Papazafeiropoulou, & Vendelø. (2009). Does Culture Affect the Adoption of Advanced Mobile Services? A Comparative Study of Young Adults' Perceptions in Denmark and the UK. *The Data Base for Advances in Information Systems*, *40*(4), 132–147.

Europa. (2007). *An EU strategy for mobile TV*. Retrieved November 24, 2007, from http://europa.eu/rapid/pressReleasesAction.do?aged=0&format=HTML&guiLanguage= en&language=EN&reference=MEMO/07/298

Fujimoto, K. (2006). The Third-Stage Paradigm: Territory Machines from the Girls' Pager Revolution to Mobile Aesthetics. In Ito, M., Okabe, D., & Matsuda, M. (Eds.), *Personal, portable, pedestrian: Mobile phones in Japanese life* (pp. 77–102). Cambridge, MA: The MIT Press.

Goldsmith, R. E., & Hofacker, C. F. (1991). Measuring consumer innovativeness. *Journal of the Academy of Marketing Science*, *19*(3), 209–221. doi:10.1007/BF02726497

Katz, E., Blumler, J., & Gurevitch, M. (1974). Uses and gratifications research. *Public Opinion Quarterly*, *37*, 509–523. doi:10.1086/268109

Kim, B. S. (2004). TV in my hand, Satellite DMB. *Kwanhun Journal*, (Winter), 118-123

Kim, T. H. (2009, June 26). Mobile TV Audience Reaches 22 million. *The Korea Times*. Retrieved March 3, 2010, from http://www.koreatimes.co.kr/www/ news/biz/2009/06/123_47547.html

Korea net (2006, December 5). Cell phones revolutionize Korean lifestyles. *Korea.net*. Retrieved Feb. 20, 2008, from http://www.korea.net/news/news/ newsview.asp?serial_no=20061204020

Lee, M. H. (2006). *A study on the type of acceptance of early adaptors of the digital multimedia broadcasting – Mainly on the realistic application of gratification-niche theory according to the media competition*. Unpublished doctoral dissertation, SungKyunKwan University, Seoul, Korea.

Lee, S. H. (2007). *Mobile TV-DMB*. Paper presented at the 2007 annual conference of International Communication Association, San Francisco, May, 2007.

Lee, S. H. (2008). *Incorporating mobile multimedia into everyday life: Diffusion and use of Mobile TV*. Unpublished doctoral dissertation, University of Wisconsin, Madison.

Leung, L., & Wei, R. (2000). More than just talk on the move: Uses and gratifications of the cellular phone. *Journalism & Mass Communication Quarterly*, *77*, 308–320.

Lewis-Beck, M. S. (1994). *Factor analysis and related techniques*. London: SAGE.

Lin, C. A. (1994). Exploring potential factors for home videotext adoption. In Hanson, J. (Ed.), *Advances in telematics* (pp. 111–124). Norwood, NJ: Ablex.

Ling, R. (2004). *The mobile connection: The cell phone's impact on society*. San Francisco, CA: Morgan Kaufmann.

Matsuda, M. (2006). Discourses of Keitai in Japan. In Ito, M., Okabe, D., & Matsuda, M. (Eds.), *Personal, portable, pedestrian: Mobile phones in Japanese life* (pp. 19–40). Cambridge, MA: The MIT Press.

May, H., & Hearn, G. (2005). The mobile phone as media. *International Journal of Cultural Studies*, *8*(2), 195–211. doi:10.1177/1367877905052417

Media Corp, T. U. (2004). *Satellite-DMB's policy issues and market prospect*. Paper presented at the Commercial DMB demonstration, Seoul, Korea. Retrieved July 3, 2005, from www.kmis.or.kr/3_sig/comm_ data/comm(may04)1.pdf

Ministry of Information and Communication (MIC) of Korea (2005 & 2006). *Dynamic U-Korea*.

Miyata, K., Boase, J., Wellman, B., & Ikeda, K. (2006). The Mobilie-izing Japanese: Connecting to the Internet by PC and Webphone in Yamanashi. In Ito, M., Okabe, D., & Matsuda, M. (Eds.), *Personal, portable, pedestrian: Mobile phones in Japanese life* (pp. 143–164). Cambridge, MA: The MIT Press.

National Information Society Agency (NIA). (2007a). *Informatization White Paper*.

National Information Society Agency (NIA). (2007b). *Korea informatized progress and status overview*.

November 4, 2007, from http://news.zdnet.com/2100-1035-5905677.html

O'Keefe, G., & Sulanowski, B. (1995). More than just talk: Uses and gratifications and the Telephone. *Journalism & Mass Communication Quarterly*, *75*, 922–933.

Okabe, D., & Ito, M. (2006). Keitai in Public Transportation. In Ito, M., Okabe, D., & Matsuda, M. (Eds.), *Personal, portable, pedestrian: Mobile phones in Japanese life* (pp. 205–218). Cambridge, MA: The MIT Press.

Pedersen, P. E., & Ling, R. (2003). *Modifying adoption research for mobile Internet service adoption: Cross-disciplinary interactions*. Paper presented at the 36th Hawaii International Conference on System Sciences (HICSS 2003), Hawaii, January.

Reardon, M. (2005, October 21). The lowdown on mobile TV. *ZDNet News*. Retrieved

Rogers, E. M. (1995). *Diffusion of innovations* (5th ed.). New York: Free Press.

Rubin, A. M. (2002). The uses-and-gratifications perspective of media effects. In Bryant, J., & Zillmann, D. (Eds.), *Media effects: Advances in theory and research* (2nd ed., pp. 525–548). Mahwah, NJ: Lawrence Erlbaum Associates.

Ruggiero, T. (2000). Uses and gratifications theory in the 21st century. *Mass Communication & Society*, *3*(1), 3–37. doi:10.1207/S15327825MCS0301_02

Shin, D. (2006). Prospect of Mobile TV: Another Bubble or Killer Application. *Telematics and Informatics*, *23*, 253–270. doi:10.1016/j.tele.2005.08.001

Stafford, T. F., & Stafford, M. R. (2001). Identifying motivations for the use of commercial web sites. *Information Resources Management Journal*, *14*, 22–30.

Suh, Y. K. (2005). *Satellite-DMB: Successful strategy for industrialization of new technology*. Paper presented at Korea IT Leaders Forum, Seoul, Korea.

Teng, R. (2005). Digital multimedia broadcasting in Korea. *In-Stat*. Retrieved July 21, 2005, from http://www.instat.com/promos/05/IN-0502469WHT.pdf

Trepte, S., Ranné, N., & Becker, N. (2003). Personal Digital Assistants—Patterns of user Gratifications. *Communications*, *28*, 457–473. doi:10.1515/comm.2003.029

White paper 2005 & 2006. Retrieved December 10, 2007, from http://www.dynamicitkorea.com/media/index.jsp?nodeId=1914

Yoon, J. H. (2007). KDDI's Unified Portal Site – au one: A brief analysis on KDDI's unified portal site for mobile phones and PCs. *Research On Asia (ROA) Group White Paper*. Retrieved December 12, 2007, from http://www.infoedge.com/product_ type.asp?product=RO-0019

ENDNOTES

1 In this paper, everyday life means routine or daily life or ordinary occasion.

2 Japanese mobile phone.

3 They were the main source to recruit actual DMB users at the time of the survey. Because this study focused on only DMB users to see their actual use behavior, a random sample of the population would rarely find actual DMB users in its early stage at the time of survey. At that time of the survey, the DMB adoption rate was below the 10 percent of the population, so it was difficult to grasp the views of DMB users through a random sample.

4 Measurements of variables:

- *Influencing factors for the adoption and use of mobile TV-DMB:* Through a pilot study, 15 items were constructed and selected to identify influencing factors for the adoption and use of mobile TV-DMB: immediacy, convenience, mobility, usefulness, stimulation, substitution, exclusiveness, innovativeness, show-off, fun, education, loneliness reduction, personal multimedia, escape, and relaxation. These items were measured, using an 11 point Likert scale, ranging from 'strongly disagree (0)' to 'strongly agree (10)' ($M=87.68$, $SD=27.93$, $\alpha=0.88$).
- *Mobile TV-DMB use behavior:* DMB use time ($M=3.19$, $SD=1.24$) was measured by asking the respondents how much time they spend on DMB on an average weekday, ranging from less than 15 minutes to more than three hours. Also, using a categorical time period, the frequency of DMB use ($M=5.24$, $SD=2.19$) was measured by asking the respondents how often they used DMB during the last week, ranging from never (0) to everyday (7). In terms of locations ($M=16.02$, $SD=11.05$, $\alpha.=0.70$), the respondents were asked how often they used DMB on public transportation and the street, in cars and public places, and at home respectively during the last week, ranging from never (0) to everyday (7). In addition, to measure DMB use in different time slots ($M=18.65$, $SD=12.23$, $\alpha.=0.80$), the respondents were asked how often they used DMB during their morning commute, lunch time, evening commute, and evening time, ranging from never (0) to everyday (7). For the frequency of DMB program use ($M=18.85$, $SD=12.29$, $\alpha.=0.80$), the respondents were asked how often they watched each of the programs during the last week. The scale ranges from never (0) to everyday (7).
- *Adoption time of mobile TV-DMB:* This variable was measured by asking respondents when they started to use mobile TV-DMB for the first time.
- *Demographic variables:* Demographic characteristics of the respondents were measured by conventional dimensions such as age, gender, education attained, and average monthly household income. Gender was coded as dummy.

5 SPSS factor analysis using principal component solution and verimax rotation was employed to better explain expected correlation among factors and to classify motivations for the use of mobile TV-DMB (Lewis-Beck, 1994). Exploratory factor analysis was performed by a principal component extraction method with varimax rotation to classify the 15 influencing items on mobile TV-DMB use. Four factors were extracted, explaining 66% of the total variance: entertainment, innovativeness, usefulness, and immediate accessibility. A reliability analysis was performed on each factor and Cronbach's Alphas for three factors were above 0.78, indicating a high level of internal

consistency. The value for factor 4 was 0.62, which was acceptable. The overall reliability score of all variables is Cronbach's Alpha (α) =0.88.

6 Japan's MBCo. launched its S-DMB service for use in vehicles, not for mobile phone use, on October 20, 2004, while TU Media launched the world's first commercial satellite DMB (S-DMB) service to mobile phone subscribers on May 1, 2005.

7 Gap filler is a technology to improve a level of radio reception by retransmitting radio waves in areas with poor reception conditions.

8 "The ultimate goal of the 'u-Korea Master Plan' is to achieve an advanced Korea, a society where all people can benefit from a safer ubiquitous society," providing advanced individual life (NIA, 2007a, p.14).

9 "When the number of individuals adopting a new idea is plotted on a cumulative frequency basis over time, the resulting distribution is an S-shaped curve" (Rogers, 1995, pp.22-23).

Chapter 17
Call Centers, India, and a New Politics: Cultural Interpretations

Maheswar Satpathy
Indian Institute of Technology Kanpur, India

ABSTRACT

The establishment of myriad Customer service Centers, or as colloquially known, Call Centers have become a much accepted reality now in India. The country known for its assimilative nature has also painfully adopted it, though not quite got adapted to its demands. This much appreciated system by the parents of young students leaves them agape at the amount of salary it promises in comparison to their children's educational qualifications. Call Centers have become a major hub for any intermediate passed out youngster, with dreamy eyes, full of ambitions, and passions to win the world at any cost. In the present chapter an attempt is made to interrogate the relative benefits and weigh them with the demerits that it has accumulated on the loyal workers. The key theme of the chapter is the analysis of subtle politics in the engagement of these young souls which steals their youthful charisma, vivacious spirit, zeal and vigor in return of some rupees which neither the beneficiary can enjoy nor utilize. A cultural interpretation of modernization, progress and development with a focus on the sustained core-periphery relation in a theoretically imagined world economy finds its place into the rubric of chapter. It interrogates the motives of western countries (or at least the cause behind such accusation) to sustain their monopolistic cultural imperialism through multifarious means, as evidenced in the case of Call centers in India.

INTRODUCTION

Call centers have become an inevitable aspect of service delivery in today's business world, and are becoming increasingly popular with organizations as their preferred mode of interaction with their customers (Mukherjee & Malhotra, 2008). Call centers are supposed to be a major customer interface for many organizations, and are often characterized by their role of providing the first, if not the only, point of contact for the customers to engage with their service provider (Dean,

DOI: 10.4018/978-1-60960-037-2.ch017

Copyright © 2011, IGI Global. Copying or distributing in print or electronic forms without written permission of IGI Global is prohibited.

2004). There has been an enormous growth in the interest of researchers to conduct research in the area with the advent and growth of services of off-shoring (Penfold, 2009). Several studies have been conducted on off-shored work from diverse perspectives like the globalization of labor (Farrell et al. (eds) 2005), decent work (Sasikumar and Varma 2004; Penfold 2008), human resource management (Chatterjee 2007), and trade union participation (Penfold 2007; Taylor et al. 2007). Most of the work has been focused on call centres within this sector (Taylor and Bain 1999, 2000, 2005; Ramesh 2004; Batt et al. 2005; Pradhan 2005; Nornoho and d'Cruz 2006, 2007). Here, we shall take a political perspective for constructing the base of the discussion.

The present chapter focuses on the burgeoning Business Process Outsourcing sector and the leading host country i.e. India, and how a new occupation for unemployed youth has simultaneously created avenues for life and vigor, while depleting it. For few years, India has been chosen by many countries, especially U.S.A. and Europe, like many other countries who have felt compelled to outsource their major portions of processes from this country. India has won its spurs as the world's outsourcing destination of choice and currently has a commanding share of the global outsourcing market. The union Communications and Information Technology Ministry in India states that the India IT-BPO sector is likely to achieve a target of USD 60 billion in export revenues by 2010.

Global organizations have always preferred outsourcing call center services from India, in comparison to other countries like China, Philippines, Malaysia and other Asian countries (www.outsource2india.com; Abraham & Manning, 2007). India has always been the sweetheart of MNCs as an outsourcing location, because call centers in India offer a wide variety of advantages that other countries can't afford to offer. This is reflected in the increase of clientele base and the mushrooming BPO sector in India, and several reports like the McKinsey Report do illustrate the veracity of this claim. Today, having call centers in India has become the norm for several global companies, considering the cheap availability of a large and skilled workforce. In recent years, India has been able to effectively meet the growing international demand for call center outsourcing services by providing cost-effective services and customer-oriented call centers. This seems to be supported by the increasing desire among many international organizations to setup call centers in India to leverage upon the easily available large qualified workforce and provision of cost-effective call center outsourcing services.

Upon investigation, we find that the reason why western world has shown sufficient interest to establish BPO units in India lies with India's ability to provide the largest number of qualified, tech-savvy, IT literate, trained, skilled and experienced professionals. India has the largest English-speaking population after the USA. India's large and well-educated workforce has been one of its main advantages over other countries. The irony of the situation is that India's large manpower is willing to work for a lesser price due to heavy unemployment. Generally, in a call center operation, manpower typically accounts for 55% to 60% of the total cost, whereas in India, manpower is available at a fraction of the cost overseas. The relative advantage of a large, well-educated, English-speaking work force has enabled India to be the recipient of much of assignments from countries such as the USA, the UK, and Australia (Penfold, 2009). India's services exports rose by over 30 per cent per year during the five years 2001 to 2006, during which time they rose from 29% to 38% of total exports (Department of Foreign Affairs and Trade 2008).

The call center industry is well situated within India's global leadership; with its offshore information technology and business process outsourcing industries increasing at an annual rate greater than 25% and generating export revenues of $60 billion by 2010 (NASSCOM, National Association of Software and Service Companies, 2005, cited by Pal & Buzzanell 2008). Holman report (Holman et al. 2007) indicates that a typical Indian call center has about 500 employees, is about four years old, services the international market (80% international customers), and focuses on outbound calling (60% revenue from outbound sales). When we look for reasons as to why it is the most preferred job by youngsters we find many, and some compelling ones. A society in which the majority of the population suffers from the pangs of hunger, destitute and unimaginable misery, and then this '*little*' sacrifice seems not so preposterous. The argument also seem to gain weight as a nation strives to produce educated souls every year out of its glorious factories known for shaping humanity, and still does not possess the potential or *know-how* to get them employed, then they have to seek refuge in an occupation which can promise their much deserved but less available sources of need satisfaction. The failure to produce adequate avenues for employment of these educated frustrated souls leads them to find their much sought after solace in these alluring ivory towers known as Call Center. Hence, one question which still hunts the minds of both parents and policy makers is 'would you rather have graduates roaming the streets jobless or would you rather have them get some interim employment, be financially independent and be able to plan a future?'

Some of the reasons attributed by youngsters for this preference of a call centre job seems to be oriented by their largely consumerist and hedonistic lifestyle and desires. Few of the most compelling reasons attributed are not getting a better job, nothing better to do, job not matching with the education level, the provision of good work environment (deceptive though), good fringe benefits, flexibility of time (!), attractive life style (new clothes, new watches, but who sees them!), and transport facility. Some of the youngsters think that call centers are very good opportunity for young Indians to get work experience and with this experience and money they saved, they can go back to school for higher education. Also, these opportunities will motivate more youngsters to attain a degree, which they thought is useless couple years ago. They opine that most of the people move on and look for new and better opportunities after a couple of years of Call center experience. The youngsters seem to be using this experience to hone their communication skills, customer handling skills and earn money for higher education.

However, this much valorized system of working for others in a novel environment, equipped with modern amenities and congenialities seems to have its subtle ways of exploiting these innocent soldiers too. These Centers work on the model of getting the work done by some other people by seeking help of an intermediate party, who seems to be there eternally '*at your service*'. This very adage, '*at your disposal sir*' requires these youngsters to get trained to sacrifice their much needed pleasures of life, during their very formative years of life. In the First half of the chapter an attempt is made to highlight the following issues: First, Call Centers and their entry into the Indian setup, and mindset. Second, the very ingrained attitude towards the call centre, and the kind of jobs they offer with the promises of lucrative remuneration packages. Third, The challenges and nuances of Call Centre jobs, their Organizational culture, and how this monstrous coordination takes place. Fourth, an Analysis will be made of the contributions of these Call Centers to the lives of people in general. Some issues like do customer care executives really care for customers, and if not, why not; and the lessening concern for the human problems and a gradually increasing indifference tainted by the training given will also be discussed.

The second half will try to focus on some of the experiential aspects revealing the trepidations and struggles that a youth faces during his every waking hour of the life. The individualistic perception of the job, evaluation and sentiments attached to it, and the collective negativistic schema emerging largely because of its impact on the health and interpersonal lives of the youngsters is discussed. A dissection shall be made to look skeptically at the apprehensions in the psyche of a layperson of India regarding the use of subtle politics by the developed countries in their vent for a new monopolistic and capitalistic regime. The apprehension surrounds the use of a series of built-in techniques used for crippling the talented youth, reminding the adage "catch them when they are young, and kill them in their womb". A revisit to the concept of growth and development as experienced in the advent of globalization which has been construed by Indian scholars and lay alike as neo-colonization in a new garb also forms one of the core arguments. This has been complemented with a rationalization through an iteration of several back ground factors behind the construal characterized by a persecutory nature of the motives of the western countries by Indians. It provides a critical assessment of (if) call centers are a reflection of a new politics emerging in India. Though, this chapter does not explicitly endorse the view, still an objective attempt is made to give a critical look at the very situation. Hence, an effort is necessary to weave together interluding issues of an Indian youth as a subaltern, waiting for the call, though the nature and gravity of these calls and its impact on him are unknown…

WHY INDIA FOR OUTSOURCING?

One of the strongest arguments in favor of effectiveness of outsourcing efficacy of India comes from a giant in this sector. Outsource2india.com advocates why India is and should be a most preferred outsourcing destination. On a surface level, the reasons cited may sound like a real boon to the Indian youngsters incessantly engaged with a struggle to come to terms with their self-image by acquiring a 'respectable job', but ultimately upon little in-depth analysis they fail to serve their purpose.

India has emerged as a key player in the BPO market, and continues to attract outsiders as a preferred BPO destination opines Dharmawat in Patni Computers Report (2003) based on his rich data and sophisticated statistical analysis. According to NASSCOM – the revenue from BPO services and related services like call centers grew by 59 percent to $2.3 billion in the financial year 2002-2003. NASSCOM estimates that the revenue from BPO and related services will grow 54 percent in the year up to March 31, 2004, to touch $3.6 billion. McKinsey & Co. predicts the global market for IT-enabled services to be over $140 billion by 2008. NASSCOM, the premier industry association of Indian IT companies, classifies Indian business process outsourcing services under IT enabled services and characterizes it as the fastest-growing segment of Indian IT exports over the past two years.

Some other reasons cited by Dharmawat (2003) for the preference of the western world for India as a rich land of labor is its user-friendly nature

Figure 1. Get the India Advantage! (Source: www.outsource2india.com)

and culture, people are high IT Savvy and skilled, open to new cultures and are having a strong work ethics. This seems to be driven by the higher level of education which teaches cosmopolitanism that every Indian is trying to get imbued with. Due to a greater emphasis on language and value education, Indian population have gradually felt compelled to accustom themselves with the usages of English, modern (western) mannerisms, professional etiquettes and values.

Indian educational system is very competent in producing educated souls who are left struggling with their fates as soon as they finish their education. Definitely, after 25 years of dependence on family members for all their expenses their morality strikes them to do something! Few graduates go abroad to pursue further education or better prospects, for those who stay behind, a career in a IT or an ITES affords them a much better standard of living than local jobs. A postgraduate, for example commonly starts at a salary of around Rs. 7,500 per month, which approximates to US $150. A comfortable salary in India is considered to be around Rs.10, 000 which translates to approximately US $222 per month. And so, it is no surprise that because the US dollar goes further in India that companies worldwide can drastically cut costs without compromising quality (Dhrmawat, 2003).

Table 1. Expected global markets for IT enabled services

Customer Interaction Services	33.0
Finance and Accounting Services	15.0
Translation, Transcription, and Localization	2.0
Engineering and Design	1.2
HR Services	5.0
Data Search, Integration and Management	44.0
Remote Education	18.0
Networking Consulting and Management	15.0
Website Services	5.0
Market Research	3.0
Total	**141.2**

Source: NASSCOM McKinsey Study - India IT Strategies

Knights and Jones (2007) opine on why people should go for BPO sectors in India, citing the urgent and existential concerns:

State socialism and policies of trade protection over many years seemed not to have benefited the poor too much and maybe the stimulation of a thriving knowledge-based BPO industry will also be a drop in the ocean but any country suffering as many problems as India can hardly kick a gift horse in the mouth.

But, the question is can this advantage which all authorities have been boisterously glorifying for all these years are really going to prove advantageous for Indians in the long term. Is this care and commitment really getting reciprocated, or even having been thought by their masters sitting miles away, enjoying the luxurious services. The issue seems to be very sensitive and a proper discussion with a foresight is certainly the need of the hour.

Issues, Controversies, Problems

One of the most attractive factor for global organizations to India and their preference for Indian outsourcing services is due to an assurance of cost-effective customer support services without compromising on quality. However, the oft ignored issue is whether this component of cost effectiveness really takes care of both party's interests and does this cost-effectiveness component and subsequent attraction for off shoring has any costs for India too. It remains to be analyzed and needs an in-depth exploration. We shall discuss the underlying issues in the light of development perspectives, and the perpetual core-periphery relations, and how does the peripheries negotiate and re-negotiate their development at an unknown cost.

Imperialism, Decolonization and Neo-colonialism have played and continue to play a powerful and decisive role in shaping the world in which we live—starting with the national borders i.e. part and parcel of our identity construction, to a perception of self and others. Al-Rodhan (2006) comments on a shifting definition of globalization and opines on the role of an individual's political ideology, geographic location, social status, cultural background, and ethnic and religious affiliation in creating a canvas for construal of globalization. Though, there is a controversy with regard to the definition of globalization, increasingly scholars have come in to terms with the homogenous features of west-rooted globalization. This is evident in Waters (2001) who define globalization as "...the direct consequence of the expansion of European culture across the planet via settlement, colonization and cultural replication. It is also bound up intrinsically with the pattern of capitalist development as it has ramified through political and cultural arenas..." It is evident that these discussions regarding the nature of globalization have equated the former to the later i.e. re-colonization or neo-colonization.

In a very popular book entitled 'Globalization or Re-colonization?' Jain (2001) comments:

"The harsh reality about globalisation is that it is nothing but 'recolonisation' in a new garb. The third world countries are being turned into economic colonies of the developed, or imperialist, countries – the same imperialist powers who had once before directly colonised them. In the name of 'free market' and 'free trade', multinational corporations and banks of the West are re-entering the third world economies to plunder their wealth and resources. The imperialists are not enforcing this new colonial order by force as they once did in the 19th and early 20th centuries. They don't need to. The ruling classes of the third world countries themselves are betraying the interests of their people and countries. They are voluntarily handing over control of the economies of their countries to Western corporations and governments and the international financial institutions controlled by them. The third world elites are of course being well rewarded for their treachery: they are getting a share of the imperialist plunder of their own countries."

The Era of Modern Laborers

Though call centers have been accepted in Indian Industry and even in the workplace, it has faced some of the most vehement criticisms ever heard about anything. The concept of 'cyber-coolies' laments on the new model of engaging young minds in developing countries and exploiting talents by making them work on some of the less preferred jobs in western worlds. In fact, the situation has worsened so much that the lives of these employees seem to be drab and monotonous, mechanical and routine enough to work for the big bosses who are eternally preoccupied in savoring the pleasure of life. These modern laborers equipped with all the major degrees from universities in metropolis like Delhi, Mumbai, Bangalore, etc. go out at odd hours from home chasing their dreams. On a flipside, the issue is can India really ensure a provision of equitable jobs for its millions of educated youngsters, craving to fulfill their basic needs? Can a concept of 'respectable job' be conceived in a developing country where people race eternally for an occupation for livelihood? And, consequently when they fall prey to a mentally corrupting system, what happens to the lives of such modern laborers who struggle with their several aspects of lives?

As though it is not enough, they are required to work in a non-ergonomic fashion in workplaces. Some critics argue that the work atmosphere in call centers is de-humanizing [EFLIWC]. Others point to the low rates of pay and restrictive working practices of some employers [Pay Scale], [HSE, 2006]. There has been much controversy over such things as restricting the amount of time that an employee can spend in the toilet [Hazards]. Furthermore, call centers have been the subject of complaints by callers who find the staff often do not have enough skill or authority to resolve problems,[Russel, 2006] while the dehumanized workers very often exhibit an attitude of apathy to even the most abusive customer (Ahmed, 2006). Owing to the highly technological nature of the operations in such offices, the close monitoring of staff activities is easy and widespread. It has been argued that such close monitoring breaches human rights to privacy [Health Canada], even though it is implemented mostly with an assumption regarding its positive outcomes.

On The Quality/Kind of Works: Are Some More Equal than Others?

The type and quality of work which has been transferred to India has mainly focused only on the front office functions, rather than the core ones, and hence the inevitable and invincible call centers are sky rocketing in the market. Knights and Jones (2007) opine that the hugely accumulating highly educated labor force has compelled the call centre organizations to recruit, and a economic considerations seem to be the underlying sole motivation(Evalueserve, 2004; Sen and Shiel, 2006). While a predominant utopian world order has been conceptualized by many theorists who believe that it's a global panacea to the mass deprivation and poverty of third world economies (Friedman, 2005), the situation seems to be reverse. It has been supported by few others who lament that it's a "dystopian nightmare in which labor exploitation associated with western capitalism is exported to the underdeveloped world at the cost of displacing labor in the west" (Mirchandani, 2005; Bain and Taylor, 2004; Taylor and Bain, 2005). Though, it is reflected from the article of Knights and Jones that economy has been transmitted from western countries to the underdeveloped Indian context, it's still under debate.

The issue rises whether does this targeting India as a BPO destination due to the free labor (liberalized one) systematically create slaves of consumerism and hedonism with the illusion of affluence and freedom? Another issue is why do developed countries conveniently take India as a country full of talents, but never nurture it? As, it's quite obvious that if they feel that India does posses a talent pool, then a mutual flow should have been encouraged for the core functions of business and services between all these countries rather than a charitable engagement in some less cared for job by their own citizens. Though it's not intended to exaggerate the issue and neither to deny the equality of jobs, but then the unresolved issue is why there is discrimination in allocation of functions in a globalized era? Are some people more equal than others, and hence be responsible for the important functions of business and those which can be "done any anyone" will be thrown developing or underdeveloped countries? These latent, but serious issues necessitate neutral stand, and self-reflexivity on the part of policy analysts.

Work Global but Pay Local!!!: Analysis of Inequitable Remuneration

It's quite clear that Western countries chose India as a hub for Call centers for the reason of economic viability and possibilities. The inherent agreement was Indians will work for foreign (International) organizations, serve their processes and functions, but will be paid locally. And, the basic reason that availability of good quality labor force (still used instead of Human Resource/capital) at an almost fragment of their currency has lured capitalists, and entrepreneurs to establish their business houses in India. Though, at initial stages, all youngsters blindly accept call center jobs because of pecuniary benefits, gradually they start feeling the subtle efforts behind strategic entrapment of young minds. A strongly held perception of pay inequity and exploitation percolate down the mass mindscape giving rise to job dissatisfaction and lack of motivation, and commitment (both affective and continuance). The problem seems to be accentuated by the low compensation (pay and bonuses).

If, we compare the call center job with other jobs that an average undergraduate student holds, then definitely on a superficial level it seems to be superior in every possible way. But, an in-depth analysis of the psychological issues (to be discussed) weighed by the pay equity measures will make it liable to be labeled as a pure exploitation, at least from the perspective of the employees. When it comes to consultancy jobs, projects at higher level (white collar jobs), then the pay is being guided by the currency of the country service is being rendered to than the service being rendered by. So, in this context this incongruity can definitely be a predictive factor of resentment among the Indian youngsters who devote their heart and soul into the "not too important" processes of business.

Accent Neutralization or Linguistic Customization: Creating Serviceable Others

Sociolinguists have gradually recognized the phenomenon of globalization which besides bringing a set of far-reaching, transnational, economic, social and cultural changes has implications for patterns of language-use, linguistic variation and change too (Cope and Kalantzis 2000; Fairclough 1992; Heller 1999). One of the interesting aspects of globalization has been the 'new work order' (Gee, Hull and Lankshear 1996) which reflects a new ('post-Fordist') ways of working making new demands on the linguistic abilities of workers. Commentators opine (e.g. Cope and Kalantzis 2000; Gee,Hull and Lankshear 1996; Gee 2000) that the demands of new capitalism necessitates newer forms of linguistic and other agency. There is another argument that these new linguistic demands on workers might lead to a control "over their linguistic behavior and thus and thus a diminution of their agency as language-users" (Cameron, 2000). Hence, a systemic entity which entails control over not only the practice of a language but its features is but obvious to lead to the death of agency among call centre employees.

After the call center business started in India, metropolis and even small towns saw emergence of some culture transition hubs known as call center training houses. The basic objective with which they surfaced was to acculturate the "uncivilized", "unmannered" Indian youth and make them to achieve a minimum level of business etiquettes. Since the beginning these centers have tried their various strategies of attracting youngsters to join these short term training programs which promise improvement in fluency of English, business etiquettes and mannerisms, etc. Agents in Indian call centers are required to learn American accents, as mostly they cater to the American customers (Ganguly, 2009). They are even expected to be conversant with day-to-day American issues to the extent that they are able to carry on casual conversations with clients (Mirchandani 2004). This cultural familiarization is an extension of locational masking reducing the chances of the identification of the agent as non-American, an essential feature of the business strategy. Taylor and Bain (2005) in their study on call centre employees have observed that these westernized pseudonyms, locational masking and forced cultural familiarization or customization exacerbate which they call as 'culturally-specific dimensions of psychological strain' subsequently leading to emotional labor. "...In a Foucaultian sense then" observes Poster (2007, p.291), "national identity management allows these firms to hide the exercise of power on the [American] consuming public".

The main purpose of accent neutralization / reduction is to help proficient English speakers speak with a more North American or British accent. The main cause of this trend towards accent neutralization / reduction is the demand created by outsourcing. The basic focus of these centers lies in creating training grounds for accent neutralization, i.e. erasing what they call "mother tongue effect", thereby diminishing the value of own mother tongues in the eyes of youngsters. Even, sometimes many young souls do curse their fates as to why are they born in India which they believe is a major reason of their linguistic deficiency (see, Ganguly, 2009). This supposed linguistic deficiency and its effect is much exaggerated in each context is likely to become unbearable for an average educated individual entering into this sector. Therefore, initial phase of screening weeds out the incompetent ones. This overemphasis on linguistic skills, especially accent (American/ British), less emphasis on any other talent, and dejection on a supposedly less adaptable accent improvement leads to a strange depression among these subalterns of modernity.

This linguistic imperialism which is reflected in the attempt by core countries to train others in their language i.e. English and thereby establish their hegemony in the world economy has been discussed by several scholars worldwide. Phillipson (1992, P.51, 47) draws on Johan Galtung's (1971) typology of imperialism and posits linguistic imperialism as a "sub-type" of cultural imperialism as he defines it as "the dominance of English...asserted and maintained by the establishment and continuous reconstitution of structural and cultural inequalities between English and other languages" (1992, pp. 51, 47). Structural inequality at the global level is reflected in the center-periphery distinction that Phillipson adopts from the neoimperialism critique proffered by dependency theory and its theoretical companion, world systems analysis (Chirot and Hall 1982). Accordingly, the "First World" center enforces a set of economic, political and military structural relations (constraints) on "Third World" nations in the periphery which makes them "dependent" on the core or center nations for economic and political survival in the global capitalist system (Sonntag, 2009). These very clearly reflect a deeply ingrained systemic modification.

Many of these guinea pigs who go through this painful process of accent reduction often find this trend distasteful, but compelling to follow. But, perhaps we might find answer to this unnerving tendency among the trainees to accommodate the changes with an analysis of a fascinating book entitled "The World is Flat" by Thomas L. Friedman. In this he explains the positive side of the trend and how a middle class youth accepts the process with a defense mechanism:

"... before you disparage it, you have to taste just how hungry these kids are to escape the lower end of the middle class and move up. If a little accent modification is the price they have to pay to jump a rung of the ladder, then so be it - they say".

Though, Friedman is able to tell us about the individual compulsions about accent reduction, he has not spent much effort in working out its deleterious effects with a restricted emphasis only on the etiology. But, some issues which have been ignored are whether this linguistic deficiency is really a deficiency? What competence does this accent neutralization and customization bring in an individual's ways of working? And, is there really anything fishy about the strategic movement of developed countries proclaiming their linguistic superiority, and efforts to gradually erase the mother tongue (value and hence use of it) from the minds of their own denizens?

Pressure of work, Noise and working Environment: The call centre industry has been infamously been lamented to be suffering from a poor reputation of low pay, monotonous work, high demands, low control, limited social support and few opportunities for participation and learning (Anderson & Jansson, 2006; Houlihan, 2000). Poor working conditions e.g. poor facilities, poor maintenance and cleaning, cramped working conditions, management interference, lack of privacy and noisy are some of the other working condition factors which affect the working patterns, performance and individual wellbeing of the employees(Taylor and Bain, 1999). These services are mostly the —high-volume, low-value, routinized end of call centre work which tends to be shifted to India (Taylor and Bain 2005). Agents are mostly expected to work at night to cater to U.S. time zones, and adjust to an altered social and family life.

Benninghoven et al. (2005) have analyzed the role of certain factors which adversely affect the health and wellbeing of individuals working in call centres. Work in open-plan offices with a high level of background noise, a high rate of work in front of computer screens, often with a limited amount of space affects the working style and performance. There is a very high demand for mental concentration due to the simultaneous requirements to speak, listen, and look as well as to use fine motor skills like typing and it compels at time to lose coordination between these faculties.

All the working pattern and load has an adverse impact on the Quality of Work Life of employees. Quality of work has been defined as "better jobs and more balanced ways of combining working life with personal life" (European Foundation for the Improvement of Living and Working Conditions, 2006). Employee's lack of control over their work and extensive monitoring, including the high emphasis on targets, accentuated by the strain of dealing with customers all day, compels them to experience work as pressurized, intensive and emotionally stressful (Richardson et al., 2000; Deery and Kinnie, 2002).

Verbal Abuse: the eternal Applaud of the enslavement: It's not obsequious that call center employees encounter monsters howling at them by their crude and irrational comments and often a series of vulgar, rude and abusive customers (Taylor and Bain, 1999). The situations become tense often and make call center employees get rebuked for nothing, just because one man had fought with his wife in his home thousands of miles away, he had a bad day at office, she does not know how to turn off the washing machine, etc. The system of call center trains the young minds to be submissive and meek, and systematically ingrains values of slave mentality by brainwashing them. The effort to undo the reason and logic is the fruitful culmination results in perpetuation of this mentality, which is so applauded in the industry.

However, it's always a possibility that the frustrations of irated consumers ventilate when he talks over phone to a person overseas, and due to the lack of restriction and sanction, all of his negative emotions gets aired out. It puts us another issue before us i.e. has anyone ever tried to understand the trauma of endless and passive reception of applauding abuses? This systematic enslavement gives rise to loss of individual freedom, self-dignity and results in lack of self-esteem and self-worth, causing permanent havoc in the development of personality.

Analyzing Possible Impacts of Rigidity in Work Style

Taylor and Bain (1999) observe that the call centers have very hazardous outcomes as reported by employees who undergo the ritualistic strict regime of call center execution few deserve mention and analysis. Firstly, close scrutiny by management (e.g. frequent random call monitoring) is one of the essential ingredients of the call center job. The monotony of the job characterized by repetitive job tasks like receiving calls from different parts of the world, telling people solutions for silly problems, and restrictive working practices (some operators are required to follow a pre-written script). Call centre customer service positions have been found to be highly repetitive and "dead-end" jobs. Most are labeled as 'electronic sweatshops', and are often criticized for their "low-cost HR practices". The work itself is often monotonous, repetitive and routine, dominated by short cycle times and the prioritization of quantitative call throughput (Connell and Harvey, 2004,).

Call centre frontline employees (FLEs) routinely engage in highly demanding scripted interactions with customers that are continuously assessed and monitored by management (Mahesh and Kasturi, 2006; Siong et al., 2006, Ashil et al., 2009). Such conditions engender demanding, repetitive and stressful work roles and produce dysfunctional selves, and adaptive problems. Indeed, research indicates that excessive scripting is positively related to emotional exhaustion (Batt & Applebaum, 1995; Deery, Iverson, & Walsh, 2002; Holman, 2003a, b, McGuire & McLaren, 2007).

Emotional Labor and Apathy

All these factors added together leads to what might be called as emotional labor, basically fanned by the process of coercive control over expression of feelings and emotions in workplace. In a study on call center employees Johnson and Spector (2007) have demonstrated how the emotional labor process can influence employee well-being. In a field study Wilk and Moynihan (2005) examined the effect of supervisory regulation of display rules--the rules about what kind of emotion to express on the job on the emotional exhaustion of subordinates. On the basis of a sample of 940 call center employees, the authors found that worker emotional exhaustion varied across supervisors within jobs, suggesting that emotion work is influenced at the supervisory, rather than job level.

In another study Goldberg and Grandey (2007) observe that "Service with a smile" is satisfying for the customer, but such rigid display rules may prove to be costly to employee and the organization. The authors used an experimental approach in contrast to previous use of self-reported and cross-sectional designs to test tenets of resource depletion theories; specifically, whether the self-regulation of emotions required by display rules depletes energy and attentional resources during a service encounter. Using a call center simulation with three "customer" interactions, the authors found that participants given positive display rules (e.g., be enthusiastic and hide frustration) reported more post-simulation exhaustion and made more errors on the order form compared to those with display autonomy. Customer hostility during one of the calls also increased exhaustion overall and the number of errors during that specific call, though proposed interactions with display rules were not supported. Surface-level emotion regulation, but not deep-level, was the mechanism for the energy depletion effect of display rules, while display rules had a direct effect on performance decrements. The efforts for emotional assuage also results in vain, and is characterized by high levels of depressogenic and negativistic schema regarding self, others and world.

There is theoretical consensus as well as empirical evidence to suggest that emotional exhaustion and depersonalization lead to undesirable job outcomes such as reduced job performance (Bakker et al., 2004; Halbesleben and Buckley, 2004; Schaufeli and Bakker, 2004; Singh, 2000; Singh et al., 1994). According to COR theories, call centre FLEs may try to minimize the loss of emotional resources by implementing coping strategies that help utilize and conserve their scarce resources (Wright and Hobfoll, 2004). Employees who detach themselves from their customers and co-workers will perform ineffectively in their organization. This self-monitored apathy results in interpersonal skill loss as a result of disuse. We can expect that both emotional exhaustion and depersonalization will negatively affect the performance of call centre FLEs in dealing with issues surrounding customer dissatisfaction, initiating a circular relationship among the two (Ashill et al, 2009).

Stress, Psychological Health and Wellbeing: call centre FLEs often suffer from burnout (Singh et al., 1994; Singh, 2000) that is a form of psychological strain resulting from persistent work stress. Taylor and Bain also observe that high stress also is a common problem associated with front-end jobs where employees deal directly with customers. Individuals generally become victimized with subjective concerns headache, muscle tension, irritation, not feeling in the mood for work and feeling worn out which may represent a variety of disease conditions (Eriksen et al., 1999). Call centers have been criticized vehemently by several researchers and management experts for their employee management practices, i.e. their "sweatshop" approach (Taylor and Bain, 1999), "sacrificial HR strategy" (Wallace et al., 2000), and the "low-cost HR Practices" (Batt, 2000), where control and efficiency are emphasized at the cost of employee stress and turnover. Hence, it is but obvious that it results in poor service quality and low customer satisfaction, sometimes reflecting a chicken-egg problem.

Several specific patho-physiological pathways have been hypothesized for the relationship between the workload, stress, and mediation of stress in affecting the physiological mechanisms/ processes in the body. For example, high mental load and job demand may increase muscle tension and decrease micro-pauses in muscle activity. Even in the case of low loads it may lead to muscle fatigue, due to continuous firing of low threshold motor units, which are not only triggered by low level physical loading, but also by mental loading (Westgaard, 1999; Sjögaard et al., 2000). High mental load and job demands may also lead to adverse changes in immune response (Theorell et al., 2000). The severity of the effect can be well imagined as it has been found that Job stress may reduce the ability to reduce the physiological activation to resting levels during breaks and after work, and thus adversely influence recovery (Lundberg and Melin, 2002). Moreover, constant attention to the flashing messages on computer and being on guard, talking continuously for 8-10 hours, and listening with the help of a headphone causes impaired vision and hearing problems (Taylor and Bain, 1999).

Sleepless night shifts are never a cakewalk for anyone, be they Indian or western men. But, it is often a complaint of employees that 'when it comes to taking service, the big bosses call you when your body has given up, struggling with yourself to give it some rest'. Daily night shifts can cause insomnia, and other sleep problems including somnambulism. This also creates what has been referred to as Jet lag, a result of imbalance of circadian rhythm. The symptoms of jet lag can be quite varied, depending on the amount of time zone alteration. They may include digestive problems, headache, fatigue, irregular sleep patterns, temporary insomnia, disorientation, grogginess, irritability and mild depression. According to COR theory, people try optimally to adapt to the situation, but gradually with resource depletion (accentuated by lack of support systems), they start losing hope, and this causes further cycling of symptoms.

Interpersonal Lives and Mechanical Lifestyle: A Process Analysis

It seems that these employees become very efficient in social skills like persuasion and communication (fluency), especially with the help of BPO training programs. But, ironically all these come at a cost of compensating against their interpersonal lives. It is well described by Knights and Jones (2007):

...the ethical attention has been diverted to the offshore BPO service sector in India and the work intensive, high pressure call centres where frequent night shifts disrupt family or social life and where well educated staff are expected to remain in dead end routinized low level software, back office or call centre service and sales with little prospects of career development (see Bhaghat, 2005 for some fictional realism on these issues). In these critical ethical judgements, the positive sides are usually underplayed.

Jagdish Seth (2009) talks about one of the related theme i.e. interpersonal relations and lifestyle of Call centre Couples (CCC), which he very vividly describes, and objectively shows the loopholes in the systemic flaws which affect their lifestyles. The rise of this New Modern Class (NMC) will not seem to be as rosy after reading his theses. He explains as to how the lifestyle of these classes has been systematically aborted with the illusion of affluence and prosperity. Their lifestyle includes a variety of new trends which not only challenges many traditional societal institutions like "friendship replaces kinship", "blurring of responsibility", "when wants become needs" and introduce a highly dependent consumerist and hedonistic worldview. The implications of these are huge and require proper deliberation.

Growth as an Individual and Society: Are They Really Growing?

Many Indians in the country and Diaspora population feel that *"India should not fall prey to such low paid and valueless jobs from the west. I see no growth in the local Indian economy with the creation of jobs that exploit manpower. Moreover these conditions are going to short live and not in the far future that many Indian workers in the call centers would be let without jobs. Job from the west be it for any purpose is not a reliable and sustainable way to development in India or any Asian country for that matter" (Kannan Pasupathiraj, Germany)*. They are often characterized by low status, poor pay and few career prospects.

In a BBC report (2003) Mr Bidwai said that call centers were exploiting young English-speaking Indians who "have an undergraduate degree and nowhere else to go." They work extremely long hours badly paid, in extremely stressful conditions, and most have absolutely no opportunities for any kind of advancement in their careers. This blind acceptance of these jobs has been attributed by him to the acute joblessness in India in spite of GDP growth. In spite of some people's strong support for Call centre industry as a vital part of India's economic development, Mr Bidwai said it had little value to India's wider economy. "It's a couple of billion dollars a year at the moment, which isn't a hell of a lot, even by Indian standards," he argued. "Our exports of textiles, for instance, are much, much bigger. *"There is a kind of irrational exuberance about the prospects of this kind of low-paid, low-value-addition work which creates this content both here and in the West, where people are losing jobs. As it has been seen, a large numbers of employees suffer from a range of ailments - for instance sleeplessness, serious depression, ear infections - so you have a very high rate of employee turnover"*. He has criticized to the Government system in India saying *"Some of our leaders are looking for shortcuts to the sort of hard task where you actually address the basic needs of the millions who are in a state of depravation. They're looking at call centers as a kind of magic wand, and I think that's actually evading and ducking the real issues that confront the country - those are to do with poverty, equal opportunity, illiteracy"*.

Besides national identity and growth call centers seem to have an impact on the individual identity, or at least the perception and construction of it. Basi (2009) examines the concept of identity in a globalized world and the diversified exercise of agency over identity construction by women working in India's transnational call centre industry. She provides very incisive accounts and nuanced analysis of the experiences of Indian women call centre workers based on a rich empirical data. This gives a thorough treatment and an in-depth analysis regarding the role of women's participation in the global labor market. It seems that Identity is floundering under the realms of growth and sacrifice with a flame of self-purging perseveration. Pal and Buzzanel (2008) through a case study based analysis have also revealed us important facts about how an individual negotiates his personality and identity and relatedness to the world. They have tried to explore the effects of call centers in creating grounds for negotiating and expanding "an array of possible, especially new, identities and identifications and the ways that these choices affected changing social discourses". They observe that call center employees engage in constructions and reconstructions through a process of negotiation with "an array of discourses that bracketed opportunities for particular identities and identifications", which definitely emerged out of their experiential import.

They opine that these negotiation processes led to an engagement in what they call as "strategic identity (ies) invocations" and revamped several arenas of life like work, career, and familial intercourse and practices.

There is large scale change in the transactions and interactive modalities between countries as soon as we get into a commercial relation. This has been supported to some degree by Friedman (2005) who observes that:

"The flattening of the world will reshape political identities, recast political parties and redefine political activists. Tensions among our identities as consumers, employees, citizens, taxpayers and shareholders will come into sharper conflict. This will all have to be sorted out. Who will look out for different communities that have never connected before? How will we determine whether or not a business partnership is a true collaboration or if someone is being taken advantage of?"

This definitely will arouse anxieties of various kinds in the collective psyche of both developing and developed nations. But, few issues which Friedman could not do justice with are dealt by Goldman (2009) who asks very interesting and incisive questions related to the scope for development of India to be called on by the call centers. Though he is exclusively focusing on the city of Bangalore, some of the issues raised by him have got potential implications on a large scale. Few of the questions asked are pertinent to our discussion: " Do investments from the World Bank and India's IT sector generate the type of urban social relations that transforms poverty-level incomes and living conditions into the socially equitable and healthy "world city" of the 21st century? Or, at the very least, will these dynamics overcome some of the "mega-city" problems that seem to plague large global South cities?" As his discussions which profusely complimented with examples illustrate that in fact these call centers neither contribute in substantial fashion neither towards to the economic development nor the creation of equality or improve of standards of living and wellbeing index.

GLOBALIZATION AND NEO-COLONIZATION: IMPACT ON CULTURE AND CULTURAL IDENTITY

"The comparison of cultures requires not that we reduce them to platitudinous similarity but that we situate them apart as equally significant, integrated systems of differences", says James Boon in his book other tribes, other scribes. However, it is very evident that one of the biggest achievements of the process of globalization has been reflected in among the culmination of cultural heterogeneity giving rise to a coercive cultural hybridity among subjugated countries. And, it has been basically due to the capitalist processes of neo-colonization governed by the core countries with modern archery as evidenced by an economic submissiveness and despondency instilled among developing and underdeveloped countries (periphery and semi-periphery). It has also been observed that globalized work processes have substantially and fundamentally changed the ways in which labor markets are organized around the world (Mirchandani 2004). Ganguly (2009) opines that at a macro-level, the emergence of call centre industry is a representation of the shift in globalization. He observes that *"unlike silicon chip production in Taiwan, maquiladoras in Mexico, or McDonald's in France, transnational customer service employment represents a shift from exporting the production of material goods or culture to a full scale reproduction of identity and culture"*. It's not limited to the identity reproduction only, rather as Patel complements that the call centre employees are being trained to embody an American identity and cultural cues (Patel 2006).

In our present context of call centres, the issue with which we grapple is: are call centers and all the mercantile activities which goes on in the name of outsourcing and big business of engaging young unemployed denizens are really designed with noble intentions! Are the 'beleaguered souls' really resting in peace and have found ultimate solace? Or, is it just the beginning of a culture customization which has started with getting imbued with the values and patterns of another dominant culture in order to satiate their physical needs. Though, expressing the desire to be open to others' cultures, values and ethos is purely rational, what seems to be irrational ironically is the mocking diminution of emphasis on one's language and culture, and failing (forceful) attempts to ape others since the very childhood. This seems to create a gap in the mindset of an individual struggling to achieve an image in the eyes of his/her reference group which devalues them eternally. This culture shock gives rise to excessive engagement with the accumulation and assimilation of others' cultures as reflected in accent neutralization, which generally ends up with imitation of British/American accent (still making a mess of it!).

The 'melting pot' ideology suggests that each cultural group loses some of its dominant characteristics in order to become mainstream through a process of assimilation, or what Triandis (1994) calls subtractive multiculturalism. In contrast, when people from a cultural group add appropriate skills and characteristics of other groups, it may be called integration, or additive multiculturalism (Leung et al., 2005). But, in this case what seems to be happening is that learning or adapting another's culture though is a propositionally very beneficial one, has resulted in loss for the adoptee. This seemingly glorified imposition of melting pot approach on the minds of Indians and a compulsion to learn to adapt to western culture(s) is likely to produce a considerable amount of umbrage.

Some researchers also opine that the efforts expedited at the synchronization of agents' shifts with western customer servicing hours is both symbolic of the subordination of Indian workers to the interests of western capital. Moreover, it induces nocturnal work patterns and hence significantly exacerbates the generic pressures associated with routinized call-handling (Taylor and Bain 2005; Ganguly, 2009). Taylor and Bain (2005) have also gone to the extent of categorizing these practices as neo-imperialist and indeed racist. They have further argued that it cannot be assumed that glamorized "globalization goes unchallenged and that Indians become eager agents – in both literal and metaphorical senses – willing to re-model themselves at the behest of American or British corporate clients".

English as a language has been instilled into the mainstay of political and social sphere since the colonization phase of Britishers in India. In an interesting piece of analysis concerning a long standing problem of the divisive nature of affiliation with English in India between the so-called elites and the subalterns, Sonntag (2008) provides a refined analysis of how today's call centers perpetuate the issues of status quo in India. She concludes that the appropriation of American identities by Indian call center workers through a deceptive process of identity management and actualization is for the express purpose of "cultivating linguistic capital within the Indian context, and does not entail a loss of authenticity or reveal cultural insecurity". But, the question is what actually leads to this passive acceptance of a cultural loss, or which is otherwise referred to as cultivation of cultural capital? And do they always accommodate to the demands of lingual and identity appropriation positively. Perhaps the evidence present indicates to the contrary.

Winifred (2004) has explored the globalization of service work through Indian call centers that involve a process of identity management or deception to be more appropriate. He observes that there is difference of views with regard to the role of development of ICT on the cultural homogeneity and hybridity. He comments that "while some scholars argue that the development of ICTs will lead to a homogenization (especially an "Americanization") of identities, others see increasing global disjuncture and renegotiation of identities". He concludes that though acting American, as a part of identity management reflects global power of American firms, it does actually lead to assimilation of American identity for most of the employees. The side effects of this coercive identity actualization are they are increasingly "...accommodation, ethical or political objection, or active resistance". It very clearly indicates that the oft presented rosy picture of the situation in call centers is not perceived very positively by Indians.

The herculean effort of developed nations to develop the underdeveloped by providing assistance and employment opportunities have been highly publicized in recent years. The first assumption of these countries of an order (basically of their own), and creating criterion points against them for other countries is quite systematic and deliberate. It is quite possible under such situations that an Indian psyche will be reminded of an egoistic logic settled by developed countries dividing the world into pluralistic entities (sometimes two binary entities) as reflected in a song "we are the world and its counterparts". It raises interesting issues like the strengthening of a status quo existence in relation to the boundary definition of core and periphery. And, if the situation is so, then does not it warrant folks of India to smell something fishy!

SOLUTIONS AND RECOMMENDATIONS

The situation seems to be very sensitive due to the interaction of several factors which leads to worsening of situation. The perception of call centers and exploitation is mediated by the cultural factors, and the long standing systemic mental captivity among Indians, boiled with inferiority complex elicited by these anti-cultural practices. The issues are very sensitive and it creates problems for both the employers and employee. But, what seems to be very evident is the dense haze of sense of animosity resulting from the perception of injustice and a perception regarding the unfair treatment at work, and belief of a strategic design of developed countries for crippling the engendered. There have been various attempts made by scholars to resolve issues and concerns of the call center employees in India. But, it does not seem to be such a cakewalk to vanish these feelings, and reaction just with swish and flick. But, can we think that the solutions to this seem to be a figment of imagination and give up? As, the business has to run, and millions of young Indians need to engaged in some fruitful work, so the need for call centers can't be denied. Therefore, some of the suggestions which might be provided to call center organizations to engage in value-centric employee treatment and develop an employer brand image:

First, Organizations need to identify the key traits of Individuals, not just the English fluency as the core competence for these Jobs. Secondly, those who are competent in technical skills, but lack in communication skills should be trained in communication skills (not just English fluency!). It is of paramount importance to emphasize on empowering people and develop human capital in true terms, rather than creating serviceable others for satisfaction of petty needs. Thirdly, provide advice on how to integrate working shifts with home life, which covers, in particular, eating properly and how to promote good sleep during the day. Ensure car parks are well lit and consider employing a security person to patrol car parks or installing closed circuit television to increase security. Four, Telephone for taxis to collect at the call centre rather than staff catching them on the street. Fifth, make bus timetables available so employees can wait in the call centre rather than on the street if they miss the bus they had intended to catch.

Some other suggestions which may be provided for the health maintenance of employees and creating a humane environment based on ergonomic model:

It should be ensured that there is a clear chain of command is identified out of hours and adequate maintenance and technical services are available overnight. It should be made very clear that working hours are not too long and employees get enough rest between shifts, neither employees work too many night shifts in a row. Managers need to negotiate with staff about overtime (including a reference to double shifts appears to condone it – double shifts should be avoided where ever possible) Managers should try to avoid planning critical jobs for the ends of night shifts or at times of the day and night when staff are naturally less alert, e.g.2-4am.

There should be opportunities to express real concerns like anyone can report fatigue problems to management and the company will make improvements. Enough care should be taken so that the environment does not cause drowsiness (it is light with visual interest, not too hot and there is variation in the level of sound). Incidents or accidents where fatigue may be responsible should be thoroughly investigated.

Enough opportunities should be provided to enhance interpersonal relations at work, rather than following a rigid regime. Communication between shifts should be ensured, for example by a 'buzz session' at the start of a shift; communication whiteboards, the Intranet, Email and newsletters can also be used. It has been a complaint of temporary workers that they are treated very whimsically by other superior staff. Temporary workers should be given the same treatment as permanent workers in terms of workplace characteristics, demands, expectations etc.

Call centers should have a formal, defined and documented career development plan in place for all staff. Emphasis on career planning of youngsters should be a concern of organizations too, as a symbol of true socially responsible agency. Career development plans should be regularly discussed for each individual employer, where goals and sub-goals are altered in line with recent achievements. Call centers should allow staff a certain amount of time for career progression training. Call centers should actively encourage their staff to progress in their careers, and provide funding and/or specified training allowance for each employee. Hence, Call centers should use external training providers where necessary, and to relieve demand of time pressures on management staff.

FUTURE RESEARCH DIRECTIONS

Though, a review of literature suggests a presence of several suggestions, known as ideal practice, it is to be taken with a pinch of salt that the results have been largely marginal and still the perception of unfairness permeate the mental space of Indians. It is expected from researchers, scientists, as well as policy makers to steer the future direction for research as well as policy imperatives to facilitate the lives of these individuals. More individually garnered attention needs to be paid in order to enhance upon the systemic flaws. The strong feelings of inferiority complex among Indian call center employees fueled by the strange irritating behavior of abusive customers can only be reduced by creating more sensitivity among the western citizens to other cultures and lessening animosity among cultures. More efforts are to be initiated at grassroots level to make people feel closer to each other as fraternity than feel someone, or some groups as inferior to others on the basis of some falsely based criteria.

The future efforts need to be oriented towards creating more harmonious relations among cultures encouraging multiculturalism and respect for each others' values and ethos. More efforts should be initiated to create better work environment at the workplace, by fostering ethical values, by inculcating amore relationship-oriented culture than a task oriented rigid work culture. The needs and concerns of youngsters (call center employee) need to be taken into consideration.

CONCLUSION

The concerns of call center employees are manifold, and in a country suffering from pangs of hunger and problems of severe nature, this small husk of support (by call centers) can't be ignored by youngsters who have been fooled by the educational system. But, at the same time the subtle politics of engaging young minds of India in a manual task (abandoned by others) does not seem to be quite rational too. So, the employers of west should realize that people of this developing country do posses a great deal of talent, and hence can also be considered for more engaging tasks. So, a combination of cultural sensitivity techniques equipped with multicultural HR management, strong focus on the development of young minds (career development) than just '*using*' them for petty business benefits, and more humane ways of dealing will create better systems of business enterprise and better civic society.

REFERENCES

Abraham, M., & Manning, E. (Aug 12, 2005). *Service on the Line: Indian Telephone Call Centers and Customer Care.* Paper presented at the annual meeting of the American Sociological Association, Available at http://www.allacademic.com, retrieved dated 2010-05-07.

Advice regarding call centre working practices. (2006, October). *Health and Safety Executive* retrieved 2008-06-05

Ahmed, Z. (22-02-2006). Abuse rattles Indian call centre staff. *BBC News*. Available at www.bbc.co.uk, retrieved 2008-06-05.

Al-Rodhan, N.R.F. (19th June, 2006). *Definitions of Globalization: A Comprehensive Overview and a Proposed Definition*. Geneva, Switzerland: Geneva Centre for Security Policy.

Anderrson, E., & Jansson, A. (2006, May 23-25). *A Call for Learning - A model for Developmental Learning in Call Centres*. Paper presented at the 6th UFHRD/AHRD Conference on Human Resource Development Research and Practice across Europe, Tilburg.

Ashill, N. J., Rod, M., Thirkell, P., & Carruthers, J. (2009). Job resourcefulness, symptoms of burnout and service recovery performance: an examination of call centre frontline employees. *Journal of Services Marketing*, *23*(5), 338–350. doi:10.1108/08876040910973440

Bain, P., & Taylor, P. (2004). *No passage to India? UK unions, globalisation and the migration of call centre jobs*. Work, Employment and Society Conference, 1-3 September, Manchester.

Bakker, A. B., Demerouti, E., & Verbeke, W. (2004). Using the job demands-resources model to predict burnout and performance. *Human Resource Management*, *43*(1), 83–104. doi:10.1002/hrm.20004

Basi, J. K. T. (2009). *Women, Identity and India's Call Centre Industry*. New York: Routledge.

Batt, R. (2000). Strategic segmentation in front-line services: matching customers, employees and human resource systems. *International Journal of Human Resource Management*, *11*(3), 540–561. doi:10.1080/095851900339756

Batt, R., & Applebaum, E. (1995). Worker Participation in Diverse Settings: Does the Form Affect the Outcome, & If So, Who Benefits? *British Journal of Industrial Relations*, *33*(3), 353–378. doi:10.1111/j.1467-8543.1995.tb00444.x

Benninghoven, A., Bindzius, F., Braun, D., Cramer, J. Ellegast, R. Flowerday, U. Genz, A., Heyden, T. von der, Pfeiffer, W., Schittly, D., Schweer, R. and Roger Stamm(2005). CCall—Healthy and Successful Work in Call Centres. *International Journal of Occupational Safety and Ergonomics* (*JOSE*), *11*(4), 409–421.

Bhaghat, C. (2005). *One Night@ the Call Center*. New Delhi, India: Rupa & Co.

Boon, J. A. (1982). *Other Tribes, Other Scribes: Symbolic Anthropology in the Comparative Study of Cultures, Histories, Religions, and Texts*. Cambridge, UK: Cambridge University Press.

Call centers in India. (n.d.). Available at www.outsource2india.com, retrieved dated 15th December, 2009.

Call Centre Monitoring. (n.d.). *Management*. Available at www.callcentrehelper.com. Retrieved 06-05-2008.

Cameron, D. (2000). Styling the worker: Gender and the commodification of language in the globalized service economy. *Journal of Sociolinguistics*, *4*(3), 323–347. doi:10.1111/1467-9481.00119

Chatterjee, S. R. (2007). Human resource management in India: "Where from" and "where to?". *Research and Practice in Human Resource Management*, *15*(2), 92–103.

Chirot, D., & Hall, T. D. (1982). World-system theory. *Annual Review of Sociology*, *8*, 81–106. doi:10.1146/annurev.so.08.080182.000501

Connell, J., & Harvey, H. (2004). Call centres and labour turnover: do HRM practices make a difference? *International Employment Relations Review*, *10*(2), 49–66.

Deery, S., & Kinnie, N. (2002). Call centres and beyond: a thematic evaluation. *Human Resource Management Journal*, *12*(4), 3–14. doi:10.1111/j.1748-8583.2002.tb00074.x

Deery, S. J., Iverson, R. D., & Walsh, J. T. (2002). Work Relationships in Telephone Call Centres: Understanding Emotional Exhaustion & Employee Withdrawal. *Journal of Management Studies*, *39*(4), 471–496. doi:10.1111/1467-6486.00300

Department of Foreign Affairs and Trade. (2008). *Australia-India FTA Feasibility Study Industry Consultations. Background Paper*. Australia: DFAT, Canberra.

Dharmawat, N. (September, 2003). *BPO-Destination India*. Patni Computer Systems Ltd India. Available at www.patni.org, retrieved dated 25th December, 2009.

Eriksen, H. R., Ihlebaek, C., & Ursin, H. (1999). A scoring system for subjective health complaints (SHC). *Scandinavian Journal of Public Health*, *27*, 63–72. doi:10.1177/14034948990270010401

European Foundation for the Improvement of Living and Working Conditions. (2006). *Industrial Relations Dictionary*. Retrieved from www.eurofound.eu.int., accessed 30 November 2009.

Evalueserve. (2004). *The next big opportunity – moving up the value chain – from BPO to KPO*. Retrieved from www.evalueserve.com, retrieved dated 5th January, 2010.

Farrell, D., Laboissiere, M., Pascal, R. et al. (2005). *The emerging global labor market: Part 1--the demand for offshore talent in services*. McKinsey Global Institute (MGI) Report, June.

Friedman, T. (2005). *The World Is Flat: A Brief History of the Globalized World in the 21st Century*. London: Allen Lane.

Galtung, J. (1971). A structural theory of imperialism. *Journal of Peace Research*, *8*(2). doi:10.1177/002234337100800201

Ganguly, K. (December 15, 2009). *Managing Customer Hostility in Transnational Call Centers in India*. Retrieved from http://ssrn.com, retrieved dated 1st April, 2010.

Goldberg, L. S., & Grandey, A. A. (2007). Display rules versus display autonomy: emotion regulation, emotional exhaustion, and task performance in a call center simulation. *Journal of Occupational Health Psychology*, *12*(3), 301–318. doi:10.1037/1076-8998.12.3.301

Goldman, M. (2009). *Getting Bangalored: Struggles over Urbanized Space in the Making of Asia's World Cities*. ASA Annual Meetings, January 2009.

Halbesleben, J. R. B., & Buckley, M. R. (2004). Burnout in organizational life. *Journal of Management, 30*(6), 859–879. doi:10.1016/j.jm.2004.06.004

Hazards 81 extended briefing: Toilet breaks: Give us a break! (n.d.). *Hazards*. Available at www.hazards.org, Retrieved 2008-06-05.

Holman, D. (2003a). Employee Well-being in Call Centres. *Human Resource Management Journal, 12*(4), 35–51. doi:10.1111/j.1748-8583.2002.tb00076.x

Holman, D. (2003b). Phoning in Sick? An Overview of Employee Stress in Call Centres. *Leadership and Organization Development Journal, 24*(3), 123–130. doi:10.1108/01437730310469543

Holman, D., Batt, R. & Holtgrewe, U. (2007). *The global call center report: International perspectives on management and employment.*

Houlihan, M. (2000). Eyes Wide Shut? Querying the Depth of Call Centre Learning. *Journal of European Industrial Training, 24*(2/3/4), 228-240.

Hourly Rate Survey Report for Industry. Call Center. (n.d.). *Pay Scale*. Available at www.payscale.com, retrieved dated 15th December, 2009.

Jain, N. (2001). *Globalisation or recolonisation?* (1st ed.). Pune, India: Elgar.

Johnson, H. A., & Spector, P. E. (2007, October). Service with a smile: do emotional intelligence, gender, and autonomy moderate the emotional labor process? *Journal of Occupational Health Psychology, 12*(4), 319–333. doi:10.1037/1076-8998.12.4.319

Knights, D., & Jones, B. (2007). Outsourcing (the) economy to India: utopian and dystopian discourses of offshoring. *The International Journal of Sociology and Social Policy, 27*(11/12), 433–446. doi:10.1108/01443330710835792

Leung, K., Bhagat, R. S., Buchan, N. R., Erez, M., & Gibson, C. B. (2005, July). Culture and International Business: Recent Advances and Their Implications for Future Research. *Journal of International Business Studies, 36*(4), 357–378. doi:10.1057/palgrave.jibs.8400150

Lundberg, U., & Melin, B. (2002). Stress in the development of musculoskeletal pain. In Linton, S. J. (Ed.), *New Avenues for the prevention of chronic musculoskeletal pain and disability* (pp. 165–182). Amsterdam: Elsevier Science Publishers.

Mahesh, V. S., & Kasturi, A. (2006). Improving call center agent performance. *International Journal of Service Industry Management, 17*(2), 136–157. doi:10.1108/09564230610656971

McGuire, D., & McLaren, L. (2007). The Impact of Physical Environment on Employee Commitment in Call Centres: The Mediating Role of Employee Well-Being. *Team Performance Management, 14*(5/6).

Mirchandani, K. (2004a). Webs of Resistance in transnational call centers: strategic agents, service providers and customers. In Thomas, R., Mills, A., & Helms Mills, J. (Eds.), *Identity Politics at Work: Resisting Gender, Gendering Resistance* (pp. 179–195). London: Routledge.

Mirchandani, K. (2004b). Practices of global capital: Gaps, cracks and ironies in transnational call centers in India. *Global Networks, 4*, 355–374. doi:10.1111/j.1471-0374.2004.00098.x

Mirchandani, K. (2005). Webs of resistance in transnational call centres. In Thomas, R., Mills, A., & Helms-Mills, J. (Eds.), *Gender, Organization and the Micro-politics of Resistance*. London: Routledge.

Mukherjee, A., & Malhotra, M. (Guest Editorial. (2009). Call centre services: the good, the bad, and the ugly. *Journal of Services Marketing, 23*(5).

News, B. B. C. (11 December, 2003). *Call centers 'bad for India.'* Retrieved from www.bbcnews.co.uk

Noronha, E., & d'Cruz, P. (2007). Reconciling dichotomous demands: Telemarketing agents in Bangalore and Mumbai, India. *Qualitative Report, 12*(2), 255–280.

Pal, M., & Buzzanel, P. (2008). The Indian Call Center Experience: A Case Study in Changing Discourses of Identity, Identification, and Career in a Global Context. *Journal of Business Communication, 45*(1), 31–60. doi:10.1177/0021943607309348

Patel, R. (2006). Working the Night Shift: Gender and the Global Economy. *ACME: An International E-Journal for Critical Geographies, 5*(1), 9–27.

Penfold, C. (2007). Trade union responses to white-collar off-shoring. *Economic and Labour Relations Review, 18*(1), 53–77.

Penfold, C. (2008). Off-shoring and decent work: Worlds apart? *International Journal of Comparative Labour Law and Industrial Relations, 24*(4), 573–594.

Penfold, C. (July, 2009). Off-shored services workers: labor law and practice in India. *Economic and Labor Relations Review*.

Phillipson, R. (1992). *Linguistic imperialism*. Oxford: Oxford University Press.

Pradhan, J. P., & Abraham, V. (2005). Social and cultural impact of outsourcing: Emerging issues from Indian call centres. *Harvard Asia Quarterly, 9*(3), 22–30.

Ramesh, B. P. (2004). *Labour in Business Process Outsourcing: A Case Study of Call-Centre Agents*. Noida, India: V. V. Giri National Labour Institute.

Russell, S. (30-01-2006). Tone-deaf to customer complaints, Dell opens yet another call center in India. *ZDNet*. Retrieved 2008-06-05.

Sasikumar, S. K., & Varma, U. K. (2004). *Information and Communication Technology and Decent Work: Study of India's Experience*. Noida, India: V. V. Giri National Labour Institute.

Schaufeli, W. B., & Bakker, A. B. (2004). Job demands, job resources, and their relationship with burnout and engagement: a multi-sample study. *Journal of Organizational Behavior, 25*(3), 293–315. doi:10.1002/job.248

Sen, F., & Shiel, M. (2006). From business process outsourcing (BPO) to knowledge process outsourcing (KPO): some issues. *Human Systems Management, 25*, 145–155.

Sheth, J. (2009). The call center couple: India's new middle class. *Journal of Indian Business Research, 1*(1), 10–13. doi:10.1108/17554190910963163

Singh, J. (2000). Performance productivity and quality of frontline employees in service organizations. *Journal of Marketing, 64*(2), 15–34. doi:10.1509/jmkg.64.2.15.17998

Singh, J., Goolsby, J. R., & Rhoads, G. K. (1994). Behavioral and psychological consequences of boundary spanning burnout for customer service representatives. *JMR, Journal of Marketing Research, 31*(4), 558–569. doi:10.2307/3151883

Siong, Z. M. B., Mellor, D., Moore, K. A., & Firth, L. (2006). Predicting intention to quit in the call center industry: does the retail model fit? *Journal of Managerial Psychology, 21*(3), 231–243. doi:10.1108/02683940610659579

Sjogaard, G., Lundberg, U., & Kadefors, R. (2000). The role of muscle activity and mental load in the development of pain and degenerative processes at the muscle cell level during computer work. *European Journal of Applied Physiology, 83*, 99–105. doi:10.1007/s004210000285

Sonntag, S. K. (2008). Appropriating Identity or Cultivating Capital? Global English in Offshoring Service Industries. *Anthropology of Work Review*, *26*(1), 13–20. doi:10.1525/awr.2005.26.1.13

Sonntag, S. K. (2009). Linguistic globalization and the call center industry: Imperialism, hegemony or cosmopolitanism? *Language Policy*, *8*, 5–25. doi:10.1007/s10993-008-9112-9

Taylor, P. & Bain, P. (1999). An assembly line in the head: work and employee relations in the call centre. *Industrial Relations Journal.*

Taylor, P., & Bain, P. (2000). Entrapped by the "electronic panopticon"? Worker resistance in the call centre. *New Technology, Work and Employment*, *15*(1), 2–18. doi:10.1111/1468-005X.00061

Taylor, P., & Bain, P. (2005). India calling to the far away towns: the call centre labour process and globalization. *Work, Employment and Society*, *19*(2), 261–282. doi:10.1177/0950017005053170

Taylor, P., Scholarios, D., Noronha, E., & d'Cruz, P. (2007). *Union formation in Indian call centres/BPO--the attitudes and experiences of UNITES members*, University of Strathclyde Business School and Indian Institute of Management, Ahmedabad. Retrieved from http://www.union-network.org

The Call Center Answer Team reaches out to the industry for to crack a tough nut. (2003, July 30). Q&A: How Many Calls Should I Monitor. Retrieved from www.callcentermagazine.com. Retrieved 2008-06-05.

Theorell, T., Hasselhorn, H. M., Vingård, E., & Andersson, B. (2000). Interleukin 6 and cortisol in acute musculoskeletal disorders: results from a case-referent study in Sweden. *Stress Medicine*, *16*, 27–35. doi:10.1002/(SICI)1099-1700(200001)16:1<27::AID-SMI829>3.0.CO;2-#

Wallace, C. M., Eagleson, G., & Waldersee, R. (2000). The sacrificial HR strategy in call centres. *International Journal of Service Industry Management*, *11*(2), 174–184. doi:10.1108/09564230010323741

Waters, W. (2001). *Globalization,* (2nd ed., p. 6). London: Routledge.

Westgaard, R. H. (1999). Effects of physical and mental stressors on muscle pain. *Scandinavian Journal of Work, Environment & Health*, *25*(4), 19–24.

Who's on the Line? Women in Call Centres Project. (n.d.). Atlantic Centre of Excellence for Women's Health. *Health Canada.* Available at www.acewh.dal.ca, Retrieved 2008-06-05

Wilk, S. L., & Moynihan, L. M. (2005, Sep). Sep Display rule "regulators": the relationship between supervisors and worker emotional exhaustion. *The Journal of Applied Psychology*, *90*(5), 917–927. doi:10.1037/0021-9010.90.5.917

Winifred, P. (2004). *Who's On the Line? Indian Call Center Agents Pose as Americans for U.S.-Outsourced Firms*. Paper presented at the annual meeting of the American Sociological Association, Hilton San Francisco, Available at www.allacademic.com retrieved dated 2010-05-07.

Wright, T. A., & Hobfoll, S. E. (2004). Commitment, psychological well-being and job performance: an examination of conservation of resources (COR) theory and job burnout. *Journal of Business and Management*, *9*(4), 389–406.

ADDITIONAL READING

Ankerl, Guy. Global communication without universal civilization. INU societal research. Vol.1: *Coexisting contemporary civilizations: Arabo-Muslim, Bharati, Chinese, and Western.* Geneva: INU Press. ISBN 2-88155-004-5.

Constantino, R. (1978). *Neocolonial identity and counter-consciousness: Essays on cultural decolonization*. London: Merlin Press.

Farrell, D., et al. (2005) Reforming India's financial system, *The McKinsey Quarterly* special edition: Fulfilling India's promise, p.103

Farrell, D., et al. (2005). Ensuring India's offshoring future, *The McKinsey Quarterly* special edition: Fulfilling India's promise, p.75, www.mckinseyquarterly.com.

Farrell, D. (2005). India's lagging financial system. *The McKinsey Quarterly*, (2): 11.

Rajan, R. G. (2005). Making India a global hub, *The McKinsey Quarterly* special edition: Fulfilling India's promise, p.113.

Sartre, J. P. (2001). *'Colonialism and Neocolonialism*. Translated by Steve Brewer, Azzedine Haddour, Terry McWilliams, republished by Routledge, France.

Sharma, P., Mathur, R., & Dhawan, A. (2009). Exploring customer reactions to offshore call centers: toward a comprehensive conceptual framework [Emerald]. *Journal of Services Marketing*, *23*(5), 289–300. doi:10.1108/08876040910973404

Taylor, P., & Bain, P. (2003). 'Subterranean Worksick Blues': Humour as Subversion in Two Call Centres [SAGE: London]. *Organization Studies*, *24*(9), 1487–1509. doi:10.1177/0170840603249008

Wegge, J., van Dick, R., Fisher, G., Wecking, C., & Moltzen, K. (2006, January). Work motivation, organisational identification, and well-being in call centre work. *Work and Stress*, *20*(1), 60–83. doi:10.1080/02678370600655553

KEY TERMS AND DEFINITIONS

Accent Neutralization: It is the process of reducing the accent of your native language, and using the accent of your target language. Most times this target language is English. The use of accent neutralization has become common as the economy transitions from local to global. With call centers and offices for American companies scattered all over the world, a method of teaching people to successfully communicate with American businesses and consumers became critical.

Business Process Outsourcing (BPO): Business process outsourcing (BPO) is a form of outsourcing that involves the contracting of the operations and responsibilities of specific business functions (or processes) to a third-party service provider.

Conservation of Resources Theory (COR): (Hobfoll, 1988, 1989): is a comprehensive theory of stress based on the central tenet that people strive to obtain, build, and protect that which they value (e.g., resources), and psychological stress occurs when these resources are lost, threatened with loss, or if individuals fail to replenish resources after significant investment.

Emotional Assuage: in the present context, it means an attempt made by the call centre employees to alleviate their own pain and affective suffering though several efforts to undo and compensate for the interpersonal loss, which ultimately they fail to achieve.

Jet Lag: Jet lag, medically referred to as "desynchronosis," is a physiological condition which is a consequence of alterations to circadian rhythms; it is classified as one of the circadian rhythm sleep disorders. Jet lag results from rapid long-distance transmeridian (east-west or west-east) travel, as on a jet plane. But, colloquially it used to refer to the sleep disorder caused by frequent sleep loss.

Negativistic Schema: The evaluative tendencies of self, others and world by basically depressed individuals is known as negativistic schema. Severely depressed individuals are examined in the areas of judgments of contingency, attributions of causality, expectancy estimates, and self-schemata/self-reference. The available evidence is suggestive of negativistic evaluative tendencies in severely depressed individuals, with this bias being most clearly manifested in the area of self-schemata/self-reference.

Slave Mentality: Slave mentality refers to the mindset of individuals to accept meekly supremacy of a particular group of society without reacting to their actions and a compelling tendency to be guided by others' commands rather than self-conscious freewill. The implications and after-effect of slavery does a lot of damage to both the minds and the economy of colored people even years after slavery. Racism still flourishes in society.

Subalterns of Modernity: those individuals who are being deprived of growth and development even though being in a globalized modern era. So, in a way they are subalterns or disadvantaged sections as a result of the modernization.

Systemic Mental Captivity: A subtle and systematic attempt made by a group of society to capture the mental freedom of another group in order to make them serviceable others. This may have a crippling impact on the mindset of people who have been victimized with this mental perpetration.

Cultural Hybridity: Cultural hybridity is used to describe societies that emerge from cultural contacts of European "explorers" and those "explored". It does not view these contacts and exposures as impostures of a dominant culture over the non-dominant one rather emphasizes on a mutual intermingling.

Neo-Colonialism: Neocolonialism is used by post-colonial critics of developed countries' involvement in the developing world. The proponents of this framework argue that existing or past international economic arrangements created by former colonial powers were or are used to maintain control of their former colonies and dependencies after the colonial independence movements of the post-World War II period.

Multiculturalism: The acceptance or promotion of multiple ethnic cultures, for practical reasons and/or for the sake of diversity and applied to the demographic make-up of a specific place, usually at the organizational level, e.g. schools, businesses, neighborhoods, cities or nations. In this context, multiculturalists advocate extending equitable status to distinct ethnic and religious groups without promoting any specific ethnic, religious, and/or cultural community values as central.

Section 4
Games

Chapter 18
Mobile Games in China:
Formation, Ferment, and Future

Zixue Tai
University of Kentucky, USA

Haifang Zeng
Shanghai University, China

ABSTRACT

This chapter presents a critical analysis of China's nascent but fast-evolving mobile game market. Through a penetrating examination of the historical evolution of the market from the late 1990s to the dawn of the 3G era, it offers insight on the changing dynamics of interplay as well as major trends of concentration and internationalization among key sectors of the Chinese mobile market structure. By casting findings in the comparative lenses of major global mobile markets in the United States, Japan and South Korea, the chapter contributes to an in-depth understanding of the universalizing features and the peculiarities of China's mobile games and gamers.

INTRODUCTION

China boasts the largest mobile market in the world. According to official statistic from the Ministry of Industry and Information Technology (MIIT), the number of cell phone users exceeded 756 million as of January 2010,[1] and it is expected to maintain its track of fast growth in the coming years. Meanwhile, three years after the formal announcement of TD-SCDMA as the national 3G mobile standard in 2006, and one year after the commercial trial run in over a dozen major cities across the country, China Mobile Limited (CML) formally started its commercial 3G services to users in most areas in China in early 2009, and its competitor, China United Network Communications Group Co. Ltd (China Unicom), followed suit in October 2009. As of February 2010, MIIT reports that the 3G user base reached 160 million (with 1.5 million additions in the month of February alone).[2] Against this backdrop of the latest 3G upgrade is the much-publicized anticipation of the blossoming of a variety of 3G-based applications and platforms to the world's largest user base thereafter. One hot area of global attention is the fledgling but fast-evolving field of mobile gaming.

DOI: 10.4018/978-1-60960-037-2.ch018

Copyright © 2011, IGI Global. Copying or distributing in print or electronic forms without written permission of IGI Global is prohibited.

The Asia-Pacific region, spearheaded by powerhouses such as Japan, South Korea and China, has been the undisputable global hub of online video game production, circulation and consumption (Hjorth & Chan, 2009; McLelland, 2007). In the array of expanding mobile applications and services from text and multimedia messaging to display advertising to gaming, Asia-Pacific has been consistently projected to the world's leading market for the near future by industrial research services such as JupiterResearch[3] and BuddeComm[4] in their data-driven analysis and trend studies. Although China lags behind many countries in the rate of penetration by a variety of emerging m-services (i.e., mobile services) in the country, its absolute size of mobile population puts it far ahead of the rest of the world, and the trajectory of growth shows strong signs of accentuation in the forthcoming years. Specifically, Analysys International, a leading advisor of technology, media and telecom (TMT) industries in China, reports that China's mobile game market size reached RMB ¥383 million and ¥526 million for the first and second quarters of 2009, registering a revenue growth of 32.4% and 21.3% over the previous quarters respectively.[5]

The youth-driven cultures of mobility rendered by the widespread use of variegated cellphone-based applications and content are well on the path to (re)shape the social, cultural, and political landscapes across the globe (Castells, 2007; Goggin, 2006; Ling & Donner, 2009). At the same time, however, the "socio-cultural differences and contextual specificities" of online gaming cultures tend to be neglected by prevalent scholarships (Hjorth & Chan, 2009; p. 4). As the "next big thing" on the horizon, mobile games are destined to be a pivotal driver in China's ongoing telecommunications revolution and in redefining China's burgeoning technospaces. Nonetheless, as McLelland (2007) astutely observes, academic discussions of mobile communication have been marked by the primacy of European/North American models, and the particularities of the Asia-Pacific region tend to be elided or neglected. In a similar vein, there is an unfortunate dearth of scholarly attention to (both online and mobile) network games as well as the shifting dynamics of the gameplay culture in China. With this in mind, this chapter offers a critical, in-depth overview of the fast-changing and exploding mobile gaming industry in China in the broader context of globalization and convergence. Specifically, it analyzes the burgeoning mobile game culture as exemplified in player data, use patterns, gamer attitudes and content genres by integrating results from recent surveys and industrial reports, and then looks at the evolving gaming platforms as manifested in popular technologies and applications developed both in and outside of China. This is followed by an analysis of the evolving market structure, the official regulatory framework, and emerging trends of industry concentration as well as interplay of domestic and global corporations on multiple fronts.

BACKGROUND: THE BUSINESS AND CULTURE OF MOBILE GAMES

Proliferating mobile technologies have profoundly transformed every aspect of daily life. Many rich veins of inquiry, however, are yet to be explored among the global community of social scientists (Katz, 2006; 2007). As a "taken-for-granted aspect of everyday life" (Ling & Donner, 2009, p. 143), mobile communication has engendered a particular type of "mobile logic" enmeshed in the routine fabrics of daily socio-cultural experiences of billions of interconnected individuals. Ling and Donner (2004) note that this mobile logic has three distinct attributes: ubiquitous and widespread access, individual addressability, and the ability to interlace interactions (i.e., the intertwining and coordinating of multiple activities involving interactions with different parties at the same time). Likewise, in their comparative study of mobile media usage in Europe, North America and Pacific Asia, Castells et al. (2007, p. 245) conclude that the "multimodal, multimedia, portable system" of mobile communication has triggered a number of social practices, cultural values, organizational patterns and behavioral trends – among them: youth culture is setting the trend all segments of society; "relentless connectivity" is blurring spatial contexts and time frames as well as redefining social and cultural norms; perpetual contact has transformed interpersonal networks and has created unplanned, mostly spontaneous communities of practice in instant time; texting via SMS and multimodal forms of communication are changing a variety of language parameters and cultural practices. Moreover, Castells et al. note the emerging trend of mobile users as creative producers of an ever-evolving list of content and service types fulfilling niche needs of different user groups. In that regard, this subgenre of mobile culture is the latest accentuation of the well-trodden practice of produsage that pervades cyberspace (Bruns, 2008).

Meanwhile, the video game industry, which includes stand-alone and networked games encompassing a variety of platforms, has been ingrained as an indispensible part of global popular cultures. In 2008, worldwide video game revenue surpassed that of DVD and Blue-Ray for the first time,[6] and for some years now, global video game sales have surpassed movie box office revenue and music sales respectively.[7] For over half a century, the history of video games has witnessed a number of technical and conceptual revolutions from stand-alone computer-based to console to arcade to networked to handheld games (Malliet & de Meyer, 2005; Williams, 2006). However, irrespective of the particular platforms and hardware features these video games are attached to, there have been a number of persistent thematic threads that interweave the different game genres together throughout their generational evolutions (Ip, 2008; Sellers, 2006). With respect to video gaming, therefore, the manifestation of the convergence culture as observed by Henry Jenkins (2006) lies in that gaming content evolves, but does not vanish, with a multitude of Black Boxes serving as niche-filling delivery tools. This overall trend continues with the expanding arena of mobile game market (Hall, 2005), particularly in the case of Japan (Chan, 2008), where old themes and formats find a renascence on the move.

As the most popular type of games in the network era, it is no wonder that massively multiplayer online role playing games (MMORPGs) have attracted considerable attention among the academic community, particularly in the context of community building, identify formation and user participation (see Chan & Vorderer, 2006 for an overview). Online gaming culture intricately reflects the material world, and represents to varying degrees recreations of the milieu of practices in the mundane world on the part of the players (e.g., Taylor). As primarily a youth-driven culture, the "spectacular space" created by online games (Crowe & Bradford, 2007) has much to tell about young people's construction of identities online (Thomas, 2007). In particular, our understanding of the cultures of mobile communication and online gaming in Asia Pacific – the most dynamic region in both areas – must be localized and particularized in order to avoid the perpetuating stereotypes and universalizations as often projected in prevalent Western theorizing (Bell, 2005; Hjorth & Chan, 2009; McLelland, 2007). Towards that end, we have seen, among others, highly revealing scholarly efforts in recent years to contextualize the so-called *Korean Wave* of online gaming and *PC Bang* (Jin & Chee, 2009), the *keitai* as a unique Japanese cultural phenomenon (Ito, Okabe & Matsuda, 2006), and the popularization of the highly location-based mobile gaming in Japan (Chan, 2008). Research on the socio-cultural practices of mobile communication and gaming in China, the biggest market of video gaming and mobile communication, unfortunately, has lagged substantially behind but has much to contribute to the expanding academic literature in these fields.

MOBILE GAMING IN CHINA[8]

Within the last decade or so, China's mobile value-added services (MVAS) have experienced staggering growth, with an estimated market value of RMB 155 billion (approximately US$ 22.7 billion) for the year 2009. While text messaging (SMS) and Color Ring Back Tone (CRBT, also known as PRBT, or Personalized Ring Back Tone in North America), the leading areas of explosion in the previous decade, have shown signs of tapering off lately, other areas of wireless content and applications such as multimedia messaging services (MMS) and mobile gaming are projected to be the next triggers of market growth.[9] In particular, mobile gaming is expected to lead the next round of China's mobile revolution as the country finally embarks on its highly-anticipated (albeit much-delayed) transition to 3G networks in 2009. The popularization of wireless video gaming will undoubtedly reconfigure China's unfolding mobile landscape, and will significantly shape the future trajectories of the ongoing mobile revolution in China.

Historical Overview of the Mobile Game Market in China

An overview of the online game industry in China, which has experienced phenomenal growth in the past decade, is in place to put things into perspective with respect to China's mobile gaming. Compared with those in South Korea, Taiwan, Japan, and the United States, the Chinese online gaming industry was off to a later start, with the first online game title the *King of Kings* (*KoK*), which was developed by Taipei-based Lager Interactive Inc., released to Chinese players in July 2000. In the first few years, online video games in China were heavily dominated by foreign titles, especially those from South Korea. For example, in 2001 and 2002, no less than 90% of popular game titles constituted imports (and pirated derivatives) from South Korea, distantly followed by productions from Taiwan, Japan and the United States (China Game Publishers Committee & International Data Corp., 2004). However, thanks to years of vigorous efforts by Chinese game developers and enthusiastic support by the Chinese government, domestic game titles made up 61.2% of China's market share, with 80 of the 115 licensed game releases being Chinese developments in 2009 (China Cultural Industry Net, 2010).

According to the latest statistical report released by the China Internet Network Information Center (CNNIC) (2010), as of December 2009, the online game population reached 265 million (out of the 384 million Internet users). In terms of market return, the 2009 China Network Game Market White Paper released by the Ministry of Culture in January 2010 indicates that subscription-based revenue alone reaches RMB¥ 25.8 billion in mainland China in 2009, registering an average annual growth rate of more than 50 percent in the ten-year old industry (China Cultural Industry Net, 2010). Mobile gaming in China has found fertile ground side by side with the explosive development of online gaming during this time.

From its inception to the present, mobile gaming in China has roughly gone through four distinct stages of development: induction (1998 - early 2002); transition (late 2002 - 2006); consolidation (late 2006 - 2008); and the 3G era (2009 onward). The induction phase (1998-2002) marks the advent of mobile games in China, mainly accomplished through the importation of foreign games into the Chinese market. Single-player casual games built-in within the cell phone dominated the playing experience. Although there were a limited number of games available for downloading via a small number of proprietary mobile networks in major metropolitan areas such as Guangzhou and Shanghai, low bandwidth and high cost of connectivity prohibited the diffusion of mobile gaming to the majority of consumers. Meanwhile, mobile service and content providers as well as commercial game developers largely ignored the nascent market of mobile gaming during this time due to the lack of a critical mass of consumers and a viable business model.

The first mobile game was brought to the Chinese players by a Western cell phone manufacturer – Nokia – through its 6110 handset released in 1998. Its single-player, built-in puzzle game *Snake* immediately caught the imagination of China's mobile population dominated by people in their 20s and early 30s, and has been one of the most-played games via cell phones in China. *Snake* introduced millions of Chinese consumers to this novel entertainment, and put Nokia 6110 and 3310 (released in 2000) among the best-sellers in the Chinese market at the time. Noticeably, this development in China paralleled the timeline of the widespread diffusion of *Snake* in other regions of the world (Hall, 2005). Other popular games at the time were Blackjack card games, Reversi and similar types of board games, as well as variants of Tetris puzzle games. These handset-specific games used minimal graphics, and had already earned recognition through other platforms such as video arcades, home PCs and online networks. So it was natural for mobile game developers to look into the old stockpiles of game ideas and products for adaptations to the emerging platforms of mobile phones. The few games that Chinese players had access to were simply spillovers from foreign game developers and cell phone manufacturers, and there was no effort to customize to the needs and tastes of domestic players within China.

Things started to change in the transition stage (late 2002 - 2006) as domestic developers and investors caught on and jumped to the mobile bandwagon. The most immediate trigger for the next round of buzz was the commercial deployment of 2.5G mobile network technologies across the country in mid-2002, started by the opening of GPRS (General Packet Radio Service) data and voice services by China Mobile in 160 cities on May 17 and followed shortly by the state-sponsored competing CDMA (Code Division Multiple Access) 1X service offerings of China Unicom in the following months. Availability of these data services not only popularized downloadable cell-phone based games (due to reduced connectivity cost) but also opened the door for network-based interactive mobile games.

In order to streamline the distribution of the expanding list of mobile games to cell phone users, China Mobile released its Treasure Box (aka Magic Box) (*baibaoxiang*) service through monternet.com in July 2003. Consumers using China Mobile wireless services can pick games of interest (along with other types of content such as animations and music) from a large pool of items for downloading to their Java-supporting cell phones; additionally, players can also choose multiplayer interactive games through the site. At about the same time, China Unicom came up with its U-Magic (also known as Uni-Magic) platform, which, through partnership with QUALCOMM, a San Diego-based company, adopts BREW (Binary Runtime Environment for Wireless) technology to offer subscribers similar services within China. Another major service provider (SP) in China's mobile game industry, DigiFun was founded in August 2003 to compete for a share of the mobile entertainment market. Since then, Treasure Box has been a leader in China's mobile game distribution market, trailed significantly by Unicom's BREW services. DigiFun, however, has successfully maintained its status as a viable competitor by cultivating a niche sector of mobile services. Two landmark events since its inception are DigiFun's release of the *Legends of the Three Kingdoms*, an MMORPG in China tailored toward the mobile platform in 2004, which garnered 15 million registered users within the first three months of its debut, and its 120-episode animation *Journey to the West* specifically for the mobile audience in 2005.

Another mobile phenomenon was started during this period by KONG.net. Founded in 2002, KONG.net specializes in WAP (Wireless Application Protocol) services targeting mobile phones, with a wide range of data and value-added applications from news, multimedia message services (MMS), instant messenger services, CRBT, to mobile games. Starting in January 2004, KONG.net added its Communication Over The Air (COTA) platform through which content from conventional print, radio, television media outlets becomes accessible to the mobile audience. Its quick rise in popularity in China earned it the nickname "The No. 1 Mobile Entertainment Portal" among Chinese cell phone users. Its prominent success led to its NASDAQ IPO on July 9, 2004, setting the record among China enterprises for the shortest amount of time taken from start-up to enlisting within 26 months. The leading area of revenue growth for KONG.net since 2002 has been mobile gaming.

The dominant business model during this period was to charge players a one-time fee for single-player games, and to offer free real-time playing capability on wireless networks but make a profit through mobile connectivity. Revenue generated from the players was shared among mobile service providers, game content providers and game developers. The sheer size of China's mobile market makes it possible for many games to attract millions of players with no time. Meanwhile, the low threshold of market entrance in the mobile game field led to the overnight rise of hundreds of developing companies or Content Providers (CPs), many of which employed no more than a few dozen game programmers in one rented office. As a result, popular service providers (SPs) were flooded with tens of thousands of low-quality, copycat games on their sites, and bedazzled users had a hard time finding the ones fitting for their playing preferences and individual tastes. At the same time, growing demands from the expanding user base continued to drive the need for newer innovative games to keep players on the hook. As of early 2006, although over 2,000 companies had taken on the business of developing mobile games for Chinese players, popular game content was still dominated by foreign imports, with 80% of the games being played from game providers mainly from South Korea, Japan, Europe and the United States. It is only natural then that many domestic developers would find themselves struggling for survival for failing to make a splash in the market.

This chaotic scenario led to the next phase of market consolidation and restructuring (late 2006 – 2008). As more developers joined the bazaar, the tens of thousands of mobile game titles proved to be marketing nightmare for both the content providers and the mobile service providers. For the average consumers, the most common complaint was the sloppy, non-differentiating quality of many games they downloaded. Yet from the consumers' perspective, nothing could be worse than being charged for a game they did not like. The existing business model, however, exacerbated the user experience because fees were primarily levied based on per download. Moreover, there was a clear shift in the mobile gaming market as a few major brands/companies had emerged as leaders while hundreds were still struggling for survival. In 2006, among the close to 300 service providers on China Mobile's Treasure Box, the most dominant mobile game platform in the country, the top three leaders, MIG China Co. Ltd, Sina, and Guangzhou Mosun, totaled 20% of the market share, while the top 20 service providers made up about 70% of the overall revenue. As a contrast, the bottom 32% of the struggling service providers scored a revenue stream of near-0 during the same period.

Leading the charge to tilt the mobile game playing field to favor quality and talent was China Mobile, whose footsteps were quickly followed by China Unicom. China Mobile's first move was the creation of a section called "Premium Game Community" on its Monternet Treasure Box in mid-2006 exclusively promoting high-quality games to Chinese players. The Premium Game Community landed over ten million registered users within the first six months, and has topped other services/sections on Monternet in both volumes of mobile traffic and transactions since its launch. A follow-up measure China Mobile adopted is to set limits on the number of titles each SP can post on its Treasure Box services. This was publicly released in a notice to all of its contracted SPs in July 2008 announcing that starting September 1, 2008, quotas will be reduced for contracted SPs on the number of mobile game titles they can publish with Treasure Box. For A-tier SPs, the allowed number goes down from 130 to 40; for B-tier SPs, from 100 to 30; for C-tier clients, from 80 to 20; normal SPs are only allowed 6 titles (from the previous 60). This step, which effectively eliminates about 70% of the game title offerings within Treasure Box, is defended by China Mobile as necessary in order to achieve its goal of supporting fewer but higher-quality games to its player base.

In a related effort to streamline its business model, China Mobile's Monternet switched to a unified two-track transaction system for game players: a points system through which players can make payments via a registered account with China Mobile, and points can be purchased with cash deposit into the account at any time; alternatively, players can add transaction charges to their monthly bills with China Mobile to be paid at the end of the month. Prior to this move, it was common for various SPs to enforce their own payment schemes for downloading/playing games under their domain, and players would have to make different transactions in accordance with specific requirements. Plus, it was not uncommon for individual SPs to add hidden fees in the processes without users' prior knowledge. The new payment system significantly improves players' experiences and facilitates game transactions through a centralized payment platform.

This phase also witnessed the diversification of revenue-generating models with mobile gaming. Previously, the dominant business model had been per download charging and revenue-sharing among mobile carriers, IPs and CPs. *Crystal Tree*, an MMORPG for the mobile platform developed by Moloon International, led mobile gaming in China to new territories by being the first mobile game to achieve profitability in the first quarter of its formal release in January 2007; it also set the new mobile game record (up till that time) to have 1,500 players in the same real-time game session. Moreover, it introduced the items-based scheme of game-playing in which users play for free while revenue is generated via selling game accessories (aka props) and virtual currencies. Moloon attributes the main source of inspiration for this concept to South Korea, where this model has a proven track of success (Jin & Chee, 2009). This model was quickly propagated as a major business model as more MMORPG titles entered the market. Meanwhile, other trends that were consolidated during this period were the further diffusion of the ARPU (Average Revenue Per User) model through which users pay a flat monthly fee (typically RMB¥ 5 per month) among a growing number of popular titles and the integration of mobile and online gamers through the same networked game. An example of the latter is the popular MMORGP game *The World of Legend*, which is a major collaborative project between Moloon and China Unicom and allows mobile users to share their playing experience with counterparts connected via the Internet in multiple episodes of the game. The convergence of online and mobile games for a unified playing experience promises to be dominant mode of networked gaming experience in China.

The game changer in the unfolding mobile market in China is the switch to 3G coverage, which took place in January 2009, with MIIT officially granting licenses to three mobile carriers - TD-SCDMA to China Mobile, WCDMA to China Unicom, and CDMA2000 to China Telecomm. This much-anticipated, albeit delayed, transition is predestined to reinforce many of the well-established aspects of the mobile culture and will most likely set new trends as mobile connectivity reaches the next level. However, the first few months after the opening of the 3G networks have not brought in the expected outpouring of switchovers from mobile phone users, as most of them are still cautiously watching over what the market has to offer in terms of enhanced values and services in the 3G packages. Nonetheless, it is clear that mobile games will continue on the track of rapid growth in the years to come. A useful reference point is that, at the current stage of development, mobile gaming only accounts for less than 1% of the revenue generated by value-added services via cell phones in China, while it makes up nearly 10% in both markets in Japan and South Korea. Improved speed and connectivity via the 3G networks will undoubtedly drive the demand for newer and better games.

Gamers and Game Genres: A Profile

While China's mobile game industry has experienced unique twists and turns in its development, it also shares a lot in common with those in most other countries. It is useful here to compare the major demographics and behavioral patterns between Chinese players and their counterparts in other national contexts. The following discussion, unless otherwise noted, is based on the survey results conducted by iResearch (2008) for Chinese players and the overview of cross-national player statistics and profiles in recent years by Katz and Acord (2008).

First of all, Chinese mobile gamers are primarily male-dominated youth in their early twenties. Unlike in many other countries where women are generally as likely as men to play mobile games, the Chinese mobile game field is heavily dominated by male players (93%); the gender distribution displays a similar but less extreme pattern among video game playing (70% male vs. 30% female) in China. In respect to age, mobile gaming in China is predominantly a youth-driven culture, with 74.5% players in the 18-24 age group, followed by 14.9% in the 25-30 range; players younger than 18 account for 10.6% while those beyond 30 only make up about 2%. While twentysomethings tend to dominate the mobile playing field in most countries, available statistics from cross-national contexts indicate no other country displays this extreme pattern as found in China. In terms of brand, Nokia is the most preferred platform, being used by 70.6% of the Chinese mobile gamers, while Sony Ericsson owns 12.7% of the player market.

As for payment, 30.2% of the players express willingness to pay less than RMB ¥10 ARPU, and 37.2% indicate they may pay more than RMB ¥10 per month on mobile gaming, whereas 32.6% don't see a need for paying to play at all. This is no surprise considering that 51.2% of the players are secondary (i.e., high and technical) school students and 33.6% of them are college students. Noticeably, there is a distinct pattern in the regional diffusion of mobile gaming, as more economically developed areas tend to have more players, with Guangdong leading the rest of the country in producing the largest player base. There is also a high level of diffusion of mobile gaming in Beijing and Shanghai. Clearly, level of economic development is closely tied to the tendency to play mobile games: on one hand, more economically advanced regions have access to more resources to build a sophisticated mobile network infrastructure; on the other hand, users in more developed areas display a higher demand for mobile connectivity, and they are better off to subscribe to mobile services.

With regard to single-player games, role-playing games are favored by 42.8% of the players, trailed by kung-fu games (16.8%) and casual games (8.1%). Other games, such as racing (6.3%), adventure (5.9%), strategy (5.5%), sports (4.2%), are much less popular. Among networked games, MMORPGs are by far the most prevalent, being favored by 78% of the mobile players. This is followed distantly by kung-fu action games, with 14.2% players reporting preferring this genre. Social games, which comprise roughly 40% of the global mobile market, are much less played in China, as they are listed by merely 1.7% of the players as the most-favored games. However, this typology is somewhat misleading since MMORPGs can be played for both purposes of competing and community-building. Moreover, mobile game titles are short-lived in China, as 32.2% of the players stay with the same title for less than a month, and only 37.6% indicate they play the same game for more than three months. This is in sharp contrast to online video games, in which 75% players report staying with the same title for over a year. This discrepancy most likely reflects the lack of quality titles in the mobile game market, and poses a great challenge to game developers in attracting loyal players.

In terms of content, three genres of mobile games are popular with Chinese players. First, casual puzzle, card, shooting, race and arcade games such as variants of Tetris, the Sims series, Bejweled, and Rocket Car are still popular among Chinese mobile gamers. The main reasons for this continuing popularity are affordability, widespread (built-in or downloading) availability, ease of playing, and familiarity with such games (often from past playing on computers and other platforms). As mentioned before, these games were the first to be introduced to the Chinese players, and were dominated by imported titles from the West, largely due to the prevalent practice of global cell phone manufacturers embedding these games to the handsets they sell in China. However, in recent years, some cell phone manufacturers have resorted to promoting handset sales based on particular gaming features and it is not uncommon for them to use customized embedded games as a sales pitch to Chinese consumers. Moreover, most mobile game developers have released their own game titles in these traditions in the past decade, and the quality has dramatically improved as content becomes more and more customized to cater to the tastes of Chinese players. Therefore, these classic game genres still provide a desirable pastime for many Chinese gamers from time to time.

Secondly, regularly hitting the most-played titles among Chinese mobile players are a variety of games produced by foreign developers (mainly from Japan, South Korea, and the United States) and primarily marketed in overseas markets. Compared with their Chinese counterparts, these companies have garnered a lot of experience and expertise in churning out high-quality game titles that are avidly sought after by Chinese players. It is precisely for their competitive edges that both China Mobile and China Unicom have for the years been enforcing a proactive policy to ensure the availability of enough titles from foreign top-notch companies in their offer list of mobile games. On the other hand, foreign companies attracted by the huge Chinese mobile market have aggressively pursued partnership opportunities in China to vie for a share of the ever-expanding Chinese mobile population.

The third type of popular games consists of a miscellany of creative productions ranging from kung-fu to fantasy to sports themes by domestic companies. While Chinese game developers started late in the video and mobile gaming fields, and they are still lagging behind in many areas compared with counterparts from countries such as Japan, South Korea and the United States, they have accumulated valuable experience and made impressive progress in recent years. Five or six years ago, the mobile market was overwhelmingly dominated by foreign imports across all game genres, and domestic productions were mainly sloppy imitates of foreign games. Today, popular hits are a mixture of Chinese and foreign productions, with the former gradually gaining an upper hand in the playing field.

We can gain some valuable insight from the historical imprint of the ChinaJoy Annual Golden Plume Awards, the most prestigious and the most coveted awards in China's network and mobile game business. Currently in its sixth year and with results tallied based on votes directly from the players, the Golden Plume Awards are dubbed "The Oscars of China's Video Game Industry." Each year, these awards offer some snapshots of the state of development and achievements of China's blooming video game industry. We hereby provide an analytical overview of the awardees in the mobile game category for the most recent two years (2008 and 2009).

Of the fifteen awardees in the mobile category in 2009, ten are networked titles and five are non-networked games. With the former, seven are MMORPGs whereas three are single-player RPGs. All networked games are promoted in the commercial framework of playing-for-free and charging-for-virtual-transactions. This is consistent with the prevalent business model of video gaming in China, and makes a distinct contrast with the practices found in countries such as Japan, South Korea and the United States where subscriptions-based model tends to prevail. As an indication of a prominent emerging front in video (mobile) gaming, one from each category features a 3D visual design (*Feng Shen Online* for MMORPG and *Jiu Zhou Online* for RPG).

Two of the ten networked mobile games are in the kung-fu domains, and six feature real-time action and art-of-war (strategy) themes; the remaining two are fantasy/adventure games, with one set in an animal kingdom and the other situated in the Kingdom of Women from the 16th-century classic *The Journey to the West*. Six of the ten game titles have developed varying degrees of story lines in specific ancient Chinese dynasties: the plots of three games are adapted from the ancient Chinese classics of *The Enfeoffment of the Gods*, *The Journey to the West*, and *Romance of the Three Kingdoms* respectively; the four remaining titles are rooted in fictional virtual settings. In particular, *Anrufen Online*, developed by Beijing-based WiSTONE for Symbian S60 devices, takes the unusual approach of setting the play in the Western cultural context of knights and wizards. The reason for this becomes obvious if we take into consideration that *Anrufen Online* was almost released simultaneously in China and the English-speaking world. There exists, however, a significant twist: in China, the game is promoted on an items-based model while it is marketed on a subscription-based approach in the West.

Unlike the networked titles which are all developed by Chinese companies, the five stand-alone single-player game titles featured in the awards are more diversified. Two of them are produced by non-Chinese developers – *Age of Empires III* by Microsoft Game Studios (United States) and *Demon Overtaken* from Game Plus (Japan). Another awardee, *Guide to Lost Souls*, is grounded in the same-titled popular kung-fu reader by famed Taiwanese Gu Long. To complete this category are two are music games, one of which is *XDancery*, a Simlife (based in Xiamen, Fujian Province) 3D production, which was nominated among the Top Ten finalists for Nokia's 2008 Mobile Games Innovation Challenge contest.

In the 2008 Golden Plume Awards mobile game category, all five networked titles are MMORPGs, with two focusing on adventure, two on real-time action/strategy, and one featuring romance. *One Piece*, developed by Moliyo Online Media & Entertainment (Japan), is based on the popular same-titled Japanese manga series (which was subsequently adapted into an anime series). The other four are local productions, with three fixed in fictional settings and one cast in the familiar thematic context of the Three Kingdoms. In particular, *Jing Ling Online* – developed by Palm World in Fujian, and situating the gaming in a fantasy world of love and romance interwoven with virtual pets – marks a special effort to recruit female players on board. This theme is echoed in *Pet Love* (*Chongwu Qingyuan*), a production of DigiFUN Mobile Entertainment headquartered in Shenzhen, that involves virtual pets in real-time strategy fighting. A similar approach is found in *Poke Elf*, an award-winning title in 2009 by Beijing-based Troodon Co., Ltd. It is obvious that female players are gradually becoming the target of mobile game marketing in recent years. With the three non-networked RPG titles in the winners' list, two are from developers in Taiwan – *Xuan-Yuan Sword* by DOMO under SoftStar, and *Legend of Ancient* by Auer Media & Entertainment Corp – and both are grounded in age-old Chinese folktales, and are trenched in the well-trodden territory of kung-fu fights. The third title, *Dragon Eyes*, a strategy game release of Greatelsoft Co., Ltd in Shanghai, has also been promoted in other Southeast Asian countries since its release in China.

It is worth highlighting here that a few Chinese literary classics have been the perennial source of popular storylines in online and mobile video games, as revealed in the above titles. Naturally, cultural resonance with the players is essential for success. Chinese game developers are certainly in a vantage position to tap into traditional cultural themes/elements to their advantage.

Regulation and Market Structure of Mobile Gaming

As growth in China's mobile game market accelerates, the mobile industry has already experienced a significant level of concentration in the last three years. As an illustration of the status of the field, the China Mobile Game Content Provider Union (CPU) was founded in April 2009 with three overall objectives: to promote a healthy industrial environment through self-regulation and self-support; to create a platform for exchange for professionals in the field; and to build a formal mechanism of coordination among practitioners.[10] Founding parties consist of nine premium and 17 ordinary members.

From the peak of over two thousand game developers in 2006 to less than 100 ones that still manage to maintain a significant presence in the mobile game market in early 2010, China's mobile gaming industry has already witnessed a considerable degree of concentration and consolidation in a rather short period. According to statistics from CPU,[11] from 2003 to 2006, the number of game mobile content providers (aka developers) increased more than tenfold, and that number shrunk by 80% from 2007 to 2008; meanwhile, transaction prices for game title distribution rights increased by 100% from 2008 to 2009, indicating intensifying market demand for quality titles. The Chinese mobile game industry remains a volatile one, and clear winners won't emerge until the dust of the 3G era has settled a few years down the road. At the same time, after years of market turbulence and waves of mergers and buyouts, a small number of developers, such as WiSTONE, Mommoth, 3GUU, MocaWorld, Moloon, and Legame, have solidified their grasp on different niche markets through successfully releasing popular game titles in recent years, while non-differentiating companies are entangled in struggles for survival. In the process of market consolidation, the three megacities of Beijing, Shanghai, and Guangzhou have established themselves as leading hubs of mobile game development and consumption in China. These major metropolises possess the technical talent, an affluent consumer base, and a robust mobile network, and thus they will remain trend-setters for the Chinese mobile industry in the future.

The peculiarities of China's mobile game market structure are presented in Figure 1. China's accession to the WTO in 2001 has incurred a number of commitments on market access to different areas of the telecommunications sector by foreign commercial interests (for Roseman, 2005 for an overview). Foreign access to value-added services such as CRBT, MMS and SMS are regulated by *the Provisions for the Administration of Foreign-invested Telecommunications Enterprises,* first promulgated by the State Council in December 2001 and revised in October 2008, which set the cap for foreign capital at 50%. Distribution services of video/mobile games, however, are exempted from the WTO obligations as they are broadly classified in the telecommunications content area. The Ministry of Culture (MOC) exercises broad oversight over online content, and the regulatory framework is specified in the *Interim Provisions on the Administration of Internet Culture* becoming effective on July 1, 2003. This is further supplemented by the *Joint Directives by the Ministry of Culture and the Ministry of Information Industry on the Development and Administration of Network Games*, which took effect on July 12, 2005.

Figure 1. Market Structure of China's Mobile Game Industry (Adapted from China Mobile Game Content Provider Union (CPU) (http://www.cpunion.org/))

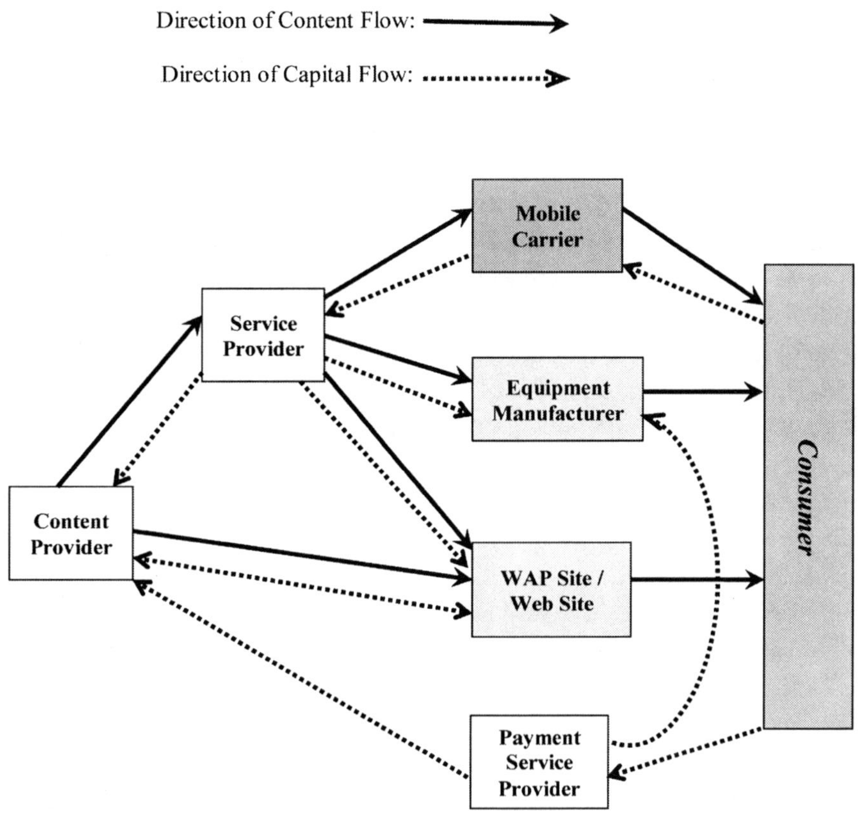

Official regulations stipulate that all service operators of networked games (classified as "Internet Cultural Service Unit") must obtain a license from the MOC, and all imported games must be inspected and approved by certified inspectors through the local cultural administrative bureaus. Furthermore, the 2005 directives require that network game services carry on "advanced" and "progressive" cultural products glorifying Chinese national culture, and that they stay away from subversive, politically sensitive, pornographic, and vulgar content. Although the MOC has not promulgated any clear rules on foreign investment in the area of online content services, officials from the General Administration of Press and Publication (GAPP), a government bureaucracy that also has regulatory responsibility over online publishing, explained in October 2009 that foreign-participated business units are not allowed to engage in network content services.[12] In essence, this means that overseas game developers, or foreign CPs as specified in the figure, can only set foot onto the Chinese mobile game market through state-licensed SPs. In addition to the license from the MOC (for cultural services), mobile service providers must also be approved with a license from the Ministry of Information Industry and Technology (MIIT) (for network information services), plus a permit from the GAPP (for online/network publishing).

The Chinese online video game industry, which includes mobile games, was elevated to the status of a state-supported and nationally-prioritized pillar enterprise in the aforementioned 2005 *Joint Directives*. Such elevation entitles the industry to state-induced financial, taxation, land lease, and other policy support from both the national and local government authorities. Other initiatives in fostering national game development include prioritizing ongoing game titles that promote "superior enduring Chinese cultural values and moral virtues," and the construction of a small number of clusters of industrial bases specializing in research and development of digital entertainment products. Topmost level strategic positioning of the game-related cultural industry in the national economy was codified in the *Cultural Industry Promotion Plan* released by the State Council in July 2009. Growth in the mobile game sector, then, has benefited greatly from meticulous state nurturing.

China's mobile carrier market is a tripartite structure, divided among China Mobile, China Unicom and China Telecomm, although China Mobile maintains a formidable presence by owning about 60% of the mobile population in China. Treasure Box of China Mobile and U-Magic of China Unicom are principal wireless retail centers for mobile games to their respective subscribers. WAP sites (i.e., mobile sites, such as wap.kong.net, adopting the Wireless Application Protocol) and Web sites (e.g. Sina.com) engaging in the business of mobile games are numbered in the thousands, and market consolidation in the near future will likely concentrate the market power to a much smaller list of players. Mobile payment service providers include online operations of major banks and Paypal-type of specialized online payment services, the most popular two of which are *Yifubao* and *Shenzhoufu*. Currently, there are over 200 Service Providers (SPs) distributing tens of thousands of mobile game titles to various commercial entities and user groups. At this point, SPs play the critical role in bridging mobile carriers and CPs, and can be broadly classified in three types.

First, there are those SPs that specialize exclusively in mobile gaming, and they have played a dominant role in promoting game titles to the mobile audience in recent years. Big names include Mig Games, Moloon International, Linktone, Pearl in Palm, D.CN, Troodon, and China Wireless Arts. They form close alliances with both mobile carriers and WAP/Web sites in marketing hit titles from commercial developers, and most of them are also engaged in game development themselves. While they will continue to dominate the service market for the near future, the biggest challenge most likely comes from finance. The second type of SPs consists of providers of multiple value-added mobile services, and mobile gaming is just one of the many areas of mobile entertainment for them. The biggest name in this category is KONG.net, whose business ranges from news, stock information, weather services, CRBT, video, social networking, music to games for mobile users. Many other sites, such as 51Momo, SOOYUU and SPCOOL, have followed this model in promoting their services. Driven by the need for revenue generation, these sites typically prioritize their services to focus on the areas that are the most profitable, and mobile games are not very likely to top their agenda for expansion at the time being. Survival is the most immediate challenge for most start-ups in this category.

The third type of SPs features the participation of comprehensive portal sites in the mobile game business. Leaders in this category are Sina, Tencent Inc. (more commonly known as QQ), Shanda and Tom.com, which have already made a solid presence in the Internet era and have recently branched off in mobile gaming as one add-on for their offerings. Compared with the other types, these SPs are in the best financial position to make new investments. However, they are more likely to be trend-followers than pacesetters with regard to mobile gaming because this is not strategically their primary focus. The most likely route for these players is to break into new ground through buyouts and mergers of promising start-ups.

There are clear trends of consolidation and integration in the Chinese mobile industry in recent years. More than one hundred mergers have already taken place within the past few years, with most deals involving SPs buying out upstart CPs. Some big names, such as Shanda, KONG.net, Moloon, and Linktone, have been involved in multiple transactions. These mergers make it possible for the same corporate entity (e.g. KONG.net) to fulfill the multiple roles of WAP site, CP, and SP at the same time. Nonetheless, formidable players such as Sina, Sohu and Netease have thus far balked from making bids on mobile game businesses, and this is likely to change once the 3G market conditions become settled in the next couple of years.

Additionally, there are signs of heightened globalization in China's mobile game market. Attracted by the world's largest mobile population, many global conglomerates have established varying degrees of presence in China since the early 2000s. The main strategy for foreign companies to break into the Chinese market has been merging with local companies or forming business alliances with localized partners. Among major corporations that have extensive business operations in mobile game development in China for the time being are SEGA (Japan), KONAMI (Japan), Sony (Japan), UbiSoft (France), G-mode (Japan), Com2Us (South Korea), Webzen (South Korea, recently acquired by US-based NHN), VODUS (South Korea) and SOFTSTAR (Taiwan) as well as major manufacturers such as Nokia and Motorola. These foreign companies have played an important role in shaping the historical evolution of mobile gaming in China. The following are but a few snapshots of landmark events ushered in by overseas developers in recent years.

California-based Mtone Wireless Corp. made a big splash with the launch of its *Multi-User Dungeon*, first in Guangzhou in 2003 and expanded to major Chinese markets in 2004. This was the first real-time MMORPG in China (due to bandwidth limitations at the time, it was SMS-based), and introduced the concept of subscription-based mobile gaming (users had to pay RMB 12 per month). Additionally, Mtone formed an alliance with the NBA and China Mobile in 2005 for a multi-year partnership to distribute NBA-related content and gaming to mobile users in China. While sports-related games have not been popular with Chinese players, basketball is one of the few rare exceptions, joined by soccer, thanks to the efforts of Electronic Arts (EA) with its FIFA Online, which was first released in 2006 and has launched annual editions since then. Not to be left behind, Disney has pursued an aggressive strategy in developing and marketing mobile games derived from its proprietary content to Chinese players since the early 2000s. Currently with three regional offices in Beijing, Shanghai and Guangzhou, Disney has maintained a prominent presence in the Chinese market through its theme-based game titles; as a matter of fact, Disney debuted one of the first popular pay-per-play (RMB¥ 8) title *Mickey Mahjongg* in 2005 specifically targeting female players.

Finally, it should be noted that this process of globalization is not unidirectional. Encouraged by the government's "Go-Out" policy in telecommunications products and services, more and more Chinese companies with the know-how and the financial capability are reaching out to global markets. Many game titles developed by Chinese companies have successfully entered overseas markets, especially those in Southeast Asia, in recent years, and major CPs are looking outside to enhance global visions in pursing partnerships and long-term strategies. Meanwhile, mobile game development in China has gained competitive edges at the global level, as indicated by the decision by Ubisoft to move its R&D center to China (through its Shanghai and Beijing operations) in the past decade and EA's strategic move to make Shanghai a key part of its Asian video and mobile game business with the founding of EA Computer Software Shanghai Co, Ltd in 2006. These trends manifest perfectly the "complex connectivity" of the global game industry in which an "ever-increasing network of interconnections and interdependencies" (Tomlinson, 1999, p. 2) pervades the developers and the products.

FUTURE RESEARCH DIRECTIONS

While arcade, console and PC games, which have significantly shaped the development of mobile gaming in South Korea and Japan, find no less enthusiasm among Chinese players, one key difference between China and South Korea/ Japan that may have contributed to the unequal development of Chinese mobile gaming is the lack of an affordable high-speed mobile data network infrastructure in China. The switch to 3G network coverage is going to usher in a new era of mobile connectivity in China. Therefore, the 3G era will lead to monumental changes in China's mobile gaming industry. Future research should examine the nature of these changes and the impact they may have on the ongoing Chinese mobile culture. Toward that end, useful frames of reference are South Korea and Japan, which installed 3G networks years ahead of China, and the trajectories of transformation in those countries can provide valuable insight for researching mobile gaming in the 3G era in China. As made clear earlier, however, mobile culture in each national context is heavily infested with local flavors. Thus, the similarities and differences between mobile gaming in China and those in other countries will offer rich terrains for research ideas and questions in the ensuing years.

To a great extent, mobile culture is youth culture. While there have been a rising stockpile of enlightening publications in recent years on youth gaming culture and identify building in a variety of national/local settings, what is acutely lacking is penetrating research on the same topic in China. This is made more urgent by the fact that Chinese mobile gaming is overwhelming dominated by youngsters in their early 20s. So youth play a much more dominant role in China than in most other countries. This opens an exciting line of research that promises to make cornerstone contribution to the global literature in this area. Additionally, in a society deeply rooted in authoritarian traditions of politics and culture, the transformative impact of mobile communication in general and mobile gaming in particular is especially worth inquiring about in future scholarship. Another closely related issue involves the nature and dynamics of the interplay of Chinese and global companies and the interaction of domestic and overseas players.

Lastly, and equally importantly, in a country where the state maintains a formidable omnipresence in everything people do in the online and mobile environment, the Chinese authorities are still coming to grips with the fast-evolving mobile networks. What new policies are being contemplated and what regulatory framework lies on the horizon to control what people can game on are worth monitoring in the years to come.

CONCLUSION

It has been widely noted across the academic literature that the Asia Pacific (especially South Korea and Japan) region has led the world in online video gaming. China, with the largest mobile market in the world, has emerged as a global leader of mobile game production and consumption. From its inception in the late 1990s to the present, mobile gaming in China has gone through four distinct stages: induction, transition, consolidation, and the 3G era. Within a decade or so, mobile games have quickly evolved from a rarity to widespread availability to an expanding player base, and the Chinese mobile industry has meanwhile transformed from one that used to lopsidedly rely on foreign titles to one that is incubating hundreds of home-grown hit titles annually that are avidly sought after not only in China but also in many parts of the world.

As for game genres, China, like many other countries, has followed the similar path of historical development from built-in casual games to downloadable sing-player titles to networked MMORPGs. However, there are also outstanding differences in terms of game content, with Chinese players resonating most comfortably with gaming settings historically situated in the rich domains of ancient Chinese dynasties and thematically cast in classic literacy works. In terms of demographics, Chinese mobile players display two tendencies that deviate significantly from counterparts in other countries: they are disproportionately dominated by male gamers (93%), and are overwhelmingly populated by people in the 18-24 age group (74.5%). Only a very small percentage of the players are older than 30. To be looked at another way, however, this may bode well for game developers and distributors, as there is a huge portion of the mobile population to be tapped into. For the time being, MMORPGs are the most-favored game types, and items-based model of playing dominates the mobile market. This is in contrast with many other countries such as the United States, Japan and South Korea, where subscriptions-based play tends to prevail. This does not necessarily mean that subscriptions-based model cannot survive in the Chinese mobile game market; it does suggest, however, that those game distributors intending to go with this model will face an enormous hurdle in the Chinese market.

There is already a considerable level of market concentration and integration in the mobile game industry. But more fundamental shape-ups will most likely take place in the forthcoming years, as pivotal players in different sectors of the market emerge in the next few waves of market consolidations in a highly globalized sector. Due to the peculiarities of its market structure, Chinese mobile carriers and service providers (SPs) have enormous leverages in shaping the rules of game-playing in the market, as do the state regulators.

REFERENCES

Bell, G. (2005). The age of the thumb: A cultural reading of mobile technologies from Asia. In Raessens, J., & Goldstein, J. (Eds.), *Handbook of computer game studies* (pp. 67–87). Cambridge, MA: MIT Press.

Bruns, A. (2008). *Blogs, wikipedia, second life, and beyond: From production to produsage*. New York: Peter Lang Publishing.

Castells, M., Fernández-Ardèvol, M., Qiu, J. L., & Sey, A. (2007). *Mobile communication and society: A global perspective*. Cambridge, MA: MIT Press.

Chan, D. (2008). Convergence, connectivity, and the case of Japanese mobile gaming. *Games and Culture, 3*(1), 13–25. doi:10.1177/1555412007309524

Chan, E., & Vorderer, P. (2006). Massively multi-player online games. In Vorderer, P., & Bryant, J. (Eds.), *Playing video games: Motives, responses, and consequences* (pp. 77–88). Mahwah, NJ: Lawrence Erlbaum Associates, Publishers.

China Cultural Industry Net. (2010, January 20). *China network game market white paper: 2009. A synopsis*. Available: http://www.cnci.gov.cn/content%5C2010120/news_55882.shtml

China Game Publishers Committee & International Data Corp. (2004). *China game industry report: 2003*. Beijing: People's University Press.

Crowe, N., & Bradford, S. (2007). Identity and structure in online gaming: Young people's symbolic and virtual extensions of self. In Hodkinson, P., & Deicke, W. (Eds.), *Youth cultures: Scenes, subcultures and tribes*. New York: Routledge.

Goggin, G. (2006). *Cell phone culture: Mobile technology in everyday life*. New York: Routledge.

Hall, J. (2005). Future of games: Mobile gaming. In Raessens, J., & Goldstein, J. (Eds.), *Handbook of computer game studies* (pp. 47–55). Cambridge, MA: MIT Press.

Hjorth, L., & Chan, D. (2009). Locating the game: Gaming cultures in/and the Asia-Pacific. In Hjorth, L., & Chan, D. (Eds.), *Gaming cultures and place in Asia-Pacific* (pp. 1–14). New York: Routledge.

Ip, B. (2008). Technological, content, and market convergence in the games industry. *Games and Culture*, *3*(2), 199–224. doi:10.1177/1555412008314128

iResearch (2008). *The investigation report of Chinese mobile phone players*. Retrieved online, November 10, 2009, at http://www.docin.com/p-18767066.html

Ito, M., Okabe, O., & Matsuda, M. (2006). (Eds.) *Personal, portable, pedestrian: Mobile phones in Japanese life*. Cambridge, MA: MIT Press.

Jenkins, H. (2006). *Convergence culture: Where old and new media collide*. New York: New York University Press.

Jin, D. Y., & Chee, F. (2009). The politics of online gaming. In Hjorth, L., & Chan, D. (Eds.), *Gaming cultures and place in Asia-Pacific* (pp. 18–37). New York: Routledge.

Katz, J. E. (2006). Mobile communication and the transformation of daily life: The next phase of research on mobiles. *Knowledge, Technology, &. Policy*, *19*(1), 62–71.

Katz, J. E. (2007). Mobile media and communication: Some important questions. *Communication Monographs*, *74*(3), 389–394. doi:10.1080/03637750701543519

Katz, J. E., & Acord, S. K. (2008). Mobile games and entertainment. In Katz, J. E. (Ed.), *Handbook of mobile communication studies* (pp. 403–418). Cambridge, MA: MIT Press.

Ling, R., & Donner, J. (2009). *Mobile phones and mobile communication*. Malden, MA: Polity.

Malliet, S., & de Meyer, G. (2005). The history of the video game. In Raessens, J., & Goldstein, J. (Eds.), *Handbook of computer game studies* (pp. 23–45). Cambridge, MA: MIT Press.

McLelland, M. (2007). Socio-cultural aspects of mobile communication technologies in Asia and the Pacific: A discussion of the recent literature. *Continuum: Journal of Media & Cultural Studies*, *21*(2), 267–277. doi:10.1080/10304310701269099

Roseman, D. (2005). The WTO and telecommunications services in China: Three years on. *Info*, *7*(2), 25–48. doi:10.1108/14636690510587207

Sellers, M. (2006). Designing the experience of interactive play. In Vorderer, P., & Bryant, J. (Eds.), *Playing video games: Motives, responses, and consequences* (pp. 9–22). Mahwah, NJ: Lawrence Erlbaum Associates, Publishers.

Taylor, T. L. (2005). *Play between worlds: Exploring online game culture*. Cambridge, MA: MIT Press.

Thomas, A. (2007). *Youth online: Identity and literacy in the digital age*. New York: Peter Lang.

Williams, D. (2006). A brief social history of game play. In Vorderer, P., & Bryant, J. (Eds.), *Playing video games: Motives, responses, and consequences* (pp. 197–212). Mahwah, NJ: Lawrence Erlbaum Associates, Publishers.

ADDITIONAL READING

Cao, Y., & Downing, J. (2008). The realities of virtual play: Video games and their industry in China. *Media Culture & Society*, *30*(4), 515–529. doi:10.1177/0163443708091180

Chehimi, F., Coulton, P., & Edwards, R. (2008). Evolution of 3D mobile games development. *Personal and Ubiquitous Computing*, *12*(1), 19–25. doi:10.1007/s00779-006-0129-9

Consalvo, M. (2009). Convergence and globalization in the Japanese videogame industry. *Cinema Journal*, *48*(3), 135–141. doi:10.1353/cj.0.0122 de Souza e Silva, A. (2009). Hybrid reality and location-based gaming: redefining mobility and game spaces in urban environments. *Simulation & Gaming*, *40*(3), 404–424. doi:10.1177/1046878108314643

de Souza e Silva, A., & Hjorth, L. (2009). Playful urban spaces: A historical approach to mobile games. *Simulation & Gaming*, *40*(5), 602–625. doi:10.1177/1046878109333723

Ernkvist, M., & Ström, P. (2008). Enmeshed in games with the government: Governmental policies and the development of the Chinese online game industry. *Games and Culture*, *3*(1), 98–126. doi:10.1177/1555412007309527

Fannin, R. A. (2008). *Silicon dragon: How China is winning the tech race*. McGraw-Hill.

Glotz, P., Bertsch, S., & Locke, C. (Eds.). (2005). *Thumb culture: The meaning of mobile phones for society*. New Brunswick, NJ: Transaction Publishers.

Heather, A., Horst, H. A., & Miller, D. (2006). *The cell phone: An anthropology of communication*. Oxford, UK: Berg Publishers.

Hjorth, L. (2007). The game of being mobile: one media history of gaming and mobile technologies in Asia-Pacific. *Convergence (London)*, *13*(4), 369–381. doi:10.1177/1354856507081955

Izushi, H., & Aoyama, Y. (2006). Industry evolution and cross-sectoral skill transfers: A comparative analysis of the video game industry in Japan, the United States, and the United Kingdom. *Environment & Planning A*, *38*(10), 1843–1861. doi:10.1068/a37205

Jenkins, H. (2006). *Fans, bloggers, and gamers: Media consumers in a digital age*. New York University Press.

Ling, R. (2004). *The mobile connection: The cell phone's impact on society*. San Francisco, CA: Morgan Kaufmann.

Ling, R., & Campbell, S. (Eds.) (2010). *The reconstruction of space and time: Mobile communication practices*. New Brunswick, NJ: Transaction Publishers.

MacInnes, I., & Hu, L. (2007). Business models and operational issues in the Chinese online game industry. *Telematics and Informatics*, *24*(2), 130–144. doi:10.1016/j.tele.2006.04.002

Ng, W. B. (2001). Japanese video games in Singapore: History, culture and industry. *Asian Journal of Social Science*, *29*(1), 139–162. doi:10.1163/156853101X00361

Okazaki, S., Skapa, R., & Grande, I. (2008). Capturing global youth: Mobile gaming in the U.S., Spain, and the Czech Republic. *Journal of Computer-Mediated Communication*, *13*(4), 827–855. doi:10.1111/j.1083-6101.2008.00421.x

Rheingold, H. (2003). *Smart mobs: The next social revolution*. Cambridge, MA: Basic Books.

Shin, D. H. (2010). The dynamic user activities in Massive Multiplayer Online Role-Playing Games. *International Journal of Human-Computer Interaction*, *26*(4), 317–344. doi:10.1080/10447310903575473

Tai, Z. (2006). *The Internet in China: Cyberspace and civil society*. New York: Routledge.

Tai, Z. (2010). Casting the ubiquitous net of information control: Internet surveillance in China from Golden Shield to Green Dam. *International Journal of Advanced Pervasive and Ubiquitous Computing*, *2*(1), 53–70.

Zhao, Y. (2008). *Communication in China: Political economy, power, and conflict*. Lanham, UK: Rowman & Littlefield Publishers, Inc.

KEY TERMS AND DEFINITIONS

Average Revenue Per User (ARPU): This is a business model of online video gaming through which the users are charged a flat monthly fee for the privilege of game playing online.

Content Provider (CP): In the Chinese regulatory framework, a content provider in the mobile game industry is a commercial entity that supplies (produces) game titles (hence content) to players. Content providers are not allowed to distribute game content to users unless they are licensed to do so.

Massive Multiplayer Online Role Playing Games (MMORPGs): These are a popular genre of online video games in which a large number of players can take on particular roles and simultaneously interact with one another in a virtual environment. MMORPGs are a special type of Massive Multiplayer Online Games (MMOGs). Other MMOG genres include MMOFPS (Massive Multiplayer Online First-Person Shoot), MMORG (Massive Multiplayer Online Rhythm Game), MMOSG (Massive Multiplayer Online Sports Game), and MMOR (Massive Multiplayer Online Racing).

Mobile Carriers: Commercial companies that provide wireless phone services in a particular region/area to paying customers. Additionally, they offer a wide range of services, such as text messaging and video streaming/downloading, through their wireless networks. In China, three mobile carriers are licensed to operate in the market: China Mobile, China Unicom, and China Telecomm, with China Mobile dominating about 60% of the market.

Mobile Value-Added Services (MVAS): In the official telecommunications classificatory scheme, mobile services are divided into basic services and value-added services. Mobile basic services encompass basic voice and data delivery services across the mobile networks, while mobile value-added services include all other non-core services (often called mobile informational services in the Chinese framework).

Multimedia Messaging Services (MMS): It is a popular way of delivering multimedia content via mobile phones. This is an enhancement from purely textual information (called text messaging, or SMS) communicated via the mobile networks.

Color Ring Back Tone (CRBT): This is a rapidly-growing area of value-added (enhanced) services across the world. Users can select a personalized song or audio file to play for incoming calls/messages. It is called Personalized Ring Back Tone (PRBT) in North America.

Service Provider (SP): In the Chinese mobile market, mobile game service providers are commercial entities that are engaged in the business of distributing game titles across the mobile networks. Chinese regulations require that SPs be licensed by the state, and current rules exclude foreign capital or corporations from this area of services. Content providers have to work through service providers in order to gain access to the mobile game market.

ENDNOTES

1. Statistic retrieved from the MIIT Website at http://www.miit.gov.cn/n11293472/n11293832/n11294132/n12858447/13059405.html.
2. MII statistics as reported at: http://www.miit.gov.cn/n11293472/n11293832/n11294132/n12858447/13106713.html.
3. www.forrester.com/
4. http://www.budde.com.au/
5. The summary of the report is available at: http://enfodesk.com/SMinisite/index/articledetail/type_id/1/info_id/3910.html
6. See the report by *Edge* Online at: http://www.edge-online.com/news/global-game-sales-overtake-dvdblu-ray
7. MC Insight: Advertising and Video Games. July 2008, available http://www.media-

contacts.com/images/common/mc-insight/mc_insight_advergaming.pdf

8 Unless otherwise noted, this section of the discussion draws extensively from news reports published from 2002 to 2009 in the Chinese press.

9 A synopsis of China's MVAS market report by China Market Research Web for 2009-2012 by can be found at http://www.cnscdc.com/22506.html.

10 More information can be found at the official Web site at http://www.cpunion.org.

11 A variety of statistics are available from http://wap.wan8.mobi/cpu/09年工作事项.rar.

12 Xinhua News Agency release, October 10, 2009, available http://news.xinhuanet.com/politics/2009-10/10/content_12203357.htm

Chapter 19
Cinema of the Gun:
Science and Technology Studies and the First Person Shooter

Sam Hinton
University of Canberra, Australia

ABSTRACT

Games, like other technological artifacts, do not fall mysteriously from orbit. They are constructed by humans within a matrix of social, economic and cultural conditions of production and consumption. Technologies including the refrigerator, the M16 machine gun and the bicycle have famously been the subject of so-called science and technology studies (STS) analyses, following MacKenzie and Wajcman, Pinch and Bijker and Latour and Callon, among others. Aside from a few initial forays, games have largely escaped this kind of in-depth and sustained analysis. This chapter argues for the use of STS in games research and uses an STS approach to explore the evolution of a 'genre' of computer games – the first person shooter, or FPS – and in particular the cinematic turn that has taken place within this genre over the past decade.

INTRODUCTION

In 1992 *Wolfenstein 3D* presented players with a virtual world on their computers. They could move through this world as if seeing through the eyes of the game's protagonist, William "BJ" Blazkowicz as he mounted a Rambo-eqsue escape from a Nazi prison. This type of game, which has come to be known as a first person shooter (FPS) has been an important part of the game industry, and is a genre that has grown with it. To illustrate the position of the first person shooter in the game market, a recent release, *Call of Duty: Modern Warfare 2* (2009) reportedly cost US$200 million to produce and grossed US$550 million in sales in its first week, with revenues of up to US$1 billion expected. (Fritz 2009)

Since *Wolfenstein 3D* first person shooters have become increasingly cinematic, both visually and in terms of narrative development. It is tempting to see this as a natural process of development, an inevitable result as the games industry matures

DOI: 10.4018/978-1-60960-037-2.ch019

Copyright © 2011, IGI Global. Copying or distributing in print or electronic forms without written permission of IGI Global is prohibited.

and as the technology becomes capable of more complex image and data. Similar technologically deterministic assumptions are often made about many technologies, and have come under increasing scrutiny from social theorists working under the broad banner of science and technology studies (STS) who emphasise the social dimensions of technological development over the scientific or technical.

This chapter has two purposes. Its primary goal is to present science and technology studies as an empirical method for analysing games. The secondary goal is to use an STS approach to examine the social reasons that underlie the cinematic turn of first person shooters. To reach these goals the chapter is organised into three sections. Section one provides an overview of science and technology studies. The second section looks briefly at how science and technology studies might be useful in the analysis of games. The third and final section acts as an exemplar, and employs a science and technology studies approach to look at the development of the first person shooter genre and the rise of cinematic verisimilitude in the genre.

BACKGROUND

Science and technology studies (often abbreviated as STS) is an approach to the study of technology that focuses on the importance of social factors in the construction of technologies. Two books in particular were important in bringing the STS approach to prominence – MacKenzie and Wajcman's *The Social Shaping of Technology* (1985); and, Pinch and Bijker's *The Social Construction of Technological Systems*. (1987) MacKenzie and Wajcman pose the fundamental question like this: "what shapes the technology in the first place, before it has 'effects'. Particularly, what role does society play in shaping technology?" (MacKenzie and Wajcman 1985, 8)

The term "technology" is problematic because it is widely used with a variety of subtly different meanings. MacKenzie and Wacjman associate three distinct meanings with the term "technology". First, technology can be said to relate to a physical object or artefact – a hammer is a technology, a computer is a technology. The second definition acknowledges the human dimension of technology. MacKenzie and Wacjman use the example of steelmaking as a technology – this emphasises that steel is made through a physical process of production that necessarily incorporates the activity of people who use the apparatus required to make steel. The third definition of technology presented by MacKenzie and Wacjman is derived from the etymology of the word technology as a form of knowledge: a "systematic knowledge of the practical arts". (MacKenzie and Wajcman 1985, 4)

A central problem in technology studies concerns the nature of the relationship between technologies and society. On one hand, technologies are often said to have an impact, or an effect, upon human societies. The microcomputer, for example, has been touted as both a positive technology that will unleash humans from the drudgery of repetitive labour, and as a negative technology that reduces humans to binary sequences and processes them like any other commodity. This perspective that technologies alone change society is an approach that is often referred to as technological determinism. (MacKenzie and Wajcman 1985, 4)

Science and technology studies is itself derived from earlier work done in the sociology of scientific knowledge by theorists like Kuhn. (1962) This work emphasised the social contingency of scientific discoveries, and argued that scientific discoveries were open to interpretative flexibility. In other words, scientific knowledge was not absolute – the results of experiments are always open to a certain level of interpretation and it was this interpretation that could be debated. Through analysis of these debates sociologists could show how different perspectives were united through a rhetorical process, and finally resolved into a single widely agreed piece of scientific knowledge. The construction of these technological artefacts is, much like the construction of scientific knowledge, also a matter of social contingency. (Pinch and Bijker 1987, 18)

Science and technology studies extends the sociology of scientific knowledge into the realm of technology, which is seen as an extension of the scientific process. In developing new technologies, inventors (or engineers, or technologists) take scientific knowledge and turn it into practical innovation. Pinch and Bijker argue that this process is not straight forward – scientific knowledge does not lead to innovation. Instead, innovators cherry-pick scientific knowledge, taking some bits and leaving others to arrive at a technological artefact. (Pinch and Bijker 1987, 20)

While there are a variety of different approaches to STS – including social shaping, social construction and actor-network theory (Brey 1997) – they all share similar underlying principles. The three main principles are symmetry, interpretative flexibility and stablisation. The principle of symmetry argues that the researcher should remain agnostic about the real properties of the technology they are examining. Interpretative flexibility maintains that different social groups will attach different meanings and properties to an artefact – meaning is not fixed by the artefact itself. Symmetry demands that the interpretations that any group attaches to an artefact should be taken at face value, without an attempt by the researcher to judge their value or truth – the artefact in question should be assumed to have no real or absolute properties. Stabilisation occurs as different interpretations of the technology are resolved and a consensus of interpretation is reached by social groups. At this point debates about the interpretation of the technology cease and the technology is said to have become black-boxed: the technology's details are neatly packaged away into the conceptual form of the technology and no longer serve as a matter of debate. There need not be any relationship between the real or physical properties of the object and the socially constructed interpretation of that technology.

This has a range of practical applications in looking at technologies and understanding the range of social factors that lead to one outcome over another. For example, why did the electric refrigerator end up becoming the conventional domestic technology when the alternative gas powered option was just as viable? By examining the debates that led to the development of the electric refrigerator, and the way these debates were resolved, Ruth Schwartz-Cowan shows quite compellingly how important social factors were in the selection of one technology over another. (Schwartz-Cowan, 1987) This challenges the common-sense belief that the technologies that come to the fore are the technically superior options. Numerous other STS analyses of technological artefacts make this point again and again.

STS AND GAMES

Games are comparable with other technological artefacts like bicycles, refrigerators and M16 machine guns (all of which have been the subject of STS analyses). Games are constructed by teams of developers, informed by the success and failure of other games (O'Donnell 2009) and the work of computer scientists. There are also a number of parallels between game developers on the one hand and technologists on the other. Where technologists take scientific knowledge and cherry-pick build technologies, game developers take computer science and build software.

The STS approach is laid out clearly by Pinch and Bijker in the introductory chapter of *The Social Construction of Technological Systems.* (1987) They provide an example of an STS study of the development of the bicycle in which they begin by describing the artefact, then move on to examine the debates about the form and function of the bicycle. They define social groups such as cyclists, anticyclists, women bikers and so on, each of whom attached different interpretations and values to the bicycle and each of whom had different problems with its design. (Pinch and Bijker 1987, 30-34) A process of debate ensued which saw various issues emerge, such as safety concerns, vibration concerns and so on. Each of these issues were dealt with through rhetoric (that is, the problem did not necessarily disappear, but people changed their minds about the problem) or through redefinition of the problem. For example, there was significant debate over the value of the inflatable tyre and the solid tyre. The solid tyre was initially preferred although it led to too much vibration when the bicycle was ridden across uneven ground. While this was seen as a problem for some, others didn't mind the vibration, so the inflatable tyre was not seen as a solution. Indeed, the inflatable tyre was seen as ridiculous and ugly, although proponents of the inflatable tyre argued that it provided a much smoother ride than the solid tyre. The resolution of this debate seemed to be when cyclists with inflatable tyres proved they were faster than cyclists with solid tyres, which redefined the problem – the problem for solid tyres was that they were too slow, so the air tyre became a solution to this new, redefined problem. The resolution of this debate among others stabilised the bicycle's design. (Pinch and Bijker 1987, 44-46)

Applying the STS approach to games can follow a similar approach. In the example in the following section the approach is applied to first person shooters, but could easily be applied to a series (for example, the Halo series of games) or indeed through close analysis of the development process, to an individual game. The goal of such a study would be to show how social choices rather than technical constraints alone influenced the construction of a game. A typical application of STS to games would begin with a description of the game, then move on to identify social groups, the ways that these groups interpret the game through its development and the problems or debates that these different interpretations engender. The study would then look at how these debates were stabilised – what was done to resolve these issues thus leading to a stable and widely agreed upon interpretation of the game?

Despite the apparent potentials of such an approach, there has not been a great deal of application of STS to games, nor very much analysis of the way that the conditions of game production shapes games and determines the kinds of games that get to market. Instead, much of the games studies literature has been focussed more heavily on describing the field of games studies or analysing player interactions and the forms of games. There is, however, growing interest in looking more critically at the work of game producers and the technology that games are built on, and STS forms one approach to these kinds of studies.

Ian Bogost and Nick Montfort's *Racing the Beam*, while not an STS study of games, outlines a field of research that they call platform studies which they see as "a kind of membrane between computing, STS, and the humanities." (Jenkins 2009) They note that while games and digital media in general have been subjected to analysis of the cultural products and the ways they are used by players "little work has been done on how the hardware and software of platforms influences, facilitates, or constrains particular forms of computational expression." (Bogost and Montfort 2009, 3) In a different but related vein, O'Donnell's work on creative collaborative practice uses what he calls a microethnographic approach to look at the way that social factors influence game development. (O'Donnell 2009) His notion of creative collaborative practice draws upon STS, but again is not an STS study.

On a larger scale, Lev Manovich has pointed to the increasing importance of these kinds of studies for the analysis of media systems. He argues that software has become "*a layer that permeates all areas of contemporary societies*" (Manovich 2008, 7, original emphasis), and deserves attention as a fundamental principle in the study of media. He has developed the term software studies to describe an approach that examines "the role of software in forming contemporary culture, and cultural, social, and economic forces that are shaping development of software itself." (Manovich 2008, 5)

These approaches are compelling and suggest a strong developing interest in the rich territory that exists between the techno-scientific field of information technology and the arts and humanities. It may be that the application of STS to games is just part of a larger program of analysis that is emerging, along the lines of Manovich's software studies (in whatever guise it takes). In some ways it doesn't matter: what matters it that all of these approaches seek to emphasise the social factors that underlie, instruct and mould the technological.

STS AND THE FIRST PERSON SHOOTER

An Overview of the Genre

Before going on to look more closely at the social construction of first person shooters from the STS perspective, it's valuable and in keeping with the STS approach to provide a brief overview of the FPS genre. To outline and describe the genre it's useful to briefly outline some of the key titles and the developments that went along with them. While the following is by no means an exhaustive list of first person shooters (and indeed, fans of the genre will note numerous omissions) my main goal is to present an overview to help the reader gain a broad understanding of the genre rather than to present a definitive history of the genre.

A first person shooter is a game that, as the name suggests, presents a virtual world as if the player were looking through the eyes of a character who is in the world. First person shooters are also based primarily around ranged (gun-based) combat. The player progresses in the game by moving through the virtual environment, encountering and defeating increasingly difficult enemies. In many first person shooters players have a health score or similar that is decremented each time they are hit by an enemy; this game mechanic ensures that the player can take a few hits before dying, which increases the playability of the game. Often this health score can be increased by finding power-ups, but in more recent games it is recovered automatically or by some other player action. In many recent first person shooters progression through the game is accompanied by the progression of a narrative, although all but the most superficial narrative was absent from early first person shooters.

Probably the first commercial first person shooter was *Wolfenstein 3D* (1992), a game produced by Id Software and sold mainly as shareware and through mail order by publisher Apogee Software. The game used a technique called raycasting to take data from a two-dimensional map and project it on the player's screen as a three-dimensional environment. This technique was not developed by Id (the technique was described by Roth in a technical paper in 1982), but was the first application of the technique in a game. While raycasting did not present a true 3D world (players could not look up or down and were limited to movement in only two dimensions), the illusion was sufficient to make the player feel that he or she was immersed in a virtual world.

The following year the same developers released *Doom* based on the same technology, which was highly successful and went on to firmly establish the genre. Also in 1993 Looking Glass Technologies released *Ultima Underworld*, a dungeon exploration game that also used raycasting techniques to create a virtual 3D world. *Underworld* was in some ways technically superior to *Doom*, allowing players to cast spells and engage in hand-to-hand combat with monsters, and to jump and swim as well as walk and run. A third game, *Pathways into Darkness* was also released in 1993 for the Apple Macintosh by a game developer called Bungie. *Pathways* also used raycasting techniques and, like *Ultima Un-*

derworld, focussed on dungeon exploration, but in a 1940s milieu. Following the success of *Doom*, all three companies released sequels based on the same technology. Notably, Bungie's sequels were titled *Marathon* and laid the foundations for *Halo* (2001), a very popular first person shooter.

In 1995 Parallax Software released *Descent*, a flight simulator that bore more resemblance to a first person shooter than a conventional flight simulator. Instead of flying through skies, the player flew through rocky mazes, shooting enemies in a zero-gravity environment. Importantly, *Descent* did not use the raycasting system of other FPSs, but drew much of the world – including opponents – in full 3D, allowing the player to move and look freely in all three dimensions. But it was the release of *Quake* by Id Software in 1996 that created a great deal of interest in the industry.

Quake used a rendering technique that allowed players to look around a completely three-dimensional environment. Players could jump (and fall) and had to contend with enemies that may be above or below them. The enemies themselves were also three dimensional unlike enemies in earlier games who were literally two dimensional. In addition to this, *Quake* introduced an internet play mode, which allowed players who had an internet connection to compete against each other in games of up to 16 players. *Doom 2* (1994) had provided network play options but this was limited to players on the same local area network.

Numerous first person shooters can be listed following *Quake*, many boasting various improvements or innovations in the quality and complexity of graphic output. An important development was the market for computer game engines as developer tools separate to the games that supported them. A game engine is the part of a first person shooter that is responsible for drawing 3D graphics on the screen, and increasingly have also come to handle other technical tasks like networking, collision detection and physics simulation. Development of these engines is technically complex, expensive (increasingly so as games become more complex) and time consuming. For many companies, licensing the use of a game engine was a more effective way of getting a game to market than developing one in-house. For example, while Id Software continued to produce games, their core business relied heavily on licensing their 3D engines to other developers. Epic, the makers of the *Unreal* series of games also provided their engine for licensing by developers.

Notable among these developers was a company called Valve, who purchased rights to the *Quake* engine and redeveloped it significantly to create their own engine (which they called Goldsource) upon which they based their FPS title *Half-life.* (1998) *Half-life* was notable as one of the first highly popular first person shooters to present the player with a narrative, replete with characters and plot arc. Another game, *Deus Ex* (2000), based on Epic's *Unreal* engine, also presented the player with an involved and engaging narrative. From at least 2000 onwards, first person shooters were dominated by two significant themes: steadily increasing graphical complexity with the goal of producing a more photoreal visual experience; and, the increasing sophistication and integration of narrative into game play.

Taken together, the increased visual complexity and greater prominence of narrative in first person shooters could be deemed something of a cinematic turn in the evolution of the first person shooter. Cinematic realism here does not refer simply to the idea of trying to reproduce or simulate reality. Instead, cinematic realism is about reproducing a cinematic verisimilitude, so that playing a game becomes like participating in a cinematic experience, complete with the conventions (or at least, references to the conventions) of contemporary cinema. The virtual camera, effects, plot and score all work together to produce this experience.

Producers, Players and Anti-Gaming

In STS terminology, the game development and design process is being black-boxed as debates about realism and the role of narrative in first person shooters become resolved and closed. Part of this black-boxing concerns the cinematic form of the first person shooter – the way that cinematic conventions are becoming a largely taken-for-granted feature of the FPS genre. From a science and technology studies perspective, then, how can we account for this cinematic turn in first person shooters? According to Pinch and Bijker (1987) the initial approach should be to identify social groups and the meanings they ascribe to artefacts, then identify problems and finally, identify how solutions to these problems led to changes in the technology. A more thorough study could identify numerous social groups and sub-groups, and numerous tensions and problems between these groups and first person shooters. However, the primary goal of this chapter is to present an approach to analysis rather than a thorough-going analysis itself, so to progress the discussion, I will identify three very broad groups: producers, players and anti-game advocates.

The making of games involves a number of different elements: programming and art production (development), marketing and finance (production), packaging, distribution and retail. While many of the early FPS games (like games more generally) were assembled by small developers with only a handful of staff and shoestring budgets, games are now big business. As noted above, a typical 'triple-A' title like *Modern Warfare 2* may cost millions of dollars to make, and involve dozens or hundreds of developers. While the failure of a title can lead to significant losses, the profit margins enjoyed by successful titles can be huge, making games a risky but potentially lucrative investment.

It is beyond the scope of this chapter to describe the structure of the game industry in any depth, and others such as Kerr (2006) have discussed this at some length. A number of commentators have also begun to examine the role that the structure of the industry plays in the development and shape of games (Hinton 2009; Dyer-Witheford and Sharman 2005; Kline, Dyer-Witheford and DePeuter 2003). The salient point in the context of an STS study of games is that the structure of the industry has a profound impact on the way games are made and the kinds of games that make it to the shelves. At the very least, as Kerr points out, the industry is largely controlled by a small number of large publishers, who enjoy a great deal of power over developers and retailers and who ultimately decide which games will make it to market and which won't.

Players are clearly those people who play first person shooters, although again, the term suggests an homogeneity that is not there. Before games moved into the mainstream, players or 'gamers' as they are often called, were simply people who played games on their computers. Because the total population of computer game players was small this served as a useful way to describe a group of people and their relationship with the technology. As games have become more popular new terms like 'casual gamer' and 'hardcore gamer' have appeared, used mainly by the industry to delineate different kinds of players. Casual gamers, as the name suggests, play games only briefly, perhaps one night a week, or on a mobile device on the train on the way to work, or while waiting in a queue. Hardcore gamers are people who spend a lot of time and/or money on their games

While players as consumers form an important part of the "circuit of culture" (du Gay et al. 1996) and feed back into the production of media devices through mechanisms like market research, the involvement of players in the development of games (and the FPS in particular) is even more direct. As Sotamaa puts it, games are "inherently 'configurative' and 'participatory'". (2007) The act of simply playing a game implies a certain level of active engagement that goes beyond the 'gaze' of the film viewer and enters what Chesher (2004) has identified as an immersive, sticky mode of viewing he calls the 'glaze'.

Beyond the participative modes of engagement that characterise game playing, some players' engagement with games sees them extending games by producing digital materials designed to enhance or change the game – a modification of the original game generally referred to as a game 'mod'. This kind of engagement has been encouraged by many games companies, and has seen the evolution of a mode of player participation that is hybrid of play and labour – something Kücklich (2005) calls 'playbour'. Kücklich makes a good argument that player-based production is fundamental to the games industry in increasing the shelf-life of games, acting as a recruitment portal for games companies, and aids innovation and the development of new game features. This is particularly true in the FPS genre, which Kücklich notes boasts both the first game mod (*smurfenstein*, a mod for *Wolfenstein 3D*) and the most successful mod (*Counterstrike*, a mod for Valve's first person shooter *Half-Life*). The impact of this kind of player engagement with games should be clear with respect to STS analyses of games: a subset of players actively influence not just sales and marketing of games, but the development of the genre.

The third group identified above can be loosely referred to as anti-game advocates, although perhaps by labelling the range of views this group encompasses suggests a greater homogeneity than it is meant to. Unlike players and producers, anti-game advocates are not part of the game industry, but criticise first person shooters for a variety of reasons, though most significantly because of the violence that first person shooters portray.

Traditionally the FPS genre has been closely linked to hardcore gaming, perhaps because mastery of games like *Quake* (especially in an online player versus player environment) require hours of play, and running the latest titles at their best requires investment in gaming technology that needs a level of technical knowledge that often goes well beyond the game itself. Whole communities and web sites have sprung up to support people who want to squeeze as much performance as possible from their computer systems. A computer system that has been customised to maximise game performance even has a name – a 'gaming rig'.

Since at least 1999 the publishers and developers of first person shooters began to realise that if the market was to continue to grow, it needed to expand its market beyond the traditional hardcore gamer demographic. As if to drive this point home harder, the response to the 1999 Columbine high school massacre pointed a long finger of blame at the game industry – particularly at makers of first person shooters, leading to a US Senate enquiry and lawsuits against developers such as Id.

Broadening the market for first person shooters was no small challenge. From its inception the first person shooter genre was irreverent and anti-social. David Kusher's book *Masters of Doom* recounts the development of Id Software and focuses large on the two key personalities in the company – John Romero and John Carmack – who are described as clever but socially challenged individuals. Romero was described in the book as "a fucked-up kid making fucked-up games to deal with the fucked-up physical and emotional violence he'd experienced when he was young." (Kushner 2003) Carmack was also depicted as a troubled youth for whom the computer was an escape from the world.

In response to growing negative publicity and the threat of regulation, the industry moved quickly to defend its markets. At E3 (a major games industry convention) in 1999, Doug Lowenstein of the Interactive Digital Software Association made the point that games were becoming a mainstream entertainment medium, and that the industry recognised they had responsibilities to the community. (Keegan 1999)

Players were also becoming increasingly diverse and no longer conformed to the hardcore gamer demographic that had defined much of the FPS market in the mid 1990s. A short article in *Esquire* magazine summarised this feeling amongst some players by suggesting that playing a first person shooter is "like being a Japanese air-traffic controller during a Godzilla movie – it's a panicky, bewildering experience where you mash buttons at random and run in circles while being blown up every eight seconds". (Stein 2008)

Attitudes to the hardcore gamer also reflect changing values within the games industry. When asked in an interview about how much influence hardcore gamers had on games, Laddie Ervin, a games marketer with more than 16 years industry experience responded negatively. He pointed out that "as games become even more expensive to create, there's going to be bigger bets placed on titles that appeal to a much wider audience because you can't target only the hardcore at that point or you'll go bankrupt". (Kuo 2006)

Developers, too, were looking to bring greater maturity to their games. According to one developer, Warren Spector, who produced the first person shooter *Deus Ex* (2000): "You have an obligation to show the consequences of choices [...] One of the biggest problems with games, especially more linear games, is they say 'kill everything that moves. Good player!'" (Sheffield 2007)

Rise of Photorealist Graphical Complexity

One of the key factors in the development of the cinematic qualities of the first person shooter is the development of the photorealistic qualities of game engine rendering, a goal that has been followed doggedly by many game developers – particularly those engaged in the development of first person shooters.

Early first person shooters like *Wolfenstein 3D* and *Doom* were not able to render full 3D environments, largely because the contemporary computer hardware was not fast enough. Because of the so-called raycasting method used to create the illusion of a 3D world, the player was unable to look up or down and had only limited movement in the up/down direction. While clever programming could produce the illusions that allowed players to walk up steps, jump or swim, it was not until 1996 with the release of *Quake* that real 3D, which gave the player full three-dimensional movement appeared.

Following *Quake* a number of 3D games were released with increasing visual complexity. Table 1 summarises some of the key developments and system requirements of first person shooters developed by Id Software. Between 1992 and 2004 the minimum system requirements changed dramatically.

Underlying the increasing photographic realism of games was a rapid development in graphics and computer hardware. Around 1995 hardware manufacturers began marketing video cards with dedicated circuitry designed to speed up the rendering of 3D graphics. These so-called 3D video accelerators allowed games programmers to cre-

Table 1. System requirements for Id Software's first person shooters

Game	Year	Memory	CPU	Graphics	Features
Wolfenstein 3d	1992	528K	386	VGA	Raycasting
Doom/Doom 2	1993/1994	8MB	486-66	VGA	Raycasting/height variation
Quake	1996	8/16MB	Pentium 75	VGA	Full 3D, pre-rendered lightmaps
Quake 2	1997	16MB	Pentium 90/133	open-gl support	
Quake 3	1999	64MB	Pentium 233	4/8MB (ha)[1]	spline curves, shaders
Doom 3	2004	384MB	Pentium IV 1.5Ghz	64MB (ha)	unified lighting, per-pixel lighting
Crysis Warhead[2]	2008	1GB	Pentium IV 2.8GHz	256MB (ha)	

ate more complex graphics because they freed up the computer's CPU from processing intensive graphics drawing operations. Because games needed to be written specifically to take advantage of video accelerators, programming interfaces were developed which hid the specifics of each graphic card's technical implementation. Using these interfaces (Glide, OpenGL and DirectX) greatly simplified the development of 3D games and allowed the development of games that could run on a wide variety of hardware.

First person shooter developers were quick to push the limits of successive generations of hardware, leading to a leap-frog process of development in both hardware and the graphical complexity of first person shooters. Gamers tinkered and updated their computers to squeeze as much performance from the hardware as possible, with a number of magazines using first person shooters as benchmark tests in comparing the performance of rival hardware vendors.

One review reflected on the perceived value of 3D acceleration in this way:

Many gamers describe playing Quake II in semi-religious terms, and when immersed in the 3Dfx version of the game, it's easy to understand why. [...] Of course, Quake II supports many graphics modes and 3D cards (it utilizes OpenGL), and players will surely find a combination that works well on their system (even if it's a now-ancient Pentium 150). But none approaches the 3Dfx version, and it is the only way through which one can fully appreciate the technological achievement of the game. (Broady 1997)

It is tempting to see a causal relationship between developing hardware capacity and the development of visual complexity. The cause of increasing visual complexity in games appears at face value to be a result of the increasing power of the hardware. The problem with this assumption is that it does not explain the predominance of photorealism (other forms of rendering are equally challenging), nor does it account for the two-way nature of this arrangement: it is just as valid to say that games drove the hardware as the other way around. There has been, however, an undeniable push towards graphical realism in first person shooters. As Henry Jenkins put it in his blog: "Game designers seem relentless in their push for more realistic graphics, often failing to explore other potentials within the medium. There is no reason why games should embrace photorealistic graphics just because they can." (Jenkins 2007)

The potential for games to move beyond a photorealist aesthetic is certainly there, although very few first person shooters have attempted to do this. Notable exceptions include *XIII* (2003) and *Red Steel 2* (2010), both of which use a cel-shading technique that creates a comic book visual style. There have also been independent experiments with alternate non-photorealistic rendering methods for *Quake*. (Mohr and Gleicher 2001)

An alternative explanation, as suggested above, is that underlying this push toward photorealism in first person shooter graphics is a more fundamental imperative for games to become more cinematic so that they appeal to a broader market of players for whom the cinema is already a well established and accepted visual medium. It is social choice, not technical requirement, that is behind the rise of photorealism in first person shooters.

Rise of Narrative

In addition to the increasing visual complexity of games, the importance of narrative has also increased. Narrative in games is an issue of some debate, not only among game developers, but among academic commentators, too[3].

Early games in the first person shooter genre did not engage narrative to any great extent. What narrative there was provided a back-story that served as the most basic explanation to support game play. John Carmack, one of the key people behind *Doom* and *Quake* said this about what he saw as the role of narrative in his first person shooters in the lead up to the release of *Quake 3 Arena*:

We're still all about the experience of the combat in that way, yes, it is very much like the fighting games where there is some back story, but who the hell cares? The game has to be fun as you're doing your actions [...] that's what has to be fun – not the fact that you are progressing through a story to some end goal. [...] Yes, there is a little back-story, but that's not the point. The point is that the game is fun to play. (Sones 1999)

Carmack did not dismiss the value of narrative in games, but stressed the importance of fun and play over narrative in his games. More recently, this approach has been challenged, even seen as old-fashioned and out of step with the modern first person shooter market. According to one commentator: "(id) has all these great franchises that may seem outdated to the current player. I hope we see them all revitalized with today's better graphics, awesome new game engines and amazing advancements in game play." (Travis Moses from GamePro magazine, cited in Snider 2009)

For game developers the debate about the role of narrative in games seems to be based on a tension between two differing views of the primary purpose of games. For some, like John Carmack, the value of games is in the fun and play aspect – a game devoid of story can still be fun and engaging. Early first person shooters like *Doom* and *Quake* exemplify this approach. In these games the emphasis is on the game play – the honing of fine motor skills between keyboard and mouse and the refinement of the twitch-click response. This kind of game play is more akin to playing a sport than watching a movie, and, especially in online player-versus player environment, rewarded players who spent time practicing and learning the idiosyncrasies of the game engine.

Half-life (1996) was seen by many in the industry as a turning point for the first person shooter, and the key factor was not the game's technical sophistication but its use of narrative. One industry commentator remarked that *Half-life* was "an evolutionary jump among first-person shooters, moving the genre away from blast-happy gib heaps to a more mature game style." (Carless 2003) This more mature style was largely about the game's narrative. The game puts the player into the first-person perspective of Gordon Freeman, a scientist who has just started work at the Black Mesa research laboratory. An accident occurs and the player finds themself swept up by a narrative – in the confusion it's not clear what is going on and a mystery begins to take form. While the hand-eye coordination of *Quake* still plays an important role in the playing of the game, progression in the game becomes about more than simply shooting enemies and getting to the final encounter – it's now largely about following the story and finding out how it ends.

Debate about the role of narrative has shifted more recently; that is, the problem of narrative has been redefined. The question of whether narrative plays a role in game development has become a discussion about the best ways for narrative to be integrated into gameplay, and ways for narrative to become more interactive and immersive. (see for example, Remo and Sheffield 2009) What's interesting about this turn in the debate is that it signals the resolution of the fundamental question about the role of narrative in the first person shooter. The question is not *whether* narrative should have a role, but *how* narrative should be constructed. This signals a resolution and stablisation of the role of narrative in the first person shooter and pushes the first person shooter into a more cinematic form.

CONCLUSION

Science and technology studies was developed as a method of analysing the social construction of technologies. Games, as technologies, are valid targets for STS analysis. As games continue to grow and become a mainstream entertainment medium, understanding the social factors driving the technological development of games will become increasingly important. Indeed, as Manovich has pointed out, studies that cross the science/humanities divide and that are sympathetic and informed by both are increasingly important in a world that is driven by software.

In this chapter STS has been applied to the analysis of one part of the game industry and has focussed mainly on what has been identified here as the cinematic turn in first person shooter game development. Rather than suggesting this cinematic turn is purely the result of the development of more sophisticated computer hardware, the analysis suggests that the growing size and diversity of the game-playing demographic coupled with demands of publishers to reach larger markets drove a push towards the realisation of a more broadly accepted format.

Publishers and developers of first person shooters had a number of problems they needed to address which related to the way that first person shooters were constructed by some groups (for example, anti-gamers, hardcore gamers, developers), and the way that other groups (publishers, casual gamers) understood games. Debates about the nature of first person shooters and their players kept the meaning of the FPS in a state of flux that has been somewhat stabilised as narrative and photorealist rendering techniques have established the FPS in a more cinematic form. The response to these challenges was to bring games more into line with another older and more established (and thus socially acceptable) media industry: the cinema. To do this, games needed to become more cinematic and less technical – they needed more compelling narratives and more cinematic graphics.

There are, of course other factors not explored here. For example, the thesis presented here does not look at the way that game developers' interactions with their creative environment influences the kinds of games they create. This provides some insight into the desire amongst developers for the development of realistic graphics prior to the growth of the games industry. As noted above, Casey O'Donnell has engaged in just such an analysis. (O'Donnell 2009) Although noted, this chapter has not engaged to any significant degree with the effects of player-led development, nor has it engaged significantly with the effects of the structure of the industry upon game development. I raise these issues here to point out the richness of the STS approach. No doubt there will be an increasing number of studies that follow these kinds of STS-informed approaches and methods.

REFERENCES

Bijker, W. E., & Pinch, T. J. (Eds.). (1987). *The Social construction of technological systems: new directions in the sociology and history of technology*. Cambridge, MA: MIT Press.

Brey, P. (1997). Philosophy of Technology Meets Social Constructivism. *Society for Philosophy and Tehnology, 2*(3-4). Retrieved December 3, 2009, from http://scholar.lib.vt.edu/ejournals/SPT/v2_n3n4html/brey.html

Broady, V. (1997). Quake II Review for PC. *GameSpot*. Retrieved December 3, 2009, from http://au.gamespot.com/pc/a ction/quake2/review.html

Carless, S. (2003). Marc Laidlaw On Story And Narrative In Half-Life. *Gamasutra*. Retrieved December 3, 2009, from http://www.gamasutra.com/fea tures/20030808/carless_01.shtml

Chesher, Chris (2004). Neither gaze nor glance, but glaze: relating to console game screens. *SCAN Journal, 1*(1).

Dyer-Witheford, N., & Sharman, Z. (2005). The Political Economy of Canada's Video and Computer Game Industry. *Canadian Journal of Communication, 30*(2).

Frasca, G. (2003). Ludologists love stories, too: notes from a debate that never took place. In *Level Up Conference Proceedings: Proceedings of the 2003 Digital Games Research Association Conference* (pp. 92-99). Utrecht: University of Utrecht. Retrieved from http://www.digra.org/dl/display_ html?chid=05163.01125.pdf

Fritz, B. (2009). Call of Duty: Modern Warfare 2 video game gets Hollywood-scale launch. *LA Times*. Retrieved December 3, 2009, from http://www.latimes.com/business/la-fi-ct-duty 18-2009nov18,0,5238209.story

Hinton, S. (2009). Gaming Nation: The Australian Game Development Industry. In Hjorth, L., & Chen, D. (Eds.), *Gaming Cultures and Place in Asia Pacific*. London: Routledge.

Jenkins, H. (2007). Applied Game Theory, R.I.P. 1: Melodrama and Realism. *Confessions of an Aca/Fan*. Retrieved December 3, 2009, from http://www.henryjenkins.org/2007/ 04/applied_game_theory_rip.html

Jenkins, H. (2009). A New "Platform" for Games Research?: An Interview with Ian Bogost and Nick Montfort. *Confessions of an Aca/Fan*. Retrieved December 3, 2009, from http://www.henryjenkins.org/book_shelf/

Keegan, P. (1999). *Culture Quake*. Retrieved December 3, 2009, from http://www.motherjones.com/ politics/1999/11/culture-quake

Kerr, A. (2006). *The Business and Culture of Digital Games*. London: Sage.

Kline, S., Dyer-Witheford, N., & de Peuter, G. (2003). *Digital play: the interaction of technology, culture, and marketing*. Montréal, Canada: McGill-Queen's University Press.

Kücklich, J. (2005). Precarious Playbour: Modders and the Digital Games Industry. *Fibreculture Journal, 5*.

Kuhn, T. S. (1962). *The Structure of Scientific Revolutions*. Chicago: University of Chicago Press.

Kuo, L. C. (2006, April 14). Do "Hardcore Gamers" Still Matter? *GameSpy*. Retrieved December 3, 2009, from http://au.gamespy.com/articles/701/701787p1.html

Kushner, D. (2003). *Masters of Doom: How Two Guys Created an Empire and Transformed Pop Culture* (1st ed.). New York: Random House.

Malaby, T. M. (2007). Beyond Play: A New Approach to Games. *Games and Culture, 2*(2), 95–113. doi:10.1177/1555412007299434

Manovich, L. (2008). *Software Takes Control*. Retrieved April 19, from http://softwarestudies.com/softbook/

Mohr, A., & Gleicher, M. (2001). Non-invasive, interactive, stylized rendering. In *Proceedings of the 2001 symposium on Interactive 3D graphics* (pp. 175-178). New York: ACM.

Montfort, N., & Bogost, I. (2009). *Racing the Beam: The Atari Video Computer System*. Cambridge, MA: MIT Press.

O'Donnell, C. (2009, March 15). The everyday lives of video game developers: Experimentally understanding underlying systems/structures. *Text.Serial.Journal*. Retrieved December 3, 2009, from http://journal.transformativeworks.org / index.php/twc/article/viewArticle/73

Remo, C., & Sheffield, B. (2008, July 18). Redefining Game Narrative: Ubisoft's Patrick Redding On Far Cry 2. *Gamasutra*. Retrieved December 4, 2009, from http://www.gamasutra.com/view/feature/3727/redefining_game_narrative_.php

Roth, S. D. (1982). Ray casting for modeling solids. *Computer Graphics and Image Processing, 18*(2), 109–144. doi:10.1016/0146-664X(82)90169-1

Schwartz-Cowan, R. (1985). How the refrigerator got its hum. In Wajcman, J., & MacKenzie, D. (Eds.), *The Social Shaping of Technology*. Buckingham, UK: Open University Press.

Sheffield, B. (2007, March 5). All For Games: An Interview with Warren Spector. *Gamasutra*. Retrieved December 3, 2009, from http://www.gamasutra.com/featur es/20070305/sheffield_01.shtml

Snider, M. (2009, August 18). Back to the future for id Software, 'Wolfenstein'. *USA Today*. Retrieved December 3, 2009, from http://www.usatoday.com/life/lifestyle/ 2009-08-17-wolfenstein_N.htm

Sones, B. E. (1999, June 30). *Interview with John Carmack*. Retrieved December 3, 2009, from http://web.archive.org/web/20030515172424/http://www.cdmag.com/articles/ 021/004/jc_interview.html

Sotamaa, O. (2007). Let Me Take You To the Movies. *Convergence, 13*(4), 383–401.

Stein, S. (2008). Six Reasons Why First-Person Shooters. *Esquire*. Retrieved December 3, 2009, from http://www.esquire.com/the-side/opinion/six-reasons-why-shooters-suck

Wajcman, J., & MacKenzie, D. (Eds.). (1985). *The Social Shaping of Technology*. Buckingham, UK: Open University Press.

ADDITIONAL READING

Bogost, I. (2006). *Unit Operations: An Approach to Videogame Criticism*. Cambridge, MA: The MIT Press.

Cowan, R. S. (1983). *More Work For Mother: The Ironies of Household Technology From the Open Hearth to the Microwave*. New York, NY: Basic Books.

Huizinga, J. (1971). *Homo Ludens: A Study of the Play-Element in Culture*. Boston, MA: Beacon Press.

Jasanoff, S., Markle, G., Petersen, J., & Pinch, T. (Eds.). (1994). *Handbook of Science and Technology Studies*. Thousand Oaks, CA: Sage.

Juul, J. (2001) "Games Telling Stories? A Brief Note on Games and Narratives." *Game Studies*. 1.1

Kline, S. (2003). *Digital Play: The Interaction of Technology, Culture, and Marketing*. Montreal, CA: McGill-Queen's University Press.

Latour, B. (1996). *Aramis, or the Love of Technology*. Cambridge, Mass: MIT Press.

Latour, B. (2005). *Reassembling the Social: An Introduction to Actor-Network Theory*. United Kingdom: Oxford University Press.

Latour, B., & Woolgar, S. (1986). *Laboratory Life: The Construction of Scientific Facts*. Princeton, NJ: Princeton University Press.

Law, J. (Ed.). (1991). *A Sociology of Monsters: Essays on Power, Technology and Domination. Sociological Review Monograph*. London: Routledge.

Poole, S. (2000). *Trigger Happy: VideoGames and the Entertainment Revolution*. New York, NY: Arcade Press.

Williams, R., & Edge, D. (1996). The Social Shaping of Technology. *Research Policy, 25*, 856–899. doi:10.1016/0048-7333(96)00885-2

Wolf, M. J. P., & Perron, B. (2003). Introduction. In *The Video Game Theory Reader*. New York, NY: Routledge.

ENDNOTES

1 (ha)–hardware accelerated graphics required to play the game

2 *Crysis Warhead* is not an Id game, but is shown here to show the continued trajectory of game development

3 It is beyond the scope of this chapter to recount the academic debate. For a summary, see: (Frasca, 2003)

Chapter 20
Playing as Producing: Convergence Culture and Localization of EA Digital Games in Taiwan

Ying-Chia H. Lin
Fu-Jen Catholic University, Taiwan

ABSTRACT

The main goal of this chapter is to explore the dynamics and interactions between foreign producers, media technologies, and local consumers in the process of globalization through a discussion of the localization of digital games. As digital games have become global entertainment products, the game industry has encountered various lingual and cultural challenges in the process. The content of digital games can be changed more easily and many game players are able to create objects or modify the game world to express their programming abilities or identities. The game industry takes advantage of digital technologies and player modding practices to overcome language and cultural barriers. Modding and other localization practices, such as translation and local packaging, imply the convergence of the global and the local cultures (i.e. glocalization) as well as the producers and consumers (i.e. prosumers).

INTRODUCTION

The global market for digital games has expanded rapidly in the last a few decades. As digital games have become global entertainment products, the game industry has encountered various lingual and cultural challenges. Yet, unlike some other cultural industries, the game industry has used localization as an important tactic in penetrating international markets. Game designers adopt different degrees of localization practices, and some have exceeded the conventional practices of translation and packaging. These new localization practices represent a new trend toward converging local and global cultures. Thus, localization is not only an industrial practice, but also an important cultural practice worthy of further research and analysis.

Digital games differ from traditional entertainment products, such as films and TV programs, in that it is easy to change their content using digital media technology (Manovich, 2000). More and

DOI: 10.4018/978-1-60960-037-2.ch020

Copyright © 2011, IGI Global. Copying or distributing in print or electronic forms without written permission of IGI Global is prohibited.

more game players are able to create or change games to increase their enjoyment, express their programming abilities, or forge their own identities. Though scholars have explored the media practices of fans and regular digital media consumers (Jenkins, 2006; Kücklich, 2005; Postigo, 2007), few of them have extended the discussion to the realm of localization or examined cross-cultural creation and consumption in participatory and convergence culture. The purpose of this chapter is to initiate a discussion about global digital game localization through the empirical study of the local strategies or localization practices of American digital game companies in Taiwan. The biggest game company in the US, Electronic Arts (EA), is the subject of this in-depth analysis. EA established its Taipei office in 1996, and has since published and localized many of its game in Taiwan. It successfully attracted local players and built a fan community. Studying the localization practices of a world-leading game developer and publisher such as EA provides several insights into the dynamics of the game industry, digital technologies, and game players involved in the process of global circulation of digital games.

BACKGROUND AND LITERATURE REVIEW

Theorizing "Localization": Localization and Global Mass Culture

The discussion of "localization" usually accompanies the greater discourse of globalization. Scholars view localization as either a part of, or a reaction to, globalization. There are three common approaches to the dynamic between the local and the global, and the overall purpose of localization: 1. Localization maintains cultural identity and resists globalization. 2. Localization is the domestication of global culture, resulting in cultural hybridization. 3. Localization is a means through which global capitalism expands.

The first of these approaches considers the local and the global to be opposite and incompatible. The local represents the *authentic*, *heterogeneous*, and *grassroots*. In contrast, the global is *commercial*, *homogeneous*, and *rootless* (Tomlinson, 1997). From this viewpoint, the global expansion of multinational cultural industry has a negative influence on local cultural production because it dominates local markets, impairs local cultural productivity, and weakens the public sphere (Schiller, 1998). To resist the threat of the global culture, local people must rediscover and revive (lost) local cultures and traditions, and reconstruct their local identities. Many indigenous cultural movements are based on this approach, including the Ainu (in Japan) and the Hawaiian cultural movement (Friedman, 1994). Some criticize the idea that globalization is a homogeneous process as an oversimplification of the dynamic cultural exchange between the local and the global. First, whether or not the local is really *authentic* and *uncontaminated* remains questionable. The idea that local cultures are authentic is unrealistic, goes the argument; cultures are always hybridized (Morris, 2002). Therefore, any claim of authenticity is only a discursive act. Furthermore, in spite of some advantages that foreign cultural products have, they cannot dominate all local markets (Chadha & Kavoori, 2000).

To some, localization represents local people partly adopting, integrating, and converging their local culture and foreign cultures. The result of this form of localization is a hybridized culture. In this case, the local culture does not disappear but finds a new space and a new way to grow and survive in the process. Recent studies in cultural hybridization explore not only cultural hybridity, but also the identity formation reflected in this process of hybridization (McMillin, 2001; Kraidy, 2002). However, some criticize the cultural hybridity theory for its lack of political position or further contextual analysis. Their main concerns include the following: What are the power relations in the process of hybridization? Who has the power to decide which cultural elements to keep? Does cultural hybridization entail resistance and subversion to colonial or global power? How are cultural identities negotiated in the process? Quoting Nederveen Pieterse, Tomlinson (1999) warned "Hence, hybridity raises the question of the *terms* of mixture, the conditions of mixing and mélange. At the same time it is important to note the ways in which hegemony is not merely reproduced but *refigured* in the process of hybridization" (p. 147). In other words, hybridity is not neutral—the historical context in which hybridization takes place, the power structure between countries, the imbalance of cultural flows, and the hegemony of western capitalism all are important factors. Hybridity is ambivalent. It can be what Bhabha (1994) called *third-space*, which is full of potential for subversion. It can be the hegemonic practice that erases cultural differences and tames the *Other*. Therefore, when hybridization and localization take place, it is necessary to recognize the major agents involved and ask how different forces interact in the process.

The two types of localization above assume that the agents of localization are local audiences or cultural consumers. However, transnational corporations and producers engage in another form of localization. This type of localization, called *commercial localization*, is not an indigenous response to global or imported cultural products, but an organized practice designed to maximize commercial benefits in specific regions and countries. While the emphasis of localism is to resist global or foreign power, the goal of commercial localization practice is to overcome that local resistance by controlling the differences between the local and the global.

The expansion of global mass culture in the late 20th century encountered complex and decentered local forces ranging from language to revived nationalism. Hall (1997) asserted that capitalism and global mass culture are still western-centered, but not centered in the same way: "in order to maintain its global position, capital has had to negotiate.... to incorporate and partly reflect the differences it was trying to overcome" (p. 32). Therefore, the global capitalism no longer simply destroys the differences of cross-national borders, but also accommodates and absorbs them. Capital "works in and through specificity" (p. 29). However, the main purpose of the negotiation and accommodation is not to reflect the values and lives of the local, but to maintain the interests of foreign capitalists in various historical and cultural contexts. In this sense, commercial localization is actually working under the logic of capitalism, creating *global localism* (Dirlik, 1996) or *glocalization* (Robertson, 1995).

Dirlik (1996) stated that global localism is the result of global capitalism that "represents a further deterritorialization, abstraction, and concentration of capital,... global capitalism represents an unprecedented penetration of local society globally by the economy and culture of capital" (p. 28). Transnational corporations do not only enter national and local markets to sell their products; many of them move their production to the local if the local can provide the necessary technology and manpower. However, the power of management and decision-making remains with the global. Robertson (1995) believed there were a variety of modes for practicing global localization, or *glocalization*. Economic glocalization represents "the tailoring and advertising of goods and services on a global or new-global basis to increasingly differentiated local and particular markets" (p. 28). Furthermore, economic glocalization "involves the construction of increasingly differentiated consumers, the 'inventions' of 'consumer tradition'." (p. 29). Market research plays an important role in localizing imported cultural products. Maxwell (1997) noted that market research continuously extracts information and meaning from popular culture and everyday life to promote new ways of consumption. Using surveys and other methods of information gathering, multinational corporations invent identities and cultivate desires that articulate their products and services. Hence, the "knowledge of local value assessments and how they differ inside and outside national territories is a geo-strategic asset" (p. 197). Similarly, Dirlik (1996) commented that the purpose of global capitalism was "...to reconstitute subjectivities across national boundaries to create producers and consumers more responsive to the operations of capital" (p. 32).

The three approaches above are different in terms of the goals and purposes of localization, the agent of localization, and the impact of localization on local subjects. The *resistance* model essentializes the local and simplifies the dynamic between the local and the global. This approach also reduces the cultural exchanges caused by consumption of imported cultural products to one-way transmissions. The *cultural hybridization* model focuses on the result of the cultural exchange and localization, and provides a more detailed view of how the local and the global interpenetrate and inform each other. However, empirical research on cultural hybridization must be placed in the social and historical context of western colonization and global capitalism to provide a more critical understanding of how localization actually occurs. Like the cultural hybridization model, *commercial localization* usually creates a form of hybridization. However, we should pay attention not only to how glocalization changes or hybridizes cultural content, but also how global capitalism and local agents interact, compete, and negotiate.

Commercial Localization of Digital Games

This chapter focuses on the commercial localization of digital games: how it is practiced, and its implications in cultural convergence and globalization. As discussed earlier, the main purpose of commercial localization is to sell products globally. Thus, the cost, means, and effects of localization are the main concerns of commercial localizers. *The Localization Industry Standards Association* defines localization as the process of "taking a product and making it linguistically and culturally appropriate to the target locale (country/region and language) where it will be used and sold" (LISA, 2006, Section I, para. 4). According to *The Game Localization Handbook* (Chandler, 2005), there are different degrees of localizing digital games—no localization, packaging and manual localization, partial localization, and full localization. However, her definition of localization is "the process of translating the game into other languages" (p. 12). In this context, localization depends on the extent of translation for in-game text and packaging and the use of voiceover dubbing. The difference between partial localization and full localization is that full localization translates everything—text, manual, voiceover and packaging— while partial localization leaves the voiceover untranslated. Game content, representations, and local preferences are not the major concern of the game industry unless these components act against local regulations or seriously violate local social customs. The localization of digital games is more of a technical job than creating games for local needs. More importantly, decisions on what, or how, to localize are based on "...financial viability. To do this (making the decisions), the publisher evaluates the time, money, and resources spent on the localization development versus the expected return on the investment in the game" (p. 46).

In Thayer and Kolko's (2004) comparison of commercial localization of digital games and productivity software, they pointed out that some game components, such as narrative and characters, are culture-specific. Therefore, localization does not always take place only on the linguistic level. In this case, game designers change other elements of the game, such as the characters, storyline, graphic user interface (GUI), and difficulty levels, according to local preferences—a process called *blending*. Successful localization or blending efforts can create positive returns for game companies in the long term, but it is easier and cheaper to skip the blending process in the short term. Thus, from an industrial perspective, the localization of digital games primarily consists of the translation and altering of game content. However, this definition does not cover all the translocal strategies that game companies implement. The practice of localization is not limited to the production stage, but extends to the consuming stage.

Digital Games: Media Characteristics and Localization

Any form of media localization practices, whether they be bottom-up, indigenous response, or top-down commercial practices, are the result of the interaction between the media, foreign producers, and local consumers. As a form of computerized media, digital games share many cultural characteristics with other media, but are unique in their relations to consumers. Digital games have visual representation, stories and narratives, music and sound effects like in other audio-visual media. Yet, because of computer algorithms, it is possible to change digital games partly and with less effort and cost (Manovich, 2000). Furthermore, players are able to interact with the computer and the game world.

Although *interactivity* is a broad and slippery term, it is probably the most commonly used term to describe the characteristic of computerized media. The concept of interactivity simply refers to what the users or players can do to and in the mediated environment or the game world. Steuer (1992) defined interactivity as "the extent to which users can participate in modifying the form and content of mediated environment in real time" (p. 86). Ryan (1994) also thought that real interactivity is not merely the ability to navigate a virtual world, but the power of the user to modify the environment. In a truly interactive system, the mediated environment must respond to the user's actions and change accordingly.

Instead of *interactivity*, Raessens (2005) used *participation* to characterize digital games. According to him, participation consists of three domains: interpretation, reconfiguration, and construction. The interpretation of digital games is not much different from the interpretation of a traditional text, reading a book, or watching a movie. Reconfiguration involves navigating through the game space, *interacting* with the game environment. By choosing different paths and the objects in the game, the player is able to create and change the narrative and a body of information. Construction in digital games refers to the addition of new elements and modification of existing elements by creating patches hacking into the game system, or using codes. Construction extends the degree of player participation and involvement in the game environment by giving players the chance to construct and reconstruct the digital game content. Construction typically requires much more knowledge and skill than a regular gamer has, and is usually a form of resistance. Nevertheless, more and more games have a preprogrammed hacker system or provide patches that allow gamers to customize their gameplay. This creates ambivalence between real resistance and controlled marketing strategy.

Interactivity (or *participation*, in Raessens's term) changes the relationships between the digital games producers and consumers. Interactivity releases the power of content production to the consumers. The traditional media industry places many limitations on what consumers can do to the media content, mostly due to technical and legal reasons. Due to the technological possibility and characteristics of digital games, there is more room for consumers to change or localize games to accommodate their needs.

CASE STUDY: LOCALIZAITION PRACTICED BY ELECTRIONIC ARTS IN TAIWAN

Localization of Content and Packaging

Electronic Arts established the *EA Taiwan* (Taipei) office in 1996, forming one of the first EA offices in the Asian-Pacific region. EA was also one the first foreign game companies to set up local offices to market and distribute their own games instead of hiring a local, third party distributor to do the job. However, in its 10 years in Taiwan, *EA Taiwan* has only released and marketed games developed by its major studios; it has never developed any games locally. This case study selects several top selling EA games for the following localization analysis: *The Sims* and *SimCity* (simulation games), *Need for Speed: Most Wanted* (racing), *MVP Baseball, 2005* (sports simulation), and *Battlefield: Vietnam* (first-person shooter). The Chinese versions of these games were purchased at the EA flagship store in Taipei between July and November, 2005. The English versions of these games were purchased in retail stores in Washington, USA, and BC, Canada during the same period.

Table 1 lists the overall summary of changes in the localized version of each game. This list includes changes not only *within* the games, but also elements *affiliated* with the games, such as the packaging and manuals. Most of the in-game elements, such as characters, rules, narrative (if there is any), objects, graphics, and symbols, remain the same in the Taiwanese localized version. The only major changes are linguistic—from English to Chinese—and the degree of translation varies. Most games focus on the textual part and leave the soundtrack and voiceover content the same. In other words, the virtual world remains basically unchanged. However, there are minor changes that bring hints of locality. For example, in the *SimCity: Formosa,* the mayor's name changes to "Ma, Xiao-Ju" (馬小九), which resembles the name of the former mayor of Taipei, "Ma, Ying-Ju" (馬英九). All games allow the player to name his or her characters. However, the interface only allows the player to type in Roman characters, not Chinese characters. The most obvious changes are in the packaging. EA Taiwan has rewritten the box texts, and applied different pictures and design. There are also differences between the English and Chinese user manuals. The translated and edited Chinese user's manual is more compact size and has fewer game instructions, while the English manual is full of examples and detailed instructions. The titles of the games are another important element designed to appeal to local consumers. All the localized versions have their own Chinese titles. These titles are partly based on the English title and partly on how the titles sound and are read in the social context. For example, the Chinese title of *MVP Baseball 2005* is *Meiguo Zheban Dailiangmon 2005* (美國職棒大聯盟2005), which simply means "*American Major League 2005*". The meaning of *MVP* in the original title is lost in translation.

The sense of locality is relatively weak in these localized games. Instead, these games provide exotic, westernized (or North Americanized) virtual spaces based on the lifestyles of North Americans. *The Sims* is a simulation of North American daily life: the buildings and daily items in the game include single detached houses, backyard BBQ stoves, and private swimming pools, which are very different from what most Taiwanese consumers live in and use. In *Need for Speed: Most Wanted,* the scenery, car models, and characters are all set in North America. *MVP Baseball, 2005* is a simulation of American big league baseball. In *Battlefield: Vietnam* the players are American soldiers by default. Even *The SimCity*: *Formosa* is not based on the landscape of Taiwan—there are only a few famous Taiwanese landmarks, such as Taipei City Hall. In spite of the foreignness, these games are still popular with local players. Thus, the issue is not whether

Table 1. EA localization check list

	The Sims	*SimCity: Formosa*	*Need for Speed: Most Wanted*	*MVP Baseball, 2005*	*Battlefield: Vietnam*
User's interface (GUI)	translated into Chinese	translated into Chinese	translated into Chinese	translated into Chinese	translated into Chinese
Graphics, objects, and symbols	no changes but the player can create his or her own objects	no major changes, with addition of localized objects	no changes, no localized objects	no changes, no localized objects	no changes, no localized objects
Rules	no changes	no changes	no changes	no changes	no changes
Difficulty level	no difficulty level	no difficulty level	no significant changes	no significant changes	no significant changes
Voiceover	no v.o.	no v.o.	original v.o.	original v.o.	original v.o.
Storyline or narrative	no major storyline	no major storyline	no changes	no major storyline	no changes
Main characters	the player can create his or her own character	in-game characters have localized Chinese names; the player can also create his or her own character	original characters, but the player can name his or her character	original characters, but the player can name his or her character	original characters, but the player can name his or her character
Title of the game	Chinese title	Chinese title	Chinese title	Chinese title	Chinese title
User's manual	Redesigned Chinese manual	Redesigned Chinese manual	Redesigned Chinese manual	Redesigned Chinese manual	Redesigned Chinese manual
Packaging	repackaged and new design	repackaged and new design	the design is similar to the original	similar to the original design but the American player is replaced by a Taiwanese player	repackaged and new design
Box texts	translated and some new text added	translated and some new text added	translated and some new text added	translated and some new text added	translated and some new text added

local Taiwanese players find a sense of locality in the games, but whether the games make sense to the players.

The penetration of American culture power and the globalization process in the past several decades has already created a *hybridized* culture that integrates many western values and ideas into the Taiwanese culture. Hollywood movies, TV programs, and advertising already broadcast the consumerism in *Sims,* sports competition in *MVP Baseball, 2005;* violence, crime, and car racing in *Need for Speed: Most Wanted;* and military and masculinity in *Battlefield: Vietnam* to Taiwan. American cultural power is so pervasive that most people do not usually feel its existence. The *westernness* or *Americanness* of the games may be visible, but in most cases it does not prevent local players from playing the games. Sometimes, this sense of westernness or Americanness even creates a sense of superiority for the products. As Kohler (2004) observed about the use of English in Japanese society: "English is used to convey a sense of foreignness, but one that is cool and chick" (p. 211). Taiwanese advertising commonly uses English, Japanese, and some western European languages to create a similar feeling and appeal (Raymond *et al.*, 1999).

With the advent of digital technology, it is easier to alter digital media than traditional media content. Nevertheless, most localized games are not so different from their original English forms. This reflects the power imbalance between different markets. With a population of 22 million, Taiwan is a relatively small market compared to larger markets like the US and Japan. Because Taiwanese regulations on game content are quite relaxed, game producers do not have to change the content to accommodate them. Under these circumstances, EA has little incentive to localize games beyond the minimal efforts of repackaging and translation. Besides, local gamers already know how to play "unlocalized" games due to the American culture influence.

The degree of localization for foreign games in the US is generally greater than that for American games in Taiwan. It is not uncommon for Japanese digital game publishers to change the title, difficulty level, graphics, and even characters to accommodate the preferences of American players. *Street Combat* is an extreme example. The original main characters in this game, based on Japanese comics, were all altered in the American version: they changed from comic-style characters to "more mature" characters[1] (Kohler, 2004). In this case, the incentive and degree of alteration depend not on technological limitations, but on the market size, cost and returns, and the consumer's consumption behaviors, which are conditioned by the contextual relations between two countries. Meanwhile, although foreign game companies in Taiwan don't extensively change the game content and appearance, consumers are able to express their own version of locality.

Game Modding and Localization

At the most basic level, digital games consist of digital bits and bytes. By adding, changing, and reproducing these bits and bytes and the underlining codes, anyone with sufficient programming knowledge can alter or create elements that influence the content, gameplay, and presentation of the game. This changes the relationship between cultural consumers and producers, and effectively blurs the lines between producers and consumers. In other words, the medium of digital games provides an opportunity for potential *player-authorship* (Poremba, 2003) and *prosumer* behavior (producer-consumer) (Herman *et al.*, 2006; Lister *et al.*, 2003; Toffler, 1980).

Herman, Coombe, & Kaye (2006) pointed to at least four practices that use the medium of digital gaming technologies. The first is the creation of artifacts, avatars, or spaces within the game using digital resources provided by the game space as it has cumulatively evolved. The second practice is the remediation of other media forms into digital games. The third practice is the modification, or *modding,* of existing games using software design engines and the underling source code provided by game developers to create entirely new games. Finally, the activity of *meta-gaming* involves the creation and dissemination of mods, patches, and gaming advice through the network interactivity of virtual gaming communities (p. 194).

The types of game mods vary. *Add-ons*, *plug-ins*, *skins,* and *patches* are the most common mods. The more extreme sort of mods are total conversion ones in which the player uses the original game engine to make a completely new game, for example, converting a first person shooter set in a modern-day military environment to a sword-fighting game set in feudal Japan (Postigo, 2003). Many mods are available on the Internet, circulated by individual hobbyists and amateur game designers. Sometimes the original game developers sell expansion packs or patches in stores or make them available for download on the Internet.

Game developers have various attitudes towards mods and the dynamics created by these participatory cultures. Some companies emphasize their copyrights on these games and discourage mods and patches, while an increasing number of companies try to incorporate these changes and capitalize on them. The most famous example may be the creation of a series of first-person shooters from two games, *Doom* and *Quake*. These games allow players or hobbyist programmers to modify the games' codes to change details and scenery. Although consumers have the chance to alter the game elements, the game companies or developers control the ownership of the programs, the game engine, and all codes under current copyright laws. For example, *Valve*, the company that created *Half-Life*, retains the intellectual property rights of all mods created using the software development kit (SDK) shipped with the game by providing the modders with rights to use the source-code. In other words, the modders can play with the codes, but do not own the rights of their creation (Kücklich, 2005). Moreover, game mods still need the original games to work.

Digital technology calls attention to the instability of the content of digital media and the significance of this instability in the international context. If the modding potential embedded in digital games allows local consumers to customize and localize foreign games to fit their needs, then why do foreign developers need to localize the games in the first place? Two important reasons are the control of ownership and the supply of stable and quality labor. The quality of the localization practiced by the local prosumers may not be acceptable to the original game developers. More importantly, the purpose of commercial localization is to attract the maximum local consumers and generate the maximum benefits with minimum efforts. Game companies must be able to control which games are localized, and when and where these games are released. However, with the emergence of the *modder culture* and a *modder-friendly* structure, local players have increased localization or customization much more than game companies. Local prosumers provide all types of mods that express their needs, preferences, and identities.

Baseball is the most popular professional sport in Taiwan. There is a professional baseball league in Taiwan and several Taiwanese players have joined Japan's professional league and the American major and minor leagues. Although the teams, players, and baseball fields in *MVP Baseball, 2005* are all American, local gamers have modified the digital game to include Taiwanese players and teams. The result is a hybridized, blended content that mixes Chinese and English text. The new localized version retains the names and symbols of the original teams, but changes the players. The original Chinese version by EA does not translate or subtitle the background announcement in *MVP Baseball, 2005*. Now, however, Chinese soundtracks (mods) are available for download on the Internet. These localization mods provide players with a gaming experience that is more familiar than the original *localized* version provided by EA.

In addition to the *MVP Baseball 2005*, EA games such as *Sims* and the *Battlefield* franchise have a large modding community. The *Sims* franchise allows players to design their own objects, furniture, buildings, and characters. Players have created Chinese-style furniture and costumes on the Internet. On the website of Taiwan's *Battlefield* modding community, bfx.tw, players can find various mods to change the soundtrack, add new special effects, maps, weapons, characters, and backgrounds. Because of the technological characteristics of digital games, the semiotic domain of the games is not fixed, but fluid.

The attitude of the EA Taipei office towards these mods is positive and encouraging. On EA's official Taiwanese site for *Battlefield 2*, there is a direct hyperlink to bfx.tw under the category of *recommended community sites*. The Chinese user's manual of *Battlefield: Vietnam* has a section about mods: "Mods can partly or largely change the content of the game, and they are usually free. They can extend the value and playability of the game infinitely."[2] The manual also lists several websites and internet-based discussion boards where the players can acquire information about mods.

Not only players but also game companies use modding to construct the sense of locality and identity. In April 2005, EA Taiwan hosted an event inviting Taiwanese players to help record voice files for the "National Special Force mods, Republic of China." The event description read as follows: "This is a modding project contracted by EA Taiwan. The project leader is the producer of *Taiwan Straight*, Kenimaru. [The purpose of this project is] mainly to modify the uniforms, radio voices, flags and symbols, etc., of the US Special Forces in the *Battlefield 2: Special Forces Expansion Pack* to the kinds used by the Special Forces of the Republic of China (Taiwan)" (Nasasaw, 2005). Moreover, among the various languages spoken in Taiwan, the project specifically requested the use of Mandarin. What is important here is how EA constructed a national identity and commodified that national identity at the same time. These gestures signify the intention to control and standardize the industry for further commodification and maximum sales. As the call stated, "submitting [the voice files] means you fully license the rights to use [them] in the *National Special Force Mods* or the future's *Taiwan Straight Mods*".

Kücklich (2005) noted that mods add to the shelf-life of the original product and increase customer loyalty. Mods can be fun both inside and outside of actual gameplay, and can expand the fan base. Although the modding possibility gives players the freedom to subvert and extend the original game world or express their identities, it is now more common for the original game developers to develop mods and sell them as expansion packs or stand-alone products. Thus, the modding community has become a source of free or semi-free labor for the game industry. This is how localization becomes a site of power struggle and configuration.

CONTRADICTIONS IN GLOCALIZATION AND CONVERGENCE CULTURE

Circulation is the main focus of globalization studies. Appadurai (1996) used the term *flows* to describe the global cultural configuration. Lee and LiPuma (2002) also proposed a concept called *cultures of circulation* and argued that the contemporary processes of globalization demonstrate that capitalism is transitioning from a production-centric system to one whose primary dynamic is circulation. The circulation of cultural products is de-territorializing cultural products from where they are produced. These products are then sold to, and consumed by, consumers in other places. When cultural products circulate beyond national borders, foreign cultural producers may have to worry about cultural differences and local or regional competition. This study does not assume that certain cultural differences cross national boundaries, or that the cultural differences only exist across national borders. The concept of differences is always comparative, relative, and, most importantly, constructed through discourses between different parties for various reasons and in different historical periods. However, at the risk of cultural essentialism, it is necessary to recognize the existence of cultural differences across national borders to further understand the dynamics of the transnational circulation of cultural products.

Because of the existence of cultural differences, transnational cultural corporations must ensure the unimpeded circulation of their products. Unlike what Lee and LiPuma (2002) argued, unimpeded circulation is not simply based on the effacement of historically discrete cultures and capitalism, but also on complex localization practices in different stages of production, circulation, and consumption. Although the corporate sector uses the term *localization*, the analysis in this chapter demonstrates that the global operation of the digital game industry is a blend of centralization and decentralization, and of universalization and particularization. On one hand, multinational corporations must decentralize to accommodate local differences. On the other hand, they must maintain dominance through control and centralization. This combination of universalization and particularization is similar to *glocalization* (Robertson, 1995). Nevertheless, various cultural industries practice glocalization in different ways. Moreover, the percentage of universalization vs. particularization differs in each domain, and remains a topic for further examination.

The digital game industry centralized original game development and design in developed countries. The text and content of many games are not particularly localized for a smaller market like Taiwan. Thus, in terms of controlling cultural capital (brands, forms, designs, etc.) and technology, the primary principle is centralization. However, product marketing focuses on particularization: marketing is based on a centralized system but operates differently on the local level. In most cases, there is a specific office handling local repackaging, advertising, and marketing activities aligned with local identities and preferences. Between the centralization in development and the decentralization in marketing, there lies a grey area of semi-centralization based on dispersed production. The main driving force of global dispersed production is that capitalists are always looking for cheaper labor and new markets (Bell, 1976; Castells, 1999; Harvey, 1989). Nevertheless, the production of the digital game hardware may be dispersed, digital game software production is fundamentally centralized in certain locations (Johns, 2005; Kline et al., 2004). There is also another type of decentralization that is not horizontal across various geographical locations, but vertically dispersed in the production-circulation-consumption cycle.

In the conventional model of cultural production, there are clear-cut boundaries between production, circulation, and consumption. The role of cultural producers is clear: they create texts and content from scratch. Then their creations circulate through various media. In the case of digital games, production exceeds the conventional stage of production by original producers, and can also happen in the stage of circulation (i.e., modification by the distributors) and consumption (i.e., modification by the consumers). Although vertically-dispersed production and decentralization helps the digital game industry broaden its consumer base, accommodate differences, and foster sales and circulation, it is not a conflict-free model. The contradiction between centralization and decentralization can create tension. Dispersed production also provides opportunities for *prosumers* to challenge the centralized cultural dominance and control based on the logic of mass production. *Piracy,* (partly) *reproduction*, and *modding* that exceed the permission of the game companies cause conflict between consumers and the game companies.

If decentralization and dispersed production facilitate the scale and velocity of circulation, a system for controlling and managing production is necessary to ensure that this is in the best corporate interests. On one hand, game players use modding activities to express themselves, and gain the pleasure of doing it. On the other hand, there are limitations to what a player can do with the games, and copyright restrictions. Game companies want players to have fun because this increases gamer loyalty. In most cases, not only are modding changes limited, but these changes are later commoditized by the game developers. In this way, modding becomes a way for game developers to accommodate and commercialize the culltural differences. Moreover, only a few devoted hobbyists and fans can do modding constantly because it requires a large investment in time and certain programming skills. Hence, although some people consider modding to be a form of resistance, the potential of modding as a form of large-scale resistance is not as strong some may claim. Corporate interests always limit the possibilities of subversion within the commercial domain.

We must acknowledge that in spite of careful calculation and operation, commercial glocalization is not always successful in facilitating product circulation. Although glocalization attempts to facilitate the circulation and consumption of digital games, it may also reinforce national boundaries. This contradiction arises from the incompatibility between the desire for centralization, the need for decentralization, and the conflicting interests of producers and consumers. In addition to the legal restrictions on modding, game companies can use tactics such as release dates, regional encoding, system standards, and hardware lockout chips (Consalvo, 2006). These tactics not only monitor sales and control the product flow, but they more importantly allocate commercial benefits to different departments involved in large commercial operations. For example, the Taiwanese distributor of the very popular South Korean online games *Lineage II* locked out all access to their Chinese servers from outside Hong Kong and Taiwan (Oyang & Lee, 2004) despite the fact that many overseas ethnic Chinese players were angry and protested this decision. As Consalvo (2006) said, "Culture can still flow, but only along well-marked paths, designed to ensure careful tracking and control" (p. 131).

Another contradiction in glocalization is the future direction of culturally specific elements in cultural products. For easier global adoption and localization, non-culturally specific products are arguably more popular in the international market. However, further decentralization and localization occurs in larger and specified markets and areas. For example, EA recently set up an office in Shanghai, China. The new office is not only responsible for local sales, marketing, and distribution, but is also hiring game producers and software engineers to develop games for the Chinese market (Feldman, 2004). This suggests that local preferences, differences, and context are not something to be obliterated, and there is no perfect formula for global cultural products. Hall commented that westernization and homogenization is never an absolutely completed process (Hall, 1997, p. 28). Similarly, glocalization is never completed, but is a continuous process in which the global and the local, the universality and the particularity, compete and negotiate.

CONCLUSION

This chapter uses the example of the localization of digital games to demonstrate the complicated dynamics between foreign producers, media technologies, and local consumers in the process of cultural convergence and globalization. Global digital game companies, such as EA, have successfully marketed games based on the logic of glocalization. Utilizing the practice of modding and other marketing strategies, EA is able to avoid expensive localization process such as content alteration and dubbing, and establish connections with fan communities. Yet, the emergence of modding and participatory culture creates new challenges for digital cultural industries. Indeed, cultural convergence and glocalization is not friction-free. It is a dialectical and dialogical process in which different actors interact and negotiate. Media researchers must pay attention to the contradictions, tensions, and conflicts stimulated by their interactivities and, moreover, how the actors solve conflicts and how power is configured in the process. For example, issues such as copyright ownership and the control of content creation and product flow have been and will continue to be a power struggle.

In our increasingly interconnected world, many are experiencing the rapid cross-border circulation of images, signs, and information carried by mass media and popular cultural products. We generate meaning, find pleasure, and even express our identities through consuming foreign cultural products. Although certain parties control and regulate these flows for various reasons, new digital technologies have blurred the boundaries between cultural producers and consumers, further empowering consumers in the realm of cultural circulation. It is this new trend of convergence between the global and local, the producer and consumer, that makes the future full of challenges and excitement.

REFERENCES

Appadurai, A. (1996). *Modernity at Large: Cultural dimensions of globalization*. Minneapolis: University of Minnesota Press.

Bell, D. (1976). *The coming of post-industrial society*. New York: Basic Books.

Bhabha, H. (1994). *The Location of culture*. New York: Routledge.

Castells, M. (1998). *End of millennium*. Malden, MA: Blackwell Publishers.

Chadha, K., & Kavoori, A. (2000). Media imperialism revisited: Some findings from the Asian case. *Media Culture & Society*, *22*, 415–432. doi:10.1177/016344300022004003

Chandler, H. M. (2005). *The game localization handbook*. Hingham, MA: Charles River Media Inc.

Consalvo, M. (2006). Console video games and global corporations: Creating hybrid culture. *New Media & Society*, *8*(1), 117–137. doi:10.1177/1461444806059921

Dirlik, A. (1996). The global in the local. In Wilson, R., & Dissanayake, W. (Eds.), *Global/local: Cultural production and the transnational imaginary* (pp. 21–45). Durham, NC: Duke University Press.

Feldman, C. (2004, October 8) Q&A: EA's man on the ground in China, Erick Hachenburg. *GameSpot*. Retrieved May 1, 2006, from http://www.gamespot.com/news/6110098.html

Friedman, J. (1994). *Cultural identity & global process*. London: Sage.

Hall, S. (1997). The local and the global: Globalization and ethnicity. In King, A. D. (Ed.), *Culture, globalization and the world system: Contemporary conditions for the representation of identity*. Minneapolis: University of Minnesota Press.

Harvey, D. (1990). *The Condition of Postmodernity*. Cambridge, MA: Blackwell.

Herman, A., Coombe, R. J., & Kaye, L. (2006). Your second life? Goodwill and the performativity of intellectual property in online digital gaming. *Cultural Studies*, *20*(2-3), 184–210. doi:10.1080/09502380500495684

Jenkins, H. (2006). *Fans, bloggers, and gamers: Exploring participatory culture*. New York: New York University Press.

Kline, S., Dyer-Witheford, N., & Peuter, G. D. (2004). *Digital play: The interactions between technology, culture and marketing*. Kingston, ON: McGill/Queens University Press.

Kohler, C. (2004). *Power-up: How Japanese video games gave the world an extra life*. Indianapolis: Brady Games.

Kraidy, M. M. (2002). The global, the local, and the hybrid: A native ethnographic of glocalization. In Taylor, S. (Ed.), *Ethnographic research: A reader*. London: Sage.

Kücklich, J. (2005) Precarious playbour: Modders and the digital games industry. *Fibreculture, 5*, Article 5. Retrieved March 15, 2006, from http://www.journal.fibreculture.org/issue5/kucklich.html.`

Lee, B., & LiPuma, E. (2002). Culture of circulation: The imaginations of modernity. *Public Culture*, *14*(1), 191–213. doi:10.1215/08992363-14-1-191

Lister, M., Dovey, J., & Giddings, S. (2003). *New media: A critical introduction*. New York: Routledge.

Manovich, L. (2000). *The Language of New Media*. Cambridge, MA: MIT Press.

McMillin, D. (2001). Localizing the global: Television and hybrid programming in India. *International Journal of Cultural Studies, 4*(1), 45–68. doi:10.1177/136787790100400103

Morris, N. (2002). The myth of unadulterated culture meets the threat of imported media. *Media Culture & Society, 24*, 278–298. doi:10.1177/016344370202400208

Nasasaw. (2005, April 7). *Mods of National Special Force: EA Taiwan calls the Taiwanese Players to make (ROC) Mandarin Voice Mods.* Message posted to http://www.ezhobi.net/discuz/viewthread.php?tid=8925 (In Chinese)

Oyang, Y. S., & Lee, C. H. (2004) *The Lineage II* locks out the overseas IP, the players criticize the NCSoft Taiwan as cheater. *ET Today,* Retrieved May 20, 2006, from http://www.ettoday.com/2004/03/16/10846-1600618.htm (In Chinese)

Poremba, C. (2003). *Player as author: Digital games and agency.* Unpublished master's thesis. Simon Fraser University, Vancouver, Canada.

Postigo, H. (2003). From pong to planet quake: Post-industrial transitions from leisure to work. *Information Communication and Society, 6*(4), 593–607. doi:10.1080/1369118032000163277

Postigo, H. (2009). Of mods and modders: Chasing down the value of fan-based digital game modifications. *Games and Culture, 2*(4), 300–313. doi:10.1177/1555412007307955

Raessens, J. (2005). Computer games as participatory media culture. In Raessens, J., & Goldstein, J. (Eds.), *Handbook of computer game studies* (pp. 373–388). Cambridge, MA: The MIT Press.

Raymond, M. A., Shao, A. T., & Taylor, C. (1999). Shifting advertising appeals in Taiwan. *Journal of Advertising Research, 39*(6), 61–70.

Robertson, R. (1995). Glocalization: Time-space and homogeneity-heterogeneity. In Featherstone, M., Lash, S., & Robertson, R. (Eds.), *Global modernities* (pp. 25–44). London: Sage.

Ryan, M.L. (1994). Immersion vs. Interactivity: Virtual reality and literary theory. *Postmodern Culture, 5*(1). Retrieved March 2, 2006 from Project MUSE database.

Schiller, H. (1998). Striving for communication dominance: A half-century review. In Thussu, D. K. (Ed.), *Electronic empires: Global media and local resistance*. London: Arnord.

Steuer, J. (1992). Defining virtual reality: Dimensions determining telepresence. *The Journal of Communication, 42*(4), 73–93. doi:10.1111/j.1460-2466.1992.tb00812.x

Thayer, A., & Kolko, B. E. (2004). Localization of digital games: The process of blending for the global game market. *Technical Communication, 51*(4), 477–488.

The Localization Industry Standards Association. (2006). *Frequently asked questions about LISA and the localization industry.* Switzerland: The Localization Industry Standards Association. Retrieved, July 7, 2006 http://www.lisa.org/info/faqs.html

Toffler, A. (1980). *The third wave*. New York: Morrow.

Tomlinson, J. (1997). Cultural globalization and cultural imperialism. In Mohammadi, A. (Ed.), *International communication and globalization* (pp. 170–190). London: Sage.

Tomlinson, J. (1999). *Globalization and culture*. Chicago: The University of Chicago Press.

ENDNOTES

[1] The game's sequel was brought back to the original anime characters due to the fans' reaction.

[2] Translated from Chinese by the author.

Section 5
Corporate Convergence and Media Market

Chapter 21
TV Box on the Internet: The Interplay between Politics and Market in China

Anthony Fung
The Chinese University of Hong Kong, Hong Kong

Luzhou Li
University of Illinois at Urbana-Champaign, USA

ABSTRACT

With the advancement of media technologies, traditional media and new media converge at a pace faster than ever globally. In the People's Republic of China (PRC), nonetheless, while infrastructure is mature and audience is abundant, media convergence is not at all ready to take off. In theoretical term, the zigzag path of the convergence between television industry and the Internet in China origins from the clash between the state and the market. The chapter presents a special case in which political forces become a potential shaping force for media convergence and in the authoritarian environment of China, politics still reigns over technologies and the market. From a political economy perspective, the dominance of politics can be explained in terms of the process of regulatory spatialization.

INTRODUCTION

With the advancement of media technologies, traditional media, especially the television, and new media have been converging at a pace faster than ever globally, and sometimes we could even see that new media is driving out traditional media. In the People's Republic of China (PRC), just between 2007 and 2008, the proportion of traditional media outlets in the whole media industry has decreased to 72%, whereas that of new media has increased to 28% (Chinese Media Industries Report 2007-2008, 2008). Within the realm of traditional media, in China, except the cable television service, major categories like television advertising revenue and newspaper circulation revenue are all experiencing recession. In the meantime, new media such as Internet use has on average grown by 33.8% and 40% respectively in 2007 and 2008 (CNNIC). Ostensibly, both developments of the television industry and the Internet industry have reached a stage where media convergence becomes necessary unless the Chinese state lets its state-owned traditional media

DOI: 10.4018/978-1-60960-037-2.ch021

Copyright © 2011, IGI Global. Copying or distributing in print or electronic forms without written permission of IGI Global is prohibited.

fade out (which is quite unlikely). Nonetheless, although there is an imminent need, the infrastructure is mature and audience is ready to change, the trend of television-Internet convergence is not seen at all to take off in China.

The zigzag, or even stagnant, path of the convergence between the television and Internet in China can be explained from the perspective of political economy. In the concrete context of China, political factor instead of market force constitutes a potential shaping force for media convergence. Contrary to the Western view that regulating the communication industry is a government reaction to market problems (Mosco, 1996), in China, state activities emanate from political agendas. Presumably, this centralized regulation would largely shape the ways in which television-Internet convergence takes place. There are various forms of television-Internet convergence in China including telephony, Internet television, Internet Protocol Television (IPTV), digital TV, cable TV, and online video. This chapter will present cases of both online video industry and IPTV in China to see how political factors influence the process of television-Internet convergence in China. It will conclude with the argument that the barrier in the process of television-Internet convergence originates from the clash between politics and market, or rather, between the political control of the state and the profit motive of the media and Internet industry in the sense that market-oriented content and programs may go beyond the political limit of the authorities. The total control of both the media and the Internet could enhance a more seamless control over the information, opinions and culture, and that is going to be explained by the process of regulatory spatialization. Yet, the situation is more complex than just a pure political economy analysis. More than that, ironically, the communist state itself cannot resist the temptation of money and profits from the increasingly expanding market. Thus, the state is also seen having the ambition, besides the political mission, to embrace the big market. The empirical data in this chapter comes from interviews with industry practitioners and observers, and interpretations of major polices concerning the media and Internet industry in China.

BACKGROUND

The Internet and "Online Television" in the PRC

With the online video technology, the Internet has become an alternative site to television for watching visual images. In China, the growth of the video industry is characterized both by a rapid pace of development and high diversity of programs online. First, online video has become one of the top five most frequently used Internet services among Chinese Internet users. By June 2009, online video users have reached more than 200 million and increase with a rate of 10% (within the half year). Second, according to Vobile's data, the Chinese online video websites are home to as much as 65% of online copyright-infringing content.[1] A lack of copyrighted materials and regulations in online video precisely increases the content diversity. Alternative "television" then appears in three forms online. The first of this kind is the YouTube-style streaming technology driven video hosting/sharing websites such as tudou.com, youku.com, and 56.com whereby both professional/traditional videos from TV stations and Film/TV/music production companies and user-generated content (UGC) are provided. The second type of online audio and video service provider is the peer-to-peer (P2P) technology driven platforms offering access to hundreds of live streaming video channels as well as on-demand shows. In practice, users can download a small application program to their own local computers and watch full-length television programs that are available from terrestrial broadcasters as well as movies and sports events. These hugely popular P2P TV platforms in China include PPLive, PP-

Stream, and UUSee. Needless to say, a significant amount of content provided by these platforms is pirated. The third type is the similar video website created by leading Chinese portal websites like Sina, Sohu, and Tencent.

However, despite the diversity of programs, content and a relatively loose control of the online video industry, it does not guarantee a high level of freedom. This is due to the fact that the information infrastructures in China on which the online video industry operates are stated-owned, regulated and controlled.

Television Industry Going Online

The same kind of control has also been adopted in the television industry for years, although the rapid commercialization of the industry has made the issue of control more complicated. Television in China used to function effectively as an ideological state apparatus. However, since the mid 1990s, Chinese television industry has experienced an accelerated process of commercialization. This trend resulted partly from China's media reform, which in essence, was an official initiative with which the Chinese authorities could terminate subsidies for stated-owned national and terrestrial television stations, thereby forcing these stations to rely on market for self-finance. Of course, the prevalence of cultural globalization and China's entrance into the World Trade Organization (WTO) later legitimated and accelerated the process. After more than a decade's market operation, the state-monopolized television industry in China has already been proven a very profitable business. But the industry is not satisfied with the existing profit. When the Internet is perceived as another large cake of the market, television stations also forge effective strategies attempting to occupy this new market and to maximize the economies of scale. Therefore, despite the unsolved copyright problems online, keen competition among the new media companies, major state televisions also want to get their feet wet in the new media market.

China's two media giants, Shanghai Media Group (SMG) and China Central Television (CCTV) are the major players. Since 2005, SMG, the mother corporation which controls all Shanghai media, has established four new media subsidiaries, namely "Oriental Broadband", "BestTV", "SiTV", and "Oriental Dragon", to extend their arms to Internet television, IPTV, digital TV, and mobile multimedia services respectively. As for CCTV, it integrated its existing online resources and created a new company called "CCTV International Networks" in charging all online operations. Apparently, this kind of development increasingly blurs the boundaries between the commercial role of the television industry and the political role of the station. And as the new online terrain is relatively "free", it further deepens the Party's paranoia lest that the Party could lose its ideological frontline upon television's fusion with the new technology. Thus, the extension of the television industry to the Internet has foretold a new wave of control over the Internet, an area otherwise the official would not have paid attention.

Parallel Control between the Television and Internet Industry

In China, with this specific reason, what can be observed is a parallel control between television and the Internet, whereas, in most Western states, the regulation over Internet could be relatively looser than that over traditional media outlets. Ways of controlling and disciplining applicable to traditional media in China include "the imprisonments of journalists, the closure and forced reorganization of media outlets, editorial censorship, the imposition of an official line (*koujing*) in media reporting of major events, regular media monitoring by a specialized team of Party elders, the appointment of key management personnel in major media outlets, and the maintenance of ad hoc rules regarding media operations, or 'propaganda disciplines' (*xuanchuan jilu*)" (Zhao, 2004, pp. 179-180). To coordinate and strengthen the state control over the content of television, State Administration of Radio, Film, and Television (SARFT), under the directive of the State Council, was created in March 1998 to control the broadcasting and film industries in the form of censorship.

The Internet once gave hope to more freedom. Optimistic and idealistic, some Chinese scholars believed that the Internet could eventually breakdown the Communist Party's control over information and the authorities would give up its rein over the state media (Zhang, 2006). But there are more and more studies tending to suggest that Chinese government has come up with practical resolution to control the Internet (Hartford, 2000; Kalathil, 2003; Tsui, 2001). Now it has become increasingly evident that the state's anxiety to control over new communication and technologies never wanes in as much as the Chinese government has been skilful in controlling the Internet. Some studies (e.g. Kalathil, 2003) have listed out various ways of control from restricting Internet access, filtering content, monitoring online behaviors, to blocking Internet use entirely.

Lessig (2001) summarized four general modalities of Internet control: (1) the law that sanctions by the (perceived) threat of legal actions; (2) the market that regulates by price; (3) social norms that restrain the community by social pressure; (4) the Internet architecture that is built-in with control. Applying this model to the Chinese context, Tsui (2003) analyzed how the Internet control actually takes place in China. His discussion of the law and the market is particularly relevant to this chapter. With the set of laws and regulations, written in very ambiguous terms, promulgated in 1996, the authorities interpret it in their own way whenever they want to tighten the control. The law becomes the basis for the authority to jail any deviants and in the long run, forces market players to keep guessing and steering to the safe side. On the market side, as Tsui argued, the PRC is learning to control the operators with market rule. To do online business, Internet service and content providers are required to apply a license from the government office which serves as gatekeepers to prevent these companies from straying from Party ideology. Thus, there is no surprise why these Internet operators do self-censorship to stay out of political blunder.

An evolving parallel control between television and Internet implies that it might necessitate a centralized unit like SARFT to efficiently regulate the Internet and the content that circulates in it. If there is going to be a bureau, the philosophy of control should also be based on political agendas. But until now, a full regulation over the Internet in China has not been observed.

Internal Competition among Government Units

Internet control in China seems to be partially regulated, as opposed to a full mode of regulation which requires a single comprehensive legal and regulatory framework (Garcia-Murillo & Maclnnes, 2003). Since 1996, the authorities have implemented a set of regulations to manage different aspects of the Internet: content, online reporting, e-commerce, and many other Internet-related aspects. The characteristic of such control is that multiple—and to certain extent chaotic—regulatory bodies exist and their functions are overlapping, and sometimes even contradictory to each other. There are at least six government bodies trying to extend their arms to the Internet industry. The Ministry of Industry and Information Technology (MIIT)[2] is the office that overseas China's telecommunication and software industry, and it is responsible for the licensing and registrations of all "Internet service providers" (ISPs) in the PRC. The State Council Information Office (SCIO) is responsible for overseeing the country's Internet sites engaging in news publishing and postings. The Ministry of Public Security (MPS) serves to patrol, monitor and filter the "undesirable" elements on the Internet. There are also various other government bodies drafting various regulations and staking out their claims over the still undefined Internet industry in China. For instance, the Ministry of Culture began drafting regulations extending their jurisdiction over online games. The State Copyright Bureau also claimed their territory over online copyright protection. An industry observer[3] quoted a Chinese proverb to describe this chaotic situation: "It just looks like nine dragons managing one river at the same time. Every relevant ministry tries to involve itself in the Internet industry so as to take a share of the profits, SARFT, MIIT, the General Administration of Press and Publication, the Ministry of Culture…." Apparently, the central government remains cautious of permitting any one particular government body to wall off the whole market. To put it in another way, under this struggle of power, the authorities resort to ensure a "controlled competition" between different state organs over the control of the Internet sector.

It is true that being regulated differently by different parties without a unifying structure means that the political control over the Internet is less effective (at least than that over television). Theoretically speaking, given that MIIT is the department designated to license ISPs, it should have become the authority over the Internet industry including the online video business. However, in the case of online video and audio industry, SARFT, gains the upper hand. It is a situation in which SARFT, the perceived political gatekeeper of the broadcasting industry, goes into the management of the telecommunication or Internet-related industry under the pretext of convergence. After formally and informally interviewing personnel in different departments, the study found that one major reason for this "balanced state" is that each of these official units carries the political mission, that is, to help the government to control the Internet and hence the ideology. At this moment, the authorities might not need all of them to exercise their power, but in times of critical political moments or certain political campaigns, the government has to rely on these agents to centralize the control.

Internal competition among different official units should not simply be perceived as maladministration of the government. It reflects the legacy of control and the constant fear of losing control by the administration. In retrospect, in July 1998, the State Council authorized MII (later renamed as MIIT) to oversee the administration of the nation's television broadcasting network, but in the same year in March, SARFT was established and given a secondary role to involve in "doubly" monitoring the industry.

As nowadays in China television-Internet convergence will most likely to result in large multiplication of online video, the existing online audio and video industry has become an important strategic ground for both SARFT and MIIT to take control of. SARFT's rising status in the ongoing political battle is certainly backed up by support from the State Council, which obviously would like to advance its control of cultural content—not just media content, and at the same time increase its financial revenues and resources. The latter is a precondition for more political control. From the point of a view of the Party, content control is surely a more important consideration and therefore the power will go to SARFT which has been more experienced than MIIT to censor video content.

There is of course the possibility that the television-Internet convergence can produce other forms of content. Based on interviews with some anonymous sources, China is now able to manufacture television sets that could view online video from certain websites such as youku.com (one of the largest online video websites with a large number of illegal video resources). But such move was immediately banned by SARFT; otherwise SARFT would soon lose control over the content.

However, except the content, SARFT has very minimal role with the Internet industry. A pragmatic solution for the authorities is therefore to let SARFT take the primary role in regulating online video and MIIT take the secondary and supporting role to regulate other elements apart from the video content. In December 2007, SARFT and MII (later renamed as MIIT) co-released the Regulations on Online Audio and Video Service, effective from January 31, 2008. According to the third act of this regulation, "SARFT is the authority to administer, monitor, and regulate the industry's development, while the MII, with authority over the internet and mobile industry, will take related monitoring responsibilities and provide a set of service guidelines".[4] This clarifies the dominant role of SARFT in regulating online audio and video industry. The fact that SARFT plays a fundamental role in regulating online video and audio industry largely shapes the nature of television-Internet convergence in China. SARFT is not against the trend of media convergence. Early this year, it published the general requirements to guide the work of the whole radio, television, and film industry, which clearly showed SARFT's intention to encourage and promote media merger. However, to SARFT, the television-Internet convergence is better to be directed to a way in which traditional television and radio can occupy the dominant and leading role in the wave of new media booming.

In this way, it can both enhance its authority and revenue and extend the stringent and effective censorship over the new media. Under this circumstance, not surprisingly, policies made by SARFT are in fact partial to the television industry and discourage the development of new media per se, in particular, the online audio and video business.

THE POLITICAL ECONOMY OF ONLINE TELEVISIN AND IPTV IN CHINA

Political Measures, Economic Consequences

Political control over online video hosting websites encompasses two aspects: regulation and censorship. Policy can regulate capital investments and directly influence the development of telecommunications and media industries. The impact of policy is particularly significant in East Asian countries whereby the telecommunication and media industry has little or no regulatory experience (Hukill, 2000). Oba and Chan-Olmsted (2005) examined the development of cable television in ten East Asian countries by looking at country-specific environmental factors including economic factor and regulation/policy factor. They concluded that cable-related policies instead of economic conditions play a direct and significant role in the development of cable industries in these countries. To consider regulation over online audio and video industry in China, major policies concerning Internet business were investigated and interpreted in this section.

According to the seventh act of the Regulations on Online Audio and Video Service, all online audio and video service providers will be required to apply for an online audio-visual broadcasting license through local broadcasting authorities and the license should finally be authorized and issued by SARFT. Without this license, no organization and individuals are allowed to operate the online audio and video business. Besides, the eighth act of the regulation stipulates that key qualifications of online video and audio service providers should fulfill two conditions: first, being a state-owned enterprise, second, building in itself a sound program censorship system and conveying legal content.

This regulation appears to be an attempt to fix the unregulated situation as the introduction of online audio-visual broadcasting license signals the authorities' determination to lay down the rule of the game. However, the imposition of state ownership and program censorship as key qualifications for online audio and video websites in fact reveals SARFT's intention to limit the private players' maneuver and exercise political control. On April 1, 2008, SARFT issued the first batch of online audio-visual broadcasting license. Twenty-three new media enterprises received the qualification but none of the online video hosting websites can be found on the list. Absentees included the two leading players in the field: youku.com and tudou.com. If the regulation is drafted just to regulate the unregulated instead of exerting political control, online video hosting websites would not have been all expelled out of the license competition.

The consequence of market interference by the authorities is that it directly threatens the survival of the online video hosting websites. What we can see is that while the market of online video hosting websites is increasing in an unprecedented high rate, ironically, none of the companies involved in the industry can redeem any profit. The primary market obstacle is the operation cost of online video companies with most expenditure going to bandwidth maintenance. Currently, youku.com spends around RMB24 million every month on bandwidth cost and the yearly hosting expense exceeds RMB290 million. Second, since Internet users do not have to pay for the viewing cost, online video websites primarily rely on advertising revenue for profitability. However, half of the total advertising revenues in China go to television industry (that provides the content) and only around 6% goes to Internet business. Publicly released report shows that online advertising revenue of a medium size video portal 56.com is RMB1 million per month, and that of a large size, ku6.com, is RMB3.5 million (BDA China). The advertising revenue can hardly meet the expenses. Up to now, thanks to the investment by overseas venture capitals including Softbank, Carlyle, IDG, Sequoia Capital, and SIG, these online video businesses still survive. Since 2005, the investment just on the five major online video hosting websites in China has reached USD 173 million. However, due to perceived political risks, since Jan 2008, 50% of the venture capital investing in this industry, especially those investing in middle-to-small size video hosting websites, has began to cease pumping in money and even withdraw. More and more investors chose to wait for a clearer government signal (Liang, 2008). This directly led to an industry-wide survival crisis. Censorship constitutes another important factor that hinders private investment in the video hosting industry, thereby putting the new state-owned media comers in a more advantageous position. Censorship encompasses both external censoring and internal censoring. In terms of external censoring, the ways in which Chinese government has exercised to control the online video websites are more or less the same with what have been applied to the traditional media. First, online video hosting companies will receive "notifications" or political orders from higher organizations now and then. This usually happens during politically sensitive time like Anniversary of the June 4 Incident, NPC and CPPCC sessions, and Beijing Olympics. Second, regular media monitoring usually results in warnings and penalties. For instance, on March 20, 2008, all of a sudden, SARFT publicly released a report detailing its covert examination over the entire online audio and video industry, and in it, 32 websites including tudou.com, one of the most popular video portals in China, were warned and penalized. Third, serious violation leads to temporary/partial suspension of online video services or even a closure. On March 14, 2008, tudou.com abruptly suspended its services for one day. The reason offered by tudou.com was that the website was undergoing technological maintenance and expansion. However, one interviewee told that tudou.com was forced to shut down its websites as a penalty for its violating the Regulations on Online Audio and Video Service which was made effective from January 31, 2008. The industry-wide self-censorship has become another noticeable phenomenon. As one of the interviewees pointed out, "Actually, they (video websites) have been very self-disciplined. Their extent of self-censorship is even wider than censorship from without." Self-censoring mechanism rely both on technology and censors (personnel). The leading players of the industry like youku.com and 56.com have all developed a special technology to detect whether or not the content of the video is "healthy" in the eyes of the authorities. In general, the industry practitioners think that the accuracy of this automatic censoring is very high. Despite this, to play safe, these institutions also hire staff to screen and clean politically dangerous content. Tudou.com's biggest department is for censoring user-uploaded content. Youku.com has established a 24-hour on-duty content censoring team. 56.com recruits more than 50 people entirely and exclusively dedicating to video-censoring. Ku6.com is said to have the most well-established program censoring system within the industry. Its monitoring team reaches nearly 100 staff. As the volume of uploaded content increases and the information control tightens, the hosting expenses on content filtering will inevitably increase. This adds the operation cost of the industry. The uncontrollability of the user-generated and uploaded content makes the occurrence of politically sensitive content possible. For example, before 2008, a lot of user-uploaded videos in major video websites like tudou.com and youku.com were perplexed by the fact that some of the videos were 'tainted' with Falungong information. The only way to tackle them is to fence itself with more filtering devices to ensure that they have fulfilled their "responsibility" lest that their license could be revoked.

Media-Dominated Convergence

Under the control of SARFT, when online video companies suffer from both political and economic pressure, traditional media has the vintage points to enter the new media market. To some extent, it can be seen as SARFT's strategy to restrict online video companies for traditional television stations' sake. In the 15th act of the regulation, SARFT explicitly encourages state-owned strategic investors to invest in online video and audio industry. It also encourages TV and film production companies to produce content suitable for online broadcasting. In March 2008, Beijing Radio and Beijing TV Station simultaneously announced the operation of their own video websites and aimed to become the first-class online video websites. Prominent examples also include Oriental Broadband hosted by SMG and i-Phoenix hosted by Phoenix TV (a pan-Chinese satellite TV with its headquarter in Hong Kong). These websites are owned by traditional media groups or television stations and are eligible and responsible for the online circulation of all the copyrighted audio-visual programs possessed by the real world organizations. The primal advantage of traditional broadcasting organizations is their large amount of copyrighted videos and programs. For instance, CCTV has several-million hours' original visual programs. This will become a great asset if they decide to march into the new media industry. Apparently, traditional television presents a different logic of profit-making from new media companies. Traditional television stations make profit through maximizing their existing resources; however, online video companies reap profits mainly by means of illegally capitalizing on the content of Western and domestic copyright holders. Traditional television appears to be more legitimate and in an advanced position to integrate international standards. From the point of view of the authorities, as mentioned earlier, traditional media is more experienced in filtering and censoring content and accustomed to following the political order from the authorities. In this way, SARFT kills two birds with one stone. In sum, SARFT's extension of power and control over the Internet has reshaped the nature of television-Internet convergence in China. Although the media convergence stimulated in this way may not necessarily be unsuccessful, this process is highly catered for a more centralized political control and hence reflects the political complexity in China.

IPTV as the Site of another Struggle

A similar political influence of the convergence can also be witnessed in the trajectory of development of the IPTV in the PRC. The IPTV, a crossover between the traditional television industry and the telecommunication industry, is illustrated here to further reconfirm the earlier argument of SARFT's ambition on not only political control but also economic profit. In the development of the IPTV, SARFT and telecom offices (MIIT as the authority) strategically compete for organizational extension and control. As a legacy of control, the State Council has refrained MII from stepping into the radio and television industry and SARFT from stepping into the telecom industry. However, the document *Concerning 2009 in-depth Economic System Reform Working Plan* released by the State Council in May 2009 has allowed SARFT and MIIT to enter into each other's own territory. As the Regulations on Online Audio and Video Service defines SARFT as the authority in guiding the development of IPTV and issuing IPTV broadcasting license, SARFT succeeds in excluding MIIT from the control of the IPTV. SARFT's dominant role in the IPTV business compels telecomm operators to actively seek collaboration with television stations, which ultimately moulds a media-dominated-over-telecom IPTV business. However, initially, SARFT did not have a landslide victory. The disadvantage of the television industry in developing IPTV lies in its lack of a well-established technological infrastructure. SARFT-led television industry players still have to depend on telecom operators unless SARFT installs its own fix-line and broadband fiber optics for its own interactive television. To offer value-added service, these players should also upgrade the existing cable networks to high-speed two-way communication networks, which require both financial investment and cooperation from local government. It is a project that cannot be accomplished in the short run. In this regard, although television operators would be very reluctant to see their own interests being shared by telecom operators, they still need to consider the cooperation with the latter. At the same time, telecom operators have to comply with the "unfair" game because they simply see no room for further expansion without the approval of the SARFT. The traditional telecom industry is a saturated market and government basically dictates the growth of the industry. To explore IPTV is perceived as a way for, at least in the short term, expansion. Lack of content is the Achilles' heel of the industry. Collaborating with tradition media that serves as program providers not only solves the content problem. It also means that telecom operators get a license to operate the IPTV business because, currently, all the six IPTV license holders are media players. It seems that the cooperation between SARFT and MIIT is conducive for the development of IPTV in China. For instance, the success of Hangzhou (a coastal tourist city) model in developing IPTV in China lies in the deep collaboration between the local radio and television network operators and local telecom operators. Another case is the IPTV service launched by SMG and local China NetCom and China Telecom operators. At this moment, players all acknowledge that such cooperation is necessary. But considering that the concrete political settings vary from province to province and in the central-government level the competition and tension between SARFT and MIIT have not been finally resolved, how to promote this cooperation nation-wide remains an empirical question. In sum, to some extent, political forces still limit the development of IPTV in China.

TOWARD A SPATIALIZATION OF POLIITCAL CONTROL

The state's dominance over the entire path of convergence can be explained by the concept of spatialization. Spatialization refers to the institutional extension of organizational activity in both horizontal and vertical forms (Mosco, 1996). From a political economy perspective, spatialization maximizes both political control and economic profit because of economies of scale. Horizontal concentration or "cross-media concentration"—in which an older media line buys out a firm in a newer line—is a particularly useful concept here (Mosco, 1996). The extension of traditional television industry into the Internet-related business in China is a horizontal spatialization. However, what we observe in China's unique situation is not just a process of spatialization in which the media conglomerates are made dominating the market. Dupagne and Garrison (2006) suggested that, besides economic convergence or cross-media mergers and acquisitions, there is also regulatory convergence. Such a convergence, apart from an organizational or financial monopoly of certain territory or space by the state-owned media or state, can be conceived as a means of the state to streamline its regulatory regimes and extend the political control through the legal means over a wider scope of institutions. In fact, Mosco (1996) has pointed out that the process of spatialization of communication industries involves both the logic of production and the logic of power over economic decisions. While the former is well documented in traditional political economy analysis, the latter is often not highlighted, but in the case of China, a state or politics driven convergence which improvises the state's power and regulatory territory exemplifies the logic of power in the complex interplay between market and politics. In such expansion of political space, unlike its Western counterparts, the communication industry in China is not operated under the rule of the invisible hand of the market. The Chinese media is owned outright by the state and is always deeply embedded into the wider state functions through processes of control and regulation. The existing entire national institutional makeup and setting were established in accordance with the original configuration of media outlets. Now, the rise of new media can be regarded as an issue disrupting the original media or state configuration. Under this circumstance, in order to restore its controlling ability over the communication industry, the state is forced to stretch its activities over this uncharted new media terrain. In the process of television-Internet convergence, a media-dominated situation can help the state in maintaining its political control. The reality of television-Internet convergence in China is basically an internal reshuffle and restructuring of power relations to restore the national order. Politics therefore is naturally the shaping force—and sometimes a hindering force—for convergence, and at this moment, politics still reigns over technologies as well as the market. However, there is a by-consequence out of this convergence that is originally aimed for political purpose. As illustrated in the competition of television station and online video companies over the terrain of Internet, SARFT's favor to the television station on the one hand strengthens political control, and on the other hand, boosts the economic resources of the state-owned media. That is, state is a direct driving force for media commercialization. The state's recent expansion of regulatory power over the Internet can be seen as the direct involvement of state in the capital accumulation process. It is quite overt that the state now realizes the importance of restructuring internal organizations and expanding organizational activities to acquire more economic benefits. In a nutshell, the spatialization of political control and the subsequent media-dominated convergence can be viewed simultaneously as the state's attempts to occupy a wider political space to exert ideological control more efficiently and to command more economic resources and benefits.

REFERENCES

Cui, B. G. (2008). *Zhongguo chuanmei chanye baogao (Chinese media industries report)*. Beijing: Social Sciences Academic Press.

Dupagne, M., & Garrison, B. (2006). The meaning and influence of convergence: A qualitative case study of newsroom work at the Tampa news center. *Journalism Studies*, *7*(2), 237–255. doi:10.1080/14616700500533569

Garcia-Murillo, M., & Macinnes, I. (2003). The impact of technological convergence on the regulation of ICT industries. *International Journal on Media Management*, *5*, 57–67.

Hartford, K. (2000). *Cyberspace with Chinese characteristics*. Retrieved January 20, 2010, from http://www.china-wired.com/pubs/ch/home.htm

Hukill, M. (2000). Telecommunications and development policy in Southeast Asia: Foundations for convergence. In Hukill, M., Ono, R., & Vallath, C. (Eds.), *Electronic communication convergence: Policy changes in Asia* (pp. 48–76). New Delhi: Sage.

Kalathil, S. (2003). *Open network, closed regimes: The impact of the Internet on authoritarian rules*. Washington, DC: Carnegie Endorsement for International Peace.

Lessig, L. (1999). *Code and other laws of cyberspace*. New York: Basic Books.

Liang, S. N. (2008). Shipin wangzhan de kunjing ji yunying moshi tanxi (An exploration of the dilemma and business model of the online video website). [Youth Reporter]. *Qingnian Jizhe*, *33*, 95.

Mosco, V. (1996). *The political economy of communication: Rethinking and renewal*. London; Thousand Oaks, CA: SAGE Publications.

Oba, G., & Chan-Olmsted, S. M. (2005). The development of cable television in East Asian countries. *Gazette: The International Journal for Communication Studies*, *67*(3), 211–237. doi:10.1177/0016549205052225

Tsui, L. (2001). Big mama is watching you: Internet control and the Chinese government. *The economist*, (January): 10–14.

Tsui, L. (2003). The Panopticon as the Antithesis of a Space of Freedom. *China Information*, *17*(2), 65–82. doi:10.1177/0920203X0301700203

Zhang, L. L. (2006). Behind the "Great Firewall": Decoding China's internet media policies from the inside. *Convergence: The International Journal of Research into New Media Technologies*, *12*(3), 271–291. doi:10.1177/1354856506067201

Zhao, Y. Z. (2004). The state, the market, and media control in China. In Thomas, P. N., & Nain, Z. (Eds.), *Who owns the media?: Global trends and local resistances* (pp. 179–212). London, New York: Zed Books.

ADDITIONAL READING

Castells, M. (1996). *The rise of the network society*. Oxford and Cambridge, MA: Blackwell.

Donald, S. H., & Keane, M. (2002). Media in China: New convergences, new approaches. In Donald, S. H., Keane, M., & Yin, H. (Eds.), *Media in China: Consumption, content and crisis* (pp. 3–17). London, New York: Routledge Curzon.

Dupagne, M., & Garrison, B. (2006). The meaning and influence of convergence: A qualitative case study of newsroom work at the Tampa news center. *Journalism Studies*, *7*(2), 237–255. doi:10.1080/14616700500533569

Fung, A. (2009). Globalizing televised culture: The case of China. In Turner, G., & Jay, T. (Eds.), *Television studies after TV* (pp. 178–188). Oxon: Routledge.

Gripsrud, J. (2004). Broadcast television: The chances of its survival in a digital age. In Spigel, L., & Olson, J. (Eds.), *Television after TV* (pp. 210–223). Durham: Duke University Press.

Jenkins, H. (2006). *Convergence culture: Where old and new media collide*. New York: New York University Press.

Kalathil, S. (2003). *Open network, closed regimes: The impact of the Internet on authoritarian rules*. Washington, DC: Carnegie Endorsement for International Peace.

Keane, M., & Donald, S. H. (2002). Responses to crisis: Convergence, content industries and media governance. In Donald, S. H., Keane, M., & Yin, H. (Eds.), *Media in China: Consumption, content and crisis* (pp. 200–211). London, New York: Routledge Curzon.

Kendall, L. (1999). Recontextualizing cyberspace: Methodological considerations for online research. In Jones, S. (Ed.), *Doing Internet research* (pp. 57–74). London: Sage.

Lee, T. V. (2000). The media and the legal bureaucracy of the People's Republic of China. In Lee, C. C. (Ed.), *Money, power, and media: Communication patterns and bureaucratic control in cultural China* (pp. 208–244). Evanston, IL: Northwestern University Press.

Lynch, D. (2000). The nature and consequences of China's unique patterns of telecommunications development. In Lee, C. C. (Ed.), *Money, power, and media: Communication patterns and bureaucratic control in cultural China* (pp. 179–207). Evanston, IL: Northwestern University Press.

Mann, C., & Stewart, F. (2000). *Internet communication and qualitative research*. London: Sage.

Mosco, V. (1996). *The political economy of communication: Rethinking and renewal*. London; Thousand Oaks, CA: SAGE Publications.

Pan, Z. D., & Chan, J. M. (2000). Building a market-based party organ: Television and national integration in China. In French, D., & Richards, M. (Eds.), *Television in contemporary Asia* (pp. 233–263). New Delhi; Thousand Oaks, Calif.: Sage Publications.

Parks, L. (2004). Flexible microcasting: Gender, generation, and television-Internet convergence. In Spigel, L., & Olson, J. (Eds.), *Television after TV* (pp. 133–156). Durham: Duke University Press.

Qiu, J. L., & Chan, J. M. (2004). China Internet studies: A review of the field. In Nissenbaum, H., & Price, M. E. (Eds.), *Academy & the Internet* (pp. 275–307). New York: Peter Lang.

Ross, S. M. (2008). *Beyond the box: Television and the Internet*. Malden, MA: Blackwell.

Tsui, L. (2003). The panopticon as the antithesis of a space of freedom. *China Information*, *17*(2), 65–82. doi:10.1177/0920203X0301700203

Waldfogel, J. (2009). Lost on the web: Does web distribution stimulate or depress television viewing? *Information Economics and Policy*, *21*(2), 158–168. doi:10.1016/j.infoecopol.2008.11.002

Zhang, L. L. (2006). Behind the "Great Firewall": Decoding China's Internet media policies from the inside. *Convergence: The International Journal of Research into New Media Technologies*, *12*(3), 271–291. doi:10.1177/1354856506067201

Zhao, B. (1999). Mouthpiece or money-spinner? The double life of Chinese television in the late 1990s. *International Journal of Cultural Studies*, *2*(3), 291–305. doi:10.1177/136787799900200301

Zhao, Y. Z. (1998). *Media, market, and democracy in China: Between the party line and the bottom line*. Urbana: University of Illinois Press.

Zhao, Y. Z. (2004). The state, the market, and media control in China. In Thomas, P. N., & Nain, Z. (Eds.), *Who owns the media?: Global trends and local resistances* (pp. 179–212). London, New York: Zed Books.

Zhao, Y. Z. (2008). *Communication in China: Political economy, power, and conflict*. Lanham, Md.: Rowman & Littlefield.

Zhao, Y. Z., & Guo, Z. Z. (2005). Television in China: History, political economy, and ideology. In Wasko, J. (Ed.), *A companion to television* (pp. 521–539). Malden, MA: Blackwell Pub. doi:10.1002/9780470997130.ch29

ENDNOTES

1 See http://www.forbes.com/2009/03/20/video-piracy-china-technology-internet-video.html

2 MIIT was previously known as the Ministry for Information Industry (MII).

3 This Shanghai-based industry observer was interviewed in September 2009 for this research.

4 See http://www.marbridgeconsulting.com/marbridgedaily/2007-12-29/article/7063/sarft_mii_co_issue_online_video_regulation

Chapter 22
Corporate Strategies in Media Convergence:
A Comparative Study of Sony vs. Samsung as Transnational Cultural Industries

Dal Yong Jin
Korea Advanced Institute of Science and Technology, South Korea

ABSTRACT

This study is a historical documentation of the recent trends of Sony's and Samsung's engagement in the cultural industries mainly by examining convergence between their own hardware and software to ascertain whether this trend is confirmation that Sony and Samsung play pivotal roles in the cultural market. This chapter investigates the cause of the changes and growth of Sony and Samsung, and it discusses the similarities and differences between them, with a focus on corporate policies and business strategies in convergence. In particular, it articulates the ways in which changing corporate policies have played a critical role in the growth of local-based transnational cultural corporations, because what makes them different from other firms is their approach to corporate policies of convergence.

INTRODUCTION

Over the past ten years, media convergence has been a business mantra in the global cultural industries. The enthusiasm of these industries for converging their hardware with digitalized software has been popular due to synergy effects. The ubiquity of the term synergy during the 1990s and until recent years has resulted in it becoming a must-take-business model for the cultural industries with digitalization of signals, which make the signals themselves equal, regardless of what kind of information or communication they represent (Fagerjord and Storsul, 2007). Due to synergy effects and economics of scale, those involved in the production and distribution of diverse cultural industries—both traditional media such as network broadcasting, newspaper, and computer corporations, and the new media sector, including the Internet, mobile, and cable industries—have vehemently sought convergence (Chambers & Howard, 2005; Chan-Olmsted, 1998).

Although several important factors have driven media convergence, two major causes for the

DOI: 10.4018/978-1-60960-037-2.ch022

Copyright © 2011, IGI Global. Copying or distributing in print or electronic forms without written permission of IGI Global is prohibited.

rapid growth in the process of convergence are technological advance in transmission techniques (Wolf, 2000) and the processes behind neoliberal media and telecommunication policies (Hesmondhalgh, 2008). In the context of communication technologies, one talks about the coming together of several discrete technologies to create a hybrid technology. Interrelated waves of convergence during the late 20th and the early 21st centuries around digital technology provided the interoperable systems crucial for transferring content from one medium to another with relative ease and speed. Because digital technology recognizes text, visuals, and audio all as binary code, previously incompatible media formats could increasingly be thought of as variant embodiment of the same governing technology—just as different dialects of the language of computers (Wolf, 2000; Miller, 1996). Convergence is challenging for many traditional forms of media which attempt to step into the new media sector, including the Internet market, because current media convergence has been fuelled by increasingly pervasive digital technologies (Huang and Heider, 2008; Bar and Sandvig, 2008; Schiller, 2007). In other words, the digitization of information technology and media as well as the development of media technology have contributed to media convergence (Lawson-Borders, 2003).

Meanwhile, the liberalization of national communication markets leads to individual national markets becoming part of a global market in communications (Mckenna, 2000). Although convergence started several decades ago, the transactions have been expedited with the rapid employment of neoliberal communication policies, mainly during the 1990s. Privatization of public broadcasting and telecommunications and liberalization of domestic markets have blurred the distinction between the old media and the new media, and media companies have become multi-media firms through consolidations in the midst of neoliberal reform worldwide (McChesney, 2001; McPhail, 2006). The loosening of regulations in cross-ownership among cultural industries has come to play one of the most significant roles in expanding their business areas beyond the traditional core business areas in many media companies (Thussu, 2006). In the midst of pro-media and cultural policies and advanced digital technologies, media and cultural corporations have pursued the convergence process in order to acquire multi-media and multi-functional networks starting with Western-based cultural industries (D. Schiller, 2007).

The centrality of multi-platform content strategies to contemporary cultural conglomerates also resonates in the very architecture of non-Western based cultural industries. In particular, two Asian-based cultural conglomerates—Sony in Japan and Samsung in South Korea—have turned themselves into convergence cornucopia. While advancing digital technologies, they have developed software such as film and television programs and have pursued the integration of software with hardware, which would benefit from already established global corporate brands. Sony and Samsung's recent growth as participants of the cultural industries is largely a result of both convergence and globalization, which have been key to both firms' strategies through commercial and technological integrations (Chang, 2008).

Regardless of the rapid involvement of Sony and Samsung in the global cultural market, there is little consensus in views on their roles in the global cultural industries. Using the example of the Sony Walkman, Paul du Gay et al. (2001), which is almost only book, analyzed how and why cultural practices and institutions have come to play such a crucial part in our lives and indicated Sony as the global cultural firm. There is no single work dealing with Samsung as a cultural corporation. Indeed, as a reflection of the recent growth of Sony and Samsung as cultural industries, there is no significant literature on the story of Sony and Samsung, although several previous works mainly focus on them as electronics corporations only (Nathan, 1999; Lee, 2006; Chang 2006). This study is therefore a historical documentation of the recent trends of Sony's and Samsung's engagement in the cultural industries mainly by examining convergence between their own hardware and software to ascertain whether this trend is confirmation that Sony and Samsung play pivotal roles in the cultural market. This paper investigates the cause of the changes and growth of Sony and Samsung, and it discusses the similarities and differences between them, with a focus on corporate policies and/or business strategies in convergence. It particularly articulates the ways in which changing corporate policies have played a critical role in the growth of local-based transnational cultural corporations, because what makes them different from other firms is their approach to corporate policies of convergence.

This paper employs comparative and historical approaches as methodological frameworks for examining social and economic differences and the specificity of the two cultural corporations. The historical approach is very useful to find out the causes behind the changing processes of Sony and Samsung towards becoming cultural firms because it gives a tool for explaining structural continuity and change in the cultural sector. The timeframe was divided into three periods; the 1980s when Sony turned into the cultural industry with the development of the Walkman; the mid-1990s when Samsung rapidly invested in the cultural sector; and in the early 21st century when they have changed their corporate strategies in convergence towards a networked-based integration. While the paper investigates corporate policies in the cultural industries based on the historical context, it primarily compares and contrasts their structures and institutions, because "the analysis of media cultures can develop through combining two types of comparison; one, a comparative analysis of the more intangible, more subtle, aspects of the media environment in different places; the other, the degree to which media rituals are embedded in wider belief structures, whether economic, political, religious or social," as Nick Couldry points out (2007, 248-249). However, because their culture and management systems are different, any comparison of Sony to Samsung is inevitably difficult.

THE EARLY CORPORATE STRATEGIES OF ASIAN CULTURAL CORPORATIONS

Sony and Samsung have expanded their involvement in the global cultural market over the last several decades. They had originally focused on hardware, such as TV sets and semi-conductors, but the major focus has changed and cultural sectors are not included. The two Asia-based transnational corporations (TNCs) have developed their own unique corporate strategies in the media and information sectors. As Johnson and Scholes claim (2002, 10), "strategy is the direction and scope of an organization over the long term, which achieves advantage for the organization through its figuration of resources within a changing environment and to fulfill stakeholder expectations." Sony and Samsung have deliberately expanded the long term strategy to the cultural sector, which means traditionally non-media corporations have turned into media conglomerates. While expanding their business areas through consolidation, they have substantially developed their own software based on their hardware sectors, which are different from other global media giants, including Disney and Time-Warner because these latter media giants do not have significant hardware sectors. Sony and Samsung's key businesses are very different: while Sony has developed significant resources not only in the electronics business but also in music, movies, and software, Samsung has centered on the electronics industry. However, Samsung has rapidly developed its cultural sector in the midst of globalization over the last ten years (Chang, 2008).

Sony and Samsung have indeed been compelled to grapple with change in the cultural sector, because they acknowledge that audiences have been migrating away from conventional radio and television, thereby eroding the mass-market, or broad-reach, media model that once predominated (Jenkins, 2006). They also learn that average hours per person per year spent with cable, satellite television, video and online games, and the Internet enormously increased (D. Schiller, 2007, 112). As with many global media giants, for Sony and Samsung, networks are the basis of new media and information technology which are powerful entities and they believe those who control them influence content providers' access to people and people's access to content (Winseck, 2002). In such circumstances, Sony and Samsung were well equipped to assimilate emergent areas of commodified cultural production, such as video and computer games, that had been spearheaded and cultivated by independent capital (D. Schiller, 2007, 114). They have advanced new technologies particularly in electronics for decades, and they have recently invested in cultural sectors including film, broadcasting, and gaming in different ways.

Sony, a Convergence Mania in Digital Technologies

Sony currently has five major corporate areas, including as electronics, music, games, movies, and financial services, and its primary goal is the integration of these different sectors through convergence processes. Headquartered in Tokyo, Japan, Sony began in 1946 and had 163,000 employees as of March 2007 (Sony, 2007). Sony as with many Japanese electronics companies had established its empire with originally several electronics products, including TV sets, home audio, LCD and semi-conductors. However, the proportion of sales of electronics has decreased over the last several years, while game and movie sectors are increasing, although Sony's sales have recently decreased due to factors such as the slowdown of the global economy and the appreciation of the Japanese yen (Sony, 2009). Sony first entered the cultural market in the 1980s with their Walkman and Play Station 2 and 3, and Sony's products and cultural goods penetrated the global cultural market. From hardware, such as the Walkman and console games, to software, including television, film, music, and game programs Sony has become one of the fastest growing Asia-based cultural industries. The corporation believes, "Sony should remain one company, given its far-flung business interests, arguing that the conglomerate's entertainment brand actually buttresses its hardware brand" (Laporte, 2006, 2). Sony indeed seems to have a clear idea: to become a global leader and make massive profit in the cultural industries through the convergence of hardware and software. Sony has developed software power, which would utilize Sony's global hardware platforms.

Sony's destiny in regards to the cultural industry particularly changed with the invention of the Walkman, which became a symbol of Sony and was released in 1979. This portable personal stereo player, which originated due to a personal request for a highly portable player that would allow him or her to listen to stereo recordings on long international flights, introduced a shift in music listening habits, allowing people to carry their own choice of music with them (Nathan, 1999). The invention and development of this cultural artefact particularly expedited Sony's role as the cultural industry in integrating the artefact with cultural content. Sony wanted to be an integrated producer of consumer electronics products. As Akio Morita, former CEO of Sony, stated in 1989, software would have to be produced to match the growth of new electronics products such as video tape, video disks, high-definition television, and satellite broadcasting (du Gay et al, 2001). Sony has expanded its major areas of production to cultural sectors in order to enjoy the synergy effects of the integration between hardware and software produced within the same company:

"Sony boasts a vast range of content, including games, motion pictures and music. By seamlessly integrating such content with our hardware, we make it possible to enjoy motion pictures,

music, games and other entertainment content on products like LCD television, Blu-ray Disc players, audio players and PlayStation 3, thereby creating innovative products and services that offer new ways to excite and amuse" (Sony, 2007, 8).

Regardless of some criticism of this software driven corporate policy in and out, Sony believes that the integration of its hardware and software is the most profitable and valuable business model to be continued. In other words, Sony has developed the cultural sectors as major commodities to be sold and bought in a form of convergence between software and hardware. Sony as a profit-driven private corporation, motion pictures and music are no longer art. Art could act as a form of critique for the rest of life, and it could provide a utopian vision of how a better life might be possible. However, culture had almost lost this capacity to act as utopian critique because it had become commodified (Hesmondhalgh, 2002). Commodity denoted the economic, commercial, and industrial logic and activities associated with capitalist market places (Gunster, 2004). In this regard, Horkheimer and Adorno pointed out (1969, 121):

"movies and radio need no longer pretend to be art. The truth that they are just business is made into an ideology in order to justify the rubbish they deliberately produce. They call themselves industries."

For Sony, cultural products are nothing but commodities, which exemplify the process of transforming use values into exchange values (Mosco, 2009), and the integration of hardware and software is not optional, but a must do business model to grow. Digital convergence continues to be one of the most significant strategies in the cultural industries, and Sony has begun to integrate hardware power and software power.

Samsung Benchmarked Sony towards Transnational Cultural Industries

Samsung Corporation encompasses almost every profitable sector, including semi-conductor, electronics, advertising, film and theme park. Samsung has become one of the world's distinctive brands in several areas, producing a 3G phone with satellite television and making the world's biggest plasma television (Lee, 2007). As of September 2007, the Samsung group had as many as 229,000 employees in more than 40 subsidiary companies (Samsung, 2007).

Sony used to be a larger corporation than Samsung until 2004. At that time, Sony was ranked 30th out of the world's largest companies with $66 billion of revenue, while Samsung Electronics was 54th with $54 billion; however, Samsung surpassed Sony in 2005 and ranked 39th ($71 B), compared to Sony (47th, $66 B) (*Fortune*, 2004; *Fortune*, 2005). In 1997, Sony took little notice of the Samsung Electronics Co., a Korean television maker fighting a life-or-death battle to survive the Asian currency crisis. Samsung Electronics has brought a large number of components for synergy creation, including semiconductor, telecommunication, and TV sets. For a decade, Samsung Electronics has been a top competitor in the memory chip business and cellular phone sets, and its competitiveness is the result of strategic investments and aggressive expenditures in R&D in these areas (Lee, 2006). Less than a decade later, Samsung has doubled the market capitalization of Sony, and Samsung has become what Sony could once claim. The invincible factory of ideas founded almost six decades ago by Akio Morita, the company that brought the world the transistor radio, the Walkman and the Trinitron television tube, seems to have lost its way (Brooke and Hansel, 2005), although it is premature to predict the future of Sony with only a recent business performance.

What is certain is that Samsung has rapidly jumped into the cultural sectors, including cable, film, advertising, and online games since the mid-1990s,, and it has emerged as one of the largest global cultural industries. Samsung has expanded its investment for its high profit potential with its convergence with hardware companies and technologies. As digital becomes the rule rather than the exception in the global market, Samsung wanted to develop digital content and to integrate them with hardware they have had. The consequence of the convergence process is the concentration of the cultural sector, including production, distribution, and exhibition industries, into already busted corporate giants. Samsung's holdings around the 1997 economic crisis consisted of the followings: newspaper (*Joong-ang Ilbo*); theater (Hoam Art Hall); film (Dreambox, a film importer and home video producer); Cable TV (Catch One and Q Channel); advertising (Cheil Communications); and music (Audio Soft), although some of them have been separated from the mother company in recent years (Shim, 2002; Jin, 2006).

Samsung has surely emphasized the cultural industry as one of the most significant new business areas. As Herbert Schiller argued (1989), since policies towards the cultural industries can no longer be separated from information and communication technologies (ICTs), included in the cultural industries are publishing, the press, film, radio, television, photography, recording, advertising, sports, and most recently, the many components that now make up the information industry. As such, Samsung as a cultural firm has developed efforts not only to produce cultural products but also to sell their products directly to consumers on a national scale and to the global market at the same time. The stress on involvement in the global production processes of Sony and Samsung indicates that Japanese and Korean cultural industries have tried to engage with global dynamics differently than other Asian local cultural industries, which primarily emphasize the local market

GLOBAL STRATEGIES OF CONVERGED CONGLOMERATES

Sony and Samsung have undoubtedly attained visibility in the global market and a degree of notoriety in recent years as corporate policy objects (Pratt, 2005). They have developed, changed, and intensified their corporate policies in the cultural sector as transnational cultural industries by mapping out their capital expansion.

Sony's Corporate Policies Promoting Software Power amid Globalization

Sony has vehemently expanded its penetration in the global market. In order to best exploit their global profits, Sony and later Samsung as capitalist leviathans are eager to precede foreign direct investment (FDI) in potential cultural markets (Gershon, 2002). As is well known, one of the most significant trends of Sony's cultural penetration beginning in the late 1980s was direct investment in Hollywood. Although Sony has emphasized internal integrations between hardware and software, it has invested its capital in order to secure its distribution and exhibition apparatus by owning commercial cultural corporations, as in the case of major media giants, including Disney and News Corporation. Sony's move has been part of the integration of its globalized hardware and software, which are cultural products.

Sony's globalization began in the U.S. market by purchasing Columbia Pictures Entertainment for $3.4 billion in November 1989, because Sony wanted to be an integrated producer of consumer electronics products (Jin, 2003). Morita moved his entire family to the U.S. in 1963. In that way he would understand Americans, their market, customs and regulation (Ohmae, 1998). Although Sony was ranked at the top in electronics hardware products worldwide, it needed strong software, and it found the solution through globalization (Du Gay et al., 2001). Sony's effort to break into the film and television industry in the U.S was arguably unsuccessful. Sony's purchase of Columbia aroused strong passions among politicians and media in the U.S., who condemned Sony's bid as an invasion. Many Americans saw the proposed deal as negative and claimed that the acquisition of one of the symbols of American culture was the equivalent of buying a part of America's soul (Jin, 2003).

Nevertheless, Sony's global investment has continued into the 21st century. Sony bought MGM for $5 billion in 2004, which would not only give the company an enormous film library but also considerable power in its fight to set the format for the next generation of digital video discs (Belson and Sorkin, 2004). Howard Stringer, who at the time headed Sony's U.S. arm, figured the library's 4,000 film titles would give Sony more clout with DVD retailers and fuel the growth of its high-definition Blu-ray technology, which is the system developed by Sony to view next-generation high-definition DVDs, as part of Sony's strategy. Stringer states, "MGM's valuable catalog grows more attractive with Sony's ability to distribute it to a huge international audience on multiple devices—think DVDs, PlayStations, cell phones and more" (Goldsmith, 2004). As such, the convergence of software and hardware (or content and gadgets) has been Sony's mantra for years, and that fuzzy concept seems to be starting to acquire a firm edge.

Sony lost its control of MGM in 2006, when MGM's board voted unanimously to dismiss Sony Pictures Entertainment as its domestic DVD distributor after it failed to meet performance goals (Eller, 2006). The right of MGM's DVD distribution has gone to News Corporations' 20th Century, instead. This setback has been recorded as another miscalculation for Sony. Sony took only a minority stake in the studio, then lost control of the prize it had fervently sought, meaning the potential profitable combination of software and hardware failed to materialize (Eller, 2006).

As such, Sony planned to have software partially by purchasing U.S. cultural companies; however, the content side of the business was still a Hollywood studio, and studios regard America as the center of the world (*Economist,* 1999, 56). Of course, Sony has never given up its globalization process, and Sony has turned its focus to the production and distribution for network TVs through its own Sony Pictures Television. Sony Pictures Television as a division of Sony of America in the U.S. owns its own studios, for example. As one executive of the company states,

"we don't own a broadcast network, we don't own TV stations, and we don't own a general interest cable network. Networks and their in-house studios can provide all the standard issue fare they need for themselves. We have to focus on creative and different ideas because that's what breaks out, that's what we need for a network to put (a Sony show) on the air" (Adalian, 2004, 14-15).

Meanwhile, Sony's corporate strategies have diverted its focus to other countries. The company decided to internationalize its production because American programming was being edged out in favor of locally produced programs. Sony now has 24 TV channels in 62 countries and production facilities all over Asia and Latin America, while it is launching channels in Europe. Some Sony channels are proving to be remarkably successful. Sony Entertainment Television in India, for example, is neck-to-neck in the ratings with Zee, India's most popular commercial channel. In 2005, Holding company SET Satellite acquired Hindi-lingo channel SAB TV from Sri Adhikari Brothers Television Network for around $13 million. The deal, which includes 20% of the SAB library, would give SET access to local award-winning sitcoms, a genre SAB TV has made its own (Lyer, 2005).

Sony also bought the BMG unit of Bertelsmann to create the world's second largest recorded music company after Universal Music in 2004 (Clark and Meller, 2004). Sony BMG claims an estimated 23% of worldwide music sales through the merger. Unlike its buyouts in the film and television sectors, those in the music industry have been successful in the U.S., which is the world's biggest music market. The company accounts for more than one out of every three new releases, making it the nation's biggest music company (Leeds and Meller, 2004). Sony has continued its clear corporate policy of the integration of hardware and software and its global investment as an Asia-based transnational cultural industry, regardless of occasional setbacks. Sony has also continued to extend its glocalization strategy, known as "think globally, act locally." From the start, Sony has actively promoted localization by recruiting local managers in order to respond to an antidumping lawsuit against Sony since Sony began trying to grow in the U.S. market (Chang, 2008). In particular, Sony's cultural sector, including its music and film business has been controlled by a U.S. subsidiary and has been managed almost entirely by local managers, which is the major glocalization strategy originated by Sony.

Samsung's Corporate Policies amid Globalization

Samsung's interest in the cultural industries has been intriguing, in particular, its follow-up globalization strategy of Sony. Early in the 1990s, the Korean government applied the globalization logic to the cultural industry as well as to other businesses to compete with the global cultural empire. For the Korean government, Sony's buyouts of Columbia Pictures and CBS Records in 1989 as well as Matsushita's purchase of MCA Studios in 1991 were seen as models for Korean electronics companies in entering new business ventures (Shim, 2002). The government and large companies emphasized that cultural software was to be more important than electronic hardware in the future. They also believed that it was time for Korea to invest in Hollywood, including production studios and sound recording studios (Han, 1994).

Several large conglomerates, particularly Samsung, followed faithfully the logic of investment in the cultural industries, including television programming, film production, and music. Samsung needed to invest in the cultural industries not only for high profit but also for transforming their industrial system from heavy-chemical industries to service and content businesses. As in the case of Sony, Samsung has used its strong base in home electronics as a springboard for aggressively participating in the cultural sector since the early 1990s. In other words, based upon its existing media and hardware industries, Samsung invested in cultural industries in the 1990s.

Most of all, the establishment of and withdrawal from Samsung Entertainment Group (SEG), whose major function was to achieve its goal of the integration of hardware and software power, is critical to understand Samsung's corporate strategy in the cultural sector. In 1995, Samsung launched the SEG, which had been responsible for managing its communication arm, including film, as a new company incorporating previously dispersed film, music, and cable businesses into a single unit (Kim, 1995). The SEG produced several films, including the first Korean blockbuster movie 'Shiri' in 1999 and imported several Hollywood movies, such as 'Leon' in 1995 and 'The Fifth Element' in 1997, respectively.

Regardless of Samsung's investments in the cultural sector, Samsung still considered itself as a hardware-driven corporation, which has partially influenced the failure of the film business. One episode during the meeting between Lee Kun Hee, CEO of the Samsung Group and Spielberg in 1995 explains what Samsung wanted with its plan in the film business, and how it failed. At that time, Lee desired to collaborate with Spielberg to invest in DreamWorks. A February business meeting then was supposed to clinch the deal between Spielberg and Samsung, whose proposed stake in DreamWorks had grown from $500 million to $900 million, as the talks progressed. However, Spielberg was suspicious of Samsung's true intention, because it focused on semiconductors too much:

"How are they going to know anything about the film business when they're so obsessed with semiconductors? The word semiconductors must have been used about 20 times during that 2-1/2-hour encounter" (Corliss and Bonfante, 1995, 54).

Spielberg's recollection clearly states what Samsung had pursued. On the one hand, Samsung wanted to integrate its hardware and software by investing in the film business; on the other hand, Samsung strongly continued to focus on hardware, unlike Sony. DreamWorks finally rejected Samsung's proposal because Samsung asked for a one-third share in the studio and all distribution rights across Southeast Asia (Park, 1995). Samsung's corporate policy in the film industry has not been smooth, and only a few years later, Samsung disbanded the SEG mainly because it did not create enough profits. Since Samsung has been successful in many areas, this withdrawal from the film business was considered as a rare failure of Samsung's business model as the cultural industry. The hardware/software synergy Samsung had pursued was viewed as a failure as Sony experienced it in the 1990s.

Of course, Samsung did not entirely withdraw from the cultural sector, and the convergence of hardware and software for profits has continued. Samsung has recently launched a film download store featuring full length versions of films including Iron man and The Dark Knight in an attempt to take ownership of the mobile movie market. Samsung is keen to take advantage of the phone's large touch screen and surround sound features to meet the rising consumer demand to view content online and via mobile devices (Farber, 2009). Samsung also established Samsung Venture Investment Co. and it still invests in the film business, although the scope of the business in money and workers is smaller than SEG's (Kim, 1999, 30). Samsung Venture Investment particularly imported about 20 films from New Regency Production in the U.S. until 2005, because this distribution right has been transferred from the SEG. In September 2007, Samsung also announced its support of independent movie production along with Landmark Theaters, which is America's largest theatre circuit dedicated to independent film (Park, 2007). Samsung claims that it wants to use the film business for the marketing of Samsung hardware. The company sees another opportunity to work in the content area, which is one of the most profitable and therefore most wanted businesses by TNCs.

NEW CORPORATE STRATEGIES: SONY AND SAMSUNG'S NETWORKED-BASED CONVERGENCE

Sony and Samsung have continued their long-term corporate strategy of digital convergence to develop their corporate value. The extension of the corporate power of Samsung and Sony in the cultural industries indeed continues. However, their corporate strategy has dramatically changed again.

The starter is again Sony. When Nobuyuki Idei became Sony's president in 1995, he began advocating regeneration for Sony. With the advent of the digital and network age, Idei attempted to develop network services through models such as the Internet, e-commerce, and broadcasting in order to help Sony create greater synergy between its hardware and software assets (Chang 2008). Sony highlighted information technology because it became easier to create synergy between hardware and software content as Internet and broadband technology developed. This synergy was difficult to achieve in the analog age because music and film content were packaged and delivered separately in different media such as CDs or VCRs. In the digital age, however, all such content is composed, stored, and shared real-time in digitalized formats

(Chang 2008). Yet Idie's vision went unrealized until recent years; for example, Sony did fail to develop a digital music player similar to Apple's iPod, with a few exceptions including PlayStation 2 and 3. Regardless of the fact that some critics doubt whether Sony could ever create synergy between its electronics and content business as the falling revenue and sales of Sony prove in the digital age, Howard Stringer recently made a clear statement of the importance of software and its convergence with hardware, and in particular, the network-enabled entertainment devices and digital delivery of content:

"I have always been enthusiastic about hardware. Sony has a long and proud history of developing groundbreaking hardware that has excited consumers. However, consumers now expect product interconnectivity and the convenience that networking can provide, and we believe that we are uniquely positioned to make great, innovative hardware with these capabilities. Our networked products and services will differentiate us and make Sony products more valuable to the consumer, and will also provide more reasons for consumers to buy and use many Sony products together, with an interoperable Sony solution. The switch from making pure hardware to making networked products where the software and user experience are as important as the hardware itself is not a short-term effect" (Sony, 2009, 13).

What he envisions is the direct connection of hardware and software. Sony wants to connect more and more of its product groups to the network and to each other, and develop networked content. A notable example is the Sony-created PlayStation Network, which has been operational since the November 2006 launch of PlayStation 3 and is used by 24 million users (Sony 2009).

Samsung is not much different. Samsung has rapidly utilized network-based convergence as seen in the online game and music industries in recent years. Online gaming is one of the most significant sectors in Korea, so in 2000 Samsung organized Samsung KHAN, the first online game competition company in order to lead and formulate youth digital culture, which has accelerated the boon of online gaming as electronic sports (Ministry of Culture and Tourism, 2006). This new business is not the same as Sony, which has invested in the console game business; however, the integration of hardware with software is the financial promise for Samsung in a digital society as the latest attempt to commodify the new media and cultural sector. In fact, many people buy Samsung's LCD TV set to see online game competitions between pro-online gamers. Samsung's latest entry into the cultural industries is Samsung Media Studio (SMS), which Samsung created in May 2006 for users to download music and music videos. Samsung's new service, known as Yepp, connects its MP3 player and Yepp to download music; however, Samsung plans to use Yepp for digital contents and eventually integration with its hardware, as Apple's iPod continues (Jung, 2006).

The trend of commodification of new cultural areas has been a confirmed business strategy of Samsung, and this corporate model is more likely to perform in diverse ways. In other words, Samsung alongside Sony focuses on the ways in which it commodifies their hardware through the integration with software. Cultural content and production must serve a corporation's needs for functional integration and market demand for mass production (Mosco, 2009). What the contemporary cultural industries symbolize is that they have moved closer to the center of the economic action in many countries and across much of the world. Moreover, the cultural industries can no longer be seen as secondary to the real economy (Hesmondhalgh, 2002). Cultural industries have developed their efforts to sell directly to consumers on a national scale likewise accelerated.

Sony and Samsung have had vastly different portfolios during their historical development as the cultural industries. They have expanded their functions in the production of messages and imagery, while circulating their productions through the digital convergence of hardware and software. Sony and Samsung are not completely overlapping each other in their product lines, so that gives both an advantage. Sony continues to develop film, music, cable television areas along with video games, while Samsung emphasizes new media hardware such as mobile and semiconductor, because they have their strengths in commodifying these products.

On the flip side of the same token, Sony has slightly ignored the importance of electronics while emphasizing software, such as music and movie. In the case of Samsung, it has a little emphasis on software, not much behind, though. Their strategic emphasis has brought similarities and differences in their businesses. Sony has gone ahead in several areas, and Samsung has benchmarked what Sony did; however, now Sony considers what Samsung has focused on, although Sony does not want to change its corporate policies entirely.

CONCLUSION AND DISCUSSION

Sony and Samsung, two Asia-based cultural corporations, have attained visibility and legitimacy in the global market. While the U.S. - based culture industry has long been the paramount purveyor to the global market, these local cultural productions have increased their presence throughout the world. Although, at the capillary level, but notably accommodating some movement by domestic capital into the ranks of the second-tier transnational companies, a scattering of conglomerates domiciled in countries such as Korea, China, and India have thereby joined those based (D. Schiller, 2007).

Sony and Samsung have continuously developed corporate strategies to multiply the functions performed by their distribution channels—cable systems, broadcast systems, satellites, and telecommunications networks. They have also advanced digital technologies as their hardware, and they have developed new hybrid services; on one hand, based on their hardware power, on the other hand, focused on foreign markets. In fact, Sony and Samsung prove that national capital has often claimed a significant role in implanting and reorganizing the enlarged transnational complex. Sometimes freshly constituted national capital has sought and gained substantial access to its home market and, on a more selective basis, even expanded into foreign markets. National capital's importance is as much political as economic: under the sign of neoliberal reconstruction, it typically allies with the leading transnationals behind a politics stressing convergence and industrial consolidation (D. Schiller, 2007). Sony and Samsung have invested in several countries, including the U.S., which has made them as global transnational cultural industries, although their approaches toward the global market have been different.

The global cultural industries have entered a period of unprecedented change which raises issues of fundamental importance for our society. Under this circumstance, Sony and Samsung develop their unique corporate policies in the cultural industries in the midst of globalization. Whatever their major activities are, one clear thing is that they are pursuing digital convergence, and Sony and Samsung possess their own communications facilities for message making and transmission to national, regional, and global consumers, as well as commodification of culture.

The major focus has fundamentally changed. Unlike in the 1980s and early 1990s when they emphasized chemical integration of hardware/software, the two conglomerates have experienced unprecedented failure or at least less synergy effects than expected; therefore, they are heading towards a new business corporate strategy, which is the information-based cultural industry. As Howard Stringer claims, Sony [and Samsung] is not pursuing a simple integration between hardware and software, but seeking networked software and content as in the cases of the Sony-operated PlayStation Network and Samsung's mobile-based movie download services. Conglomeration clearly entails an increase in the scope and power of individual cultural-industry corporations, in that the same corporation can have stakes in many different forms of communication. However, Sony and Samsung have experienced the decreasing value of convergence as in many U.S.-based corporations, and they have changed their corporate strategy accordingly.

Of course, their goal of convergence and change in corporate strategies is not mainly because of their utopian vision of how a better life might be possible through their cultural industries, but mainly because of their vision to produce cultural commodities. For Sony and Samsung, cultural industries are primarily businesses in accordance with a corporate strategy based on economic considerations. They expand their interest from hardware to software, and the integration between these two sectors is their business motto. In somewhat different ways, the two Asian-based electronics corporations have focused on cultural sectors through the convergence of software and hardware, of course, based-on their strong hardware bases. The changing nature of the integration of hardware and software has become a must do business model to create more commodities, and therefore, more profits.

REFERENCES

Adalian, J. (2004). Sony's back and ready for action. *Variety*, *393*(9), 14–15.

Bar, F., & Sandvig, C. (2008). US communication policy after convergence. *Media Culture & Society*, *30*(4), 531–550. doi:10.1177/0163443708091181

Belson, K., & Sorkin, A. (2004, 15 September). Buying MGM may give Sony more leverage to set a new DVD standard. *The New York Times*, C6.

Brooke, J., & Hansell, S. (2005, 10 March). Samsung is now what Sony once was. *The New York Times*, C1.

Chambers, T., & Howard, H. (2005). The economics of media consolidation. In Albarran, A. B., Chan-Olmsted, S. M., & Wirth, M. (Eds.), *Handbook of media management and economics* (pp. 363–386). Mahwah, NJ: Lawrence Erbaum.

Chan-Olmsted, S. (1998). Mergers, acquisitions, and convergence: the strategic alliance of broadcasting, cable television, and telephone services. *Journal of Media Economics*, *11*(3), 33–46. doi:10.1207/s15327736me1103_4

Chang, S. J. (2008). *Sony vs. Samsung: the inside story of the electronics giants' battle for global supremacy*. Hoboken, NJ: Wiley.

Clark, N., & Meller, P. (2004, 16 June). Europe hears objections to proposed music merger. *The New York Times*, W1.

Corliss, R., & Bonfante, J. (1995). Hey, let's put on a show. *Time*, *145*(13), 54.

Couldry, N. (2007). Researching media internationalization: Comparative media research as if we really meant it. *Global Media and Communication*, *3*(3), 247–250. doi:10.1177/1742766507082569

Du Gay, P., Hall, S., Janes, L., Mackay, H., & Negus, K. (2001). *Doing Cultural Studies: The Story of the Sony Walkman*. London: Sage.

Eller, C. (2006, 20 October). MGM deal a bold miscalculation for Sony. *Los Angeles Times*, C1.

Fagerjord, A., & Storsul, T. (2007). Questioning Convergence. In Storsul, T., & Stuedahl, D. (Eds.), *Ambivalence towards Convergence: digitalization and media change*. Goteborg, Sweden: Nordicom.

Farber, A. (2009, 19 March). Samsung makes bid for Mobile Movie Market with Download Service. *New Media Age*, 3.

Gershon, R. A. (2002). The Transnational media corporation and the economics of global competition. In Yahya, R. K. (Ed.), *Global Communication* (pp. 51–72). Belmont, CA: Wadsworth Publishing.

Goldsmith, J. (2004). How did Sony do it. [October.]. *Variety*, *396*(7), 4.

Gunster, S. (2004). *Capitalizing on Culture: Critical Theory for Cultural Studies*. Toronto: University of Toronto Press.

Han, G. J. (1994). Promises and Myths of Cable Television and Telecommunications Infrastructure in Korea. In *Elite Media amidst Mass Culture: a Critical Look at Mass Communication in Korea* (pp. 156–157). Seoul: Nanam.

Hesmondhalgh, D. (2002). *The Cultural Industries*. London: Sage.

Hesmondhalgh, D. (2008). Neoliberalism, Imperialism and the Media. In *Hesmondhalgh, D. and J. Toynbee (2008). The Media and Social Theory* (pp. 95–111). London: Routledge.

Horkheimer, M., & Adorno, T. (1969). *Dialectic of Enlightenment (trans. E. Jephcott)*. Stanford, CA: Stanford University Press.

Huang, J. S., & Heider, D. (2008). Media Convergence: a case study of a cable news station. *The International Journal on Media Management*, *9*(3), 105–115.

Jenkins, H. (2006). *Convergence Culture*. New York: New York University Press.

Jin, D. Y. (2003). Globalization of Japanese Culture: Economic Power vs. Cultural Power, 1989-2002. *Prometheus*, *21*(3), 335–345. doi:10.1080/0810902032000113479

Jin, D. Y. (2006). Cultural Politics in Korea's Contemporary Films under Neoliberal Globalization. *Media Culture & Society*, *28*(1), 6–23. doi:10.1177/0163443706059274

Johnson, G., & Scholes, K. (2002). *Exploring Corporate Strategy* (6th ed.). Harlow, UK: Pearson Education.

Jung, J. Y. (2006). Samsung Electronics open Multi-media Service Samsung Media Studio. *Electronics Newspaper*, 16 May.

Kim, M. H. (1995, February). Samsung launched Samsung Entertainment Group. *Chosun Ilbo*, *7*, 21.

Kim, T. S. (1999). Venture Investment Companied Inroad in the Film Business. *Kookmin Ilbo*, *30*(November), 30.

Larpote, N. (2006). Traveling Man. *Variety*, *402*(3), 2.

Lawson-Borders, G. (2003). Integrating New Media and Old Media: seven observations of convergence as a strategy for best practices in media organization. *The International Journal of Media Management*, *5*(2), 91–99.

Lee, D. Y. (2006). *Samsung Electronics: the global inc*. Seoul: YSM.

Lee, K. S. (2007). Samsung Corporation. In Downing, J. (Ed.), *The International Encyclopedia of Communication: International Communication*. Oxford: Blackwell Publishing.

Leeds, J., & Meller, P. (2004). Europe Clears Merger Plan of 2 Giants in Recordings. *The New York Times*, 20 July, C1.

Lyer, S. B. (2005). Sony buys up sitcom powerhouse. *Variety*, *398*(5, 21 March), 20.

McChesney, R. (2001). Global Media, Neoliberalism, and Imperialism. *Monthly Review (New York, N.Y.)*, *52*(10), 1–19.

Mckenna, A. (2000). Emerging Issues Surrounding the Convergence of the Telecommunications, Broadcasting and Information Technology Sectors. *Information & Communications Technology Law*, *9*(2), 93–128. doi:10.1080/713670487

McPhail, T. (2006). *Global Communication: theories, stakeholders, and trends* (2nd ed.). Oxford: Blackwell.

Ministry of Culture and Tourism. (2006). *2006 Korea Online Game White Paper*. Seoul, Korea: Ministry of Culture and Tourism.

Mosco, V. (2009). *The Political Economy of Communication* (2nd ed.). London: Sage.

Nathan, J. (1999). *SONY*. New York: Houghton Mifflin Company.

Ohmae, K. (1998). Akio Morita. *Time*, *152*(23), 100.

Park, S. H. (1995). Negotiation between DreamWork and Samsung Failed. *Chosun Ilbo.*, *24*(March), 9.

Park, S. H. (2007) Samsung's reinvestment in the film business it ended in 9 years ago, *Chosun Ilbo*, 18 September.

Pratt, A. C. (2005). Cultural Industries and Public Policy. *International Journal of Cultural Policy*, *11*(1), 31–44. doi:10.1080/10286630500067739

Samsung Group. (2007). Retrieved from http://www.mind-advertising.com/kr/samsung_kr.htm

Samsung. (n.d.). *Samsung Introduction.* Retrieved from http://www.samsung.co.kr/about/ir_annual.jsp

Schiller, D. (2007). *How to think about Information*. Urbana, IL: University of Illinois Press.

Schiller, H. (1989). *Culture Inc*. New York: Oxford University Press.

Shim, D. B. (2002). South Korean Media Industry in the 1990s and the Economic Crisis. *Prometheus*, *20*(4), 337–350. doi:10.1080/08109020210000233336

Sony. (2007). *Annual Report 2007*. Tokyo: Sony.

Sony. (2009). *Annual Report 2009*. Tokyo: Sony.

Sony Corporation. (2007). *About Sony*. Retrieved from http://www.sony.net/SonyInfo/CorporateInfo/

The Weakling Kicks Back. (1999, July 3)... *Economist*, *352*, 56.

Thussu, D. (2006). *International Communication: change and continuity* (2nd ed.). London: Arnold.

Winseck, D. (2002). Netscapes of power: Convergence, consolidation and power in the. Canadian mediascape. *Media Culture & Society*, *24*(6), 795–819. doi:10.1177/016344370202400604

Wolf, M. J. (2000). *The Entertainment Economy: how mega-media forces are transforming our lives*. London: Penguin.

World's largest companies. (2004). *Fortune, 150*(2), 163-180.

World's largest companies. (2005). *Fortune, 152*(2), 119-139.

Chapter 23
Fostering Innovation in Converging Audiovisual Media Services:
Do Mergers and Acquisitions Really Help?

Sergio Sparviero
Dublin City University, Ireland

ABSTRACT

The digitalization of media content and communication processes has been used as the main justification for an important turn in the design of regulations affecting various audiovisual media services and information industries, more in general. A brief case study presented in this chapter will testify that from the 1950s until the 1990s the United States' regulator facilitated the introduction of fundamental innovations by preventing vertical mergers between different audiovisual media services. On the contrary, nowadays regulators in different parts of the world seems to agree that differentiating between different technological platforms can hinder the emergence of new multimedia services based on technology convergence as well as the establishment of inter-platform competition. Alternative economic principles and the rationale of the old regulatory regime are used here to challenge these mainstream arguments, which justify the raise of the new and 'converged' information conglomerates.

INTRODUCTION

This chapter focuses on the benefits and shortcomings of mergers and acquisitions (M&As) in audiovisual media services, defined as all activities participating in the creation, distribution and delivery of audiovisual media content to end-users. In particular, this chapter describes the viewpoints of two types of stakeholders: the companies and the regulator. As the examples used to illustrate the relevance of different theories and claims regard the information sector of the United States, the regulator featuring in this account is the Federal Communication Commission (FCC).

In order to describe the current socio-economic climate and to define and explain the relationship between digitalization, convergence and the emergence of information conglomerates, this writing starts with the relevant literature review. Then the evolution of the audiovisual media services,

DOI: 10.4018/978-1-60960-037-2.ch023

Copyright © 2011, IGI Global. Copying or distributing in print or electronic forms without written permission of IGI Global is prohibited.

described through the most important technical and regulatory changes, is used as a case study of the 'old' regulatory regime dominating the information sector. From the history of this sector it emerges that, from the 1950s until the 1990s, the FCC and the large companies that dominated cinema box offices (i.e. the 'majors') had different viewpoints concerning the consequences of M&As. While many new activities emerged as a consequence of new ideas, new technologies and other innovations, the majors tried to own and/or control companies carrying out different types of activities within the sector in order to secure the distribution and exhibition of their products. The FCC, on the other hand, introduced many regulations aimed at avoiding the dominance of large companies and aimed at favoring the multiplication of independent media outlets.

The convergence of media and communication technologies seems to have also created a convergence of opinions between the FCC and the information conglomerates. While some communications scholars suggest that the recent wave of regulatory changes is determined by many factors, including the dominant ideology (see Dal Yong, 2008) or the pressure exerted by key corporations (e.g. McChesney, 1999, 2001), this chapter focuses on describing and challenging the rationale provided by the FCC to explain the benefits that society (and not just the corporations involved) draws from the emergence of larger and more powerful information conglomerates. This paper will argue that the FCC uses arguments underpinned by mainstream economic theories to justify M&As, and as a result, it implicitly compares information conglomerates to Fordist organization and manufacturing industries, which become more capable of introducing important innovations as they become larger. As explained in the closing sections of this paper, alternative economic principles can provide new and different arguments explaining that, on the contrary, fragmentation and interdependence between activities can foster the rate of innovation in the information sector. Furthermore, these arguments can bring to light new solutions to what has been considered so far a trade-off between favoring innovation in the information sector and preserving or expanding the range of diverse sources of information that users-citizens have access to.

BACKGROUND

The digitalization of media and communication technologies is radically transforming information industries and the way in which they affect the life of users-citizens. The swarm of changes that has involved digital communication technologies can be divided into four clusters of innovation. The first of these clusters concerns the digitalization of media and relates to the creation, storage, delivery and use of media products and services. Thanks to this cluster of innovations, media content is now de-materialized, separated from its physical form so that it can be manipulated far more easily than analogue forms. Moreover, the data generating content can be compressed in very small spaces and accessed at very high speeds and in non-linear ways (Lister, Dovey, Giddings, Grant, & Kelly, 2003; Flew 2007). The innovations responsible for the digitalization of media (i.e. the improvements in semiconductor technologies) are also at the origin of the much-debated information, communication and media industries' convergence, as communications and media content can now be routed and delivered using different combinations of tools and networks.

In the literature, there are at least three main strands of convergence studies and each one of them has a particular focus (Küng, Picard & Towse 2008). A first approach focuses on computers and their increasing role as versatile communication tools, while a second approach focuses on the rise of new networks and their capacity to offer different communication services. A third approach looks at the consequences of digitalization on the organization of the communication sector and the rise of information conglomerates that incorporate content (i.e. media), computing (i.e. information technology) and communication (i.e. telecoms and broadcast distribution) industries (Küng et al., 2008).

The second innovation clusters concerns the production of content in communication industries and how users' experience is changing with the advent of new media. Scholars investigating these innovations highlight that many-to-many communication, interactivity and virtuality are the most important features of new communication tools (e.g. Silverstone, 1999), or simply, that user experience is now multidimensional (e.g. Schroder, Drotner, Kline, & Murray, 2003). The increased possibility of many-to-many communications is given particular attention because of the potential consequences arising from the opportunities that users have, first, to distribute their own content on (new) mainstream media, and second, to access more specialized sources of information (van der Wurff, 2008). The third innovation cluster concerns the new opportunities that media channels have to expand and access a global audience (e.g. Flew, 2007), while the fourth and last cluster identified here concerns the 'de-massification' of media. From a user's viewpoint, innovations of this cluster have enabled a more personalized/individual consumption of media (Schroeder et al. 2003), while one of the most important aspects of a media de-massification, from a provider's viewpoint, is a better audience targeting (Benson, 2006).

Notably, it is not the first time that a set of key inventions triggers a wave of changes that profoundly affects the industrial structure of capitalist economies, by favoring the emergence of new production processes, new industries and new regulations, and by causing the retirement of 'old' industries and 'old' ways of making things. Notably, before microprocessors and computer technologies, the productions of steam engines, electricity and steel have also been at the origin of radical economic changes (see Freeman & Soete, 1997).

These lengthy periods of change, which were formerly known as 'Kondratiev waves', are now more precisely defined by some 'alternative economists' (i.e. neo-Schumpterian) as techno-economic paradigms or long waves of economic development. These alternative economists also explain that the current set of transformations are part of a fifth, unfolding, long wave of economic development triggered by the introduction of semi-conductors and microchips, and the consequent development of computer and telecommunication technologies (e.g. Freeman & Louçã, 2001). They also argue that regulations play a fundamental role in the development of techno-economic paradigms. Long waves of economic development are characterized by an economic upswing, during which new technologies and new industries emerge and old production processes are re-shaped as a consequence of the adoption of these increasingly cheaper and pervasive technologies. After reaching a peak, the downswing is characterized by the gradual exhaustion of the potential of new technologies and by investors' increasing interest in finding new and different types of innovations, some of which are responsible for the following (and new) techno-economic paradigm. In total, a long wave of economic development can last between 40 and 60 years and, although short term crisis and recessions still occur, the upswing of a techno-economics paradigm is characterized by strong economic growth, while the downswing is characterized by a modest rate of economic growth (see Freeman & Soete, 1997). The concept of long wave of economic development provides an interesting framework for understanding the 'regulatory turn' of the 1990s that is responsible for transforming the information sector. As stressed by scholars writing about long waves of economic development, the existence of a techno-economic paradigm depends, on the one hand, from the introduction and potential of radical innovations, and on the other hand, from the emergence and existence of a complementary socio-institutional climate. The socio-institutional climate is comprehensive of public policies and regulations; international investment, production and trade patterns; economic theories and statistical methods informing choices of public policies; and the emergence of new social activities (see Perez, 1983, for the original concepts).

The digitalization of communication and media technologies is an important aspect of the unfolding techno-economic paradigm. Also, it has been used as the main argument for the US 1996 Telecommunications Act and more recent changes promoting the information industries' de-regulation or liberalization in the United States, but also, for similar regulatory changes implemented by governmental institutions from all over the world (Freedman, Henten, Towse, & Wallis, 2008). The basic principle of these regulatory changes is that it seems increasingly meaningless to differentiate in terms of regulations between different technological platforms, which were installed originally for different purposes. Not only it seems meaningless to differentiate between technological platforms and their providers, but according to the mainstream argument it can even be counterproductive to do so, as this could represent a barrier to the development of inter-platform competition and the marketing of services across platforms, including all the new multimedia services based on technology convergence (Freedman et al., 2008).

However, there are different reasons for questioning the 'fitness' of this new regulatory regime and its efficiency in sustaining the unfolding techno-economic paradigm and these arguments represent the core substance of this writing. In essence, as argued in this chapter, new economic principles suggest that innovation can be fostered by regulations aiming at increasing the number of interdependent stakeholders in a sector. Generating the conditions for more innovations, of course, would also mean changing the socio-institutional climate so that the positive effects of the unfolding techno-economic paradigm can be sustained. Also, the recent history of the evolution of the audiovisual media services in the US –i.e. during the upswing of the current techno-economic paradigm - seems to confirm these principles, as regulations that aimed at increasing the number of interdependent stakeholders were implemented and innovations that still represent important characteristics of this sector were introduced following (and partly, as a result of) these rules.

INNOVATION IN AUDIOVISUAL MEDIA SERVICES BEFORE AND AFTER THE 'REGULATORY TURN': A CASE STUDY

The Fragmentation of the Audiovisual Media Industries (1950s to 1985)

Notably, before the diffusion of television systems in the 1970s (Winston 1998), cinema theatres were the main outlets for audiovisual media products and a handful of production companies were responsible for producing a very large share of all the films in circulation. These production companies were 'the majors', the Hollywood companies that dominated the box office revenues, that were initially conceived, owned and run by key innovators and entrepreneurs of this sector (e.g. Marcus Loew or the Warner brothers). These media companies were not only responsible for the production of films, but also for their distribution and exhibition. Most of these companies, in fact, owned a sizeable portion of cinema theatres, and those who did not own theatres were able to secure exhibition outlets for their production by block booking the venues and/or by other types of tight commercial agreements with theatre-owners (De Vany & Mc Millian, 2004).

At that time, the majors were Fordist organizations characterized by reutilization of the activities, task fragmentation and mass production (Storper,

1989). Actors, writers and other employees with less creative tasks were employed by the studios on a long term basis and were involved in the production of several movies (DeFillippi & Arthur, 1998).[1] A Fordist organization is characterized by a large production, but also by the standardization and/or the relatively low quality of its output, which is clearly a peculiar feature for creative industries. The majors, however, were forced to evolve from the Fordist mode of production as a result of two main factors. The first factor was the introduction and the diffusion of television broadcasting, while the second factor, which is more important for this analysis, is the set of consequences derived from the Paramount litigation.

The Paramount litigation started as an antitrust case between the Department of Justice and Paramount, and then led to a flow of Consent Decrees and decisions, issued and taken between 1949 and 1952, which aimed at breaking the vertical integration between film distribution and exhibition. As a result of this litigation the majors who owned cinema theatres were forced to divest: for example, Paramount lost some 1,395 theatres and a further 993 that it owned jointly with other operators; Fox was forced to divest 636 of its own theatres plus about 20 jointly owned; MGM (Loews) lost 135 and 20 jointly owned; Warner Brothers 501, and RKO 109 of its own theatres and 187 jointly owned (De Vany & Mc Millian, 2004). Furthermore, franchises and master contracts governing general terms were dissolved; some forms of contractual arrangements between exhibitors and film distributors, including block booking and forms of 'forward contracting', such as 'blind selling' (exhibition deals signed before the movie was produced) and season contracts (the purchase by an exhibitor of a studio's entire production of a season) were made illegal. In place of these tight contractual arrangements which restricted competition in the movie industry, standard spot contracts or single licenses for each individual film, theatre, and play date, became the only admissible form (De Vany & Mc Millian, 2004).[2]

Between the 1950s and the 1970s, AVMSs experienced some radical changes. First, the majors made 'blockbuster movies' their staple product, by reducing the number of films produced and by increasing the amount of resources invested in each production. This was a consequence of many factors: the reduced control that the majors had over the exhibition windows for their products as a result of the consequences of the Paramount litigation, the necessity to provide a viable alternative to the entertainment value of television programmes and to satisfy an audience with an increasingly selective taste (Aksoy & Robins, 1992; Brown, 1996; Scott, 2004).

In order to adapt to this new staple product, the majors aligned to the business model of independent producers and were forced to modify their production processes and organizational settings. Therefore, they separated the vertical stages of their production process and each blockbuster movie was delivered by creative and non-creative labor grouped together under a temporary organization. However, the new type of organization presented a shortcoming: as studios no longer had 'creative' personnel under contract on a long term basis, for each new project, producers had to recruit writers, directors, stars, and support crew (Lampel & Shamsie, 2003). However, despite these difficulties at the pre-production stage, organizing the production of a feature film as a temporary project was more sustainable and, as a result, it has became the norm since then.

Moreover, as part of their 'post-Fordist' strategy, financing and/or co-producing movies with independent producers or simply to distribute movies for them became a more important activity for the majors; in this way, the majors were able to deliver distinctively and qualitatively superior products (Robins, 1993) or, more simply, to diversify and to partially compensate from the reduced level of production (Aksoy & Robins, 1992). Therefore, from the beginning of the 1970s, audiovisual media *distribution* became an even more crucial activity of the complex networks of business led by the majors.

Also, by the 1970s, Hollywood studios were not only producing feature films for cinema theatres, but they were also involved with providing content for television. Therefore they were also affected by the introduction of other regulatory changes aimed at increasing competition at the production and distribution end of the audiovisual service industries; these Financial Interest and Syndication (or 'Fin-Syn') rules were adopted in 1970 and affected the sector's organizational structure for the following thirty years. As a result of these rules, the Networks were forced to purchase all their prime time programming from third party producers, with the exception of news and sports programmes (Bielby & Bielby, 2003; Christopherson, 2006).

The majors' television subsidiaries and new independent production companies benefited from a stronger demand for media entertainment which followed the introduction of the Fin-Syn rules and, throughout the 1980s and also as a result of these changes, variety between television programmes increased even if the Networks' strategies converged, to a certain extent, as they introduced the same *types* of successful formats (Christopherson, 2006). Second, thanks to the appearance of new technologies, but more importantly, given the increased competition and the consequent greater need for more diversity and distinction, film shooting outside of studios became more accessible and independent productions moved more regularly outside of Los Angeles for producing their movies. Therefore, the rationale behind these 'Runaway Productions' was twofold: artistic and economic. From an artistic viewpoint, shooting on location led to more distinctive and original productions, while from an economic viewpoint, moving productions away from Hollywood could help reduce production costs.

Therefore, while cinema and television converged in the sense that they became even more interdependent, cable television established itself as an ancillary service to terrestrial television from the start. Established in the early 1950s, Community Antenna Television (CATV) was the first form of cable television; its main business purpose was to capture local television signals using reception facilities and redistribute them to households using recently developed wideband amplifiers and coaxial cable technologies (Maule, 2003; Parsons, 2003; Bates & Chambers, 2004). Therefore, CATV services exploited the television stations' inability to provide the community within their Designated Market Areas (DMAs) with good reception, by rebroadcasting their signals and collecting monthly fees from their users. This industry was partly regulated by municipalities and local governments to whom they paid fees to buy rights of way (Strover, 2005). However, CATV was not under the FCC's jurisdiction as it did not use radio airwaves to broadcast radio signals. Moreover, although initially CATV did not pay any fees for retransmitting radio signals, the Networks benefited from the additional revenue they collected from advertisers as a consequence of the extra audience, therefore they viewed the expansion of these types of services very favorably.

However, the relationship between the Networks and the cable industry changed from 1961 when a San Diego-based cable company started to distribute the signals of four stations of the Los Angeles area in addition to the three local VHF stations (Corn-Revere & Carveth, 2004). As more cable providers followed this example, Multichannel Video Programming Distribution (MVPD) services became a reality and the cable industry started to create value added from audiovisual media content's variety rather then purely from its delivery. This new development in the cable industry, however, found the opposition of over-the-air (OTA) television channels, which were enterprises regulated and, most importantly, protected by the FCC. Therefore, in its First Report and Order of 1965, the FCC issued Syndicated Exclusivity (or 'Syndex') rules. These rules forced cable companies not to air programmes already aired by local stations, for a period of at least thirty days (Winston, 1998). Additionally, from 1969, cable television channels were allowed to show only movies that had been in circulation for at least ten years and sporting events that had been in circulation for at least five years (Strover, 2005); while from 1972 and by virtue of the 'must-carry' rule, cable companies were obliged to distribute the signal of the local television stations in the area where they operated (defined as a 50 mile radius) (Kassel, 2005). These regulations essentially aimed at preserving the local and rural dimension of the cable industry, however, they also weakened it; therefore provision was made to restrain or prohibit television and telephone companies from owning cable companies in the same local area (Strover, 2005).

However, from the 1970s, the cable industry was reinvigorated by the introduction of a new distribution method. As the cable industry and OTA television channels were separate industries (although interdependent) the cable industry had an interest in developing its own content, particularly if this content could represent an alternative to OTA television channels. As the transmission of television signals through satellite technologies became cheaper, some media entrepreneurs saw an opportunity to launch the new type of television station. This new type of station presented two advantages: first, it did not require a license from the FCC to use the radio spectrum, and second, it could reach virtually any household in the national territory that was using a (rapidly expanding) cable service. Home Box Office (HBO) is believed to have been the first of the non-broadcast networks. It was conceived by the Sterling Manhattan Cable owner, Charles Dolan, who launched the service in 1972 (Winston, 1998); before being distributed to cable companies through satellite communication from 1975, it was delivered by using the American Telephone & Telegraph (AT&T) long distance network (Strover, 2005). Another pioneer in the creation of non-broadcast channels was Ted Turner, who started leasing satellite time in 1977 to transmit thematic channels for cable companies (Winston, 1998). However, although HBO and other non-broadcast networks had the opportunity to expand and to be distributed to hundreds of cable systems over the national territory, they were still severely affected by the restrictions imposed on their choice of audiovisual media content introduced by the FCC in the previous decade. Therefore, in 1977 HBO brought the Commission to court claiming that, by limiting its programming options, the latter had exceeded its jurisdiction. Finally, the District of Columbia Court of Appeals supported HBO's position and recognized that it deserved greater First Amendment protections because of its new editorial role (Strover, 2005).

Therefore, from their conception, non-broadcast networks differentiated themselves from broadcast networks and they introduced new types of programming formats, which are still in use today. Overall, cable television increased the demand for cultural products overall and for different types of content (Feigenbaum, 2004). These included infomercials, 24-hour news and weather services, music video services, home shopping channels, arts channels, and a host of other narrowly targeted programming (Strover, 2005; Maule, 2003). With the increasing number of channels and the expanding cable systems, the number of bilateral negotiations and agreements between content providers and deliverers grew exponentially. Therefore, the sector greatly benefited from the standardization of the 'rules of the game' or of the relationship between cable providers and channel distributors that was introduced with the Cable Compulsory License Provisions of the 1976 Copyright Act. This act eliminated the need for bilateral negotiations as it provided a simplified procedure to allow transactions between the parties involved by assigning the Copyright Office and the Copyright Royalty Tribunal the role of interfaces. This law allowed cable systems to retransmit television broadcast signals containing copyrighted programs without the express consent of the copyright owners, upon the payment of a statutorily fixed royalty fee. The Copyright Office and the Copyright Royalties Tribunal collect, distribute, and adjust the royalty fees paid by cable companies, while the courts supervise collection, distribution, and adjustment decisions (Dobal, 1998).[3]

The New Regulatory Paradigm and the Rise of Information Conglomerates (1980s To 2006)

This era of the evolution of the AVMS is characterized by two trends: the first trend concerns the multiplication of interdependent activities, particularly in the 'delivery and exhibition' phase, and the consequent increase in the number of relationships between these activities. The most important of these new activities are: (1) the packaging and distribution of home video products, (2) the delivery of home video products through video tape and disk rental outlets; (3) the delivery of audiovisual media products through Internet Protocol (IP)-based communication systems – using the Internet or mobile phones private networks, and (4) the delivery of audiovisual media products directly to households using analogue and digital signals routed by satellite technologies. The second trend concerns the establishment of a new regulatory paradigm, which aims at strengthen media industries by reducing regulatory barriers that limit the extent of mergers and acquisitions in the sector.

Therefore, if on one hand, the emergence of new activities opened up more possibilities to increase the number of independent establishments within the AVMS, on the other hand, the wave of deregulatory changes undertaken since the end of the 1980s had the opposite effect, as it enabled more frequent and larger mergers, consolidations and takeovers. As a result of the new regulatory paradigm, in the early 1980s the consequences of the Paramount decisions were offset and feature film distributors were allowed to own cinema theatres (Christopherson, 2006). The process of dismantling the Fin-Syn rules, on the other hand, was only initiated at a later stage, in 1991. More specifically, at least officially, the FCC was prompted to change these rules due to three factors. First, a study by the National Institute of Statistical Sciences (NISS) concluded that despite twenty years of Fin-Syn rules, diversity or competition in the program supply market had not increased. Second, the Justice Department and the Federal Trade Commission stated that there was no sound evidence that the Fin-Syn rules were limiting the Networks' market power. Finally, as suggested by the FCC chairman, the video marketplace of the early 1990s was very different from the time when the rules were originally adopted and the rationale for introducing the rules at that time no longer applied (Tamber, 1995). The FCC chairman's suggestion referred to, and fed into, a more general and diffused opinion about new technologies and the changing video marketplace. According to this popular argument, new media technologies made deregulation possible, as the increased number of media outlets reduced the importance of the scarcity of airwaves and the need for an intervention from the central authority (see Louw, 2001).[4]

Alternatively, some communication scholars play down the role of technical changes and explain the rise of this new regulatory paradigm mainly as a consequence of the increasing popularity of neoliberal thinking (e.g. Blevins & Brown, 2007). However, despite the opinions of communications scholars and in light of the increase in the number of media outlets, the FCC argued that the Networks were no longer in the position of monopolizing the production of audiovisual media content and, as a consequence, the rationale for the Fin-Syn rules was dismissed. Thus, given the FCC's position, these rules were further relaxed in 1993 and finally eliminated with the changes brought about by the comprehensive revisions of the regulations governing the information sector imposed by the 1996 Telecommunication Act.[5]

However, if on one hand, the Fin-Syn rules had seemingly failed to *increase* competition in the program supply market, their elimination clearly favored the concentration of AVMSs. In fact, first, it is more profitable to own television programmes than to buy syndication rights (Fergusson, 2004), and second, bigger players have a competitive advantage over smaller players, as they can afford to exploit expensive studios and produce high-quality programmes. As a result, the largest audiovisual media content producers (i.e. the majors) acquired the Networks and created a restraint oligopoly in the distribution of television content.

The cable industry was also invested by the new regulatory paradigm. In 1985, or one year after Congress officially invested the FCC with regulatory power over the cable industry, the Commission adopted, as part of its Cable Act, a 'three-signal standard' rule. According to this rule, any cable television company servicing customers in an area covered by at least three OTA television signals (and such was the case of the very large majority of them) was considered to be in 'effective competition', even though, effectively, such a company could represent a local monopoly of the specific service of cable television provision. As a consequence of the existence of effective competition as defined by the Act, a cable company was freed from any regulation regarding the price of its basic tier (Corn-Revere & Carveth, 2004), which, as a result of this change, grew on average by 25 to 30%, between 1986 and 1988 alone. However, this de-regulation generated consumer dissatisfaction and consumers' opposition led to a redefinition of effective competition and the consequent reregulation of the rate as imposed by the Cable Television Consumer Protection and Competition Act of 1992 (Strover, 2005).

This new trend of regulatory changes, which was further fostered by the 1996 Telecommunication Act, facilitated both vertical and horizontal integrations among different AVMSs. It relied on the assumption that, because of the convergence to digital formats and the interoperability of networks, more systems could provide substitute services and compete against each other. Therefore, for example, Congress established with this Act the basis for an enlargement of the audiovisual content media services by introducing the Open Video System (OVS) framework. This allowed the local telephone companies to benefit from a different regime than cable providers and to provide video services in their own area (FCC Media Bureau, 2006). In addition, it modified the current media concentration rules and allowed any single company or network to own TV stations that reached as many as 35% (against the previous 25%) of the nation's television households (Corn-Revere & Carveth, 2004).

Overall, communication industries were the largest and most active in the M&A market at a global level, as a result of a de-regulation process initiated in the 1980s and culminating with the 1996 US Telecommunication Act and the 1997 World Trade Organisation's (WTO) agreement (Dal Yong, 2008). As a result of this process that favored the integration of different activities (not only audiovisual, but media in general), the once family-run film production companies that evolved into the Hollywood majors, are now part of larger media conglomerates. At the time of writing, there are six of these conglomerates: (1) Viacom, (2) Time Warner, (3) General Electric / Vivendi, (4) Sony, (5) News Corporation, and (6) Disney. These conglomerates own and run companies carrying out many different activities within the AVMS sector. Moreover, together they control 98% of the programs that carry commercial advertising during prime time television (broadcast and non-broadcast networks) and 96% of total US film rentals (Christopherson, 2006). They control 75% of commercial television in non-prime-time slots, 80% of subscribers to Pay TV, and 65% of advertising revenues in commercial radio (Epstein, 2005, p. 83, as cited in Christopherson, 2006).

NEW APPROACHES FOR UNDERSTANDING THE EFFECT OF REGULATORY CHANGES ON INNOVATION IN THE INFORMATION INDUSTRIES

The recent deregulatory trend documented in this paper clearly demonstrates that the FCC, which is the main regulator of the US information industry, tends to be in favor of M&As. In reality, the FCC is aware that mergers in the sector are 'likely to create potential for both good and ill' (FCC, 2001, p. 4), therefore they are granted if the companies involved can convince the FCC that 'the identifiable potential public interest benefits outweigh any potential public interest harms' (FCC 2001; 120). This type of analysis of positive versus negative effects, however, is normally carried out within an economic framework even when these effects have mainly an impact on diversity, localism or other social spheres that are different than economic welfare (see Blevins & Brown, 2006).

Furthermore, the FCC seems to embrace the viewpoint that concentration of assets in the communication sector is in line with its current policy objectives, as this supports innovation by providing sufficient capital to take the necessary risks and/or by reducing transaction costs (see FCC, 2001). Nonetheless, as there are negative consequences from the concentration of activities, the Commission attempts to limit their influence with 'finely tuned' regulations; for example, specific sets of rules prevent any single broadcasting company or network from reaching more than 35% of the nation's households (Corn-Revere & Carveth, 2004).

However, the case study illustrated in this paper suggests the need for a more complex analysis of the relationship between concentration of assets and innovation than the viewpoint embraced by the FCC. In more detail, the analysis presented in this chapter suggests that it is important to consider the specific features of activities belonging to different value-chain stages. As illustrated in the case study, the fragmentation of AVMS supported the introduction of many important changes, such as blockbuster movies becoming mainstream products or new types of television channels emerging. Therefore, vertical integrations seem to bear an opportunity cost (and a source of potential harm to the public interest) that horizontal merger do not bear and that the FCC does not consider in its reports (e.g. FCC, 2001). The differential of effects between vertical and horizontal mergers can be explained with simple arguments: first of all, the effects of the realization of economies of scale are likely to be more important when the companies merging are similar because they offer similar services. This is because companies providing the same services or producing the same goods are likely to have the same fixed capital assets and a pool of labor with similar skills and engaged in the same types of business operations. Consequently, the more similar two activities are, the larger is the number of 'opportunities' resulting from the realization of economies of scale. For example, the history of the AVMS illustrates that concentrating assets has been a successful strategy for cable companies in the 1990s, as companies exchanged systems in order to group them and take advantage of scale economies, not only in areas such as personnel, marketing and advertising sales, but also in emerging technologies such as fiber ring topologies and digital servers (see Parsons, 2003). Also, mergers between companies that own cinema theatres were responsible for the diffusion of the multiplex-cinemas business model (see Scott, 2004).

On the other hand, companies providing different types of services and operating in two different vertical value-chain stages, are likely to own some common types of capital, such as office buildings, or employees engaged in the same 'general' functions (i.e. non-specific to the business carried out, such as accounting or human resources). However, at least for the most part, the resources owned are likely to be specific to their function and objectives and therefore different (e.g. film studios as opposed to distribution facilities). As a result, the potential of innovating as a consequence of the realization of *economies of scale* is likely to be more limited. Consequently, two companies operating in two different value-chain stages and willing to concentrate their assets are likely to stress to the FCC the potential for synergies and *economies of scope* as the most important benefits.[6]

For example, in the year 2000 important economies of scope, i.e. synergies, were expected and promised from the (vertical) merger between an important content provider such as Time Warner - a corporation that included newspapers, broadcast and non-broadcast television channels and feature film production companies - and a company like America On Line (AOL) that, at the time, provided access to the internet and its portal page to over 20 million people (Economist, 2000; BBC, 2000). The FCC approved the transfer of licenses to the new company as it accepted the claims made by the applicants about the potential synergies (i.e. new service solutions, accelerated deployment of broadband content and technologies, and accelerated transformation of traditional media products to digital platforms) and the fact that these synergies and other efficiencies would have been realized only with a merger and not with a joint venture (FCC 2001). Retrospectively, one could say that, a joint venture would have been better than a merger in this case. As the recent de-merger between these two companies demonstrates (see America On Line [AOL] Corporate, 2009), the promised synergies were not fully realized and besides the attempt to create synergies, this controversial union did not create the basis for a new media giant. If these companies were kept separate, at least the costs linked to the merger and the demerger would have been saved.

Moreover, a set of alternative economic principles suggests that, under certain conditions, some of the same benefits that can be realized by M&As can be obtained with joint-ventures. For example, two companies merging are expected to reduce their 'transaction costs', i.e. the cost generated by their collaboration. However, at least to a large extent, the same result can be obtained by establishing detailed contracts or 'standard rules of the game', which are contracts that minimize the number of interactions between the parties involved (see Anand & Daft, 2007; Brusoni & Prencipe, 2001). The Cable Compulsory License Provisions of the 1976 Copyright Act discussed above, is an example of 'standard rules of the game'. This law allowed cable systems to re-transmit television broadcast signals containing copyrighted programs without the express consent of the copyright owners, upon the payment of a statutorily fixed royalty fee (Dobal, 1998). This eliminated the necessity for (at least, most) bilateral, contingent negotiations, between cable signal distributors and the providers of television channels.

Moreover, when a sector or an industry is characterized by the existence of many *interdependent* activities, which are carried out by companies owned by different stakeholders, and, whose relationship is determined by 'standard rules of the game', the sector can be said to be modular. In a modular system, the overall rate of innovation is higher than a more 'integrated system', which, in this case, is a sector where activities depend on each other (e.g. if their relationship is based on tacit knowledge and/or contingent negotiations). This higher rate of innovation that characterize modular systems when they are compared to integrated systems, is explained by the existence of independent and parallel innovation efforts characterized by a narrower and more specialized focus (see Ethiraj & Levinthal, 2004). In the case of 'integrated activities', on the other hand, the extent and scope of their innovation efforts can be determined (and limited) by key decision makers at the top of the organizational hierarchy.

The personal computer (PC), also known as IBM compatible, is the most common example of a modular industry, where different companies, mostly without collaborating, simultaneously innovate and create new peripherals that increase the utility of this tool. The peripherals produced by different companies can work together or simply at the same time, because of the existence of standard interfaces which function according to rules that are available to the general public. Therefore, component makers need to comply with the rules of the interfaces, instead of having to test their products in different PC environments (see Chesbrough, 2005).

Moreover, if on the one hand, M&As can provide new and meaningful ways of combining different types of capital and expertises, on the other hand, they also *reduce the design space* of each merging activity and/or company. In other words, after becoming integrated, the two companies do not innovate only by keeping into account their own specialized focus, but they have to consider each other requirements and capabilities. As companies adjust in different ways to the market, smaller and more focused activities are likely to adapt better and more easily to fast-changing environments.

This concept is best explained with an anecdotal example: one can imagine that in its efforts to transform its traditional media content to new media content, Time Warner was affected, during the duration of the merger, by AOL's innovation and technical capabilities. Alternatively, during the time of the AOL-Time Warner merger, other media content companies similar to Time Warner, benefitted from more freedom to establish several and multiple (and possibly, temporary) partnerships with different companies providing digital distribution and exhibition services. More importantly, these media content companies, which are not committed to find synergies with a particular partner that is specialized in a particular distribution technology, had more freedom to establish new and multilateral partnerships and, by doing so, to adapt more quickly to the fast-evolving markets and technological environments.

Furthermore, there are at least three important examples from the case study presented above that can be used to support this argument, as they show that regulatory or managerial changes that led to the separation and 'standardization' of the relationship between activities placed in different vertical stages of the value-chain, provided the basis for important innovations. The blockbuster movie is certainly an important innovation, as demonstrated by the fact that, after sixty years of industry's evolution, they still represent the most important 'venue' for attracting key innovators capable of creatively combining new stylistic ideas, technologies and talents; and regulations introduced as a result of the Paramount litigation at the end of the 1940s which aimed at breaking the tight (vertical) relationships between the Hollywood studios, the distributors and the exhibitors, clearly favored the diffusion of this type of media projects (see Robins, 1993). The second example regards the introduction of the Fin-Syn rules in the 1970s, which prohibited broadcast networks from owning most of their own prime time programming. As a result of the introduction of these rules, demand for audiovisual content was stimulated, a new pool of independent producers was established and (also because of the increased horizontal competition between producers) many stylistic innovations (e.g. new formats) in programs' content were introduced (see Christopherson, 2006). The third example concerns the establishment of non-broadcast channels, i.e. infomercials, 24-hour news and weather services, music video services, home shopping channels, arts channels, and other narrowly targeted programming, which was possible because of the existence of the cable industry, and more importantly, its independence from the television networks which was preserved thanks to appropriate regulations.

FUTURE RESEARCH DIRECTIONS

Mainstream economic theories support the principle, accepted by policy makers, suggesting that the largest companies have the potential to be the most innovative. One could argue that this principle applies to Fordist organizations (e.g. car makers, energy distributors, etc…) which innovate mainly through R&D, as the latter is an expensive activity that can benefit from economies of scale. However, the current economic environment defies this principle as companies like Amazon, Google or Microsoft, which certainly represent some of the most important success stories of this unfolding long wave of economic development, were very innovative and became leaders of their respective markets, before becoming very large and powerful corporations.

The alternative economic concepts that undermine the analysis presented in this chapter belong to neo-Schumpeterian (Freeman & Louca, 2001; Perez, 1983), 'modularity' (e.g. Anand & Daft, 2007; Brusoni & Prencipe, 2001; Ethiraj & Levinthal, 2004; Chesbrough, 2005) and complexity economic approaches (Beinhocker, 2007). Modularity theories, for example, are inspired by the dynamics of modern information industries (e.g. computer and semiconductor industries, see Baldwin & Clark, 1997; 2000), and therefore they should provide more useful suggestions for understanding the evolution of modern, complex dynamic industries than mainstream economic principles, which, on the other hand (as explained by Gallouj, 2002), are originally designed to understand the dynamics of agriculture and manufacturing activities. Therefore, this chapter should suggest to communication scholars that alternative economic principles can be inspiring and worthy of integration in analysis assessing innovation and policy changes of the media and information industries.

CONCLUSION

The dominance of the neoclassical (i.e. mainstream) economics as the main and most important approach in economics has been the subject of a debate that emerged in the early 1980s (see Nelson & Winter, 1982). This debate created the conditions for a transformation process aimed at promoting a 'pluralist economic viewpoint', i.e. the idea that many different economic approaches complement each other and that economic analysis based on different approaches are better for understanding the economic matters than analysis based on a single approach. Nowadays, this transformation process is led by the post-autistic economic movement, which aims at breaking the almost exclusive presence of the neoclassical economic theories in teaching, university books and economic policies by organizing alternative economic conferences and seminars and by publishing alternative economic books and journals (see Fullbrook, 2004). A pluralist economic viewpoint and alternative economic theories in general, are also ideal ingredients of multidisciplinary theoretical frameworks, and these types of approaches are considered as the most suitable to understand the dynamics of media and other creative industries (see Küng et al., 2008; United Nations [UN], 2008).

Thanks to a pluralist economic viewpoint, new arguments for old debates were introduced in this chapter. According to the main argument presented here, the 'tradition' of informing policies and regulations purely from mainstream economic theories needs to be challenged. As the history of audiovisual media services in the United States illustrates, the separation of the value-chain's vertical stages and the establishment of standard relationships or 'rules of the games' played a very important role in the emergence of the blockbuster movie format and the establishment of non-broadcast television channels.

On the contrary, M&As between companies operating in different stages of the value-chain can limit the range of opportunities that these companies can benefit from; opportunities that are destined to increased along with the growing complexity and diversity of the sector (i.e. the increasing number of different activities), which are the result of the 'swarm' of changes and innovations. In other words, particularly at times when technologies and other innovations emerge on a frequent basis (and such is the case during the upswing of a long wave of economic development), although they can benefit from mergers and acquisitions with similar companies (i.e. those operating in the same value-chain stage), information companies should keep 'their options open' and establish different vertical relationships with several companies specialized in either the production, distribution or delivery/ exhibition of media content (i.e. those operating in different value-chain stages).

However, media and information companies do not converge towards a modular organization of the sectors without the intervention of the regulator. These companies act as Fordist organizations and these types of organizations, as history demonstrates, have engaged in horizontal mergers to gain more effective control of output, price, and markets, and in vertical integrations to ensure a steady (in terms of quantity and price) supply of inputs (see Chandler & Hikino, 1990). Moreover, there are many potential negative consequences on the democratic life of societies, as many communication scholars would argue, when information companies grow and increase the control over their inputs, output and markets. Therefore, rules are required and media and communication sectors' regulators have a double incentive for fragmenting the information sector: to attempt to facilitate innovation in the sector and to create the conditions that favor the co-existence of different types of media content and formats and, more importantly, the emergence of a broader range of voices and opinions that users-citizens can have access to.

REFERENCES

Aksoy, A., & Robins, K. (1992). Hollywood for the 21st century: Global competition for critical mass in image markets. *Cambridge Journal of Economics*, *16*, 1–22.

America On Line [AOL] Corporate. (2009). *AOL Celebrates Day One as an Independent Company*. Retrieved March, 9, 2010, from: http://corp.aol.com/newsroom/press-releases

Anand, N., & Daft, R. L. (2007). What is the right organization design? *Organizational Dynamics*, *36*(4), 329–344. doi:10.1016/j.orgdyn.2007.06.001

Baldwin, C. Y., & Clark, K. B. (1997). Managing in an age of modularity. *Harvard Business Review*, (Sep-Oct): 84–94.

Baldwin, C. Y., & Clark, K. B. (2000). *Design rules: The power of modularity*. Cambridge, MA: MIT Press.

Bates, B. J., & Chambers, T. (2004). The economics of the cable industry. In Alexander, A., Owers, J., Carveth, R., Hollifield, A. C., & Greco, A. C. (Eds.), *Media economics: Theory and practice* (3rd ed., pp. 173–192). Mahwah, NJ: Lawrence Erlbaum Associates Inc.

Beinhocker, E. D. (2007). *The origin of wealth*. London: Random House Business Books.

Benson, R. (2006). News Media as a "Journalistic Field": What Bourdieu Adds to New Institutionalism, and Vice Versa. *Political Communication*, *23*, 187–202. doi:10.1080/10584600600629802

Bielby, W. T., & Bielby, D. D. (2003). Controlling prime-time: Organizational concentration and network television programming strategies. *Journal of Broadcasting & Electronic Media*, *47*, 573–596. doi:10.1207/s15506878jobem4704_6

Blevins, J. L., & Brown, D. H. (2006). Political issue or policy matter? the U.S. Federal Communications Commission's third biennial review of broadcast ownership rules. *The Journal of Communication Inquiry*, *30*(1), 21–41. doi:10.1177/0196859905280951

Blevins, J. L., & Brown, D. H. (2007). *The political economy of media ownership rulemaking within the U.S. federal communications commission: An analysis of media ownership studies*. Paper Presented to the Political Economy Section of the International Association for Media and Communication Research (IAMCR) for its 50th Anniversary Conference, Paris, France.

British Broadcasting Corporation [BBC]. (2000). *AOL, Time Warner confirm merger*. Retrieved January, 10, 2007, from: http://www.bbc.co.uk

Brown, A. (1996). Economics, public service broadcasting and social values. *Journal of Media Economics*, *9*(1), 3–15. doi:10.1207/s15327736me0901_2

Brusoni, S., & Prencipe, A. (2001). Unpacking the black box of modularity: Technologies, products and organizations. *Industrial and Corporate Change*, *10*(1), 179–205. doi:10.1093/icc/10.1.179

Carson, D. O. (2004). Satellite Home Viewer Extension Act, Statement by David O. Carson, General Counsel, United States Copyright Office before the Committee on the Judiciary, United States Senate, 108th Congress, 2nd Session. Retrieved January, 5, 2007 from http://www.tvb.org

Chandler, A. D., & Hikino, T. (1990). *Scale and scope: the dynamics of industrial capitalism*. Cambridge, MA: Belknap Press.

Chesbrough, H. (2005). Towards the Dynamics of Modularity - A cyclical model of technical advance. In Prencipe, A., Davies, A., & Hobday, M. (Eds.), *The Business of Systems Integration* (2nd ed., pp. 174–198). Oxford, UK: Oxford University Press. doi:10.1093/acprof:oso/9780199263233.003.0010

Christopherson, S. (2006). Behind the scenes: How transnational firms are constructing a new international division of labor in media work. *Geoforum*, *37*, 739–751. doi:10.1016/j.geoforum.2006.01.003

Corn-Revere, R., & Carveth, R. (2004). Economics and media regulation. In Alexander, A., Owers, J., Carveth, R., Hollifield, A. C., & Greco, A. N. (Eds.), *Media economics: Theory and practice* (3rd ed., pp. 49–67). Mahwah, NJ: Lawrence Erlbaum Associates Inc. Publishers.

Dal Yong, J. (2008). Media, Neoliberal restructuring of the global communication system: mergers and acquisitions. *Cultura e Scuola*, *30*(3), 357–373.

De Vany, A., & McMillan, H. (2004). Was the antitrust action that broke up the movie studios good for the movies? Evidence from the stock market. *American Law and Economics Review*, *6*(1), 135–153.

DeFillippi, R. J., & Arthur, M. B. (1998). Paradox in project-based enterprise: The case of film making. *California Management Review*, *40*(2), 125–139.

Dobal, C. H. (1998). A proposal to amend the cable compulsory license provisions of the 1976 copyright act. *Southern California Law Review*, *61*, 699–732.

Epstein, E. J. (2005). *The big picture, the new logic of money and power in Hollywood*. New York: Random House.

Ethiraj, S. K., & Levinthal, D. (2004). Modularity and Innovation in Complex Systems. *Management Science*, *50*(2), 159–173. doi:10.1287/mnsc.1030.0145

Federal Communication Commission. (2001). *Memorandum Opinion and Order*. Applications for Consent to the Transfer of Control of Licenses and Section 214 Authorizations by Time Warner Inc. and America Online, Inc., Transferors, to AOL Time Warner Inc., Transferee. CS Docket No. 00-30. Retrieved September, 2, 2008, from: http://www.fcc.gov

Federal Communication Commission Media Bureau. (2006). *Twelfth annual report* [online]. MB docket, 05-255. Retrieved January 2, 2007, from http://www.fcc.gov

Feigenbaum, H. (2004). Is technology the enemy of culture? *International Journal of Cultural Policy, 10*(3), 251–263. doi:10.1080/1028663042000255763

Flew, T. (2007). *Understanding global media.* Basingstoke, UK: Palgrave McMillian.

Freedman, D., Henten, A., Towse, R., & Wallis, R. (2008). The Impact of the Internet on Media Policy, Regulation and Copyright Law. In Küng, L., Picard R. G., & Towse, R. 2008. (Eds.), *The internet and the mass media,* (pp. 102-124). Oxford, UK: Sage Publications.

Freeman, C., & Louçã, F. (2001). *As time goes by - from industrial revolutions to the information revolution.* Oxford, UK: Oxford University Press.

Freeman, C., & Soete, L. (1997). *The Economics of Industrial Innovation* (3rd ed.). London, New York: Continuum.

Fullbrook, E. (2004). *A guide to what's wrong with economics.* London: Anthem Press.

Gallouj, F. (2002). Innovation in services and the attendant old and new myths. *Journal of Socio-Economics, 31*, 137–154. doi:10.1016/S1053-5357(01)00126-3

Hampton, B. B. (1970). *History of the American film industry: From its beginning to 1931.* New York: Dover.

Kassel, M. B. (2005). Must carry rule. *The Museum of Broadcast Communications.* Retrieved March 2, 2007, from: http://www.museum.tv/home.php

Küng, L., Picard, R. G., & Towse, R. (2008). *The internet and the mass media.* Oxford, UK: Sage Publications.

Lampel, J., & Shamsie, J. (2003). Capabilities in motion: New organizational forms and the reshaping of the Hollywood movie industry. *Journal of Management Studies, 40*(8), 2189–2210. doi:10.1046/j.1467-6486.2003.00417.x

Lister, M., & Dovey, J. Giddings, S. Grant, I. & Kelly, K. (2003). *New Media: A Critical introduction.* New York: Routledge.

Louw, E. (2001). *The media and cultural production.* Oxford, UK: Sage Publications.

Maule, C. (2003). Television. In Towse, R. (Ed.), *A handbook of cultural economics* (pp. 458–475). Brookfield, VT: Edward Elgar.

McChesney, R. W. (1999). Noam Chomsky and the struggle against neoliberalism. *Monthly Review (New York, N.Y.), 50*(April), 40–47.

McChesney, R. W. (2001). Global media, neoliberalism, and imperialism. *Monthly Review (New York, N.Y.), 52*(19), 1–19.

Motion Picture Association of America. (2007). Theatrical Market Statistics. Retrieved November 1, 2009, from http://www.mpaa.org

National Cable Telecommunications Association (NCTA). (2001). *Cable television developments.* Washington, DC.

Nelson, R., & Winter, S. G. (1982). *An evolutionary theory of economic change.* Cambridge, MA: The Belknap Press of Harvard University Press.

Parsons, P. R. (2003). Horizontal integration in the cable television industry: History and context. *Journal of Media Economics, 16*(1), 23–40. doi:10.1207/S15327736ME1601_3

Perez, C. (1983). Structural Change and the Assimilation of New Technologies in the Economic and Social System. *Futures, 15*(4), 357–375. doi:10.1016/0016-3287(83)90050-2

Rifkin, J. (2005). When markets give way to networks… everything is a service. In Hartley, J. (Ed.), *Creative industries* (pp. 361–373). Oxford: Basil Blackwell.

Robins, J. A. (1993). Organization as strategy: Restructuring production in the film industry. *Strategic Management Journal, 14* (Special Issue: Corporate Restructuring), 103-118.

Schroder, K., Drotner, K., Kline, S., & Murray, C. (2003). *Researching Audiences*. London, UK: Hodder Arnold.

Scott, A. J. (2004). Hollywood and the world: The geography of motion-picture distribution and marketing. *Review of International Political Economy, 11*(1), 33–61. doi:10.1080/0969229042000179758

Silverstone, R. (1999). What's New about New Media?: Introduction. *New Media & Society, 1*(10), 10–12. doi:10.1177/1461444899001001002

Storper, M. (1989). The transition to flexible specialisation in the US film industry: External economies, the division of labour, and the crossing of industrial divides. *Cambridge Journal of Economics, 13*(2), 273–305.

Strover, S. (2005). United States: Cable television. March 1, 2007, from: http://www.museum.tv/home.php

Tamber, C. (1995). The financial interest and syndication rules - take two. *CommLaw Conspectus, 3*, 107–120.

The Economist. (2000, 23 September). Solving the paradox: Productivity growth and information technology. *Economist Newspaper, US Edition*, pp. 15.

United Nations. (2008). *Creative Economy Report 2008*. Retrieved April, 1, 2008 from: http://www.unctad.org/creative-economy

van Der Wurff, R. (2008). The impact of the internet on media content. In Küng, L., Picard, R. G., & Towse, R. (Eds.), *The internet and the mass media* (pp. 65–85). Oxford, UK: Sage Publications.

Winston, B. (1998). *Media technology and society. A history: From the telegraph to the internet*. New York: Routledge.

ADDITIONAL READING

Blair, H., Culkin, N., & Randle, K. (2003). From London to Los Angeles: A comparison of local labour market processes in the US and UK film industries. *International Journal of Human Resource Management, 14*(4), 619–633. doi:10.1080/0958519032000057619

Gomery, D. (2004). The Economics of Hollywood: Money and Media. In Alexander, A., Owers, J., Carveth, R., Hollifield, A. C., & Greco, A. N. (Eds.), *Media Economics: Theory and Practice* (3rd ed., pp. 193–206). Mahwah: Lawrence Erlbaum Associates Inc. Publishers.

Hesmondhalgh, D. (2002). *The cultural industries*. Oxford, UK: Sage Publications.

Maltby, R. (2003). *Hollywood Cinema*. London: Blackwell.

Moul, C. C. (2005). *A Concise Handbook of Movie Industry Economics*. Cambridge, New York: Cambridge University Press. doi:10.1017/CBO9780511614422

Porter, M. E. (1985). *Competitive advantage. creating and sustaining superior performance*. New York: Free Press.

KEY TERMS AND DEFINITIONS

Economies of Scale: economies of scale are the cost advantages that a business obtains by expanding its production capabilities (e.g. increasing its capital assets or increasing labor) and reducing the average cost per unit produced at the same time.

Economies of Scope: economies of scope are the cost advantages that a business obtains by expanding the range of products and services offered and at the same time reducing the average cost per unit produced.

Horizontal Merger: a merger between companies that operate in the same value chain stage of the same sector or industry.

Mergers and Acquisitions: A merger refers to the collection of steps that two or more companies need to carry out in order to join their assets and start operating as a single company. An acquisition refers to the collection of steps that a company has to carry out in order to buy another company.

Synergies: Synergies refer to the extra energy, innovation or effectiveness that businesses create when they combine their efforts.

Value Chain Stages: before being offered to final customers, goods and services go through different processes and are modified by the contribution of different types of labor and/or companies and/or simply by adding inputs. These phases are referred to as value chain stages, as each stage increases the value of the good or service.

Vertical Merger: is a merger between companies that operate in different value chain stages of the same sector or industry

ENDNOTES

1 Notably, a Fordist organization typically aims at maximising profits by producing the largest amount of output with the lowest possible level of investments. As a result, during the Golden Age (i.e. 1920s to 1940s), each major studio was producing a larger number of movies than any production company does today. For example, in 2007 the total number of feature films released in the U.S. was 590 (Motion Picture Association of America, 2007) while during the Golden Age, a single studio like Universal was capable of producing 250 films over one year (Hampton 1970 cited in Rifkin 2005).

2 Part of the rationale for the Paramount litigation was to preserve and expand the parallel and smaller independent film industry, which was already present in the Golden Age of the Hollywood studios and differentiated itself from the majors in many aspects. Already at this time, for example, independent production companies used to create temporary associations and contract employees on a short term/ project basis (DeFillippi and Arthur 1998).

3 However, although the Cable Compulsory License Provisions certainly simplified the majority of relationships between programmes' packagers and distributors, it does not apply to all these relationships. In fact, there are at least three exceptions to this rule: first, in virtue of the must-carry rules and before the introduction of the Cable Television Consumer Protection and Competition Act of 1992, cable operators did not remunerate local stations for the retransmission of their signal (Strover 2005). Second, cable operators did not pay royalties to retransmit 'significantly viewed stations' in a particular community (Carson 2004). Third, small nonbroadcast networks, in order to reach a certain audience target, already at the time, but also nowadays, might pay to be carried and delivered by large cable systems (Bates and Chambers 2004).

4 Louw also explains that however, other and new forms of interventions were needed to allow the expansion of the new technologies,

such as regulations regarding satellite orbital spots (Louw 2001).

5 In 1993 the FCC had to redefine the rules to include the following points: 1) Network Acquisition of Back-End Rights; 2) Network Syndication of Off-Network Programming; 3) Network Participation in the First-Run Programming Market; 4) Entities that Qualify as a Network; and 5) Reporting Requirements Imposed on Networks (Tamber 1995).

6 Note that economic concepts such as 'economies of scale' and 'economies of scope' are defined in the Key Terms and Definition's section.

Chapter 24
The Three Cs of Key Music Sector Trends Today:
Commodification, Concentration and Convergence

Paschal Preston
Dublin City University, Ireland

Jim Rogers
Dublin City University, Ireland

ABSTRACT

Drawing upon a recent examination of contemporary trends in the music industry, this chapter explores the evolving relationships between new digital media technologies, socio-economic factors and media cultures as we enter the second decade of the twenty-first century. We examine the implications of these trends with regard to three fundamental concepts in the analysis of culture, namely commodification, concentration and convergence. We draw upon these concepts to guide our study of a music industry that is widely perceived as a leading site for new media developments. In this study we question the extent to which the music industry is experiencing transformations or significant disruptions resulting from technological innovations, or whether it is actually much more a case of 'business as usual' in the commercial music industry. Thus, this chapter interrogates and challenges the dominant framing of current debates around the notion of 'crisis' in the music industry. Furthermore, it considers how the concepts of commodification, concentration and convergence remain crucial to an informed and thorough understanding of current trends in the media and cultural industries.

INTRODUCTION

In this chapter we will explore the complex relationships unfolding between technology, socio-economic factors and media cultures in the early twenty-first century by examining some key recent trends in the music sector and their implications for three key, if long-established, concepts in the analysis of culture: concentration, commodification and convergence. In doing this we will avoid the transformative rhetorics of the 'technological sublime', and opt instead for a social shaping perspective to enable us to interrogate the forms

DOI: 10.4018/978-1-60960-037-2.ch024

Copyright © 2011, IGI Global. Copying or distributing in print or electronic forms without written permission of IGI Global is prohibited.

and extent of the crisis and restructuring processes unfolding in the recorded music industry over the past decade or more. Such an approach, as we will illustrate, challenges and undermines the dominant framing of current debates around the notion of 'crisis' in the music industry and ultimately suggests that the current situation and longer term outlook for the established music companies is much more positive than many current and recent commentaries and analyses imply.

In this chapter, we first consider the changing role and scope of the music sectors, its increasingly dense and complex relations with other media sectors. It will also outline why and how the music sector provides a suitable site for exploring the evolving meanings and relevance of the concepts of concentration, commodification and convergence. It will also indicate how the music industry comprises, in many respects, the leading edge sector in terms of the opportunities and challenges posed by radical technological change for the media and cultural industries more broadly. We then proceed to introduce and outline these three key concepts, their meanings and uses, and how they may be usefully applied as conceptual frames when it comes to understanding the nature and scope of major changes unfolding in the music sector and key cultural and social stakes involved in those developments.

This chapter then moves to a more detailed, empirical level investigation that focuses on recent trends in the music sector. Here, we are primarily concerned with investigating whether or how new digital technologies may have induced a certain 'crisis' for the prevailing models and practices of the recorded music industry. Our study notes and explores how these have also prompted key actors in this sector to mobilize their considerable resources to promote institutional, policy, regulatory and other kinds of innovations to meet these challenges. In essence, we explore how the interplay of technological, managerial, institutional and policy/regulatory innovations have provided new opportunities for restructuring, if not expansion, of the music industry's scope, practices and operations over the past decade. We also apply the concepts of commodification, concentration and convergence, indicating how they help us locate and map recent restructuring processes in this particular media sector.

Ultimately we consider how our chosen three (long-established, etc) concepts in the analysis of culture are still relevant to grounded analyses of the key trends unfolding in the music sector over the past decade and for understanding the scope and meaning of these changes.

MUSIC: NO LONGER CINDERELLA AT THE MEDIA INDUSTRIES BALL

Here, let us first clarify what we mean by the term 'music industry'. This is essentially an umbrella term covering four core industrial sub-sectors: the record industry; the music publishing industry; the live music industry; the music merchandise industry. While the live music industry and merchandise industry have increased in economic significance considerably over recent years [e.g. US concert industry revenues more than doubled between 2002 and 2008 to a value of US$ 4.2 billion (*Pollstar, 2009*)], it is the record and music publishing sub-sectors that are the primary focus of our attention in this chapter. Internationally, these sub-sectors are dominated by just four 'major' music companies: Universal, Warner's, Sony and EMI. In the record industry, the music industry's largest sub-sector, these companies accounted for 74% of the global record sales market in 2007 (IFPI, 2008).

While concentration has increased in the record industry, so too did corporate profits until 1999 when global sales revenues peaked at US$ 38.7 billion (IFPI, 2008). It is now a decade since compact disc sales reached saturation point in the majority of advanced capitalist societies. Since then, the recorded music industry appears to have experienced a significant and continuing decline, at least according to the sales statistics produced by the major record companies. Such data, combined with extensive staff cuts and the closure of many significant retail outlets, paints a decidedly unhappy picture for the record industry as a whole. All of this contrasts greatly with the decade prior to the turn of the millennium when year after year, this same industry boasted remarkable growth.

The past decade has seen developments, primarily associated with digital technologies that are widely viewed as key causes of the decline of this industry and the diminishing power of the major record companies. Developments in hard-drive recording have greatly reduced the costs associated with generating recordings and greatly increased accessibility to the whole recording process. The development of the MP3 file and subsequently peer-to-peer file sharing software has enabled the duplication and circulation of music among consumers to proliferate, and provide the major companies in the industry with questionably the most significant challenge of their history. Much media and journalistic commentary has detailed the collapse of the recorded music industry as peer-to-peer file-sharing is widely perceived to have decimated the market for physical recordings (e.g. Liebowitz, 2002). Many others (e.g. Barfe, 2005; Napier-Bell, 2008) argue that such developments have consigned the record industry as we know it to its deathbed. The record industry is the first major cultural industry sector to encounter the threats and opportunities associated with the internet. Peter Gabriel exemplifies this popular view as follows:

In some ways we are the canary down the mine, the first battle ground, but behind us goes anyone who creates anything that can be turned into data whether its software, films, pictures, music. (www.petergabriel.com)

As such, the outcome of the record industry in the light of this changing technological environment is of immense interest to all other media and cultural industry sectors. Thus, analysing how this sector has responded to new technological developments (the internet, mobile media, etc) can shed light on the more general trends of the shifting global media, culture and technology. Section four below is thus dedicated to examining some of the core record industry strategies in responding to the challenges associated with the internet. First, however, section three will provide an outline of the three key concepts derived from prior studies of the media industries – commodification, concentration and convergence – which we believe, help guide and inform a robust understanding of recent changes in the music sector.

FROM COMMODIFICATION, CONCENTRATION AND CONVERGENCE TO 'CRISIS'?

Here, we provide an outline of each of our three key concepts in turn. In doing so, we will consider how they may be usefully applied as conceptual frames when it comes to understanding the nature and scope of major changes unfolding in the music sector and key cultural and social stakes involved in those developments.

Commodification

In essence, the term commodification refers to the processes (and values) involved in valuing things primarily as marketplace products rather than goods and services to meet social uses or people's needs. It refers to producing and valuing things primarily in terms of the price they may raise in the marketplace, their exchange value rather than their use-value. This distinction was well known to classical political economists such as Adam Smith whose seminal book, *The Wealth of Nations*, is often viewed as a celebration of the huge productive power of the capitalist system and the 'invisible hand' of the market. Nevertheless, Smith offered some pointed observations on the 'commodity' as the specific form that products take when their production is largely organized through the process of exchange. He distinguished between products whose value is centred on satisfying specific human wants or needs, use value, and those products whose value is based on the price that they can command in the marketplace, exchange value. He also observed that use values are not fixed nor solely determined by biological factors, rather they evolve and are reshaped in accordance with cultural settings as well as socio-economic developments (Preston and Flynn, 2000). Whilst Smith is widely regarded as the founding father of modern political economy and the later field of economics, his observations on the commodity form are largely ignored within mainstream economics today where commodification tends to be taken as natural and universal good or given process.

In contrast, a critical approach to commodification and the commodity form has remained as a central concern in most strands of political economic, institutional or heterodox economic thought, not least when applied to the fields of media and cultural production. One of the first to draw and expand on Smith's earlier insights was Karl Marx who commenced his major work, *Capital*, with an analysis of the commodity form. He did so because it comprised the most visible and powerful representation of the capitalist system of production, especially as capitalism appears or presents itself first of all as an enormous collection of commodities. Marx fully recognized that, even in the mid-nineteenth century, the capitalist mode of production possessed an enormously expanded productive potential compared to all earlier modes, including the capacity to produce an ever-expanding array of commodities to meet various kinds of use values, cultural and informational as well as physical. Thus, the processes and analysis of commodification formed a core theme in Marx's critique of capitalism. This addressed not only the exploitation of labour power (attendant on its commodity form) but also the specific forms of rationality or instrumental reason that commodification engendered across all major domains of social, economic and cultural life. Indeed, Marx was stridently critical of the expanded sway of commodification process and the calculative ethic, 'the naked cash nexus', it fostered at the core of all kinds of personal, economic and cultural relationships.

A close engagement with commodification processes and their effects had prominent role in political economy and in most other critical work focused on the media and cultural sectors which emerged as specialized fields of academic inquiry from the early decades of the last century. Of course, this field of academic work has continued to grow in parallel with the expanding role of the media and cultural industries, telecommunications and the newer electronic communication services in recent decades.

The very term political economy itself suggest a certain distancing from any specialised or autonomous 'economic' science approach such as that which unpins the boundary walls of mainstream neo-classical economics. Indeed it marks a refusal to separate out the political from economic concerns and favours alternative, holistic approaches to the study of any phenomena in line with the methodological traditions of institutional, evolutionary or heterodox economics established by earlier writers such as Smith, Marx, Veblen, Innis and the Frankfurt School. It also borrows from that long, if often subsidiary stream of cultural studies which has been attentive to the multiple processes by which the expression and flows of symbolic goods (including oral, print, and audiovisual stories or 'content') have been shaped by socio-economic and political as well as technological developments. The pioneering work of the Frankfurt School is emblematic of this particular approach applied to the study of the media sector. Indeed, the very term, '*The Culture Industry*', was first coined and used by Adorno and Horkheimer in 1947 in their book *Dialectic of Enlightenment*, published in Amsterdam in 1947].

A core concern with commodification animates not only political economic, but also a broad spectrum of critical cultural studies work, addressing the changing structural role and characteristics of the media and the systemic features of communication and information in advanced capitalist economies (Golding and Murdock, 2000). Much of this seeks to advance an explicitly holistic approach to the study of the changing features and role of the media and other communication or information activities and how these in turn are linked to broader patterns of change in social, economic and political processes. Some have sought to develop a regional political economy theory of communication and information. Following the lead of Harold Innis (1950, 1951) and the Frankfurt School, they seek to develop a more adequate institutional political economy model of contemporary capitalism which takes appropriate account of the changing, indeed growing, role of mediated communication and other information services in capitalist industrialism. Thus, much of that work is focused on exploring how technological, economic and political factors, including commodification processes, have brought about fundamental but interrelated sets of changes in both (i) the media of public communication and (ii) in the broader/overall fabric of socio-economic and political structures in contemporary capitalist industrialism (Preston, 2001).

Contrary to most of the technocratic and economistic approaches focused on the idea of a knowledge economy or network society, concerns about the sources, forms and implications of the extended commodification of information have been prominent in many contributions to the political economy of communication (e.g. Schiller, 1984; Miege, 1989; Garnham, 1990, 2000; Mosco, 1996, 2009; Preston, 2001, 2008). In its approach to these issues, this body of work has also been strongly and clearly influenced by critical theory and some contributors explicitly acknowledge the importance Habermas's work (e.g. Garnham, 1990). Certainly, Habermas's notion of the public sphere, with its attendant emphasis on the factors shaping the quality and availability of public information and its modes of communication has influenced many recent contributions, both cultural studies and political economy approaches (Mosco, 2009). It has also served as a resource for empirically orientated research on changes in specific communication services, whether in the realm of print media, broadcasting, the Internet, public libraries, government information services, museums and art galleries. Indeed, Habermas's work, itself influenced by writings the Frankfurt School, constitutes an influential resources for contemporary work concerned with both understanding and potential alternatives to the singular logics of commodification processes (Mosco, 2009).

In contrast to information society theorists' celebrations of the quantitative growth in the economic significance of information and communication services (i.e. as sources of wealth and profits) many political economy and critical media researchers provide alternative concepts and more critical perspectives, including a stress on final users, defined as citizens, not merely consumers. Here too, many contributors tend to be highly critical of the commodification process and growing tendency to treat information as a commodity or 'thing' or mere 'tradable' commodity, to be enclosed behind property rights as a result of successive shifts in national and international industrial and trade policies (Miege, 1989; Preston, 2001; Hesmondhalgh, 2007; Mosco, 2009).

Many also provide empirically-based explorations of the specific economic characteristics and changing institutional contexts of information production, distribution and control processes. This includes a specific sub-set of work focused on the onwards march of intellectual property rights (such as copyright) and their implications for creative expression and the appropriation of the full potential of digital media for more democratic and participatory forms of expression (Lessig, 2001; Benkler, 2006). Here, it is worth considering the specific case of the music industry. While much contemporary commentary points to the decline in recorded music sales, the growth of revenue streams for music companies courtesy of proliferating 'intangible' sources is rarely acknowledged. The exploitation of recording and music publishing copyrights in spaces such as broadcasting, film, advertising and other is nothing new, however, the range of spaces through which such exploitation occurs has widened and the potential for revenue generation from these sources has become more significant. At the core of such trends is the dominance of the record and music publishing industries by a small handful of major music companies and their ownership of vast recorded music catalogues and music publishing repertoire – i.e. music copyrights.

Another related concern within the research tradition engaging closely with commodification processes centres on the patterns of media ownership, funding, control, and regulation (in the broad sense), and their relation to competing conceptions of democracy, citizenship versus consumer identities in contemporary societies (Preston, 2001). One sub- theme has been the growing concentration and centralisation of ownership of media and communication service attendant on market-driven and commodification logics (see 'concentration' section). This links to the dominant forms of structural transformation in the media and communications sectors, such as tendencies towards increasing integration of communication business and control functions across traditionally separate industrial and technological boundaries and tendencies towards outsourcing of certain (often high-risk) activities and functions to smaller 'independent' networks of firms. This stream of work often overlaps with that addressing successive forms of industrial restructuring processes in media sectors (Askoy and Robbins, 1992; Hesmondhalgh, 2007).

One other issue centres round the interplay of commodification and related institutional factors in shaping the diversity of information and/or the more qualitative aspects of the forms and flows of content associated with different funding, ownership and regulatory regimes (Schiller, 1984; McChesney, 2004). Such studies have addressed the implications of the growing concentration and centralisation of ownership and control of media at the national and international spatial scales and on the erosion of the public service regimes of regulation in the case of radio and television broadcasting. Others still have examined the impact and implications of the ever growing role of advertising, sponsorship and related public relations activities in re-shaping the forms and content of the media, including the role and effects of 'information subsidies' on news media (Gandy, 1982; Preston, 2008).

Concentration

Let us now turn to a brief consideration the role and meanings of the concept of *concentration*, especially as it applies to the media and cultural studies fields. For most mainstream economists' models since Adam Smith, the core image of many producers competing vigorously to supply various types of commodities explains not merely the efficiency of the capitalist market system. It also serves to underline the very legitimacy of that system as consumers are seen or constructed as 'sovereign' agents in the market place, free to choose from the multiple offerings of goods and services – including media and cultural products such as 'news', 'ideas' and recorded music.

In practice, however, the structure, operations and associated welfare benefits of capitalist markets and their attendant industrial structures have all evolved in ways which are increasingly discordant with this particular (if still dominant) representation. Instead of many small producers competing rigorously, a growing range of industrial sectors have been marked by the domination of a few, very large producers. In essence, rather than vigorous competition between many small producers involved in the supply of particular categories of commodities, more and more industries have become marked by the *concentration* of ownership and control in relatively few hands. This has been an evolving process and one that is uneven across different industrial sectors, but it first emerged as a significant feature of the economic landscape in the decades surrounding the turn of the twentieth century (the 1890s-1920s period). At that time, some contemporaries, including leftist German political economists such as Hilferding, viewed those shifts in industrial and market structures as denoting a new phase of 'monopoly capitalism'. That same period also witnessed the first emergence of truly mass media industries and from the early decades of the twentieth century. Those relative 'new' media industries were soon marked by relatively high levels of concentration, not least in terms of the dominant role of a very few, very large firms in key sub-sectors. It's hardly surprising then that such developments – the relatively high levels of concentration of ownership and control – featured as key elements in the seminal critique of the media and cultural industries advanced by the Frankfurt School in the mid-twentieth century.

In subsequent decades, many studies within the now specialist fields of media and communication studies have emphasized the relatively high, if not growing, levels of concentration in the media and cultural industries (e.g Garnham, 1990). In the fifth edition of his book, *The Media Monopoly*, Ben Bagdikian (1996) reported that over the preceding years, 'a small number of the country's largest industrial corporations has acquired more public communications power – including ownership of news – than any private businesses have ever before possessed in world history' (Bagdikian, 1996: xxix; cited in Croteau and Hoynes, 2001: 100). In more recent times, despite the apparent explosion in the numbers and variety of media services and the shifts towards deepening globalization, some researchers report distinct trends towards a media sector still marked by growing levels of concentration in terms of ownership and control of key media services or products (e.g. Bagdikian, 1996, 2000, 2004; McChesney, 2004). Whilst the various efforts to empirically measure the levels of concentration of ownership and control may differ somewhat, especially in terms of their findings on the extent to which it is growing (Croteau and Hoynes, 2001; Hesmondhalgh, 2007), there is less debate over the facts that the contemporary media landscape remains marked by relatively high levels of concentration. As Hesmondhalgh observes, 'bigger-than-ever corporations tower over the new terrain of cultural production in the 1990s and 2000s' (Hesmondhalgh, 2007: 187).

The reasons for the relatively high levels of concentration in the media and cultural industries have been well explored in recent decades, especially in political economy studies. One key factor here comprises the specific economic characteristics of media or cultural products (Preston and Grisold, 1995; Picard, 2005). This includes the relatively high first-copy costs (and the relatively low costs of subsequent copies of media products compared to most other products) and associated economic characteristics such as the high incentives towards scale economies and the crucial role of distribution and marketing functions (relative to production) within the media sector. Another factor here concerns the role advertising, an important source of revenue for many media industries. Its major agents, advertisers and advertising agencies, comprise a small number of large or 'lumpy' consumers of media products whose operational requirements favour concentration of ownership and control (Preston and Grisold, 1995). But political economy studies reveal a range of explanatory factors that extend well beyond any purely 'economic' factors. Thus, a second key factor facilitating concentration trends comprises the role of state policies and regulation. The includes the general role of neo-liberal policy and regulatory regimes in loosening the prior constraints on cross-media ownership and control, as well as their role in extending the onward march of intellectual property rights in the areas of copyright and branding (Preston, 2001; Hesmondhalgh, 2007).

So much for the facts and explanations of trends towards relatively high levels of concentration in the media and cultural industries. But why and how does concentration matter? In very basic terms, the relatively high levels of concentration and centralisation of ownership (and thus, ultimate control functions) in media industries are seen as imposing real barriers to, or brakes on, the potential diversity of stories, values and content produced in the major media of public communication. For example, in the media concerned with public communication of news and current affairs genres, ownership plays a key role in shaping or framing the patterns of news coverage (e.g. Preston, 2008). The trends towards media concentration tend to shape and frame (or naturalise) mainstream news coverage in ways which undermine the salience of a deliberative public sphere, for example by avoiding any fundamental questioning of corporate power and influence in the wider society. Beyond such shared concerns about the negative impacts of media concentration, critical researchers often differ in terms of more specific analytic points as well as in relation to the precise implications for particular media genres or sub-sectors (Picard, 1998; 2002; Croteau and Hoynes, 2001; Hesmondhalgh, 2007). For many scholars following the Frankfurt School tradition, issues related to media concentration are intricately bound up with the general critique framed around the issues posed by unrestrained commodification. However, we may also note that some have suggested that within a capitalist society, it is not surprising (or inevitably the case) that the mass media are marked by high levels of concentration since they are structured by powerful drivers towards economies of scale and scope). For example, Nicholas Garnham has argued that the critique of concentration on its own may not matter very much, or go very far, suggesting that 'the critique of concentration per se is more of a liberal than a left critique' of the media structures and power (Garnham, 2004: 100; cited in Hesmondhalgh, 2007 173).

Below, we will return to the idea of concentration, exploring some of the relevant issues posed by evolving trends in the music industry in recent times and their implications for the diversity of musical expression, performance, and cultures.

Convergence

Compared to our other key terms, the idea of convergence is relatively new as it only features in media studies fields since the early 1980s. It refers to the idea that new ICTs are leading to the erosion of the boundaries between previously separate communication services and functions (e.g. Pool, 1983). Here, the basic idea is that the increasing shift towards a common digital mode at the level of information processing or distribution technologies is leading to a significant erosion or blurring of sectoral boundaries between communication services or 'information' markets which were previously distinct. Since then, the concept of convergence has featured in many academic, industrial and policy studies of change in the media and communication sector, playing an especially prominent role in successive generations of technology-centred analyses (e.g. Techno-Z FH 1997).

This basic convergence idea proved to be a fashionable and highly influential concept in informing decision-making in industrial, financial and policy circles during the dot.com boom era of the late 1990s (Preston, 2001). The following description from a UK government green paper can be treated as typical as well as illustrative of the policy appropriations of the convergence idea during the 1990s and in more recent years:

Digital technology is rapidly being adopted for the reproduction, storage and transmission of information in all media. This means that any form of content (still or moving pictures, sound, text, data) can be made available via any transmis-

sion medium, eroding the traditional distinctions between telecommunications and broadcasting. Already it is possible to receive television or radio over the Internet, while digital televisions will have some of the capabilities we currently find only in computers. (UK, 1998, Green Paper on 'Convergence', pp. 12).

The convergence idea has proved to be another one of these particularly resilient notions, despite the fact that it has been subject to much criticism from researchers in the communication studies field. The convergence idea also played a central role in a whole series of technology-centred new communication innovation initiatives and pilot projects which have proved to be remarkably unsuccessful in commercial terms. These range from the various optimistic initiatives around videotex services in the 1980s to the many subsequent 'interactive' services trials in Europe and the USA in the early 1990s. The list here also includes the many 'video-on-demand' initiatives in which the major telecommunications services operators invested so much hope and research funds from the early 1990s as well as their major investments pursuing a convergence-centred vision that 'content is king' during the dot.com boom years of the late 1990s (Bouwman and Christoffersen 1992; Preston, 1993, 1997, 2001).

For example, the convergence idea underpinned the public announcements of Microsoft when justifying that company's major investments in new Internet-based information content services fields, especially in the latter half of the 1990s. The basic idea was the company could utilise its dominant position in the areas of computer operating systems and applications software as leverage for establishing a prominent role in the emerging domains of on-line digital content services. But despite its vast investment resources derived from it monopoly position and related technology rents in the computer operating systems and software tools markets, even this major corporation was forced to revise its strategy rather dramatically. Thus far, Microsoft has been unable to successfully realise its particular take on the convergence idea in practice, apart from the partially fruitful outcome of its large-scale investments in the digital games space (Preston, 2001).

The lessons of the costly investments and experiences during the fashion for 'convergence' during the late 1990s clearly point to the limits of techno-centric analytical perspectives. The lessons also imply that the boundaries between the various sub-sectoral components of the media sector (or other primary information services) are not fundamentally eroded by the trends towards digitalisation at the level of information engineering or processing. The core competencies, distinctive industrial structures and organisational capabilities or cultures which favour success in one domain of the media or other primary information sectors do not automatically translate into other domains. In essence, there remain many important institutional and capabilities factors which operate as barriers to the successful production and sale of new-ICT based media (content) services, however 'convergent', interactive or customisable these may seem at the level of technical delivery systems or technology-centred analyses (Preston, 2001). Yet, despite the long history of failure and criticisms of such technology-driven visions of new media developments, the convergence idea remained very much alive and operated as a core concept underpinning 'expert' financial analyses of investment strategies and major corporate investment decisions as well as policy initiatives related to new media 'content' industries during the dot.com boom of the late 1990s (Preston, 2001). However, the effect, if not intent, of many industrial and policy appropriations of the convergence idea, was to facilitate further concentration and centralisation of ownership and control in the media sector.

During the early years of the first decade of the twenty-first century, 'convergence' centred investment and industrial strategies went all out of fashion, a shift no doubt directly related to the clear and spectacular financial failure of so many such ventures undertaken in the late 1990s, especially the most prominent case – the merger Time-Warner and America Online. But if appropriations of the convergence idea to justify cross-sectoral investments and industrial mergers (and media concentration) have gone out of fashion, the term continues to be used for other purposes. More recently, the idea of convergence has been frequently invoked to refer to changes in the production practices and industrial routines in news media. Many studies have invoked the convergence concept in studying changes in news production practices within media organisations (and their 'multimedia newsrooms') seeking to produce and distribute integrated news services across different media platforms, including their implications for the professional roles and status of journalists (e.g. Preston, 2008). Other recent applications of the convergence idea include efforts to identify the specific characteristics of the new kinds of texts or textuality afforded by digital media and network systems (e.g. Fagerjord, 2003; Storsul and Stuedahl, 2007).

In what follows, we explore the convergence idea further by paying particular attention to the case of the music industry of the early 21st century. Here we observe that technology convergence trends have generally been defined in rather negative terms. In essence, they are generally represented in terms of a fundamental 'crisis' for the music industry, especially the recorded music sub-sector.

'Crisis' and the Recorded Music Industry

Much discourse within and surrounding the music industry over the past decade has been predicated on two major themes:

1) A significant decline in recorded music sales and with this, questions surrounding the long-term viability of the music industry.
2) The concomitant transition to a digital milieu which has generated several salient problems for the record and music publishing industries, and how they make money.

The notion of a 'crisis of digitalisation' in the music industry has consistently echoed from various corners of the industry itself, and from the political establishment. For example:

We run the risk of witnessing a genuine destruction of culture... The internet must not become a high tech wild west, a lawless zone where outlaws can pillage works with abandon, or worse, trade in them in total impunity. And on whose backs? On artists' backs. (Nicolas Sarkozy, President of France, November 23rd 2007)

Drawing upon these comments, Paul McGuinness, the manager of U2, appealed to national governments around the world to force internet service providers to remove music file-sharers from their networks:

I believe President Sarkozy truly caught the spirit of the age with that statement... It is a good rule of thumb that when it is the manager and not the artist getting the headlines, something is out of kilter. Well there is certainly something out of kilter with the music business today... The record industry is in crisis. (Paul McGuinness, speech delivered at Music Matters conference, Hong Kong, 4th June 2008)

Such crisis rhetoric has also consistently flowed from many quarters of the media. Headlines such as 'The music industry... knee deep in a downloading crisis' (*The Sunday Business Post*, April 6th 2008) and 'Industry crisis as album sales drop' (*The Irish Independent*, January 14th 2008) have become commonplace in news stories, features and opinion columns. Magazines such as *Globe and Mail* have carried features entitled '"Want a snapshot of an industry in crisis? Take a look at the music business right now' (*Globe and Mail*, January 31st 2008). Perhaps this doomsday scenario is captured most succinctly by *Irish Times* journalist Conor Pope who asks:

Has music had its day?...Of the all the upheavals wrought by the internet revolution over the last fifteen years, the shake up in the world of music has been amongst the most profound. The consequence of free music downloads could end up destroying not just the shops that used to sell music, but an entire industry. (The Irish Times, Monday April 27th 2009, p. 15)

Moreover, such sentiments are also to be found echoing in the chambers of academia, most notably through the comments of those who emphasized the transformative power of digital technologies. As far back as the mid-1990s, Nicholas Negroponte, a professor at MIT and celebrated guru of the information age, argued that copyright would 'disintegrate', with everything capable of being reduced to streams of ones and zeros being potentially 'up for grabs' (Negroponte, Wired.com, February 1995)). Society's digital future would revolutionise traditional social and economic structures. Such an outcome could, potentially, collapse the established music industries. According to Negroponte's intellectual peer, Kevin Kelly, the associate editor of high-tech publication *Wired* magazine: 'The recording industry as we know it is history...[with] digital file-sharing technologies...undermining the established economics of music' (*New York Times Magazine*, 17th March, 2002: 19-21). For Kelly, concomitant with the downfall of the major music company is the empowerment of the individual recording artist. Free from the shackles of entertainment corporations, those 'musicians with the highest status are those who have a 24-hour net channel devoted to streaming their music' (ibid). Indeed for Kelly (1999), the 'new economy' deemed to arise from digital technologies means the demise of the 'old' economic laws that have characterised the modern capitalist era.

Taking these various accounts at their word, we may deduce that from a lay person's perspective, digital technologies pose a grave threat to the established music industry as a whole. Furthermore, this threat is being realised and the music industry is experiencing radical upheaval in the wake of the 'internet revolution'. In the more extreme cases, these accounts spell out changes that are leading to the destruction of an entire industry. These notions of change have become common-sense assumptions in much discourse surrounding the recent evolution of the music industry.

The core thrust of the above arguments appears to be supported IFPI statistics which report how the global recorded music market has largely been in decline throughout the first decade of the new millennium. They indicate an initial drop of 17.3% from a record high of US$38.7 billion sales 1999 to US$32 billion in 2003. While there was a modest recovery of 4.7% throughout 2004 and sales held (almost) steady throughout 2005, the subsequent period to the end of 2006 saw a further drop of 5.4% to an overall global retail value of US$31.8 billion. By the end of 2008 this figure had dropped to US$26.5.

However, while recent digital technologies are commonly perceived as inducing a certain 'crisis' for the prevailing models and practices of the recorded music industry, it is crucial to examine the core response strategies of the established music companies to these digital developments. As we will see below, these actors have mobilised considerable resources to promote institutional, policy, regulatory and other kinds of 'matching' innovations to meet the new challenges and threats enabled by technological innovations (Preston, 2001). When the dominant industrial narratives are interrogated and examined in this light, the notion of crisis is given an altogether different complexion, as we will indicate below.

Table 1. Retail value of recorded music sales 1999-2008 (combined physical and digital formats in $US billions)

	1999	2000	2001	2002	2003	2004	2005	2006	2007	2008
Total	38.7	36.9	33.7	32.2	32.0	33.6	33.5	31.8	29.9	26.5
US	14.3	14.8	13.4	12.6	11.9	12.2	12.3	11.5	10.4	8.4
Rest of world	23.4	22.1	20.3	19.6	20.1	21.4	21.2	20.3	19.5	18.1

Source: Author, compiled from IFPI Recording Industry In Numbers reports 2000-2009

FROM 'CRISIS' TO NEW FORMS OF COMMODIFICATION, CONCENTRATION AND CONVERGENCE

The major music companies' pursuit of software suppliers, file-sharers and internet service providers through the courts is well-documented and comprises but one high-profile aspect of their responses to the disruptive implications of new media. The extension and application of copyright law in recent times has been greatly facilitated by neo-liberal policy regimes. Clearly, it also represents a fundamental response strategy of major industry actors keen to the potential threats to their dominant positions posed by digital technological innovation. However, it is their less-documented exploitation of recording and publishing copyrights across a proliferating array of platforms that is of greater interest here. While much emphasis is placed on the downturn in the record sales market during the first decade of the new millennium in many accounts, it is necessary to recognise the growing range of additional revenue streams now opened to music companies in order to understand more thoroughly and holistically the economic significance of the contemporary music industry.

In essence, we observe efforts to move beyond the 'defensive' use of the large and concentrated resources of the major music corporations to address how these may be mobilised to undertake 'offensive' strategies to open up entirely new positions and revenue streams in 'convergent' and neighbouring industrial spaces.

Growing the Online Music Market

In recent years, digital platforms for music have proliferated, as have digital music sales. There are, in 2009, over 500 legal online music services worldwide with more than 13 million tracks licensed to the major services (IFPI, 2009). Since the launch of iTunes in October 2003, digital sales have, according to IFPI statistics, grown to account for 20% of overall recorded music sales, globally. In 2008, the value of digital sales to the global music economy was just over US$3.7bn (IFPI, 2009). Furthermore, digital music has evolved far beyond the initial download services of 5 years ago to a variety of models that includes not only the a-la-carte download digital music stores, but also subscription services, mobile services, streaming services, social networking sites, brand partnerships and direct to consumer sites. Since 2007, major record companies have been entering into joint initiatives or partnerships with ISPs for the first time. For example, Sky in the UK, Neuf Cegetal in France and Eircom in Ireland, amongst many others internationally, have started offering music services to consumers. Thus, the record industry has increased the number of channels through which music can access its final users/consumers exponentially over the past number of years.

Thus, whilst there has been a substantial decline in the volume of global physical sales since the end of the last decade, the subsequent and phenomenal growth of digital sales paints a significantly brighter picture for the major companies

Table 2. Global digital music market value 2003-2008 (US$ billions)

	2003	2004	2005	2006	2007	2008
Total	0.02	0.4	1.1	2.1	2.9	3.7

Source: Author based on IFPI Digital Music Reports 2004-2009

that have been reaping the benefits of emerging and proliferating digital formats. For example the 2006 release of Justin Timberlake's *Future Sex/ Love Sounds* was released in 115 products or formats including physical CD, a variety of ringtones, mobile full-track downloads, video, bundled album digital music store downloads and other. It sold a total of 19 million units, only 20% of which were CDs (IFPI, 2008). According to the IFPI *Digital Music Report* which is published each spring since 2004, the value of digital music sales has grown from approximately US$20 million in 2003 to in excess of US$3.7 billion by the end of 2008. Digital revenues now account for 15% of overall global sales revenues. In the US, the world's largest music market, digital sales represent 30% of the overall market, in South Korea that figure is 60%. All of this serves to emphasise the recent success of the digital music store model and their growing importance to the major companies.

It should be noted that the above IFPI figures include recorded music sales in physical and digital formats including online, mobile and subscriptions. However the growing revenue stream generated by ringtones is not included, nor are revenues generated for record companies through licensing, synchronisation or from advertising-supported business models (such as Limewire). Even more promising news for the record industry came in 2008 when Juniper Research published a report predicting that the total value of the global mobile music market alone is expected to increase to more than US$17.5bn by 2012, driven primarily by rental music services and full-track downloads.

Licensing Revenues and Promotion from New Online Platforms

Furthermore, we observe that the major music companies are involved in an ongoing process of forging alliances and pacts with established and emerging social media networks and also other online and mobile content platforms. Outlined below is a brief overview of some of the key licensing rights agreements that have been struck, primarily on the back of legal actions regarding copyright infringement taken against the social media networks. These provide new avenues for the generation of revenue primarily for major recording and music publishing companies.

Vevo

April 2009 saw the Universal Music Group announce the launch of Vevo, a music video service in collaboration with Google. Effectively, Universal own the site, with Google/YouTube providing the technology. Users will be able to access and play music videos for free, with revenue generated from advertising being shared by Universal and Google. June 2009 saw Sony Music Entertainment sign up to this venture. The other major companies also entered negotiations over potential licensing agreements regarding the use of their content on the site (CNET News, Thursday 4th June, 2009).

Vevo was officially launched in December 2009 when many of the music industry's most senior executives and several major stars attended a lavish launch party, all expressing confidence that 'online video could be the salvation for stars and labels after a decade of digital mis-steps' (Edgecliffe Johnson, 2009). The chief executives of Universal Music, Sony Music, EMI Music and Google attended to declare that 'their online music video joint venture, as the best thing for the business since MTV began in 1981' (Edgecliffe Johnson, 2009). Doug Morris, chairman and chief executive of Universal Music told attendees that 'Vevo is a chance, perhaps the chance to please stop playing defence' and that 'Vevo is a huge platform, and you know what's best of all? It's our platform' (cited in Edgecliffe Johnson, 2009).

The CEO of Universal Music also said his company had already secured "millions of advertising dollars" from brands including AT&T, MasterCard and McDonald's, which are keen to reach "the most soughtafter demographic" of young music fans. (cited in Edgecliffe Johnson, 2009). It appears that 'Vevo is pushing for high CPMs - the cost per thousand advertising impressions - of $25 to $40' (Edgecliffe Johnson, 2009).

MySpace

September 2008 saw the recording and publishing arms of all four major music labels launch partnerships with social networking site MySpace. These partnerships involve MySpace providing on-demand and ad-supported streaming, music downloads, a subscription plan and a variety of other music related features such as the sale of concert tickets and merchandise. In effect, these deals cover the entire 360-degree spectrum of potential revenue streams. Users are also enabled to assemble and share playlists using the cast catalogues of Universal, EMI, Warner's and Sony-BMG. The pay-for-download and music streaming aspects of this service directs users to purchase tracks and albums via Amazon. As part of this overall deal, the music companies receive an equity stake in MySpace Music, which, according to the MySpace website boasts 68.6 million users in the US alone.

One of the key early success stories of this service was the album *Chinese Democracy* by US artists Guns'n'Roses. During its initial twenty-four hours of availability for streaming on the bands MySpace site, the three most popular tracks (*Shackler's Revenge*, *Chinese Democracy* and *Better*) achieved 1.3m plays (*Osorio, 2008a*).

YouTube

In late 2007 the Warner Music Group entered a licensing deal with YouTube which cleared all of its recorded music catalogue and music video catalogue for use by the consumer media giant that enables users to watch and share video content through a web experience. At that stage Sony Music Entertainment, had already signed a similar deal with YouTube. According to a YouTube press release upon the signing of this licensing agreement, the company promised to work with record companies 'to expeditiously remove certain copyrighted materials which are not available for exhibition on the site'. YouTube employs a content identification architecture that enables content holders to identify their content on the site and thus the opportunity to authorise and monetise the use of their works within the user generated content on the site.

What is also important to remember is that these deals only take into account music publishing copyrights; separate license fees are paid by social networking sites to the same music companies in respect of recording copyrights. For example, during 2008 all of the major labels forged relationships with social media network Imeem, making audio and video content available via on-demand streaming in return for a stake in advertising revenues generated by the site.

Other Ad-Supported Streaming Partnerships

Since 2006, ad-supported music streaming sites have also evolved as a significant source of finance for music companies (as the Vevo music video project indicates). Following a spate of lawsuits taken by the record industry against such sites as

Bolt and Grouper, revenue sharing models have emerged whereby these companies that were formerly streaming audio and music video content online have now entered alliances and partnerships with the four major music companies.

Since 2006 Spiralfrog has provided major label catalogues via its ad-supported download service. Universal became the first label to sell licensing rights to Spiralfrog for its entire catalogue. Downloads from the site are free of charge to the user, but come accompanied with advertising. License fees are thus paid for through advertising revenues. Similar deals have now been struck between all of the major labels and lesser profile sites such as Amie Street, Sellaband and Magnatune.

February 2008 saw the Universal Music Group enter a partnership with Dutch-based online live music platform Fabchannel.com, whereby Fabchannel acquired multi-territory rights for the streaming of recorded concert performances by Universal recording artists.

In August 2008 the Warner Music Group became the first of the major music companies to license content to We7, the first European-based ad-supported music service. Based on the Spiralfrog model, We7 grafts a series of short adverts into the start of music tracks based on consumer demographics such as location, age and gender. Users may stream content or take full-track downloads. September 2008 saw Warner's and online music jukebox Deezer announce a global content and revenue partnership. Warner's full catalogue of audio content is available for on-demand streaming as part of Deezer's free to access service. Under the terms of the agreement both companies will share revenues generated by Deezer's ad-supported platform.

In summer 2009, Universal partnered with UK-based Virgin Media to provide a streaming and download service. As with the other major partnerships in this sphere, the other major labels, publishers and stakeholders are likely to follow.

Other Internet and Mobile Alliances

2007 saw EMI Music Publishing finalise a deal with global internet communications company Skype. Under the agreement Skype acquired licensing rights to song copyrights from the EMI catalogue for sale as downloads or mastertones via the sites online retail store. EMI Music Group has also signed the first ever pan-European download agreement for a peer-to-peer music service with one of Europe's biggest mobile entertainment provides, Arvato Mobile. The deal enables Arvato to offer retailers the ability to provide users with music downloads or a subscription package from a catalogue of 300,000 recordings (EMI press release, June 2007).

2007 also saw the Warner Music Group enter a global strategic partnership with mobile operator Telenor. In a deal that encompasses nine out of Telenor's thirteen mobile phone companies, users can now access Warner music content in a variety of formats including full-track downloads, ringtones, ringback tones, mobile music videos and wallpapers.

Sony Music Entertainment, the Warner Music Group and the Universal Music Group announced in April 2008 that they had completed deals with a Danish telecoms operator TDC which opened digital catalogues of over one million tracks to mobile and broadband users as part of their monthly contract/subscription.

EMI has also forged links with various digital platforms. In 2007 Microsoft and Apple both reached agreements to sell EMI catalogue without DRM protection via their respective Zune and iTunes platforms. UMG, Sony and WMG subsequently removed DRM from their products also.

2009 saw the continuation of such trends with Universal entering partnerships with digital music distribution service Tunecore, and also an MP3-based download service with media corporation Virgin Media. The former pact also makes provision for new physical retail space for Universal CD catalogue in the 214 Tunecore-owned Guitar

Centre outlets throughout the US. 2009 has also seen Sony Music Entertainment enter a global partnership with digital music distributor IODA. The IODA website already possessed a catalogue of approximately two thousand independent-owned music tracks to which it will now add the catalogues of Sony's RED label, and other former 'indie' companies now owned by Sony.

Music and Digital Games

By the middle years of this decade, the digital games industry had evolved to form a significant component of the broader cultural industries with revenues comparable to the box office intakes of the Hollywood film industry and global recorded music industry (Kucklich, 2004; Thomas, 2003). The estimated value of the global digital games industry in 2008 was US$44 billion (NPD Group/Retail Track, 2009; Morgan Wedbush Securities estimates, 2009). Within this, the US digital games market had grown to in excess of double the combined value of the US movie box office and DVD market. The value of the global online games market was estimated to be US$7 billion. While these statistics may suggest that the growing digital games market is providing growing competition for the music industries in terms of providing alternative forms of entertainment for consumers, in practice the digital games industries also provides revenue streams for music companies on various levels. Games have grown to provide not only marketing and promotion for the music content they carry, this content generates licensing revenues for record and music publishing companies via the sale and use of the digital games themselves, and many games generate direct music sales by offering exclusive music and artist related content.

Recent years has seen traditional content owners such as Universal and Warner's move into mobile games courtesy of licensing agreements with games providers. The convergence of content providers with digital games hardware and content providers works to their mutual benefit ultimately bringing profits from both sectors back under the one roof. Sony is a key example in that it is not only a key player in the music industries, it also produces and supplies the various generation of the Playstation consoles and games.

While almost all games use licensed music, games such as *Song Pro*, *Rock Band*, *Guitar Hero* and *Singing Star* are of particular relevance. Song Pro is a cartridge that can download music to be played on a Gameboy console. The *Song Pro* package includes a flash memory card, headphone, music management software and a USB line to enable users to access the internet for music downloads. Song lyrics and artwork can also be downloaded from the *Song Pro* website. The significance of such music-based games is illustrated by metrics released by Microsoft in October 2008 which show that the computer company sells 3.8 million song downloads a month via the Xbox for the games *Rock Band* and *Guitar Hero* (Osorio, 2008). Figures for the previous twelve months indicate music download sales of 45 million tracks. The figures don't take into account song download sales for the same two games on Sony Playstation format. October 2008 also saw Harmonix, the company behind *Rock Band*, gain licensing rights to The Beatles catalogue for use in the game.

In April 2008 US rock band Motley Crüe released their *Saints of Los Angeles* single exclusively as a downloadable track via the *Rock Band* game on both Xbox and Playstation formats. Furthermore the 2008-09 Motley Crüe US tour was used as a platform for promoting the game. Concert-goers were invited up on stage during the intervals between support acts in order to play the game on a giant screen and compete with each other to become the *Rock Band* champion.

Beyond music-based digital games, all four major music companies are providing music content for games. In some instances popular back catalogue material by established recording artists is used as a constituent element of the digital game experience – for example artists like Blur on *FIFA Soccer* and Run-DMC and Ozzy Osbourne on *Grand Theft Auto*. In these three examples, all three recording acts are Sony Music artists generating direct recording and music publishing revenues for the company, as well as receiving promotion via their use on Sony Playstation. Equally digital games are used for promoting exclusive material and releases that are only accessible through digital games consoles. For example a limited edition version of *Payable on Death* by the band P.O.D. (issued on the Atlantic label, that is in turn owned by the Warner Music Group) included an accompanying exclusive Sony Playstation 2 DVD that contained music not available on the standard album release. The game *NFL Street* features eleven exclusive previously unreleased tracks by various Sony recording artists that only play on Sony Playstation 2, Microsoft Xbox and Nintendo GameCube consoles.

Other examples of music-digital game cross-promotion include a deal between Universal and Electronic Arts (EA), one of the world's largest producers of digital games, which has been in existence since 2003. The deal sees music recordings owned by Universal's Def Jam label incorporated into EA games. For example, the first version of the game *Vandetta* featured twelve Def Jam recording artists including Method Man, Scar Face, Ghost Killah and others. This game allowed users to play recordings form selected artists in a wrestling ring. This was the chosen method for initially releasing new music by these artists. New singles would only be released in other physical and digital formats, and for radio broadcast, subsequent to being released via the digital games. Snoop Dogg, The Black Eyed Peas, Vanessa Carlton and Marilyn Manson all had new music previewed exclusively through the games. Chicane, Beyond this, The Bare Naked Ladies, Metallica, Aerosmith, Blink-182, Papa Roach, Apollo 440, Paul Oakenfold, Naughty By Nature and Nelly provide just some examples of artists whose repertoire is currently (or recently was) being used in games.

In addition to all of this, revenue streams are also generated for artists and music companies by incorporating the personas/characters of popular recording artists into characters in the games themselves. For example, the voice of Geffen (a Universal label) recording artist Eve is used in the Ubi Soft game *XIII*; the character of Interscope (also a Universal label) recording artist Marilyn Manson features in the game *Area 51*; Atlantic Records (a subsidiary of Warners) artist Cupid features in the Acclaim Games online game *Dance*. This latter game enables users to 'virtually' dance to music tracks they choose from a catalogue that is exclusively licensed from the Warner Music Group. This particular game is also used to promote and directly sell a variety of artist-branded virtual items as well as clothing, accessories and other physical merchandise. Other similar dance games include Roxio's *In The Groove*, Nexon America's *Audition Online* and Disney's *High School Musical Dance*.

Other games are in themselves brand-products of specific artists. For example *Britney's Dance Beat* allows users to guide a representation of the character of Britney Spears through a variety of dance sequences and scenarios.

Many digital games have, based on their success, spawned music soundtrack releases. For example the EA game *Sims 2* audio soundtrack released in 2008 features such artists as Natasha Beddingfield, Datarock, They Might Be Giants and Great Northern. The soundtrack also features an 'in-game' which features Natasha Beddingfield singing her chart hit *Pocketful Of Sunshine* in the language of the Sim characters, Simlish. Also, in addition to incorporating music into the game experience itself, the various versions of the Rockstar Games game *Grand Theft Auto* have spawned no fewer than thirteen soundtracks on the Epic and Interscope labels. These soundtracks are designed around a variety of musical genres: dance/techno, rock/pop, jazz, classical.

Ongoing Content-Technology Convergence and Conglomeration

In examining the apparent crisis in the record industry a further matter for consideration relates to conglomeration, the heightened levels of concentration and convergence that are characteristic of major music and electronics companies. Here we observe that some music industry corporations actually gain from the pervasive 'crime' of internet downloading and CD burning via the sales of various electronic devices and software programmes used for these activities. In effect, the extended forms of industrial concentration prevailing in our contemporary economy means that music recordings, publishing repertoires and the electronic devices as well as the software that enables them are increasingly falling under the ownership of the same corporations.

We observe that each of the big four music companies also forms part of a larger conglomerate corporation which owns and controls various other media companies that generate products and content across a variety of media formats. With such structures of vertical and horizontal integration, the processes of production, marketing and distribution are integrated throughout various media holdings in efforts to reduce costs and increase revenues. The parent companies of UMG, WMG, Sony Music Entertainment and EMI all own holdings in other areas associated with media entertainment and electronics. For example the Sony Corporation has interests in electronics, CD manufacturing, battery production, movies, television, digital games and mobile phones. Warner has interests in movies, television, the internet, book publishing and mobile phones. Universal has interests in film, the internet and telecommunications. While EMI is now in the hands of financier Guy Hands, and as such can claim to be the largest stand-alone music company, it had until recently interests in consumer electronics and information technology.

The parent companies of the big four music companies have created new media and hardware in attempts to urge music consumers to repurchase music recordings on fresh formats. For example with CDs, consumers have ended up buying content they already owned on vinyl and cassette formats. From the launch of the CD onto international music markets in 1983 until the global peak of record sales in 1999, record industry sales units and revenues increased consistently. In terms of full-length formats, annual units sold from 1.5 billion in 1983 to 3.3 billion in 1999 (IFPI, 2007) at which point the global retail value of record sales was US$38.7 billion (IFPI, 2000). The big music companies have more recently looked to the internet, video game consoles and new audio formats to generate revenues.

Equally the parent companies of the music majors may be seen to have a foot in both camps when it comes to CD burning. While the music companies may claim to lose money through CD burning technologies enabling users to duplicate music recordings quickly and easily, their parent companies generate revenues from sales of the same CD burning technologies. Some of the biggest music companies in the world operate under motherships that sell a combination of hardware enabling CD copying, blank CDs or software for CD duplication. For example, Matsushita, which part-owns the Universal Entertainment Group (which is the parent company of UMG) also owns Panasonic, a manufacturer of CD burners, DVD burners as well as various storage and reproduction devices. Matsushita also manufactures JVC CD burners. Also, Sony Music Entertainment now operate the RCA Music Group which in turn manufactures electronic hardware enabling CD burning. Sony manufactures blank CDs. Sony and UMG also own stocks in Roxio who, according to the company's own webbite in 2008, claimed to have a share of approximately 70% of the global market for CD burning products. Furthermore BMG, who despite selling their recording interests to Sony Music Entertainment still remain one of the five largest music publishers on the planet, have a stake in software company RealNetworks which enables CD burning amongst other digital services. We must also remember that blank CDs are taxed within several states around the world, with subsequent monies providing revenues for the major record companies.

With regard to arguments that the sharing of digital audio files is contributing to a crisis in the record industry, the afore-mentioned hardware manufacturers with links to major music companies also produce and distribute MP3 players. While the iPod by Apple has been the biggest selling MP3 player, operating only in conjunction with the iTunes digital library, it serves not only to converge two technologies in order to make one product more appealing and functional for users, it also works to ensure that neither content nor hardware enable 'unauthorised' copying or transferring of files.

Such convergences of ownership, control and interests between manufacturers and suppliers of technology and music companies also enable the cross-promotion of products, services and content, e.g. recording artists are used to promote hardware, software and services; technologies are used to promote recording artists and content. For example, Dire Straits, who were signed to the (then) Philips-owned Polygram label released a 1985 album entitled *Brothers In Arms* album which, along with their subsequent world concert tour was used to promote the CD internationally. Some six years later, the bands next album and tour – *On Every Street* – was used as a key promotional vehicle for Philip's somewhat less successful Digital Compact Cassette (DCC). More recently, artists such as U2 have allied with Apple in cross-promotional exercises around the iPod. Also, Hewlett Packard offer exclusive designed tattoos featuring Universal recording artist's latest releases to place on the Apple iPod.

STILL COOL: COMMODIFICATION, CONCENTRATION AND CONVERGENCE

In the late 1990s, some authors confidently embraced techno-centric analytical frames to declare that the Internet and related new digital technologies were likely to produce deeply disruptive effects which would transform the established rules and norms that had characterised the capitalist economic system since the industrial revolution (e.g. Kelly, 1999). Others suggested that the new digital devices and networks would dislodge the prevailing order within the media sector, including the activities, position and power of the established 'mass' media corporations (Negroponte, 1995). The recorded music industry, and its unfolding crisis and restructuring processes have often been viewed as the leading-edge manifestations of such disruptive or transformative developments, if not exemplars of the fate awaiting other sectors of the established media industries.

What we find in this study is that it is very much more a case of 'business as usual' in the commercial music industry, rather than any signs of transformation or deep disruptions determined by technological forces. Music is a major and growing 'business', and increasingly subject to the logics of commodification. We have demonstrated that the major music corporations remain 'major' players, indicating how notions of crisis must be accompanied by attention to restructuring processes whereby those corporations have mobilised their concentrated capital and human resources to devise new kinds of institutional, marketing and organisational innovations. In the process, they have learned to live with and adapt to the challenges posed by external technological developments, developing both defensive and offensive strategies to protect and expand their revenue streams. We have demonstrated that our core concepts, derived from old economy days, are still 'cool', in the sense that they are relevant and useful conceptual steers for grounded analyses of the key trends unfolding in the music sector over the past decade and for understanding the scope and meaning of these changes.

Beyond this, we have indicated that another concept of capitalist property and ownership – the ownership copyright – is increasingly relevant to gaining a deeper understanding of the outcome of technological innovation in media and cultural spheres. The ownership of copyrights is the lifeblood of the recorded music industry. The processes of commodification, concentration and convergence outlined above provide the industry's established actors with increased scope to exploit the rights that they own and control.

We noted at the outset how new digital technologies are widely perceived as inducing a certain 'crisis' for the prevailing models and practices of the recorded music industry. Our approach has provided a grounded empirical analysis of recent key shifts in the record industry's relationships with other media sectors. We examined how various new business models launched by the music industry have opened up many new revenue-generating opportunities from music culture despite the relative decline in recorded music sales. As we have seen, the application of the concepts of commodification, concentration and convergence illuminate how the music industry is mobilising its considerable resources to meet these challenges. The concept of convergence has also been particularly helpful in conducting and describing this grounded empirical analysis.

We conclude that this study serves to interrogate and challenge the dominant framing of current debates around the notion of 'crisis' in the music industry and ultimately, it suggests that the current situation and longer term outlook for the established music companies is far from gloomy. Furthermore, we conclude that the concepts of commodification, concentration and convergence remain as central supports for an informed and thorough understanding of contemporary trends in the media and cultural industries. The industrial concentration of cultural distribution and exchange represents real power and control over the production and reception of music across increasingly global markets. The increasingly global flow of cultural products is shaped and channelled by an oligopoly of major corporations. Their continuing concentrated control over the means of distribution remains a major cause of concern for critical research concerned with the forms and extent of diversity mediated culture in the early twenty-first century.

REFERENCES

Adorno, T. W., & Horkheimer, M. (1947/1997). *Dialectic of Enlightenment*. London: [orig. published in Amsterdam, 1947]

Aksoy, A., & Robins, K. (1992). Hollywood for the 21st century: global competition for critical mass in image markets. *Cambridge Journal of Economics*, *16*, 1–22.

Bagdikian, B. H. (1996). *The Media Monopoly* (5th ed.). Boston: Beacon Press.

Bagdikian, B. H. (2000). *The Media Monopoly* (6th ed.). Boston: Beacon Press.

Bagdikian, B. H. (2004). *The New Media Monopoly*. Boston: Beacon Press.

Barfe, L. (2005). *Where Have All the Good Times Gone?: The Rise and Fall of the Record Industry*. London: Atlantic Books.

Benkler, Y. (2006). *The Wealth of Networks: How Social Production Transforms Markets and Freedom*. New Haven, CT: Yale University Press.

Bouwman, H., & Christoffersen, M. (Eds.). (1992). *Relaunching Videotex*. Dordrecht: Kluwer Academic Press.

Croteau, D., & Hoynes, W. (2001). *The Business of Media: Corporate Media and the Public Interest*. Thousand Oaks, CA: Pine Forge Press.

Edgecliffe Johnson, A. (2009). Vevo could be Saviour of Music Video. [th.]. *Financial Times (North American Edition)*, (December): 10.

Fagerjord, A. (2003). Rhetorical convergence: Studying web media. In Liestøl, G., Morrison, A., & Rasmussen, T. (Eds.), *Digital Media Revisited: Theoretical and Conceptual Innovations in Digital Domains* (pp. 293–325). Cambridge, MA: MIT.

Gandy, O. H. Jr. (1982). *Beyond agenda setting: Information subsidies and public policy*. Norwood, NJ: Ablex.

Garnham, N. (1990). *Capitalism and Communication*. London: Sage.

Garnham, N. (2000). *Emancipation, the Media & Modernity*. Oxford, UK: Oxford University Press.

Golding, P., & Murdock, G. (2000). Culture, Communications and Political Economy. In Curran, J., & Gurevitch, M. (Eds.), *Mass Media and Society* (3rd ed., pp. 70–93). London: Arnold.

Hesmondalgh, D., & Toynbee, J. (Eds.). (2008). *The Media and Social Theory*. London: Routledge.

Hesmondhalgh, D. (2007). *The Cultural Industries* (2nd ed.). London: Sage Publications.

Hickey, S. (n.d.). Industry crisis as album sales drop by 10pc. *The Irish Independent* [online], Monday January 14th. Retrieved from http://www.independent.ie/national-news/industry-crisis-as-album-sales-drop-by-10pc-1265540.html

Ingram, M. (2008). Disharmony and the business of music. *The Globe and Mail* [online]. Retrieved from http://v1.theglobeandmail.com/servlet/Page/document/v5/content/subscribe?user_URL=http://www.globeinvestor.com%2Fservlet%2FArticleNews%2Fstory%2FGAM%2F20080131%2FTWINGRAM31&ord=27610503&brand=globeinvestor&force_login=true

Innis, H. (1950). *Empire and Communications*. Toronto: University of Toronto Press.

Innis, H. (1951). *The Bias of Communication*. Toronto: University of Toronto Press.

International Federation of Phonographic Industries (IFPI) – Market Research Department. (2000). *Recording Industry in Numbers*. London: International Federation of Phonographic Industries.

International Federation of Phonographic Industries (IFPI) – Market Research Department. (2007). *Recording Industry in Numbers*. London: International Federation of Phonographic Industries.

International Federation of Phonographic Industries (IFPI) – Market Research Department. (2008). *Recording Industry in Numbers*. London: International Federation of Phonographic Industries.

International Federation of Phonographic Industries (IFPI) – Market Research Department. (2009). *Recording Industry in Numbers*. London: International Federation of Phonographic Industries.

International Federation of Phonographic Industries (IFPI) – Market Research Department. (2009a). *Digital Music Report 2009* [online]. Available at: http://www.ifpi.org/content/library/DMR2009.pdf

Kelly, K. (1999). *New Rules for the New Economy: 10 Ways the Network is Changing Everything*. London: Fourth Estate.

Kelly, K. (2002). Towards a post-napster music industry. *The New York Times Magazine*. March 17th, 19-21.

Kucklich, J. (2004). Precarious Playbour: Modders and the Digital Games Industry. In *Fibre Culture: Internet Theory, Criticism and Research, 5,* [online]. Available at: <http://www.fibreculture.org/issue5/kuckligh_print.html>

Lessig, L. (2001). *The Future of Ideas*. New York: Vintage.

Liebowitz, S. (2002). *Record Sales, MP3 Downloads, and the Annihilation Hypothesis*. [online] Working Paper, University of Texas at Dallas. Retrieved from http://www.dklevine.com/archive/leibowitz.pdf

McChesney, R. W. (2004). *The Problem of the Media*. New York: Monthly Review Press.

Miege, B. (1989). *The Capitalisation of Cultural Production*. New York: International General.

Mosco, V. (1996 & 2009). *The Political Economy of Communication*. London: Sage.

Napier-Bell, S. (2008). The Life and Crimes of the Music Biz. *The Observer – Music Monthly*, Sunday 20th January, 41.

Negroponte, N. (1995). Being Digital: A Book (P)review. *Wired.com* [online], *3.02*, February. Retrieved from http://www.wired.com/wired/archive/3.02/negroponte.html?pg=2&topic=

Negroponte, N. (1996). *Being Digital: The Road Map for Survival on the Information Superhighway*. London: Hodder & Stoughton.

O'Regan, N. (2008). Heed The Music Lessons. *The Sunday Business Post* [online], Sunday April 6th. Retrieved from http://archives.tcm.ie/businesspost/2008/04/06/story31725.asp

Osorio, A. (2008). Xbox Song Downloads Topping 3.8 Million Monthly. *Digital Music News* [online], Tuesday 21st October 2008. Retrieved from http://www.digitalmusicnews.com/stories/102108msft

Osorio, A. (2008a). MySpace Educating Masses on Chinese Democracy. *Digital Music News* [online], Thursday 20th November. Retrieved from http://www.digitalmusicnews.com/stories/112008myspace/?searchterm=Chinese%20Democracy%20on%20MySpace

Picard, R. (1998). Media Concentration, Economics, and Regulation. In Graber, D., McQuail, D., & Norris, P. (Eds.), *The News of Politics* (pp. 193–217). Washington, D.C: Congressional Quarterly Press.

Picard, R. (2002). Delusions of Grandeur: The Real Problems of Concentration in Media. In Demers, D. (Ed.), *Global Media News Reader* (pp. 33–48). Spokane, WA: Marquette Books.

Picard, R. (2005). Unique Characteristics and Business Dynamics of Media Products. *Journal of Media Business Studies*, *2*(2), 61–69.

Pool, I. de Sola. (1983). *Technologies of Freedom: On Free Speech in an Electronic Age*. Cambridge, MA: Belknap Press/Harvard University Press.

Pope, C. (2009). Has Music Had It's Day? *The Irish Times*, Monday April 27th, 15.

Preston, P. (2001). *Reshaping Communication: Technology, Info & Social Change*. Thousand Oaks, CA: Sage.

Preston, P. (2008). *Making the News*. London: Routledge.

Preston, P., & Flynn, R. (2000). Rethinking Universal Service: Citizenship Consumption Norms and the Telephone. *The Information Society*, *16*(2), 91–98. doi:10.1080/01972240050032843

Preston, P., & Grisold, A. (1995). Unpacking the Concept of Competition in Media Policy Making: The Case of Ireland & Austria. In Corcoran, F., & Preston, P. (Eds.), *Democracy and Communication in the New Europe: Change & Continuity in East and West* (pp. 67–96). Creskill, NJ: Hampton Press.

Preston, P., Kerr, A., & Cawley, A. (2009). Digital media sector innovation in the knowledge economy: rethinking knowledge inputs and policies. *Information Communication and Society*, *12*(4), 994–1014. doi:10.1080/13691180802578150

Sandoval, G. (2009). Sony joins YouTube and Universal on Vevo video site. *cnet* [online]. Retrieved from http://news.cnet.com/8301-1023_3-10257359-93.html

Schiller, H. (1984). *Information and the Crisis Economy*. Norwood, NJ: Ablex Publishing Corporation.

Storsul, T., & Stuedahl, D. (Eds.). (2007). *Ambivalence Towards Convergence: Digitalization and Media Change*. Nordicom.

Techno-Z, F. H. (1997). *The Content Challenge: Electronic Publishing and the New Content Industries - Research report commissioned and published by EC*. DG XIII/E.

Thomas, A. (2004). *The Dynamics of Games* (4th ed.). London: Informa Media.

UK Department of Trade and Industry. (1998). *Regulating Communications: Approaching Convergence in the Information Age* [Green Paper on "Convergence"]. London: Department of Trade and Industry.

ADDITIONAL READING

Bakker, P. (2005). "File-sharing – Fight, Ignore or Compete: Paid Download Services vrs. P2P Networks" in *Telematics and Infomatics*. Vol. 22. *Issues (National Council of State Boards of Nursing (U.S.)), 1-2*(Feb).

Berkman Centre for Internet and Society. (2004), iTunes: How Copyright Contract and Technology Shape the Business of Digital Media - A Case Study. Berkman Centre for Internet and Society (Digital Media Project) [online]. Available from: <http://cyber.law.harvard.edu/publications/2004/iTunes_whitepaper >

Bhattacherjee, S., Gopal, R., Lertwachara, K., Marsden, J. R., & Telang, R. (2007), "The Effect of Digital Sharing Technologies on Music Markets: A Survival Analysis of Albums on Ranking Charts" in *Management Science*, Vol. 53 No. 9, pp.1358-1374.

Dowd, T. J. (2002), "Culture and commodifictation: technology and structural power in the early US recording industry" in *International Journal of Sociology and Social Policy*, Vol.22, No. 1/2/3, pp. 106-140.

Frith, S. (2002). Illegality and the Music Industry. In Talbot, M. (Ed.), *The Business of Music*. Liverpool: Liverpool University Press.

Gillespie, T. (2009). *Wired Shut: Copyright and the Shape of Digital Culture*. Cambridge: MIT Press.

Hesmondhalgh, D. (2007), "Digitalisation, Music and Copyright", ESRC Centre for Research on Socio-Cultural Change, Working Paper number 30.

Hesmondhalgh, D. (2009). The Digitalisation of Music. In Pratt, A. C., & Jeffcut, P. (Eds.), *Creativity and Innovation in the Cultural Economy. Abingdon and New York*. Routledge.

Jones, S. (2000), "Music and the Internet", in Popular Music. Volume 19/2, Cambridge: University Press, 217 – 230.

Jones, S. (2002a), "Music That Moves: Popular Music, Distribution and Network Technologies", in Cultural Studies. Vol. 16 No. 2, March, pp. 215 – 234.

Laing, D. (2004). Copyright, Politics and the International Music Industry. In Frith, S., & Marshall, L. (Eds.), *Music and Copyright* (2nd ed., pp. 70–85). Edinburgh: Edinburgh University Press.

Mansell, R., & Javary, M. (2004). New Media and the Forces of Capitalism. In Calabrese, A., & Sparks, C. (Eds.), *Toward a Political Economy of Culture. Lanham*. Maryland: Rowman & Littlefield.

McCourt, T., & Burkart, P. (2003). When Creators, Corporation and Consumers Collide: Napster and the Development of On-line Music Distribution. In *Media, Culture and Society* (*Vol. 25*, pp. 333–350). London, Thousand Oaks, New Delhi: Sage Publications.

Oberholzer-Gee, F., & Strumpf, K. (2009), "File-Sharing and Copyright" - Harvard Business School Working Paper 09-132 [online]. Available from: http://www.hbs.edu/research/pdf/09-132.pdf

Sparks, C. (2004). The Impact of the Internet on Existing Media. In Calabrese, A., & Sparks, C. (Eds.), *Toward a Political Economy of Culture. Lanham*. Maryland: Rowman & Littlefield.

Wikstrom, P. (2009). *The Music Industry*. Cambridge: Polity Press.

Compilation of References

A day in the life of Miss McDonald. (2004). Retrieved July 22, 2007, from http://miss-mcdonald.livejournal.com/

Abraham, R. (2007). Mobile phones and economic development: Evidence from the fishing industry in India. *MIT Press Journal*, *4*(1), 5–17.

Abraham, M., & Manning, E. (Aug 12, 2005). *Service on the Line: Indian Telephone Call Centers and Customer Care.* Paper presented at the annual meeting of the American Sociological Association, Available at http://www.allacademic.com, retrieved dated 2010-05-07.

Acland, C. (2009). Curtains, cart, and mobile screens. *Screen*, *50*(1), 148–166. doi:10.1093/screen/hjn071

Adalian, J. (2004). Sony's back and ready for action. *Variety*, *393*(9), 14–15.

Adorno, T. W., & Horkheimer, M. (1947/1997). *Dialectic of Enlightenment*. London: [orig. published in Amsterdam, 1947]

Adu-Febiri, F. (2006). The destiny of cultural diversity in a globalized world. *Review of Human Factor Studies*, *12*(1), 30–64.

Advice regarding call centre working practices. (2006, October). *Health and Safety Executive* retrieved 2008-06-05

Africa Monitor. (2007, May). Inflation subsides. *Africa Monitor: Southern Africa, 12.*

Agence France Presse (AFP). New technology triggers battle for information from Tibet. (2008, March 18). Retrieved from http://afp.google.com/article/AleqM5gyuONfqd4fnQB61MMrrD-PyjMDjw

Ahmed, Z. (22-02-2006). Abuse rattles Indian call centre staff. *BBC News*. Available at www.bbc.co.uk, retrieved 2008-06-05.

Akin, D. (28 January 2010). Capitalism has gone 'badly wrong'. *The Ottawa Citizen*, d-3.

Akiyoshi, M., & Ono, H. (2008). The diffusion of mobile internet in Japan. *The Information Society*, *24*, 292–303. doi:10.1080/01972240802356067

Aksoy, A., & Robins, K. (1992). Hollywood for the 21st century: Global competition for critical mass in image markets. *Cambridge Journal of Economics*, *16*, 1–22.

Alger, D. (1998). *Megamedia: How giant corporations dominate mass media, distort competition, and endanger democracy*. New York: Rowman & Littlefield.

Alleyne, M. D. (2009). Global media ecology: Why there is no global media ethics standard. In Lee, W., & Clifford, G. C. (Eds.), *The handbook of mass media ethics*. New York: Routledge.

Al-Rodhan, N.R.F. (19th June, 2006). *Definitions of Globalization: A Comprehensive Overview and a Proposed Definition*. Geneva, Switzerland: Geneva Centre for Security Policy.

Altheide, D. (1995). *An ecology of communication: Cultural formats of control*. Hawthorne, UK: Aldine de Gruyter.

America On Line [AOL] Corporate. (2009). *AOL Celebrates Day One as an Independent Company*. Retrieved March, 9, 2010, from: http://corp.aol.com/newsroom/press-releases

Copyright © 2011, IGI Global. Copying or distributing in print or electronic forms without written permission of IGI Global is prohibited.

Anand, N., & Daft, R. L. (2007). What is the right organization design? *Organizational Dynamics, 36*(4), 329–344. doi:10.1016/j.orgdyn.2007.06.001

Andén-Papadopoulos, K. (2008). The Abu Ghraib torture photographs: News frames, visual culture, and the power of images. *Journalism, 9*(1), 5–30. doi:10.1177/1464884907084337

Anderrson, E., & Jansson, A. (2006, May 23-25). *A Call for Learning - A model for Developmental Learning in Call Centres*. Paper presented at the 6th UFHRD/AHRD Conference on Human Resource Development Research and Practice across Europe, Tilburg.

Anderson, B. (1991). *Imagined communities: Reflections on the origin and spread of nationalism*. London: Verso.

Ang, I. (1985). *Watching Dallas*. London: Routledge.

Appadurai, A. (1990). Disjuncture and difference in the global cultural economy. [). New York: Sage.]. *Theory, Culture & Society, 7*, 295–310. doi:10.1177/026327690007002017

Appadurai, A. (1991). *Modernity at large*. Minneapolis: University of Minnesota Press.

Appadurai, A. (1996). *Modernity at Large: Cultural dimensions of globalization*. Minneapolis: University of Minnesota Press.

Appfrica (2008). *The current state of Internet penetration in Africa.* Retrieved June 15, 2009, from http://appfrica.net/blog/archives/248

Armstrong, C. (2008). Korean wave: A view from America's shores. *Korea Herald*, 5 February. LexisNexus. Consulted on October 31, 2009.

Arnold, M. (2007). The Concept of Community and the Character of Networks, *Journal of Community Informatics*, 3 (2). Retrieved 20 October 2008 from http://www.ci-journal.net/index.php/ciej/article/view/327/355

Ashill, N. J., Rod, M., Thirkell, P., & Carruthers, J. (2009). Job resourcefulness, symptoms of burnout and service recovery performance: an examination of call centre frontline employees. *Journal of Services Marketing, 23*(5), 338–350. doi:10.1108/08876040910973440

Atkin, D. J., Jeffres, L. W., & Neuendorf, K. A. (1998). Understanding Internet adoption as telecommunications behavior. *Journal of Broadcasting & Electronic Media, 42*, 475–490.

Auerbach, A. (2006, February). Interpreting urban screens. *First Monday, 4*. Retrieved October 20, 2009, from http://www.uic.edu/htbin/cgiwrap/bin/ojs /index.php/fm/rt/printerFriendly/1546/1461

Austrade (2009). *Republic of Korea fact sheet*. Retrieved March 11, 2010, from http://www.dfat.gov.au/geo/fs/rkor.pdf

Australian Department of Foreign Affairs and Trade. (2010). *Republic of Korea country brief.* Retrieved March 11, 2010, from http://www.dfat.gov.au/geo/rok/brief_index.html

Australian Department of Innovation. Industry, Science and Research (2010). *Australia's exports fact sheet*. Retrieved March 11, 2010, from http://www.innovation.gov.au/Section/ AboutDIISR/FactSheets/Pages/Australia%27sExportsFactSheet.aspx

Bachen, C., Raphael, C., & Lynn, K. M. (2008). Civic engagement, pedagogy, and information technology on Web sites for youth. *Political Communication, 25*(3), 290–310. doi:10.1080/10584600802197525

Bagdikian, B. H. (1996). *The Media Monopoly* (5th ed.). Boston: Beacon Press.

Bagdikian, B. H. (2000). *The Media Monopoly* (6th ed.). Boston: Beacon Press.

Bagdikian, B. H. (2004). *The New Media Monopoly*. Boston: Beacon Press.

Baidu Encyclopedia. (2010). *NetEase company*. Retrieved April 3, 2010 from http://baike.baidu.com/view/149363.html (In Chinese).

Bain, P., & Taylor, P. (2004). *No passage to India? UK unions, globalisation and the migration of call centre jobs*. Work, Employment and Society Conference, 1-3 September, Manchester.

Baker, G., & Megler, V. (2001). *The semi-walled garden: Japan's "i-mode phenomenon*. New York: IBM.

Bakker, A. B., Demerouti, E., & Verbeke, W. (2004). Using the job demands-resources model to predict burnout and performance. *Human Resource Management*, *43*(1), 83–104. doi:10.1002/hrm.20004

Baldwin, C. Y., & Clark, K. B. (1997). Managing in an age of modularity. *Harvard Business Review*, (Sep-Oct): 84–94.

Baldwin, C. Y., & Clark, K. B. (2000). *Design rules: The power of modularity*. Cambridge, MA: MIT Press.

Balfour, F. (2008, February 11). Catching the eye of China's elite. *Business Week, 470*(4070), 55-56.

Baliamoune-Lutz, M. (2003). An analysis of the determinants and effects of ICT diffusion in developing countries. *Information Technology for Development*, *10*, 151–169. doi:10.1002/itdj.1590100303

Bandura, A. (1977). *Social learning theory*. Englewood Cliffs, NJ: Prentice-Hall.

Bandura, A. (1986). *Social foundations of thought and action*. Englewood Cliffs, NJ: Prentice-Hall.

Bann, S. (1984). *The clothing of Clio: A study of the representation of history in nineteenth-century Britain and France*. New York: Cambridge University Press.

Baoill, O. A. (2004). Weblogs and the Public Sphere. In L.J. Gurak, S. Antonijevic, L. Johnson, C. Ratliff, & J. Reyman (Eds.), *Into the Blogosphere: Rhetoric, Community, and Culture of Weblogs*. Retrieved December 2009 from http://blog.lib.umn.edu/blogosphere/ weblogs_and_the_public_sphere.html

Bar, F., & Sandvig, C. (2008). US communication policy after convergence. *Media Culture & Society*, *30*(4), 531–550. doi:10.1177/0163443708091181

Barak, S. (2008, July 2). Mobile users are all fingers and thumbs. *Inquirer*. Retrieved Aug. 9, 2008, from http:// www.theinquirer.net/gb/inquirer /news/2008/07/02/ mobile-users-similar

Barfe, L. (2005). *Where Have All the Good Times Gone?: The Rise and Fall of the Record Industry*. London: Atlantic Books.

Barnett, C. (2004). Neither poison nor cure: Space, scale and public life in media history. In Couldry, N., & McCarthy, A. (Eds.), *Mediaspace: Place, Scale and Culture in a Media Age* (pp. 58–74). London: Routledge.

Barry, A. (2001). *Political machines: Governing a technological society*. New York: Athlone Press.

Barthes, R. (1981). *Camera Lucida: Reflections on photography, (Trans. R. Howard)*. New York: Hill and Wang.

Barthes, R. (1977). *The photographic message. In Image-Music-Text*. New York: Hill and Wang.

Basi, J. K. T. (2009). *Women, Identity and India's Call Centre Industry*. New York: Routledge.

Bates, B. J., & Chambers, T. (2004). The economics of the cable industry. In Alexander, A., Owers, J., Carveth, R., Hollifield, A. C., & Greco, A. C. (Eds.), *Media economics: Theory and practice* (3rd ed., pp. 173–192). Mahwah, NJ: Lawrence Erlbaum Associates Inc.

Batt, R. (2000). Strategic segmentation in front-line services: matching customers, employees and human resource systems. *International Journal of Human Resource Management*, *11*(3), 540–561. doi:10.1080/095851900339756

Batt, R., & Applebaum, E. (1995). Worker Participation in Diverse Settings: Does the Form Affect the Outcome, & If So, Who Benefits? *British Journal of Industrial Relations*, *33*(3), 353–378. doi:10.1111/j.1467-8543.1995.tb00444.x

Baudrillard, J. (1996). *The system of objects, (Trans. J. Benedict)*. London: Verso.

Baudrillard, J. (1988). *America*. London: Verso.

Baudrillard, J. (2003). *The consumer society: Myths and structures*. London: Sage. (Original work published 1998)

Bauman, Z. (2006). *Liquid fear*. Cambridge, UK: Polity.

Beck, U. (1992). *Risk society, (Trans. M. A. Ritter)*. London: Sage.

Beck, J., & Wade, M. (2003). *DoCoMo—Japan's wireless tsunami: How one mobile telecom created a new market and became a global force*. New York: AMACOM.

Beckman, K. (2007). Telescopes, transparency, and torture: Trevor Paglen and the politics of exposure. *Art Journal, 2007*(Fall), 62–67.

Beier-de Hann, R. (2006). Re-staging histories and identities. In Macdonald, S. (Ed.), *A companion to museum studies* (pp. 186–197). Malden, MA: Blackwell. doi:10.1002/9780470996836.ch12

Beilock, R., & Dimitrova, D. V. (2003). An exploratory model of inter-country Internet diffusion. *Telecommunications Policy, 27*(3-4), 237–252. doi:10.1016/S0308-5961(02)00100-3

Beinhocker, E. D. (2007). *The origin of wealth*. London: Random House Business Books.

Belay, G. (1996). The (re)construction and negotiation of cultural identities in the age of globalization. In Mokros, H. (Ed.), *Interaction and identity: Information and behavior*. New Brunswick, NJ: Transaction Publishers.

Bell, D. (1976). *The coming of post-industrial society*. New York: Basic Books.

Bell, G. (2005). The age of the thumb: A cultural reading of mobile technologies from Asia. In Raessens, J., & Goldstein, J. (Eds.), *Handbook of computer game studies* (pp. 67–87). Cambridge, MA: MIT Press.

Bellamy, R., & Walker, J. (1996). *Television and the remote control*. New York: Guilford Press.

Belson, K., & Sorkin, A. (2004, 15 September). Buying MGM may give Sony more leverage to set a new DVD standard. *The New York Times*, C6.

Benedetti, P., & DeHart, N. (Eds.). (1996). *Forward through the rearview mirror: Reflections on and by Marshall McLuhan*. Scarborough, UK: Prentice Hall Canada.

Benhabib, S. (1992). *Situating the self: Gender, community and postmodernism in contemporary ethics*. Cambridge, UK: Polity Press.

Benjamin, W. (1981). The work of art in the age of mechanical reproduction. In *Selected writings* (pp. 319–334). Cambridge, MA: Harvard University Press.

Benjamin, W. (1968). *The work of art in the age of mechanical reproduction, (Trans. H. Zohn)*. London: Jonathon Cape.

Benkler, Y. (2006). *The Wealth of Networks: How Social Production Transforms Markets and Freedom*. New Haven, CT: Yale University Press.

Bennett, T. (1995). *The birth of the museum: History, theory, politics*. London: Routledge.

Benninghoven,A., Bindzius, F., Braun, D., Cramer,J. Ellegast,R. Flowerday,U. Genz, A., Heyden, T. von der, Pfeiffer,W., Schittly,D., Schweer, R. and Roger Stamm(2005). CCall—Healthy and Successful Work in Call Centres. *International Journal of Occupational Safety and Ergonomics* (*JOSE*), *11*(4), 409–421.

Benson, R. (2006). News Media as a "Journalistic Field": What Bourdieu Adds to New Institutionalism, and Vice Versa. *Political Communication*, *23*, 187–202. doi:10.1080/10584600600629802

Bhabha, H. (1994). *The Location of culture*. New York: Routledge.

Bhaghat, C. (2005). *One Night@ the Call Center*. New Delhi, India: Rupa & Co.

Bielby, D., Harrington, C., & Bielby, W. (1999). Whose stories are they? Fans' engagement with soap opera narratives in three sites of fan activity. *Journal of Broadcasting & Electronic Media*, *43*(1), 35–51.

Bielby, W. T., & Bielby, D. D. (2003). Controlling prime-time: Organizational concentration and network television programming strategies. *Journal of Broadcasting & Electronic Media*, *47*, 573–596. doi:10.1207/s15506878jobem4704_6

Bijker, W. E., & Pinch, T. J. (Eds.). (1987). *The Social construction of technological systems: new directions in the sociology and history of technology*. Cambridge, MA: MIT Press.

Biressi, A. (2004). "Above the below": Body trauma as spectacle in social/media space. *Journal for Cultural Research*, *8*(3), 335–352. doi:10.1080/1479758042000264975

Blanchot, M. (1988). *The unavowable community*. New York: Station Hill Press.

Blankenship, L. (1986). The conscience of a hacker. *Phrack, 1*(7), phile 3 of 10, from http://www.phrack.org/archives/7/P07-03

Blevins, J. L., & Brown, D. H. (2006). Political issue or policy matter? the U.S. Federal Communications Commission's third biennial review of broadcast ownership rules. *The Journal of Communication Inquiry*, *30*(1), 21–41. doi:10.1177/0196859905280951

Blevins, J. L., & Brown, D. H. (2007). *The political economy of media ownership rulemaking within the U.S. federal communications commission: An analysis of media ownership studies*. Paper Presented to the Political Economy Section of the International Association for Media and Communication Research (IAMCR) for its 50th Anniversary Conference, Paris, France.

Bolter, J., & Grusin, R. (1999). *Remediation: Understanding new media*. Cambridge, MA: MIT Press.

Boon, J. A. (1982). *Other Tribes, Other Scribes: Symbolic Anthropology in the Comparative Study of Cultures, Histories, Religions, and Texts*. Cambridge, UK: Cambridge University Press.

Bouwman, H., & Christoffersen, M. (Eds.). (1992). *Relaunching Videotex*. Dordrecht: Kluwer Academic Press.

Boyd, D. (2008). Why Youth Social Network Sites: The Role of Networked Publics Teenage Social Life. In Buckingham, D. (Ed.), *Youth, Identity, and Digital Media* (pp. 119–142). Cambridge, MA: The MIT Press.

boyd, D. (2003). *Reflections on Friendster, Trust and Intimacy.* Presented at School of Information Management & Systems (SIMS), University of California, Berkeley.

boyd, D. (2004). *Revenge of the User: Lessons from Creator/ User Battles*. Presented in the O'Reilly Emerging Technology Conference, February, http://www.danah.org/papers/Etech2004.html

Bracha, O. (2007). Standing copyright law on its head? The Googlization of everything and the many faces of property. *Texas Law Review*, *85*, 1799–1869.

Brey, P. (1997). Philosophy of Technology Meets Social Constructivism. *Society for Philosophy and Tehnology*, *2*(3-4). Retrieved December 3, 2009, from http://scholar.lib.vt.edu/ejournals/ SPT/v2_n3n4html/brey.html

British Broadcasting Corporation [BBC]. (2000). *AOL, Time Warner confirm merger*. Retrieved January, 10, 2007, from: http://www.bbc.co.uk

Broady, V. (1997). Quake II Review for PC. *GameSpot*. Retrieved December 3, 2009, from http://au.gamespot.com/pc/a ction/quake2/review.html

Brooke, J., & Hansell, S. (2005, 10 March). Samsung is now what Sony once was. *The New York Times*, C1.

Brooker, W. (2004). Living on *Dawson's Creek*: Teen viewers, cultural convergence, and television overflow. In Allen, R. C., & Hill, A. (Eds.), *The Television Studies Reader* (pp. 569–580). London: Routledge.

Brown, A. (1996). Economics, public service broadcasting and social values. *Journal of Media Economics*, *9*(1), 3–15. doi:10.1207/s15327736me0901_2

Bruns, A., & Jacobs, J. (Eds.). (2006). *Uses of Blogs*. New York: Peter Lang.

Bruns, A. (2008). *Blogs, wikipedia, second life, and beyond: From production to produsage*. New York: Peter Lang Publishing.

Brusoni, S., & Prencipe, A. (2001). Unpacking the black box of modularity: Technologies, products and organizations. *Industrial and Corporate Change*, *10*(1), 179–205. doi:10.1093/icc/10.1.179

Buhlmann, V. (2006, February). Intelligent skin: Real virtual. *First Monday, 4*. Retrieved October 20, 2009, from http://www.uic.edu/htbin/cgiwrap/ bin/ojs/index.php/fm/rt/printerFriendly/1554/1469

Burgess, J. (2008). All Your Chocolate Rain Are Belong to Us? Viral Video, YouTube and the Dynamics of Participatory Culture. In Lovink, G. (Eds.), *The VideoVortex Reader*. Amsterdam: Institute of Network Cultures.

Burns, S. (2007). *Mobile Tech Today*. Retrieved November 7, 2007, from http://www.mobile-tech-today.com /story.xhtml?story_id=56350

Burrell, J., & Anderson, K. (2008). 'I have great desires to look beyond my world': Trajectories of information and communication technology use among Ghanaians living abroad. *New Media & Society*, *10*, 203–224. doi:10.1177/1461444807086472

Caldwell, J. (2004). Convergence television: Aggregating form and repurposing content in the culture of convergence. In Spigel, L., & Olson, J. (Eds.), *Television after TV* (pp. 41–74). Durham, NC: Duke University Press.

Call centers in India. (n.d.). Available at www.outsource2india.com, retrieved dated 15th December, 2009.

Call Centre Monitoring. (n.d.). *Management*. Available at www.callcentrehelper.com. Retrieved 06-05-2008.

Cameron, D. (2000). Styling the worker: Gender and the commodification of language in the globalized service economy. *Journal of Sociolinguistics*, *4*(3), 323–347. doi:10.1111/1467-9481.00119

Cameron, F. (2006). Beyond the cult of the replicant: Museums and historical digital objects—traditional concerns, new discourses. In Cameron, F., & Kenderdine, S. (Eds.), *Theorizing digital cultural heritage: A critical discourse* (pp. 49–76). Cambridge, MA: MIT Press.

Cameron, F., & Robinson, H. (2006). Digital knowledgescapes: Cultural, theoretical, practical, and usage issues facing museum collection databases in a digital epoch. In Cameron, F., & Kenderdine, S. (Eds.), *Theorizing digital cultural heritage: A critical discourse* (pp. 165–192). Cambridge, MA: MIT Press.

Campbell, S. W., & Russo, T. C. (2003). The social construction of mobile telephony: An application of the social influence model to perceptions and uses of mobile phones within personal communication networks. *Communication Monographs*, *70*, 317–334. doi:10.1080/0363775032000179124

Carless, S. (2003). Marc Laidlaw On Story And Narrative In Half-Life. *Gamasutra*. Retrieved December 3, 2009, from http://www.gamasutra.com/fea tures/20030808/carless_01.shtml

Carroll, N. (2007). Art and globalization: Then and now. *The Journal of Aesthetics and Art Criticism*, *65*(1), 131–143. doi:10.1111/j.1540-594X.2007.00244.x

Carson, D. O. (2004). Satellite Home Viewer Extension Act, Statement by David O. Carson, General Counsel, United States Copyright Office before the Committee on the Judiciary, United States Senate, 108th Congress, 2nd Session. Retrieved January, 5, 2007 from http://www.tvb.org

Castells, M. (1996). *The Rise of the Network Society, The Information Age: Economy, Society and Culture* (*Vol. I*). Cambridge, MA: Oxford University Press.

Castells, M., Fernández-Ardèvol, M., Qiu, J. L., & Sey, A. (2007). *Mobile communication and society: A global perspective*. Cambridge, MA: MIT Press.

Castells, M. (1998). *End of millennium*. Malden, MA: Blackwell Publishers.

CERNET. (n.d.). Retrieved from http://www.edu.cn/CERNET_1377/index.shtml

Cerulo, K. (1997). Re-framing sociological concepts for a brave new (virtual?) world. *Sociological Inquiry*, *67*(1), 48–58. doi:10.1111/j.1475-682X.1997.tb00428.x

Cerulo, K., Ruane, J., & Chayko, M. (1992). Technological ties that bind: media generated primary groups. *Communication Research*, *19*(1), 109–129. doi:10.1177/009365092019001005

Chadha, K., & Kavoori, A. (2000). Media imperialism revisited: Some findings from the Asian case. *Media Culture & Society, 22*, 415–432. doi:10.1177/016344300022004003

Chakrabarty, D. (2002). Museums in late democracies. *Humanities Report, 9*(1), 5–12.

Chambers, T., & Howard, H. (2005). The economics of media consolidation. In Albarran, A. B., Chan-Olmsted, S. M., & Wirth, M. (Eds.), *Handbook of media management and economics* (pp. 363–386). Mahwah, NJ: Lawrence Erbaum.

Chan, D. (2008). Convergence, connectivity, and the case of Japanese mobile gaming. *Games and Culture, 3*(1), 13–25. doi:10.1177/1555412007309524

Chan, E., & Vorderer, P. (2006). Massively multiplayer online games. In Vorderer, P., & Bryant, J. (Eds.), *Playing video games: Motives, responses, and consequences* (pp. 77–88). Mahwah, NJ: Lawrence Erlbaum Associates, Publishers.

Chandler, H. M. (2005). *The game localization handbook*. Hingham, MA: Charles River Media Inc.

Chandler, A. D., & Hikino, T. (1990). *Scale and scope: the dynamics of industrial capitalism*. Cambridge, MA: Belknap Press.

Chang, M. (2005). Taiwan's nationalistic politics and its difficult 'status quo'. *The Copenhagen Journal of Asian Studies, 21*, 91–124.

Chang, S. J. (2008). *Sony vs. Samsung: the inside story of the electronics giants' battle for global supremacy*. Hoboken, NJ: Wiley.

Chang, Y. (July 12, 2005). Taiwan wave! (in Chinese) *Liberty Times*. Retrieved April 17, 2010, from http://www.libertytimes.com.tw /2005/new/jul/12/today-o4.htm.

Chan-Olmsted, S. (1998). Mergers, acquisitions, and convergence: the strategic alliance of broadcasting, cable television, and telephone services. *Journal of Media Economics, 11*(3), 33–46. doi:10.1207/s15327736me1103_4

Chatman, E. (2000). Framing social life in theory and research. *New Review of Information Behavior Research, 200*(1), 3–17.

Chatman, E. A., & Huotari, M. (2001). Using everyday life information seeking to explain organizational behavior. *Library & Information Science Research, 23*, 351–366. doi:10.1016/S0740-8188(01)00093-7

Chatterjee, S. R. (2007). Human resource management in India: "Where from" and "where to?". *Research and Practice in Human Resource Management, 15*(2), 92–103.

Chen, P. H. (2002a). The role of the state in shaping Taiwan's cable industry. *Media Asia, 29*(1), 37–41.

Chen, P. H. (2002b). Who owns cable television? Media ownership concentration in Taiwan. *Journal of Media Economics, 15*(1), 41–55. doi:10.1207/S15327736ME1501_4

Chen, Y. (2008). Looking for Taiwan's competitive edge: The production and circulation of Taiwanese TV Drama. In Zhu, Y., Keane, M., & Bai, R. (Eds.), *TV drama in China* (pp. 175–187). Hong Kong: Hong Kong University Press.

Chen, P. H. (January 14, 2006). *Ban foreign programs not close-door policy* (in Chinese). Retrieved April 17, 2010, from http://www.twmedia.org/modules/ news/ article.php?storyid=228

Chen, Y. (2004). *The formation of television audience: The market of Korean dramas* (in Chinese). Unpublished master dissertation, Tankang University.

Cheng, Y. (January 11, 2006). Audience to watch Korean dramas (in Chinese). *United Daily*, p. A15.

Chesbrough, H. (2005). Towards the Dynamics of Modularity - A cyclical model of technical advance. In Prencipe, A., Davies, A., & Hobday, M. (Eds.), *The Business of Systems Integration* (2nd ed., pp. 174–198). Oxford, UK: Oxford University Press. doi:10.1093/acprof:oso/9780199263233.003.0010

Chesher, Chris (2004). Neither gaze nor glance, but glaze: relating to console game screens. *SCAN Journal, 1*(1).

Chiang, P. (2004). *The imagined cultural landscape: The Korean wave and the fans in Taiwan* (in Chinese). Unpublished master dissertation, National Chengchi University.

Chien, C. (January 11, 2006). Support the ban on foreign programs (in Chinese). *Liberty Times*. Retrieved April 17, 2010, from http://www.libertytimes.com.tw /2006/new/jan/11/today-o10.htm

Chien, C. H. (October 23, 2005). Defensive effect of culture (in Chinese). *Liberty Times.* Retrieved April 17, 2010, from http://www.libertytimes.com.tw/ 2005/new/oct/23/today-o9.htm

China Cultural Industry Net. (2010, January 20). *China network game market white paper: 2009. A synopsis*. Available: http://www.cnci.gov.cn/conte nt%5C2010120/news_55882.shtml

China Game Publishers Committee & International Data Corp. (2004). *China game industry report: 2003*. Beijing: People's University Press.

Chirot, D., & Hall, T. D. (1982). World-system theory. *Annual Review of Sociology*, *8*, 81–106. doi:10.1146/annurev.so.08.080182.000501

Chiu, D. (January 12, 2006). Taiwanese soaps need protection (in Chinese). *Taipei Times.* Retrieved April 17, 2010, from http://www.taipeitimes.com/News/ editorials/archives/2006/01/12/2003288638/wiki

Chiu, M. L. (2007). Digital services for preserving cultural heritages: Digital Tainan experience. In L. H. Chang, Y-T. Liu, & J-H. Hou (Eds.), *The Proceedings of DACH 2007: The 2007 international conference on digital applications in cultural heritage* (Vol. 2, pp.229-242). Tainan, Taiwan: National Center for Research and Preservation of Cultural Properties.

Christians, C. G. (2004). Ubuntu and communitarianism in media ethics. *Ecquid Novi*, *25*(2), 235–256. doi:10.3368/ajs.25.2.235

Christians, C. G. (2005). Ethical theory in communication research. *Journalism Studies*, *6*(1), 3–14. doi:10.1080/1461670052000328168

Christians, C. G. (2007). Utilitarianism in Media Ethics and Its Discontent. *Journal of Mass Media Ethics*, *22*(2&3), 113–131.

Christians, C. G., Ferré, J. P., & Fackler, P. M. (1993). *Good news: Social ethics and the press*. New York: Oxford University Press.

Christians, C. (2006). The case for communitarian ethics. In Land, M., & Hornaday, B. W. (Eds.), *Contemporary media ethics: A practical guide for students, scholars and professionals* (pp. 57–69). Spokane, WA: Marquette Books.

Christopherson, S. (2006). Behind the scenes: How transnational firms are constructing a new international division of labor in media work. *Geoforum*, *37*, 739–751. doi:10.1016/j.geoforum.2006.01.003

Chu, Y.-H., & Lin, J.-W. (2001). Political development in 20th-Century Taiwan: State-building, regime transformation and the construction of national identity. *The China Quarterly*, *165*, 102–129. doi:10.1017/S0009443901000067

Chua, B. (2008). East Asian pop culture: Layers of communities. In Kim, Y. (Ed.), *Media consumption in everyday life in Asia* (pp. 99–113). New York: Routledge.

Chun, A. (1996). Fuck Chineseness: On the ambiguities of ethnicity as culture as identity. *boundary 2 23*(2), 111-138.

Chung, P., & Yuan, J. (2009). Dynamics in the online game industry of China: A political economic analysis of its competitveness. *Journal of Eptic On Line, 11*(2). Retrieved March 29, 2010 from http://www.eptic.com.br/arqu ivos/Revistas/vol.XI,n2,2009/11- PeichiChungJiangPingYuan.pdf

CINIC [China Internet Network Information Center]. (2009). *Statistical Survey Report on the Internet Development in China* (January 2009). Retrieved 14th June 2009 from http://www.cnnic.cn/uploadfiles / pdf/2009/3/23/131303.pdf

Clark, H. H., & Brennan, S. E. (1991). Grounding in Communication. In L. Resnick, J. Levine & S. Teasley (Eds.), *Perspectives on Socially Shared Cognition* (127-149). Hyattsville, MD: American Psychological Association.

Clark, N., & Meller, P. (2004, 16 June). Europe hears objections to proposed music merger. *The New York Times*, W1.

Clark, T. (2001). Baghdad by the bay. *Japan Internet Report, 59*. Retrieved August 1, 2008, from http://www.jir.net/jir7_01.html

Clark, T. (2002). How to evaluate the number of "Internet users" in Japan. *Japan Internet Report, 65*. Retrieved May 2, 2009, from http://www.jir.net/jir7_02.html#4

Clark, T. (2004, October). Mobile communications and the wireless internet—the Japanese experience. *Receiver, 11*, 1-4. Retrieved May 2, 2009, from http://www.vodafone.com/flash/ receiver/11/articles/pdf/11_02.pdf

Clark, T., & Kay, C. (2005). *Saying yes to Japan*. New York: Vertical. comScore. (2007, July 9). *comScore publishes the first comprehensive review of Asian-Pacific usage*. Retrieved April 4, 2009, from http://ww.comscore.com/press/release.asp?press=1520

CMR. (2008). About the campaign for media reform. In Y.-P. Lin (Ed.), *A 21th century anthology of media criticism*. Retrieved April 17, 2010, from http://www.twmedia.org/

CMR. (December 23, 2005). Our anger at betrayal of government (in Chinese). Retrieved April 17, 2010, from http://www.mediawatch.org.tw/ modules/news/print.php?storyid=172

Cohen, R. (2008). *Global diasporas: An introduction* (2nd ed.). London: Routledge.

Coleman, R. (2003). Elmo is black! Black popular communication and the marking and marketing of black identity. *Popular Communication, 1*(1), 51–64. doi:10.1207/S15405710PC0101_7

Condit, C. (1989). The rhetorical limits of polysemy. *Critical Studies in Mass Communication, 6*(2), 103–122. doi:10.1080/15295038909366739

Confucius,. (1979). *The analects* (Lau, D. C., Trans.). London: Penguin Books.

Connell, J., & Harvey, H. (2004). Call centres and labour turnover: do HRM practices make a difference? *International Employment Relations Review, 10*(2), 49–66.

Connery, B. A. (1997). IMHO: Authority and egalitarian rhetoric in the virtual coffeehouse. In Porter, D. (Ed.), *Internet culture* (pp. 161–179). London, New York: Routledge.

Consalvo, M. (2006). Console video games and global corporations: Creating hybrid culture. *New Media & Society, 8*(1), 117–137. doi:10.1177/1461444806059921

Constantiou, Papazafeiropoulou, & Vendelø. (2009). Does Culture Affect the Adoption of Advanced Mobile Services? A Comparative Study of Young Adults' Perceptions in Denmark and the UK. *The Data Base for Advances in Information Systems, 40*(4), 132–147.

Coombe, R., & Herman, A. (2001). Culture wars on the net: Intellectual property and corporate propriety in digital environments. *The South Atlantic Quarterly, 100*(4), 919–947. doi:10.1215/00382876-100-4-919

Cooper, T. (1996). *A time before deception: Truth in communication, culture, and ethics*. Santa Fe, NM: Clear Light Publishers.

Corliss, R., & Bonfante, J. (1995). Hey, let's put on a show. *Time, 145*(13), 54.

Corn-Revere, R., & Carveth, R. (2004). Economics and media regulation. In Alexander, A., Owers, J., Carveth, R., Hollifield, A. C., & Greco, A. N. (Eds.), *Media economics: Theory and practice* (3rd ed., pp. 49–67). Mahwah, NJ: Lawrence Erlbaum Associates Inc. Publishers.

Couldry, N. (2000). *The Place of media power: Pilgrims and witnesses of the media age*. London: Routledge.

Couldry, N. (2007). Researching media internationalization: Comparative media research as if we really meant it. *Global Media and Communication, 3*(3), 247–250. doi:10.1177/1742766507082569

Croteau, D., & Hoynes, W. (2001). *The Business of Media: Corporate Media and the Public Interest*. Thousand Oaks, CA: Pine Forge Press.

Crowe, N., & Bradford, S. (2007). Identity and structure in online gaming: Young people's symbolic and virtual extensions of self. In Hodkinson, P., & Deicke, W. (Eds.), *Youth cultures: Scenes, subcultures and tribes*. New York: Routledge.

Cubitt, S. (1991). *Time-shift: On video culture*. London: Routledge. doi:10.4324/9780203358955

Cui, B. G. (2008). *Zhongguo chuanmei chanye baogao (Chinese media industries report)*. Beijing: Social Sciences Academic Press.

d'Arcy, J. (1969). Direct broadcast satellites and the right to communicate. *EBU Review*, *118*, 14–18.

d'Arcy, J. (1977). The right of man to communicate. In Harms, L. S., Richstad, J., & Kie, K. A. (Eds.), *Right to Communicate: Collected papers* (pp. 45–52). Honolulu, HI: Social Sciences and Linguistics Institute, University of Hawaii, Manoa.

Dakroury, A. (2009). *Communication and Human Rights*. Dubuque, IA: Kendall/Hunt Publishing.

Dakroury, A. (2009). The baron of the right to communicate: Jean d'Arcy (1913:1983). In Dakroury, A., Eid, M., & Kamalipour, Y. R. (Eds.), *The right to communicate: Historical hopes global debates, and future premises* (pp. 21–42). Dubuque, IA: Kendall/Hunt Publishing.

Dakroury, A., & Eid, M. (2008). Canadian Communication perspective. In Eid, M., & Dakroury, A. (Eds.), *Introduction to communication and media studies* (pp. 303–309). Boston: Pearson.

Dakroury, A. (2005). Who owns the medium owns the message? The ambiguity of the right to communicate in the age of convergence. *Reconstruction: Studies in Contemporary Culture, 5*(2).

Dakroury, A., & Birdsall, W. (2008). Blogs and the right to communicate: Towards creating a space-less public sphere? In *International Symposium on Technology and Society proceedings*. Fredericton, New Brunswick: Institute of Electrical and Electronics Engineers (IEEE).

Dal Yong, J. (2008). Media, Neoliberal restructuring of the global communication system: mergers and acquisitions. *Cultura e Scuola*, *30*(3), 357–373.

Damm, J., & Thomas, S. (Eds.). (2006). *Chinese Cyberspaces Technological Changes and Political Effects*. New York: Routledge.

Danis, C. Bailey, M., Christensen, J., Ellis, J., Erickson, T., Farrell, R., & Kellogg, W. A. (2009) Social Computing Applications for the Next Billion Users. In *Designing Future Mobile Software for Underserved Users Workshop* at CSCW 2008.

Darling-Wolf, F. (2004). Sites of attractiveness: Japanese women and Westernized representations of feminine beauty. *Critical Studies in Media Communication*, *21*(4), 325–345. doi:10.1080/0739318042000245354

David, P. A. (1985). Clio and the economics of QWERTY. *The American Economic Review*, *75*(2), 332–337.

Dayan, D. (2009). Sharing and showing: Television as monstration. *The Annals of the American Academy of Political and Social Science*, *625*, 19–31. doi:10.1177/0002716209338364

Dayan, D., & Katz, E. (1992). *Media events: The live broadcasting of history*. Cambridge, MA: Harvard University Press.

Dayan, D., & Katz, E. (1992). *Media events: The live broadcasting of history*. Cambridge, MA: Harvard University Press.

De Muguel, A., & Barbeito, R. L. (1997). *El impacto de la telefonía móbil en la sociedad española* [The impact of mobile telephony on Spanish society]. Madrid: Ericsson.

De Vany, A., & McMillan, H. (2004). Was the antitrust action that broke up the movie studios good for the movies? Evidence from the stock market. *American Law and Economics Review*, *6*(1), 135–153.

Dearing, J. W. (2009). Applying diffusion of innovation theory to intervention development. *Research on Social Work Practice*, *19*, 503–518. doi:10.1177/1049731509335569

Dearing, J. W., & Kim, D.-K. (2008). *Diffusion of information and innovation. International Encyclopedia of Communication*. London: Blackwell Publishing.

Deery, S., & Kinnie, N. (2002). Call centres and beyond: a thematic evaluation. *Human Resource Management Journal*, *12*(4), 3–14. doi:10.1111/j.1748-8583.2002.tb00074.x

Deery, S. J., Iverson, R. D., & Walsh, J. T. (2002). Work Relationships in Telephone Call Centres: Understanding Emotional Exhaustion & Employee Withdrawal. *Journal of Management Studies*, *39*(4), 471–496. doi:10.1111/1467-6486.00300

DeFillippi, R. J., & Arthur, M. B. (1998). Paradox in project-based enterprise: The case of film making. *California Management Review*, *40*(2), 125–139.

Del Rio, E. (1996). The body as foundation of the screen: Allegories of technology in Atom Egoyan's Speaking Parts. *Camera Obscura*, (Summer): 37–38.

Dennis, E., & Merrill, J. (1984). *Basic Issues in Mass Communication: A Debate*. New York: Macmillan Publishing Company.

Department of Communication and the Arts. (1994). *Creative nation: commonwealth cultural policy, October 1994*. Retrieved February 26, 2005, from http://www.nla.gov.au/creative. nation/contents.html

Department of Foreign Affairs and Trade. (2008). *Australia-India FTA Feasibility Study Industry Consultations. Background Paper*. Australia: DFAT, Canberra.

Dharmawat, N. (September, 2003). *BPO-Destination India*. Patni Computer Systems Ltd India. Available at www.patni.org, retrieved dated 25th December, 2009.

Dickin, J. (2002). The Internet as a site of citizenship. *Canadian Journal of Communication*, *27*(4), 509–534.

Dicks, B. (2003). *Culture on display: The production of contemporary visitability*. London: Open University Press.

DiNucci, D. (1999). Fragmented Future. *Print 53*(4), 32. http://www.cdinucci.com/Darcy 2/articles/Print/Printarticle7.html

Dirlik, A. (1996). The global in the local. In Wilson, R., & Dissanayake, W. (Eds.), *Global/local: Cultural production and the transnational imaginary* (pp. 21–45). Durham, NC: Duke University Press.

Dirlik, A. (2007). Global South: Predicament and Promise. *The Global South, Winter*, 1(1), 12-23.

Dittmar, H. (1992). *The Social Psychology of Material Possessions*. Hemel Hempstead: Harvester Wheatsheaf.

Dobal, C. H. (1998). A proposal to amend the cable compulsory license provisions of the 1976 copyright act. *Southern California Law Review*, *61*, 699–732.

Donald, S. H., Keane, M., & Hong, Y. (Eds.). (2002). *Media in China: Consumption, content and crisis*. London: Routledge.

Dong, D. (2003). *Resistance within Contexts: A Study of University BBS Youth Culture in China*. Paper presented at the annual meeting of the International Communication Association, Marriott Hotel, San Diego, CA, May 27.

Du Gay, P., Hall, S., Janes, L., Mackay, H., & Negus, K. (2001). *Doing Cultural Studies: The Story of the Sony Walkman*. London: Sage.

Dudley, A. (1997). A preface to disputation. In Dudley, A., & Shafto, S. (Eds.), *The image in dispute: Art and cinema in the age of photography* (pp. vii–xiv). Austin, TX: University of Texas Publication.

Duncan, C. (1995). *Civilizing rituals: Inside public art museums*. London: Routledge.

Dupagne, M., & Garrison, B. (2006). The meaning and influence of convergence: A qualitative case study of newsroom work at the Tampa news center. *Journalism Studies*, *7*(2), 237–255. doi:10.1080/14616700500533569

Dyer-Witheford, N., & Sharman, Z. (2005). The Political Economy of Canada's Video and Computer Game Industry. *Canadian Journal of Communication*, *30*(2).

Ebenkamp, B. (2007, November 12). Jollibee is on fast (food) track in virgin markets. *Brandweek*, *48*(41), 14.

Ebenkamp, B., & Hein, K. (2005, November 21). A clown around town. *Brandweek*, *46*(42), 38.

Eckhardt, G. M., & Houston, M. J. (2002). Cultural paradoxes reflected in broad meaning: McDonald's in Shanghai, China. *Journal of International Marketing*, *10*(2), 68–82. doi:10.1509/jimk.10.2.68.19532

Edgecliffe Johnson, A. (2009). Vevo could be Saviour of Music Video. [th.]. *Financial Times (North American Edition)*, (December): 10.

Editorial. (2009). Shanghai relaxes residency rules. *BBC News online*. Retrieved 14th June 2009 from http://news.bbc.co.uk/go/pr/fr/-/2/hi /asia-pacific/8107518.stm

Edwards, G. (2004). Nagoya shuts down subway cell phone usage. *Engadget*. Retrieved October, 15, 2006, from http://www.engadget.com/2004/08012/ nagoya-shuts-down-subway-cellphone-usage

Eller, C. (2006, 20 October). MGM deal a bold miscalculation for Sony. *Los Angeles Times*, C1.

Epstein, E. J. (2005). *The big picture, the new logic of money and power in Hollywood*. New York: Random House.

Erickson, T., Kellogg, W. A., Laff, M., Sussman, J., Wolf, T. V., Halverson, C. A., & Edwards, D. (2006). A persistent chat space for work groups: the design, evaluation and deployment of loops. In *Proceedings of the 6th Conference on Designing Interactive Systems 06* (pp. 331-340). New York: ACM Press.

Eriksen, H. R., Ihlebaek, C., & Ursin, H. (1999). A scoring system for subjective health complaints (SHC). *Scandinavian Journal of Public Health*, *27*, 63–72. doi: 10.1177/14034948990270010401

Ethiraj, S. K., & Levinthal, D. (2004). Modularity and Innovation in Complex Systems. *Management Science*, *50*(2), 159–173. doi:10.1287/mnsc.1030.0145

Ettoday (August 19, 2005). At the storm of handsome Bae, fans form "a united nations" (in Chinese). Retrieved April 17, 2010, from http://www.nownews.com/2005/08/19/91-1833232.htm

Etzioni, A. (1993). *The spirit of community: The reinvention of American society*. New York: Touchstone Book.

Etzioni, A. (1996). *The New Golden Rule*. New York: Basic Books.

Europa. (2007). *An EU strategy for mobile TV*. Retrieved November 24, 2007, from http://europa.eu/rapid/pressReleasesAction.do?aged=0&format=HTML&guiLanguage= en&language=EN&reference=MEMO/07/298

European Foundation for the Improvement of Living and Working Conditions. (2006). *Industrial Relations Dictionary*. Retrieved from www.eurofound.eu.int., accessed 30 November 2009.

Evalueserve. (2004). *The next big opportunity – moving up the value chain – from BPO to KPO*. Retrieved from www.evalueserve.com, retrieved dated 5th January, 2010.

Evans, G. (2009). Creative Cities, Creative Spaces and Urban Policy. *Urban Studies (Edinburgh, Scotland)*, *46*(5&6), 1003–1040. doi:10.1177/0042098009103853

Examination Yuan of Taiwan Government. (2005). *Review of the audiovisual policy and practices in Taiwan*. Taipei: Examination Yuan of Taiwan Government. (in Chinese)

Executive Yuan (November 11, 2009). Cultural and creative industries: New engine for wider economic growth. Retrieved April 17, 2010, from http://www.ey.gov.tw/ct.asp?xItem= 64011&ctNode=1334&mp=1

Fagerjord, A. (2003). Rhetorical convergence: Studying web media. In Liestøl, G., Morrison, A., & Rasmussen, T. (Eds.), *Digital Media Revisited: Theoretical and Conceptual Innovations in Digital Domains* (pp. 293–325). Cambridge, MA: MIT.

Fagerjord, A., & Storsul, T. (2007). Questioning Convergence. In Storsul, T., & Stuedahl, D. (Eds.), *Ambivalence towards Convergence: digitalization and media change*. Goteborg, Sweden: Nordicom.

Fang, N. (2009). *Shanzhai shouji chanye yanghuo zhusanjiao 300wanren* [*Shanzhai* phone industry supports livelihood of 3 million in the Pearl River Delta]. *Southern Net*, 02-17. Retrieved February 17, 2009, from http://tech.163.com/09/0217 /03/52AV0FNJ000915BE.html

Farber, A. (2009, 19 March). Samsung makes bid for Mobile Movie Market with Download Service. *New Media Age*, 3.

Farrell, D., Laboissiere, M., Pascal, R. et al. (2005). *The emerging global labor market: Part 1--the demand for offshore talent in services*. McKinsey Global Institute (MGI) Report, June.

Featherstone, M. (1995). *Undoing culture: Globalization, postmodernism, and identity*. London: Sage Publications.

Featherstone, M. (2007). *Consumer culture and postmodernism*. London: SAGE Publications.

Federal Communication Commission. (2001). *Memorandum Opinion and Order*. Applications for Consent to the Transfer of Control of Licenses and Section 214 Authorizations by Time Warner Inc. and America Online, Inc., Transferors, to AOL Time Warner Inc., Transferee. CS Docket No. 00-30. Retrieved September, 2, 2008, from: http://www.fcc.gov

Federal Communication Commission Media Bureau. (2006). *Twelfth annual report* [online]. MB docket, 05-255. Retrieved January 2, 2007, from http://www.fcc.gov

Feigenbaum, H. (2004). Is technology the enemy of culture? *International Journal of Cultural Policy*, *10*(3), 251–263. doi:10.1080/1028663042000255763

Feldman, A. (2005). On the actuarial gaze: From 9/11 to Abu Ghraib. *Cultural Studies*, *19*(2), 203–226. doi:10.1080/09502380500077763

Feldman, C. (2004, October 8) Q&A: EA's man on the ground in China, Erick Hachenburg. *GameSpot*. Retrieved May 1, 2006, from http://www.gamespot.com/news/6110098.html

Feng, C. (2003). Press freedom, to do or not to do [in Chinese]. *Journalist*, *34*, 23–25.

Feng, C. (2006). Constructing and expanding Taiwan's public television from 1990 to 2006: Roles of academic intellectuals, social campaign and political authority [in Chinese]. *Journal of Communication and Society*, *1*, 47–67.

Feng, C. (2008). Culture & economy: Comparing Taiwan and Korea. In Peng, H. (Ed.), *Two Asian dragons: Comparing Taiwan and Korea* (pp. 212–239). Taipei: Asia and Pacific Culture and Education Foundation.

Feng, C. (December 20, 2001). Private ownership unable to solve problems (in Chinese). *United Daily*, A15.

Feng, C. (January, 26, 2008). Cultural policy in election (in Chinese). *Liberty Times*. Retrieved April 17, 2010, from http://www.libertytimes.com.tw/ 2008/new/jan/26/today-o1.htm

Fish, S. E. (1980). *Is there a text in this class?: The authority of interpretive communities*. Cambridge, MA: Harvard University Press.

Fisher, P. (1991). *Making and effacing art: Modern American art in a culture of museums*. Cambridge, UK: Cambridge University Press.

Fishwick, M. (1995). Ray and Ronald girdle the globe. *Journal of American Culture*, *18*(1), 13–29. doi:10.1111/j.1542-734X.1995.1801_13.x

Fiske, J. (1989a). *Reading the popular*. Boston: Unwin Hyman.

Fiske, J. (1989b). *Understanding the popular*. Boston: Unwin Hyman.

Fiske, J. (1993). *Power plays, power works*. New York: Verso.

Fiske, J. (1987). *Television culture*. New York: Routledge.

Fitzpatrick, M. (2007, September 27). Why mobile Japan leads the world. *Guardian*. Retrieved November 1, 2009, from http://www.guardian.co.uk/technology/2007/sep/27/guardianweeklytechnologysection.mobilephones

Flew, T. (2007). *Understanding global media*. Basingstoke, UK: Palgrave McMillian.

Flynn, B. (2007). The morphology of space in virtual heritage. In Cameron, F., & Kenderdine, S. (Eds.), *Theorizing digital cultural heritage: A critical discourse* (pp. 349–368). Cambridge, MA: MIT Press.

Fortunati, L. (2002b). The mobile phone: Towards new categories and social relations. *Information Communication and Society*, *5*, 513–528. doi:10.1080/13691180208538803

Fortunati, L. (2001). The mobile phone: An identity on the move. *Personal and Ubiquitous Computing*, *5*(2), 85–98. doi:10.1007/PL00000017

Fortunati, L. (2002a). Italy. Streotypes: true or false. In Katz, J., & Askhus, M. (Eds.), *Perpetual contact: Mobile communication, Private talk, Public performance*. Cambridge, UK: Cambridge University Press. doi:10.1017/CBO9780511489471.006

Foster, H. (1996). Death in America: Shocked subjectivity and compulsive visual repetition. In A. Michelson (Ed.), *October: The second decade, 1986-1996* (349-370). Cambridge, MA: MIT Press.

Frasca, G. (2003). Ludologists love stories, too: notes from a debate that never took place. In *Level Up Conference Proceedings: Proceedings of the 2003 Digital Games Research Association Conference* (pp. 92-99). Utrecht: University of Utrecht. Retrieved from http://www.digra.org/dl/display_ html?chid=05163.01125.pdf

Freedman, D., Henten, A., Towse, R., & Wallis, R. (2008). The Impact of the Internet on Media Policy, Regulation and Copyright Law. In Küng, L., Picard R. G., & Towse, R. 2008. (Eds.), *The internet and the mass media,* (pp. 102-124). Oxford, UK: Sage Publications.

Freeman, C., & Louçã, F. (2001). *As time goes by - from industrial revolutions to the information revolution*. Oxford, UK: Oxford University Press.

Freeman, C., & Soete, L. (1997). *The Economics of Industrial Innovation* (3rd ed.). London, New York: Continuum.

Friedman, T. L. (2005). *The World is Flat: A brief history of the twenty-first century*. New York: Farrar, Straus, Giroux.

Friedman, T. (2005). *The World Is Flat: A Brief History of the Globalized World in the 21st Century*. London: Allen Lane.

Friedman, J. (1994). *Cultural identity & global process*. London: Sage.

Friedman, J. (1990). Being in the world: Globalization and localization. In Featherstone, M. (Ed.), *Global culture: Nationalism, Globalization and modernity*. New York: Sage.

Fritz, B. (2009). Call of Duty: Modern Warfare 2 video game gets Hollywood-scale launch. *LA Times*. Retrieved December 3, 2009, from http://www.latimes.com/business/la-fi-ct-duty 18-2009nov18,0,5238209.story

Froehlich, D. M., Rachovides, D., Riga, K., Bhat, R., Frank, M., Edirisinghe, E., et al. (2009). StoryBank: mobile digital storytelling in a development context. In *Proceedings of CHI 2009* (pp. 1761-1770). New York: ACM.

Froomkin, A. M. (2003). Habermas@discourse net: Toward a Critical Theory of Cyberspace. [from http://osaka.law.miami.edu/~froomkin/discourse/ils.pdf]. *Harvard Law Review*, 751–873. Retrieved December 2009.

Fujimoto, K. (2005). The third-stage paradigm: Territory machines from the girls' pager revolution to mobile aesthetic. In Ito, M., Okabe, D., & Matsuda, M. (Eds.), *Personal, portable, pedestrian* (pp. 77–101). Cambridge, MA: MIT Press.

Fujimoto, K. (2006). The Third-Stage Paradigm: Territory Machines from the Girls' Pager Revolution to Mobile Aesthetics. In Ito, M., Okabe, D., & Matsuda, M. (Eds.), *Personal, portable, pedestrian: Mobile phones in Japanese life* (pp. 77–102). Cambridge, MA: The MIT Press.

Fulk, J. (1993). Social construction of communication technology. *Academy of Management Journal*, *36*, 921–950. doi:10.2307/256641

Fulk, J., Schmitz, J., & Ryu, D. (1995). Cognitive elements in the social construction of technology. *Management Communication Quarterly*, *8*, 259–288. doi:10.1177/0893318995008003001

Fulk, J., Schmitz, J., & Steinfield, C. W. (1990). A social influence model of technology use. In Fulk, J., & Steinfield, C. (Eds.), *Organizations and communication technology* (pp. 117–139). Newbury Park, CA: Sage.

Fullbrook, E. (2004). *A guide to what's wrong with economics*. London: Anthem Press.

Fung, A. (2007). The emerging (national) popular music culture in China. *Journal of Inter-Asia cultural studies*, *8*(3), 425-437.

Gallouj, F. (2002). Innovation in services and the attendant old and new myths. *Journal of Socio-Economics*, *31*, 137–154. doi:10.1016/S1053-5357(01)00126-3

Galtung, J. (1971). A structural theory of imperialism. *Journal of Peace Research*, *8*(2). doi:10.1177/002234337100800201

Gandy, O. H. Jr. (1982). *Beyond agenda setting: Information subsidies and public policy*. Norwood, NJ: Ablex.

Ganguly, K. (December 15, 2009). *Managing Customer Hostility in Transnational Call Centers in India*. Retrieved from http://ssrn.com, retrieved dated 1st April, 2010.

García-Montes, J. M., Caballero-Muñoz, D., & Pérez-Álvarez, M. (2006). Changes in the self resulting from the use of mobile phones. *Media Culture & Society*, *28*, 67–82. doi:10.1177/0163443706059287

García-Montes, J. M., & Carretero, M. J. (2000). La psicologa social de la telefonía móvil. [Social Psychology of Mobile Telephones] In Caballero, B., Méndez, M. T., & Pastor, J. (Eds.), *La mirada psicosocial: Grupos, procesos, lenguajes y culturas*. Madrid: Biblioteca Nueva.

Garcia-Murillo, M., & Macinnes, I. (2003). The impact of technological convergence on the regulation of ICT industries. *International Journal on Media Management*, *5*, 57–67.

Garnham, N. (2000). *Emancipation, the media, and modernity: Arguments about the media and social theory*. Oxford, UK: Oxford University Press.

Garnham, N. (1990). *Capitalism and Communication*. London: Sage.

Garnham, N. (2000). *Emancipation, the Media & Modernity*. Oxford, UK: Oxford University Press.

Gartner Group. (2009). *Gartner Says Worldwide Mobile Phone Sales Declined 8.6 Per Cent and Smartphones Grew 12.7 Per Cent in First Quarter of 2009*. Press released dated May 20, 2009. Retrieved June 15, 2009, from http://www.gartner.com/it/pa ge.jsp?id=985912

Gauntlett, D., & Hill, A. (1999). *TV living: Television, culture, and everyday Life*. London: Routledge.

Gershon, R. A. (2002). The Transnational media corporation and the economics of global competition. In Yahya, R. K. (Ed.), *Global Communication* (pp. 51–72). Belmont, CA: Wadsworth Publishing.

Giddens, A. (1990). *The consequences of modernity*. Cambridge, UK: Polity Press.

Giddens, A. (1984). *The constitution of society*. Cambridge, UK: Polity Press.

Gikandi, S. (2005). Globalization and the Claims of Postcoloniality. In Desai, G., & Nair, S. (Eds.), *Postcolonialisms* (pp. 608–634). Oxford, UK: Berg.

Giulianotti, R. (2007). Popular culture, social identities, and internal/external cultural politics: The case of Rangers supporters in Scottish football. *Identities (Yverdon)*, *14*(3), 257–284. doi:10.1080/10702890601162732

Global Fair and Festival. (2009). *Incheon Metropolitan City*. South Korea: Incheon.

Goggin, G., & McLelland, M. (Eds.). (2008). *Internationalizing Internet Studies: Beyond Anglophone Paradigms*. New York: Routledge.

Goggin, G., McLelland, M., & Hjorth, L. (2009). Waiting to participate: Introduction. *Communication. Politics & Culture*, *42*(2), 1–5.

Goggin, G. (2006). *Cell phone culture: Mobile technology in everyday life*. New York: Routledge.

Goldberg, L. S., & Grandey, A. A. (2007). Display rules versus display autonomy: emotion regulation, emotional exhaustion, and task performance in a call center simulation. *Journal of Occupational Health Psychology*, *12*(3), 301–318. doi:10.1037/1076-8998.12.3.301

Golding, P., & Murdock, G. (2000). Culture, Communications and Political Economy. In Curran, J., & Gurevitch, M. (Eds.), *Mass Media and Society* (3rd ed., pp. 70–93). London: Arnold.

Goldman, M. (2009). *Getting Bangalored: Struggles over Urbanized Space in the Making of Asia's World Cities*. ASA Annual Meetings, January 2009.

Goldsmith, R. E., & Hofacker, C. F. (1991). Measuring consumer innovativeness. *Journal of the Academy of Marketing Science*, *19*(3), 209–221. doi:10.1007/BF02726497

Goldsmith, J. (2004). How did Sony do it. [October.]. *Variety*, *396*(7), 4.

Gong, C. (2002). Kuwenhua, qingnian jiazhiguan, shehui zhuanxing [Cool Culture, youth values and social transformation]. *Youth Research*, *2*, 30–35.

Goode, L. (2005). *Jürgen Habermas; Democracy and the Public Sphere*. London: Pluto Press.

Gopakumar, K. (2006). E-governance services through telecentres: Role of human intermediary and issues of trust. *Information Technologies and Development*, *4*(1), 19–35.

Gottlieb, N., & McLelland, M. (Eds.), *(n.d.). Japanese Cybercultures*. New York: Routledge.

Gottlieb, N. (2000). *World-processing technology in Japan: Kanji and the keyboard*. Richmond, VA: Curzon Press.

Government Information Office. (1995). *The case analysis for the development of the Asian-Pacific Media Center*. Taipei, Taiwan: Government Information Office. (in Chinese)

Goyder, J. (2005). *Technology and society: A Canadian perspective*. Peterborough: Broadview Press Ltd.

Greenfeld, M. (2005). *Serious play: How gossiping is accomplished in an online fandom forum*. Paper presented at the Annual Meeting of the International Communication Association, New York, NY.

Gripsrud, J. (2004). Broadcast television: The chances of its survival in a digital age. In Spigel, L., & Olson, J. (Eds.), *Television after TV* (pp. 210–223). Durham, NC: Duke University Press.

Gronow, J., & Warde, A. (Eds.). (2001). *Ordinary Consumption*. London: Routledge.

Grossberg, L. (1997). *Dancing in spite of myself: Essays on popular culture*. Durham, NC: Duke University Press.

Gunster, S. (2004). *Capitalizing on Culture: Critical Theory for Cultural Studies*. Toronto: University of Toronto Press.

Haase, U., & Large, W. (2001). *Maurice Blanchot*. New York: Routledge.

Habermas, J. (1989). *The structural transformation of the public sphere*. Cambridge, UK: Polity Press.

Habermas, J. (2006). Does Democracy Still Enjoy An Epistemic Dimension? The Impact of Normative Theory on Empirical Research. In *Communication Theory* (pp. 411–426). Political Communication in Media Society.

Habuchi, I., Dobashi, S., Tsuji, I., & Iwata, K. (2005). Ordinary usage of new media: Internet usage via mobile phone in Japan. *International Journal of Japanese Sociology*, *14*, 94–108. doi:10.1111/j.1475-6781.2005.00071.x

Habuchi, I. (2005). Accelerating flexibility. In Ito, M., Okabe, D., & Matsuda, M. (Eds.), *Personal, portable, pedestrian: Mobile phones in Japanese life* (pp. 165–182). Cambridge, MA: MIT Press.

Halbesleben, J. R. B., & Buckley, M. R. (2004). Burnout in organizational life. *Journal of Management, 30*(6), 859–879. doi:10.1016/j.jm.2004.06.004

Hall, S. (1997). The local and the global: Globalization and ethnicity. In King, A. D. (Ed.), *Culture, globalization and the world system: Contemporary conditions for the representation of identity*. Minneapolis: University of Minnesota Press.

Hall, S. (1996). Cultural identity and Diaspora. In Mongia, P. (Ed.), *Contemporary postcolonial theory: A reader* (pp. 110–121). New York: Arnold.

Hall, J. (2005). Future of games: Mobile gaming. In Raessens, J., & Goldstein, J. (Eds.), *Handbook of computer game studies* (pp. 47–55). Cambridge, MA: MIT Press.

Halvey, M., Keane, M., & Smyth, B. (2006, April). *Time-based patterns in mobile-Internet surfing*. Paper presented at the meeting of the CHI conference, Montréal, Canada.

Hamelink, C. J. (2000). *The ethics of cyberspace*. Thousand Oaks, CA: Sage Publications.

Hamelink, C. J. (2004). Did WSIS Achieve anything At All? *Gazette: The International Journal for Communication Studies, 66*(3-4), 281–290. doi:10.1177/0016549204043612

Hamelink. C. (1983). *Cultural Autonomy in Global Communications: Planning National Information Policy*. New York: Longman.

Hampton, B. B. (1970). *History of the American film industry: From its beginning to 1931*. New York: Dover.

Han, G. J. (1994). Promises and Myths of Cable Television and Telecommunications Infrastructure in Korea. In *Elite Media amidst Mass Culture: a Critical Look at Mass Communication in Korea* (pp. 156–157). Seoul: Nanam.

Hannerz, U. (1990). Cosmopolitans and locals in world culture. *Theory, Culture & Society*, (7): 237–251. doi:10.1177/026327690007002014

Harms, L. S., & Richstad, J. (1977). Right to Communicate: Human rights, Major Communication Issues, Communication Policies and Planning. In Harms, L. S., Richstad, J., & Kie, K. A. (Eds.), *Right to Communicate: Collected Papers*. Hawaii, HI: The University Press of Hawaii.

Hartford, K. (2000). *Cyberspace with Chinese characteristics*. Retrieved January 20, 2010, from http://www.china-wired.com/pubs/ch/home.htm

Harvey, D. (1990). *The Condition of Postmodernity*. Cambridge, MA: Blackwell.

Hau, K. T., & Salili, F. (1996). Achievement goals and causal attributions of Chinese students. In Lau, S. (Ed.), *Growing up the Chinese way* (pp. 121–146). Hong Kong: The Chinese University Press.

Havis, R. J. (2005). Illusory worlds: An interview with Jia Zhangke. *Cineaste*, (Fall), 58-59.

Hayles, N. (1999). *How we became posthuman: Virtual bodies in cybernetics, literature, and informatics*. Chicago: University of Chicago Press.

Hayles, N. (2002). *Writing machines*. Cambridge, MA: MIT press.

Hazards 81 extended briefing: Toilet breaks: Give us a break! (n.d.). *Hazards*. Available at www.hazards.org, Retrieved 2008-06-05.

Heidegger, M. (1971). *'The Thing.' in Poetry, language, thought, (Trans. J. G. Gray)*. New York: Harper & Row.

Heidegger, M. (1977). *'The age of the world Picture'. The question concerning technology and other essays, (Trans. W. Lovitt)*. New York: Harper Torchbooks.

Hein, H. (2000). *The museum in transition: A philosophical perspective*. Washington, DC: Smithsonian Institution Press.

Heinrichs, J., & Spielmann, Y. (Eds.). (2002). Special issue: What is Intermedia? *Convergence, 8*(4), 5–10.

Herman, A., Coombe, R. J., & Kaye, L. (2006). Your second life? Goodwill and the performativity of intellectual property in online digital gaming. *Cultural Studies*, *20*(2-3), 184–210. doi:10.1080/09502380500495684

Hesmondalgh, D., & Toynbee, J. (Eds.). (2008). *The Media and Social Theory*. London: Routledge.

Hesmondhalgh, D. (2002). *The Cultural Industries*. London: Sage.

Hesmondhalgh, D. (2008). Neoliberalism, Imperialism and the Media. In *Hesmondhalgh, D. and J. Toynbee (2008). The Media and Social Theory* (pp. 95–111). London: Routledge.

Hesmondhalgh, D. (2007). *The Cultural Industries* (2nd ed.). London: Sage Publications.

Hickey, S. (n.d.). Industry crisis as album sales drop by 10pc. *The Irish Independent* [online], Monday January 14th. Retrieved from http://www.independent.ie/national-news/industry-crisis-as-album-sales-drop-by-10pc-1265540.html

Hill, R. (2003). A mobile phone of one's own: Japan's "Generation M". *New Zealand Journal of Asian Studies*, *5*(1), 178–194.

Hine, C. (2000). *Virtual Ethnography*. London: Sage.

Hine, C. (1998). *Virtual Ethnography*. Paper from Internet Research and Information for Social Scientists Conference, University of Bristol, UK, March 25-27.

Hinton, S. (2009). Gaming Nation: The Australian Game Development Industry. In Hjorth, L., & Chen, D. (Eds.), *Gaming Cultures and Place in Asia Pacific*. London: Routledge.

Hirschman, E., & Thompson, C. (1997). Why media matter: Toward a richer understanding of consumers' relationships with advertising and mass media. *Journal of Advertising*, *26*(1), 43–60.

Hjorth, L., & Kim, H. (2005). Being there and being here: gendered customising of mobile 3G practices through a case study in Seoul. *Convergence*, *11*, 49–55.

Hjorth, L., Kim, H., & Chan, D. (Eds.). (2009). *Gaming Cultures and Place in Asia-Pacific*. London: Routledge.

Hjorth, L., & Chan, D. (2009). Locating the game: Gaming cultures in/and the Asia-Pacific. In Hjorth, L., & Chan, D. (Eds.), *Gaming cultures and place in Asia-Pacific* (pp. 1–14). New York: Routledge.

Ho, W. (2006). A historical review of popular music and social change in Taiwan. *Asian Journal of Social Science*, *34*(1), 120–147. doi:10.1163/156853106776150216

Ho, K. C., Kluver, R., & Yang, K. C. C. (2003). *Asia@com*. London: Routledge.

Hoelscher, S. (2006). Heritage. In Macdonald, S. (Ed.), *A companion to museum studies* (pp. 198–218). Malden, MA: Blackwell. doi:10.1002/9780470996836.ch13

Hollan, J., Hutchins, E., & Kirsh, D. (2000). Distributed cognition: toward a new foundation for human-computer interaction research. *ACM Transactions on Computer-Human Interaction*, *7*(2), 174–196. doi:10.1145/353485.353487

Holman, D. (2003a). Employee Well-being in Call Centres. *Human Resource Management Journal*, *12*(4), 35–51. doi:10.1111/j.1748-8583.2002.tb00076.x

Holman, D. (2003b). Phoning in Sick? An Overview of Employee Stress in Call Centres. *Leadership and Organization Development Journal*, *24*(3), 123–130. doi:10.1108/01437730310469543

Holman, D., Batt, R. & Holtgrewe, U. (2007). *The global call center report: International perspectives on management and employment.*

Hooper-Greenhill, E. (1992). *Museums and shaping of knowledge*. London: Routledge.

Horkheimer, M., & Adorno, T. (1969). *Dialectic of Enlightenment (trans. E. Jephcott)*. Stanford, CA: Stanford University Press.

Houlihan, M. (2000). Eyes Wide Shut? Querying the Depth of Call Centre Learning. *Journal of European Industrial Training*, *24*(2/3/4), 228-240.

Hourly Rate Survey Report for Industry. Call Center. (n.d.). *Pay Scale*. Available at www.payscale.com, retrieved dated 15th December, 2009.

Houser, C., & Thornton, P. (2004, March). *Japanese college students' typing speed on mobile devices*. Paper presented at the meeting of the IEEE International Workshop on Wireless and Mobile Technologies in Education, Taipei, Taiwan.

Hsieh, C. (September 6, 2005). *First Coordinating Meeting under the Executive Yuan.* Retrieved April 17, 2010, from http://info.gio.gov.tw/ct.asp?xItem=26821&ctNode=3105&mp=1

http://journal.fibreculture.org/issue14/

http://oreilly.com/web2/archi ve/what-is-web-20.html

http://www.nytimes.com/1 995/04/29/business/cellular-phones-may-affect-pacemakers.html

Huang, J. S., & Heider, D. (2008). Media Convergence: a case study of a cable news station. *The International Journal on Media Management*, *9*(3), 105–115.

Hudson, J. H., Christensen, J., Kellogg, W. A., & Erickson, T. (2002). I'd be overwhelmed, but it's just one more thing to do. In *Proceedings of the SIGCHI conference on Human Factors in Computing Systems*, (pp. 97-104). New York: ACM Press.

Hughes, C. (2003). *China and the Internet: Politics of the Digital Leap Forward (Politics in Asia Series)*. New York: RoutledgeCurzon.

Hukill, M. (2000). Telecommunications and development policy in Southeast Asia: Foundations for convergence. In Hukill, M., Ono, R., & Vallath, C. (Eds.), *Electronic communication convergence: Policy changes in Asia* (pp. 48–76). New Delhi: Sage.

Hwang, Y., & Chang, L. (1999). Overview of satellite and cable channels [in Chinese]. *Advertising Magazine*, *95*, 15–119.

IFPI. (2009). *Digital music report 2009: New business models for a changing environment*. Retrieved October 31, 2009 from http://www.ifpi.org/content /library/DMR2009.pdf

Infotrend/CAP Ventures. (2004). *Worldwide Camera Phone and Photo Messaging Forecast: 2004-2009*. London: Kluwer. Retrieved June 15, 2009, from http://store.infotrendsresearch.com/PhotoGall ery.asp?ProductCode=MobileImagingStudy10106

Ingram, M. (2008). Disharmony and the business of music. *The Globe and Mail* [online]. Retrieved from http://v1.theglobeandmail.com/servlet/Page/document/v5/content/subscribe?user_URL=http://www.globeinvestor.com%2Fservlet%2FArticleNews%2Fstory%2FGAM%2F20080131%2FTWINGRAM31&ord=27610503&brand=globeinvestor&force_login=true

Innis, H. (1950). *Empire and communication.* Oxford, UK: Oxford University Press.

Innis, H. (1964). *The bias of communication*. Toronto: University of Toronto Press.

International Federation of Phonographic Industries (IFPI)–Market Research Department. (2000). *Recording Industry in Numbers*. London: International Federation of Phonographic Industries.

International Federation of Phonographic Industries (IFPI)–Market Research Department. (2007). *Recording Industry in Numbers*. London: International Federation of Phonographic Industries.

International Federation of Phonographic Industries (IFPI)–Market Research Department. (2008). *Recording Industry in Numbers*. London: International Federation of Phonographic Industries.

International Federation of Phonographic Industries (IFPI)–Market Research Department. (2009). *Recording Industry in Numbers*. London: International Federation of Phonographic Industries.

International Organization for Migration (IOM). (2005). *HIV/AIDS, population mobility and migration in Southern Africa: Defining a research and policy agenda*. Geneva: International Organization for Migration.

Ip, B. (2008). Technological, content, and market convergence in the games industry. *Games and Culture*, *3*(2), 199–224. doi:10.1177/1555412008314128

iResearch (2008). *The investigation report of Chinese mobile phone players*. Retrieved online, November 10, 2009, at http://www.docin.com/p-18767066.html

Irwin, W. (2007, November/December). The pop culture manifesto. *Philosophy Now*, *64*, 11–15.

Ishii, K. (2004). Internet use via mobile phone in Japan. *Telecommunications Policy*, *28*(1), 43–58. doi:10.1016/j.telpol.2003.07.001

Ito, M. (2005). Mobile phones, Japanese youth, and the replacement of social contact. In Ling, R., & Pedersen, P. (Eds.), *Mobile communication and the Re-negotiation of the public sphere* (pp. 131–148). London: Springer-Verlag.

Ito, M. (2004, October). *Personal portable pedestrian: Lessons from Japanese mobile phone use.* Paper presented at the meeting of the 2004 International Conference on Mobile Communication, Seoul, Korea.

Ito, M. boyd, d. & H. Horst (2008). *Digital Youth Research.* Retrieved from http://digitalyouth.ischool.berkeley.edu/

Ito, M., Okabe, O., & Matsuda, M. (2006). (Eds.) *Personal, portable, pedestrian: Mobile phones in Japanese life*. Cambridge, MA: MIT Press.

ITU. (2003). *Promoting broadband: The case of Japan*. Geneva: International Telecommunication Union.

ITU. (2007). *ICT statistics 2007*. Retrieved July 14, 2008, from http://www.itu.int/ITU-D/icteye/DisplayC ountry.aspx?countryId=120

ITU. (2009). *New ITU ICT Development Index compares 154 countries.* (press release dated March 2, 2009).http://www.itu.int/newsroom /press_releases/2009/07.html

Iwabuchi, K. (2002). *Recentering globalization: Popular culture and Japanese transnationalism*. Durham, NC: Duke university press.

Iwabuchi, K. (2004a). *Cultural globalization and Asian media connections*. Hong Kong: Hong Kong University Press.

Iwabuchi, K. (2004b). Discrepant intimacy: Popular culture flows in East Asia. In Erni, J., & Chua, S. (Eds.), *Asian media studies* (pp. 19–36). Malden, MA: Blackwell Publishing.

Iwabuchi, K. (2005). Introduction: Cultural globalization and Asian media connection. In Iwabuchi, K. (Ed.), *Feeling Asian modernities: Transnational consumption of Japanese TV dramas* (pp. 1–22). Hong Kong: Hong Kong University Press.

Iwabuchi, K. (2008). Cultures of empire: Transnational media flows and cultural (dis)connections in East Asia. In Chakravartty, P. & Y. Zhao, (eds.), *Global communications: Toward a transcultural political economy* (pp. 143-162). Lanham, MD: Rowman & Littlefield publishers, Inc.

Jackson, T. (1999). Taste and aesthetics of simulation: Art, technology and the future. *Journal of Arts Management. Law and Society*, *28*(4), 314–319.

Jain, N. (2001). *Globalisation or recolonisation?* (1st ed.). Pune, India: Elgar.

Jameson, F. (1992). *Signatures of the visible*. London: Routledge.

japan.internet.com. (2006, June 28). 利川しやすぃが抵抗も感じるクレジットカ「ド~オンライン決「調査 (Easy to use, but still reluctant to use credit cards online). Retrieved August 1, 2008, from http://japan.internet.com/research/20060628/1.html

Japanese cybercultures (pp. 19-33). London & New York: Routledge.

Japanese Patent Office. (2002). *Kyota Sugimoto (Japanese typewriter)*. Retrieved July 26, 2008, from http://www.jpo.go.jp/seido_e/rekishi_e/kyota_sugimoto.htm

Jazwinski, C. H. (2001). Gender identities on the World Wide Web. In Wolfe, C. R. (Ed.), *Learning and teaching on the World Wide Web* (pp. 171–189). London: Academic Press. doi:10.1016/B978-012761891-3/50012-6

Jenkins, H. (1988). *Star Trek* rerun, reread, rewritten: Fan writing as textual poaching. *Critical Studies in Mass Communication*, *5*(2), 85–107. doi:10.1080/15295038809366691

Jenkins, H. (1992). *Textual poachers: Television fans and participatory culture*. New York: Routledge.

Jenkins, H. (2006a). *Convergence culture: Where old and new media collide*. New York: New York University Press.

Jenkins, H. (2006b). *Fans, gamers, bloggers: Exploring participatory culture*. New York: New York University Press.

Jenkins, H. (2006). *Fans, Bloggers, and Gamers: Essays on Participatory Culture*. New York: New York University Press.

Jenkins, H. (2006). *Convergence culture: Where old and new media collide*. New York: New York University Press.

Jenkins, H. (2007). Applied Game Theory, R.I.P. 1: Melodrama and Realism. *Confessions of an Aca/Fan*. Retrieved December 3, 2009, from http://www.henryjenkins.org/2007/ 04/applied_game_theory_rip.html

Jenkins, H. (2009). A New "Platform" for Games Research?: An Interview with Ian Bogost and Nick Montfort. *Confessions of an Aca/Fan*. Retrieved December 3, 2009, from http://www.henryjenkins.org/book_shelf/

Jin, D. Y. (2003). Globalization of Japanese Culture: Economic Power vs. Cultural Power, 1989-2002. *Prometheus*, *21*(3), 335–345. doi:10.1080/0810902032000113479

Jin, D. Y. (2006). Cultural Politics in Korea's Contemporary Films under Neoliberal Globalization. *Media Culture & Society*, *28*(1), 6–23. doi:10.1177/0163443706059274

Jin, D. Y., & Chee, F. (2009). The politics of online gaming. In Hjorth, L., & Chan, D. (Eds.), *Gaming cultures and place in Asia-Pacific* (pp. 18–37). New York: Routledge.

Johnson, H. A., & Spector, P. E. (2007, October). Service with a smile: do emotional intelligence, gender, and autonomy moderate the emotional labor process? *Journal of Occupational Health Psychology*, *12*(4), 319–333. doi:10.1037/1076-8998.12.4.319

Johnson, G., & Scholes, K. (2002). *Exploring Corporate Strategy* (6th ed.). Harlow, UK: Pearson Education.

Jones, R. S. (2007). *Income inequality, poverty and social spending in Japan: Economics Department working papers No. 556*. Paris: OECD.

Joshi, A., Welankar, N., BI, N., Kanitkar, K., & Sheikh, R. (2008). Rangoli: A Visual Phonebook for Low-literate Users. *MobileHCI 2008*, Amsterdam, The Netherlands.

Ju, H. (November 14, 2007). *The nature of nationalism in the "Korean wave": A framing analysis of news coverage about Korean pop culture.* Paper presented at the annual meeting of the NCA 93rd Annual Convention. Retrieved April 17, 2010, from http://www.allacademic.com/ meta/ p187925_index.html

Jung, J. Y. (2006). Samsung Electronics open Multi-media Service Samsung Media Studio. *Electronics Newspaper*, 16 May.

Junger, E. (1989). Photography and the second consciousness. In Philips, C. (Ed.), *Photography in the modern era: European documents and critical writings, 1913-1940* (pp. 207–210). New York: Aperture.

Juul, J. (2009). *A Casual Revolution: Reinventing Video Games and Their Players*. Cambridge, MA: MIT Press.

Kalathil, S. (2003). *Open network, closed regimes: The impact of the Internet on authoritarian rules*. Washington, DC: Carnegie Endorsement for International Peace.

Kao, L. (January 11, 2006). Lawmakers: Yao as Facist (in Chinese). *United Daily,* A4.

Kaplan, F. E. S. (Ed.). (1994). *Museums and the making of "ourselves": The role of objects in national identity*. London: Leicester University Press.

Kaplan, F. E. S. (2006). Making and remaking national identities. In Mcdonald, S. (Ed.), *A companion to museum studies* (pp. 152–169). Malden, MA: Blackwell. doi:10.1002/9780470996836.ch10

Kassel, M. B. (2005). Must carry rule. *The Museum of Broadcast Communications*. Retrieved March 2, 2007, from: http://www.museum.tv/home.php

Katz, E. (1996). And deliver us from segmentation. *The Annals of the American Academy of Political and Social Science*, *546*, 22–33. doi:10.1177/0002716296546001003

Katz, J. E., & Aakhus, M. (Eds.). (2002). *Perpetual Contact*. Cambridge, UK: Cambridge University Press. doi:10.1017/CBO9780511489471

Katz, E., Blumler, J., & Gurevitch, M. (1974). Uses and gratifications research. *Public Opinion Quarterly*, *37*, 509–523. doi:10.1086/268109

Katz, J. E. (2006). Mobile communication and the transformation of daily life: The next phase of research on mobiles. *Knowledge, Technology, & . Policy*, *19*(1), 62–71.

Katz, J. E. (2007). Mobile media and communication: Some important questions. *Communication Monographs*, *74*(3), 389–394. doi:10.1080/03637750701543519

Katz, J. E., & Acord, S. K. (2008). Mobile games and entertainment. In Katz, J. E. (Ed.), *Handbook of mobile communication studies* (pp. 403–418). Cambridge, MA: MIT Press.

Katzenstein, P. (2006). East Asia – Beyond Japan. In Katzenstein, J., & Shiraishi, T. (Eds.), *Beyond Japan: The dynamics of East Asian regionalism* (pp. 1–36). Ithaca, NY: Cornell university press.

Kay, C. (2004, March 19). One finger, 10 keys and 6,000 characters. *Globalization Insider*. Retrieved July 17, 2008, from http://www.lisa.org/archive_doma in/newsletters/2004/1.3/kay.html

Keane, M. (2004). Asia: New growth areas. In Moran, A., & Keane, M. (Eds.), *Television across Asia: Television industries, programme formats and globalization* (pp. 9–20). London: Routledge.

Keane, M. & Moran, A. (2005). (Re)Presenting local content: Programme adaptation in Asia and the Pacific. *Media international Australia: Incorporating culture and policy, 116*, 88-99.

Keegan, T. (2004, Spring/Summer). Mobilizing shame. *The South Atlantic Quarterly*, *103*(2-3), 436–447.

Keegan, P. (1999). *Culture Quake*. Retrieved December 3, 2009, from http://www.motherjones.com/ politics/1999/11/culture-quake

Keizer, G. (2007, Dec). Hackers quickly move to exploit Bhutto assassination. *Computerworld, 28*, Retrieved from http://www.netoworkworld.com/cgibin/m ailto/x.cgi?pagetosend+/export/home/httpd/h.

Kelly, K. (1999). *New Rules for the New Economy: 10 Ways the Network is Changing Everything*. London: Fourth Estate.

Kelly, K. (2002). Towards a post-napster music industry. *The New York Times Magazine*. March 17th, 19-21.

Kerr, A. (2006). *The Business and Culture of Digital Games*. London: Sage.

Ki, S. (2009). Boys over flowers gains attention overseas. *Korea Herald*, 9 February, 2009. LexisNexus. Consulted on October 31, 2009.

Kidd, D. (2007). Harry Potter and the functions of popular culture. *Journal of Popular Culture*, *40*(1), 69–89. doi:10.1111/j.1540-5931.2007.00354.x

Kiiski, S., & Pohjola, M. (2002). Cross-country diffusion of the Internet. *Information Economics and Policy*, *14*(2), 293–310. doi:10.1016/S0167-6245(01)00071-3

Kim, L. (1980). Stages of Development of Industrial technology in a developing country: a model. *Research Policy*, *9*, 254–277. doi:10.1016/0048-7333(80)90003-7

Kim, L. (2004). The Multifaceted Evolution of Korean Technological Capabilities and its Implications for Contemporary. *Oxford Development Studies*, *32*(3), 341–363. doi:10.1080/1360081042000260566

Kim, S., & Kim, H. (2009). Does Cultural Capital Matter? Cultural Divide and Quality of Life. *Social Indicators Research*, *93*, 293–313. doi:10.1007/s11205-008-9318-4

Kim, D.-K., Chitnis, K., Vasanti, P. N., & Singhal, A. (2007). Opinion leadership in Indian Villages and Diffusion of e-Choupal. *Journal of Creative Communication*, *2*(3), 335–350.

Kim, M. H. (1995, February). Samsung launched Samsung Entertainment Group. *Chosun Ilbo*, *7*, 21.

Kim, T. S. (1999). Venture Investment Companied Inroad in the Film Business. *Kookmin Ilbo*, *30*(November), 30.

Kim, Y. (2007). The rising East Asian 'Wave': Korean media go global. In Thussu, D. (Ed.), *Media on the move: Global flow and contra-flow* (pp. 135–152). London: Routledge.

Kim, B. S. (2004). TV in my hand, Satellite DMB. *Kwanhun Journal*, (Winter), 118-123

Kim, H. M. (2005). Korean TV dramas in Taiwan: With an emphasis on the localization process. *Korea Journal*, (winter 2005), 183-205.

Kim, J. (2009). *"Boys over flowers" planning director Bae Jong-Byung*. Retrieved October 31, 2009 from http://www.asiae.co.kr/news/view.htm ?sec=ent7&idxno=2009102015351081887

Kim, M. (May, 2004). *Cultural proximity and the type of Korean television programs in Asian market*. Paper presented at the 6th Media Economic Conference, Montreal, Canada.

Kim, T. H. (2009, June 26). Mobile TV Audience Reaches 22 million. *The Korea Times*. Retrieved March 3, 2010, from http://www.koreatimes.co.kr/www/ news/biz/2009/06/123_47547.html

Kirshenblatt-Gimblett, B. (1998). *Destination culture: Tourism, museums, heritage*. Berkeley, CA: University of California Press.

Klein, N. (2002). *No Space, No Choice, No Jobs, No logo*. New York: Picador.

Kleinman, A., & Kleinman, J. (1997). The appeal of experience, the dismay of images: Cultural appropriations of suffering in our times. In Kleinman, A., Das, V., & Lock, M. (Eds.), *Social Suffering*. Berkeley: University of California Press.

Kline, S., Dyer-Witheford, N., & de Peuter, G. (2003). *Digital play: the interaction of technology, culture, and marketing*. Montréal, Canada: McGill-Queen's University Press.

Kline, S., Dyer-Witheford, N., & Peuter, G. D. (2004). *Digital play: The interactions between technology, culture and marketing*. Kingston, ON: McGill/Queens University Press.

Knights, D., & Jones, B. (2007). Outsourcing (the) economy to India: utopian and dystopian discourses of offshoring. *The International Journal of Sociology and Social Policy*, *27*(11/12), 433–446. doi:10.1108/01443330710835792

Ko, S. (January 11, 2006). GIO looking to take foreign soap operas off primetime TV. *Taipei Times*. Retrieved April 17, 2010, from http://www.taipeitimes.com/News/fron t/archives/2006/01/11/2003288408

Ko, S. (January 12, 2006). Soap opera plan's critics are ignorant, Yao retorts. *Taipei Times*. Retrieved April 17, 2010, from http://www.taipeitimes.com/News/tai wan/archives/2006/01/12/2003288583

Koch, P., Koch, B., Huang, K., & Chen, W. (2009). Beauty is in the Eye of the QQ User: Instant Messaging in China. In Goggin, G., & McLelland, M. (Eds.), *Internationalizing Internet Studies* (pp. 265–284). London: Routledge.

Koh, A. (2005). "Heteroglossic" discourses on globalization: A view from the "East.". *Globalizations*, *2*(2), 228–239. doi:10.1080/14747730500202230

Kohiyama, K. (2005). The development of mobile communications in Japan. In Ito, M., Okabe, D., & Matsuda, M. (Eds.), *Personal, portable, pedestrian* (pp. 61–74). Cambridge, MA: MIT Press.

Kohler, C. (2004). *Power-up: How Japanese video games gave the world an extra life*. Indianapolis: Brady Games.

Korea Creative Content Agency (KOCCA). (2009). *Introduction*. Retrieved September 1, 2009, from http://www.koreacontent.org/ weben/etc/kocca.jsp

Korea net (2006, December 5). Cell phones revolutionize Korean lifestyles. *Korea.net.* Retrieved Feb. 20, 2008, from http://www.korea.net/news/news/ newsview.asp?serial_no=20061204020

KPMG. (2007). *The video games market in China: Moving online*. Retrieved April 3, 2010 from http://www.kpmg.com/CN/en/IssuesAndIns ights/ArticlesPublications/Docum ents/video_games_china_0712.pdf

Kraidy, M. M. (2002). The global, the local, and the hybrid: A native ethnographic of glocalization. In Taylor, S. (Ed.), *Ethnographic research: A reader*. London: Sage.

Kristeva, J. (1982). *Powers of horror: An Essay on abjection, (Trans. L. S. Roudiez)*. New York: Columbia UP.

Kristof, N. (1995, February 21). Japan's favorite import from American: English. *New York Times.* Retrieved July 30, 2008, from http://query.nytimes.com/gst/fullpage.html?res=990CE0DC1430F932A15751C0A963958260

Kshetri, N. (2009). The evolution of the Chinese online gaming industry. *Journal of technology management in China, 4*(2), 158-179.

Kucklich, J. (2004). Precarious Playbour: Modders and the Digital Games Industry. In *Fibre Culture: Internet Theory, Criticism and Research, 5,* [online]. Available at: <http://www.fibreculture.org/issue5/kuckligh_print.html>

Kücklich, J. (2005). Precarious Playbour: Modders and the Digital Games Industry. *Fibreculture Journal,* 5. Retrieved 23 July 2008 from http://journal.fibreculture.org/issue5/kucklich.html

Kuhn, T. S. (1962). *The Structure of Scientific Revolutions*. Chicago: University of Chicago Press.

Kumar, A., Rajput, N., Agarwal, S., Chakraborty, D., & Nanavati, A. A. (2008). Organizing the unorganized: Employing IT to empower the under-privileged. In *Proceedings of the International World-Wide Web Conference* (pp. 935-944). Beijing, China: ACM Press.

Kùng, H. (1998). *A global ethic for global politics and economics, (trans. John Bowden)*. Oxford, UK: Oxford University Press.

Küng, L., Picard, R. G., & Towse, R. (2008). *The internet and the mass media*. Oxford, UK: Sage Publications.

Kuo, L. C. (2006, April 14). Do "Hardcore Gamers" Still Matter? *GameSpy.* Retrieved December 3, 2009, from http://au.gamespy.com/articles/701/701787p1.html

Kushner, D. (2003). *Masters of Doom: How Two Guys Created an Empire and Transformed Pop Culture* (1st ed.). New York: Random House.

Kwan, C. (January 13, 2006). *Dae Jang Geum a local dramas* (in Chinese). Retrieved April 17, 2010, from http://www.twmedia.org/modules /news/article.php?storyid=231

Kwong, K. K., Yau, O. H. M., Lee, J. S. Y., Sin, L. Y. M., & Tse, A. C. B. (2003). The effects of attitudinal and demographic factors on intention to buy pirated CDs: The case of Chinese consumers. *Journal of Business Ethics, 47*, 223–235. doi:10.1023/A:1026269003472

Kymlicka, W. (1995). *Multicultural citizenship: A liberal theory of minority rights*. Oxford, UK: Oxford University Press.

Lampel, J., & Shamsie, J. (2003). Capabilities in motion: New organizational forms and the reshaping of the Hollywood movie industry. *Journal of Management Studies, 40*(8), 2189–2210. doi:10.1046/j.1467-6486.2003.00417.x

Land, M. (1992). Ivoirien television, willing vector of cultural imperialism. *The Howard Journal of Communications, 4*, 10–27.

Land, M. (2006). Mass media ethics and the Point-of-Decision Pyramid. In Land, M., & Hornaday, B. W. (Eds.), *Contemporary media ethics: A practical guide for students, scholars and professionals* (pp. 15–38). Spokane, WA: Marquette Books.

Landler, M. (1995, April 29). Cellular phones may affect pacemakers. *New York Times*. Retrieved March 31, 2010, from

Landow, G. (1997). *Hypertext 2.0: The convergence of contemporary critical theory and technology*. Baltimore, MD: Johns Hopkins University Press.

Larpote, N. (2006). Traveling Man. *Variety*, *402*(3), 2.

Lattas, K. (2006). Technology of visibility: The utopian politics of cameras, videos and dreams in New Britain. *The Australian Journal of Anthropology*, *17*(1), 15–32. doi:10.1111/j.1835-9310.2006.tb00045.x

Lau, T. Y. (1988). Introducing cable television into Hong Kong: Political and economic implications. *Telecommunications Policy*, *12*(4), 379–392. doi:10.1016/0308-5961(88)90060-2

Lau, T. Y. (1992). From cable television to direct broadcast satellite: Emerging policy issues in the Asia-Pacific Region. *Telecommunications Policy*, *16*(7), 576–590. doi:10.1016/0308-5961(92)90023-I

Lawson-Borders, G. (2003). Integrating New Media and Old Media: seven observations of convergence as a strategy for best practices in media organization. *The International Journal of Media Management*, *5*(2), 91–99.

Lee, K. (2007). Questioning A Neoliberal Urban Regeneration Policy: The rhetoric of "cities of culture" and the city of Gwangju, Korea. *International Journal of Cultural Policy*, *13*(4), 335–347. doi:10.1080/10286630701683201

Lee, F. L. F. (2007). Development of pay television in Hong Kong: History, audience consumption, and challenges. *Journal of Comparative Asian Development*, *4*(2), 247–272.

Lee, F. L. F. (2009). Paradigm defining events and event routinization: Hong Kong newspapers' coverage of YouTube video clips. *The Chinese Journal of Communication and Society*, *9*, 181–202.

Lee, F. L. F. (2010). The influence of family viewing preferences on television consumption in the era of multichannel services. *Asian Journal of Communication*, *20*(3), 281–298. doi:10.1080/01292981003802176

Lee, B., & LiPuma, E. (2002). Culture of circulation: The imaginations of modernity. *Public Culture*, *14*(1), 191–213. doi:10.1215/08992363-14-1-191

Lee, D. Y. (2006). *Samsung Electronics: the global inc.* Seoul: YSM.

Lee, C. (2000). State, capital and the media. In Curran, J., & Park, M.-J. (Eds.), *De-Westernizing media studies* (pp. 110–120). New York: Routledge.

Lee, K. S. (2007). Samsung Corporation. In Downing, J. (Ed.), *The International Encyclopedia of Communication: International Communication*. Oxford: Blackwell Publishing.

Lee, D. H. (2004). A local mode of programme adaptation: South Korea in the global television format business. In Moran, A., & Keane, M. (Eds.), *Television across Asia: Television industries, programme formats and globalization* (pp. 36–53). London: Routledge.

Lee, K. (2008). Mapping out the cultural politics of "the Korean Wave" in contemporary South Korea. In B. H. Chua & K. I. Iwabuchi (Eds.), *East Asian pop culture: Analysing the Korean wave* (pp.175-189). Hong Kong: Hong Kong University.

Lee, M. H. (2006). *A study on the type of acceptance of early adaptors of the digital multimedia broadcasting – Mainly on the realistic application of gratification-niche theory according to the media competition*. Unpublished doctoral dissertation, SungKyunKwan University, Seoul, Korea.

Lee, S. (August 23, 2006). The government spends 200 billion in promoting audio-visual industry (in Chinese). *United Daily. Retrieved* April 17, 2010. Retrieved from http://udndata.com/ndapp/Story? no=2... news_id=3069674

Lee, S. H. (2007). *Mobile TV-DMB*. Paper presented at the 2007 annual conference of International Communication Association, San Francisco, May, 2007.

Lee, S. H. (2008). *Incorporating mobile multimedia into everyday life: Diffusion and use of Mobile TV*. Unpublished doctoral dissertation, University of Wisconsin, Madison.

Leeds, J., & Meller, P. (2004). Europe Clears Merger Plan of 2 Giants in Recordings. *The New York Times*, 20 July, C1.

Lefkoff-Hagius, R., & Mason, C. H. (1993). Characteristics, beneficial, and image attributes in consumer judgments of similarity and preference. *The Journal of Consumer Research*, *20*, 100–110. doi:10.1086/209336

Lemish, D., & Cohen, A. A. (2005). On the Gendered nature of mobile phone culture in Israel. *Sex Roles*, *52*, 511–521. doi:10.1007/s11199-005-3717-7

Lessig, L. (1999). *Code and other laws of cyberspace*. New York: Basic Books.

Lessig, L. (2001). *The Future of Ideas*. New York: Vintage.

Lester, P. M. (2006, February). Urban screens: The beginning of a universal visual culture. *First Monday*, *4*. Retrieved October 20, 2009, from http://www.uic.edu/htbin/cgiwrap/bin/ ojs/index.php/fm/rt/printerFriendly/1543/1458

Leung, L., & Wei, R. (1999). Who are the mobile phone have-nots? *New Media & Society*, *1*, 209–226.

Leung, L., & Wei, R. (2000). More than just talk on the move: Uses and gratifications of the cellular phone. *Journalism & Mass Communication Quarterly*, *77*, 308–320.

Leung, K., Bhagat, R. S., Buchan, N. R., Erez, M., & Gibson, C. B. (2005, July). Culture and International Business: Recent Advances and Their Implications for Future Research. *Journal of International Business Studies*, *36*(4), 357–378. doi:10.1057/palgrave.jibs.8400150

Lewis-Beck, M. S. (1994). *Factor analysis and related techniques*. London: SAGE.

Liang, S. N. (2008). Shipin wangzhan de kunjing ji yunying moshi tanxi (An exploration of the dilemma and business model of the online video website). [Youth Reporter]. *Qingnian Jizhe*, *33*, 95.

Licklider, J. C. R., & Taylor, R. W. (1990). The computer as a communication device. In *Memoriam: J. C. R. Licklider 1915-1990*. Palo Alto, CA: Systems Research Center.

Liebe, T., & Curran, J. (Eds.). (1998). *Media, ritual, and identity*. London: Routledge.

Liebowitz, S. (2002). *Record Sales, MP3 Downloads, and the Annihilation Hypothesis*. [online] Working Paper, University of Texas at Dallas. Retrieved from http://www.dklevine.com/archive/leibowitz.pdf

Lieu, P. (2002). The DNA of the Korean dramas [in Chinese]. *The Journalist (Weekly)*, *800*, 50–53.

Lieu, S. (August 24, 2009). Where is Taiwan's culture industry? Kim Dae Chung has led the miracle of the Korean wave (in Chinese). *United Daily*, A13.

Lin, L. (2005). Authoritarian states and television: A comparative study between Taiwan and South Korea [in Chinese]. *Mass Communication Research*, *85*, 1–30.

Lin, L. (2008). The press reform in Korea. *Mass Communication Research*, *95*, 257–282.

Lin, A., & Tong, A. (2008). Re-imagining a cosmopolitan 'Asian Us': Korean media flows and imaginaries of Asian modern femininities. In Chua, B. H., & Iwabuchi, K. I. (Eds.), *East Asian pop culture: Analysing the Korean wave* (pp. 91–125). London: Eurospan.

Lin, C. A. (1994). Exploring potential factors for home videotext adoption. In Hanson, J. (Ed.), *Advances in telematics* (pp. 111–124). Norwood, NJ: Ablex.

Lin, S. (March13, 2005). Aggressive Yao with dreams of films (in Chinese). *Economic Daily*, A7.

Lin, S., & Yang, W. (May 25,2005). Hsieh: Taiwan dangerous, if small and rigid (in Chinese). *Economic Daily*, A1.

Lin, T. (February 16, 2008). Taiwan's brand on the global stage (in Chinese). *Liberty Times.* Retrieved April 17, 2010, from http://www.libertytimes.com.tw /2008/new/feb/16/today-o1.htm

Ling, R. (2004). *The mobile connection: The cell phone's impact on society*. San Francisco, CA: Morgan Kaufmann.

Ling, R., & Donner, J. (2009). *Mobile phones and mobile communication*. Malden, MA: Polity.

Lister, M., Dovey, J., & Giddings, S. (2003). *New media: A critical introduction*. New York: Routledge.

Lister, M., & Dovey, J. Giddings, S. Grant, I. & Kelly, K. (2003). *New Media: A Critical introduction*. New York: Routledge.

Liu, Y., & Chen, Y. (2004). Cloning, adaptation, import and originality: Taiwan in the global television format business. In Moran, A., & Keane, M. (Eds.), *Television across Asia: Television industries, programme formats and globalization* (pp. 54–73). London: Routledge.

Lo, K. C. (2001). Double negations: Hong Kong cultural identity in Hollywood's transnational representations. *Cultural Studies*, *15*(34), 464–485. doi:10.1080/095023800110046669

Lo, S. (January 15, 2006). Support the quota system! (in Chinese) *Liberty Times.* Retrieved April 17, 2010, from http://www.twmedia.org/mod ules/news/article.php?storyid=229

Lo, T. (January 11, 2006). Korean dramas in Japan (in Chinese). *Liberty Times,* Retrieved April 17, 2010, from http://www.libertytimes.com/20 06/new/jan/11/today-o11.htm.

Louw, E. (2001). *The media and cultural production.* Oxford, UK: Sage Publications.

Lovink, G. (2007). *Zero Comments: Blogging and Critical Internet Culture*. New York: Routledge.

Low, C. (2009, July 13). Japan plots one-stop public services portal. *FutureGov*. Retrieved March 27, 2010, from http://www.futuregov.net/articles/2009/j ul/13/japan-plots-one-stop-online-public-services-portal

Lundberg, U., & Melin, B. (2002). Stress in the development of musculoskeletal pain. In Linton, S. J. (Ed.), *New Avenues for the prevention of chronic musculoskeletal pain and disability* (pp. 165–182). Amsterdam: Elsevier Science Publishers.

Lyer, S. B. (2005). Sony buys up sitcom powerhouse. *Variety, 398*(5, 21 March), 20.

Lyra Research. (2007). *The Worldwide Camera Phone Forecast.* Retrieved March 3, 2010, from http://lyra.ecnext.com/coms2/ summary_0290-801_ITM

Ma, E. K. W. (1999). *Culture, politics, and television in Hong Kong*. London: Routledge.

MacCarthur, J., Crist, G., Hartoonian, G., & Stanhope, Z. (2003). Federation square: Melbourne's most significant set of new public buildings and spaces is now open. *Architecture Australia*, *92*(2), 46.

MacColl, I., & Richardson, I. (2008). A cultural somatics of mobile media and urban screens: Wiffiti and the IWALL prototype. *Journal of Urban Technology*, *15*(3), 99–116. doi:10.1080/10630730802678002

Madamombe, I. (2005). Energy key to Africa's prosperity: Challenges in West Africa's quest for electricity. *Africa Renewal*, *18*(4), 6.

Mahesh, V. S., & Kasturi, A. (2006). Improving call center agent performance. *International Journal of Service Industry Management*, *17*(2), 136–157. doi:10.1108/09564230610656971

Malaby, T. M. (2007). Beyond Play: A New Approach to Games. *Games and Culture*, *2*(2), 95–113. doi:10.1177/1555412007299434

Malliet, S., & de Meyer, G. (2005). The history of the video game. In Raessens, J., & Goldstein, J. (Eds.), *Handbook of computer game studies* (pp. 23–45). Cambridge, MA: MIT Press.

Manning, P. (1997). Media loops. In Bailey, F., & Hale, D. (Eds.), *Popular culture, crime & justice* (pp. 25–39). Belmont, CA: Wadsworth.

Manovich, L. (2001). *The language of new media*. Cambridge, MA: MIT Press.

Manovich, L. (2000). *The Language of New Media*. Cambridge, MA: MIT Press.

Manovich, L. (2006, February). The poetics of urban media surfaces. *First Monday, 4*. Retrieved October 20, 2009, from http://www.uic.edu/htbin/cgiwrap/bin/ ojs/index.php/fm/rt/printerFriendly/1545/

Manovich, L. (2008). *Software Takes Control*. Retrieved April 19, from http://softwarestudies.com/softbook/

Mariniello, S. (2003). Commencements. *Intermédialités, 1*, 47–62.

Markman, K. (2005). Star Trek, *fan film, and the Internet: Possibilities and constraints of fan-based vernacular cultures*. Paper presented at the Annual Meeting of the International Communication Association, New York, NY.

Markus, M. L. (1987). Toward a "Critical Mass" theory of interactive media. *Communication Research, 14*, 491–511. doi:10.1177/009365087014005003

Martineau, P. (1968). Social class and spending behavior. *Journal of Marketing, 23*, 274–278.

Mary Louise Pratt, M. L. (1991). Arts of the Contact Zone. *Profession, 91*, 33–40.

Matsuda, M. (2005). Mobile communication and selective sociality. In Ito, M., Okabe, D., & Matsuda, M. (Eds.), *Personal, portable, pedestrian* (pp. 123–142). Cambridge, MA: MIT Press.

Matsuda, M. (2006). Discourses of Keitai in Japan. In Ito, M., Okabe, D., & Matsuda, M. (Eds.), *Personal, portable, pedestrian: Mobile phones in Japanese life* (pp. 19–40). Cambridge, MA: The MIT Press.

Maule, C. (2003). Television. In Towse, R. (Ed.), *A handbook of cultural economics* (pp. 458–475). Brookfield, VT: Edward Elgar.

May, H., & Hearn, G. (2005). The mobile phone as media. *International Journal of Cultural Studies, 8*(2), 195–211. doi:10.1177/1367877905052417

McChesney, R. (2001). Global Media, Neoliberalism, and Imperialism. *Monthly Review (New York, N.Y.), 52*(10), 1–19.

McChesney, R. W. (1999). Noam Chomsky and the struggle against neoliberalism. *Monthly Review (New York, N.Y.), 50*(April), 40–47.

McChesney, R. W. (2004). *The Problem of the Media*. New York: Monthly Review Press.

McChesney, R. (2001). Global Media and Democracy. In Weibel, P., & Druckrey, T. (Eds.), *Net Condition: Art and Global Media* (pp. 356–363). Cambridge, MA: The MIT Press.

Mcguigan, J. (2007, June 18). Mobility: Comment. *Fifth-Estate-Online – International Journal of Radical Mass*. Retrieved from http://www.fifth-estate-online.co.uk/comment/mobility.html

McGuire, D., & McLaren, L. (2007). The Impact of Physical Environment on Employee Commitment in Call Centres: The Mediating Role of Employee Well-Being. *Team Performance Management, 14*(5/6).

Mckenna, A. (2000). Emerging Issues Surrounding the Convergence of the Telecommunications, Broadcasting and Information Technology Sectors. *Information & Communications Technology Law, 9*(2), 93–128. doi:10.1080/713670487

McLelland, M. (2007). Socio-cultural aspects of mobile communication technologies in Asia and the Pacific: A discussion of the recent literature. *Continuum: Journal of Media & Cultural Studies, 21*(2), 267–277. doi:10.1080/10304310701269099

McLuhan, M. (1964). *Understanding media: The Extensions of man*. Toronto, Canada: McGraw-Hill.

McMillin, D. (2001). Localizing the global: Television and hybrid programming in India. *International Journal of Cultural Studies, 4*(1), 45–68. doi:10.1177/136787790100400103

McPhail, T. (2006). *Global Communication: theories, stakeholders, and trends* (2nd ed.). Oxford: Blackwell.

McPhail, T. L., & McPhail, B. M. (1990). *Communication: The Canadian experience*. Toronto, Canada: Copp Clark Pitman ltd. A Longman company.

McVeigh, B. (2003). *Individualization, individuality, interiority, and the Internet: Japanese university students and e-mail* (Gottlieb, N., & McLelland, M., Eds.).

Media Corp, T. U. (2004). *Satellite-DMB's policy issues and market prospect*. Paper presented at the Commercial DMB demonstration, Seoul, Korea. Retrieved July 3, 2005, from www.kmis.or.kr/3_sig/comm_ data/comm(may04)1.pdf

Media, F. (2009, March 23). *Focus media reports fourth quarter and full year 2008 results*. Retrieved October 20, 2009, from http://ir.focusmedia.cn/phoenix.zhtml?c= 190067&p=irol-news Article_print&ID=1268983&highlight

Mellgren, D. (2007, Sept 26). Cell Phones, web spreads news of Myanmar. *ABC News*. Retrieved from http://abcnews.go.com/Intern ational/wireStory?id=3656364

Merrill, J. (2006). Utilitarianism Vs. communitarianism. In Land, M., & Hornaday, B. W. (Eds.), *Contemporary media ethics: A practical guide for students, scholars and professionals* (pp. 41–54).

Meyer, K. (2008, Jan-Mar). Yin and Yang: The Media's Role in the Burmese Peace Movement. *Peace Magazine*, 9. Retrieved from http://archive.peacemagazine. org/v24n1p09.htm

Meyrowitz, J. (1985). *No sense of place*. New York: Oxford University Press.

Meyrowitz, J. (1985). The generalized elsewhere. *Critical Studies in Mass Communication*, *6*(3), 323–334.

Meyrowitz, J. (1997). Shifting worlds of strangers: Medium theory and changes in "them" versus "us.". *Sociological Inquiry*, *67*(1), 59–71. doi:10.1111/j.1475-682X.1997.tb00429.x

Meyrowitz, J. (1985). *No sense of place: The impact of electronic media on social behaviour*. New York: Oxford University Press.

MIC. (2007). *2007 Information and Communications in Japan White Paper*. Tokyo: Ministry of Internal Affairs and Communications.

MIC. (2009, April 7). 平成*20*年「通信利用動向調査」の結果. Retrieved November 7, 2009, from http://www.soumu.go.jp/menu_new s/s-news/02tsushin02_000001.html

Miege, B. (1989). *The Capitalisation of Cultural Production*. New York: International General.

Miller, T. (2001). Introducing … cultural citizenship. *Social Text*, *19*(4), 1–5. doi:10.1215/01642472-19-4_69-1

Miller, T. (2006). *Cultural citizenship: Cosmopolitanism, consumerism, and television in a neoliberal age*. Philadelphia: Temple University Press.

Miller, D., & Slater, D. (2001). *The Internet: An Ethnographic Approach*. London: Berg.

Miller, D. (1998). *A Theory of Shopping*. Cambridge, UK: Polity Press.

Miller, T. (2004). Trainspotting the *Avengers*. In Gwenllian-Jones, S., & Pearson, R. (Eds.), *Cult Television* (pp. 187–197). Minneapolis: University of Minnesota Press.

Miller, N. at all. (2007). Snack Attack! *Wired*, 15 (3), March. Retrieved April 1, 2008, from http://www.wired.com/wired/ archive/15.03/snack.html

Ministry of Culture and Tourism. (2006). *2006 Korea Online Game White Paper*. Seoul, Korea: Ministry of Culture and Tourism.

Ministry of culture, sports and tourism Korea game industry agency (2008). *2008 White paper on Korean games: Guide to Korean game*. Seoul, South Korea: Korean Game Industry and Promotion Institute.

Ministry of Culture. Sports and Tourism & Korean Culture and Tourism Institute. (2008). *Survey report on cultural enjoyment*. Retrieved March 15, 2010, from http://www.mcst.go.kr/web/dataCourt /statisticsData/statisticsView.jsp

Ministry of Culture. Sports and Tourism, South Korea. (2009). *Cultural Industry Report*. Retrieved March 15, 2010. From http://www.mcst.go.kr/web/data Court/research/researchView.jsp

Ministry of Information and Communication (MIC) of Korea (2005 & 2006). *Dynamic U-Korea*.

Mirchandani, K. (2004b). Practices of global capital: Gaps, cracks and ironies in transnational call centers in India. *Global Networks*, *4*, 355–374. doi:10.1111/j.1471-0374.2004.00098.x

Mirchandani, K. (2004a). Webs of Resistance in transnational call centers: strategic agents, service providers and customers. In Thomas, R., Mills, A., & Helms Mills, J. (Eds.), *Identity Politics at Work: Resisting Gender, Gendering Resistance* (pp. 179–195). London: Routledge.

Mirchandani, K. (2005). Webs of resistance in transnational call centres. In Thomas, R., Mills, A., & Helms-Mills, J. (Eds.), *Gender, Organization and the Micro-politics of Resistance*. London: Routledge.

Mitchell, J. T. (1994). *Picture theory: Essays on verbal and visual representation*. Chicago: University of Chicago Press.

Miyata, K., Boase, J., Wellman, B., & Ikeda, K. (2006). The Mobilie-izing Japanese: Connecting to the Internet by PC and Webphone in Yamanashi. In Ito, M., Okabe, D., & Matsuda, M. (Eds.), *Personal, portable, pedestrian: Mobile phones in Japanese life* (pp. 143–164). Cambridge, MA: The MIT Press.

Miyata, K. (2006, June). *Longitudinal effects of mobile Internet use on social network in Japan*. Paper presented at the meeting of the International Communication Association, Dresden, Germany.

Mohr, A., & Gleicher, M. (2001). Non-invasive, interactive, stylized rendering. In *Proceedings of the 2001 symposium on Interactive 3D graphics* (pp. 175-178). New York: ACM.

Montfort, N., & Bogost, I. (2009). *Racing the Beam: The Atari Video Computer System*. Cambridge, MA: MIT Press.

Moon, I. (2010, March 1). Korean tech is losing its cool. *Bloomberg BusinessWeek*, 52-53.

Moore, K. (2003). Overlooking the visual. *The Journal of Architecture*, *8*, 26–40. doi:10.1080/1360236032000068497

Moores, S. (2004). The doubling of place: Electronic media, time-space arrangements and social relationships. In Couldry, N., & McCarthy, A. (Eds.), *Mediaspace: Place, scale and culture in a media Age*. London: Routledge.

Morgan, R. (1998). Linking brand imagery with preference. In Greig, I. (ed), *Brand Choice Modeling*. ESOMAR, 63-84.

Morley, D. (1980). *The Nationwide audience*. London: British Film Institute.

Moroka, T. M. (2005). Family economic pressure and health outcomes in low-income Francistown, Botswana: The reverse buffering role of social support. *Journal of Social Development in Africa*, *20*, 106–127.

Morris, N. (2002). The myth of unadulterated culture meets the threat of imported media. *Media Culture & Society*, *24*, 278–298. doi:10.1177/016344370202400208

Mosco, V. (1996). *The political economy of communication: Rethinking and renewal*. London; Thousand Oaks, CA: SAGE Publications.

Mosco, V. (2009). *The Political Economy of Communication* (2nd ed.). London: Sage.

Mosco, V. (1996 & 2009). *The Political Economy of Communication*. London: Sage.

Motion Picture Association of America. (2007). Theatrical Market Statistics. Retrieved November 1, 2009, from http://www.mpaa.org

Mudimbe, V. Y. (2009). In the name of similitude. In Njogu, K., & Middleton, J. (Eds.), *Media and identity in Africa, (International African Seminars)*. Edinburgh, UK: Edinburgh University Press.

Mukherjee, A., & Malhotra, M. (Guest Editorial. (2009). Call centre services: the good, the bad, and the ugly. *Journal of Services Marketing*, *23*(5).

Mulgan, G. (1997). *Connexity: How to live in a connected world*. London: Chatto and Windus.

Muller, M. J., Farrell, R., Cebulka, K. D., & Smith, J. G. (1992). Issues in the usability of time-varying multimedia. In Blattner, M. M., & Dannenberg, R. B. (Eds.), *Multimedia interface design* (pp. 7–38). New York: ACM Press.

Munakata, N. (2006). Has politics caught up with markets? In search of East Asian economic regionalism. In Katzenstein, J., & Shiraishi, T. (Eds.), *Beyond Japan: The Dynamics of East Asian Regionalism* (pp. 130–160). Ithaca, NY: Cornell University Press.

Munster, A. & Murphie, A. (2009). Editorial — Web 2.0: Before, during and after the event. *Fibreculture journal, 14*. Retrieved 10 October 2009 from

Murray, S. (2004). "Celebrating the story the way it is": Cultural studies, corporate media and the contested utility of fandom. *Continuum: Journal of Media and Cultural Studies, 18*(1), 7–25. doi:10.1080/1030431032000180978

Murray, C. (2005). Cultural participation: A fuzzy cultural policy paradigm. In Andrew, C., Gattinger, M., Jeannotte, M. S., & Straw, W. (Eds.), *Accounting for Culture: Thinking through Cultural Citizenship* (pp. 32–56). Ottawa: University of Ottawa Press.

Naim, M. (2007, Jan-Feb). The YouTube effect: How a technology for teenagers became a force for political and economic change. *Foreign Policy*, *158*, 104–106.

Nancy, J.-L. (1991). *The inoperative community*. Minneapolis, MN: University of Minnesota.

Napier-Bell, S. (2008). The Life and Crimes of the Music Biz. *The Observer – Music Monthly*, Sunday 20th January, 41.

Napoli, P. (2003). *Audience economics*. New York: Columbia University Press.

Nasasaw. (2005, April 7). *Mods of National Special Force: EA Taiwan calls the Taiwanese Players to make (ROC) Mandarin Voice Mods*. Message posted to http://www.ezhobi.net/discuz/viewthread.php?tid=8925 (In Chinese)

Nathan, J. (1999). *SONY*. New York: Houghton Mifflin Company.

National Cable Telecommunications Association (NCTA). (2001). *Cable television developments*. Washington, DC.

National Information Society Agency (NIA). (2007a). *Informatization White Paper*.

National Information Society Agency (NIA). (2007b). *Korea informatized progress and status overview*.

Negroponte, N. (1996). *Being Digital: The Road Map for Survival on the Information Superhighway*. London: Hodder & Stoughton.

Negroponte, N. (1995). Being Digital: A Book (P)review. *Wired.com* [online], *3.02*, February. Retrieved from http://www.wired.com/wired/archive/3.02/negroponte.html?pg=2&topic=

Nelson, R., & Winter, S. G. (1982). *An evolutionary theory of economic change*. Cambridge, MA: The Belknap Press of Harvard University Press.

Newman, S. (2005). Virtual virillo. *IM Issue*, 1. Retrieved from http://nass.murdoch.edu.au/docs /S%20%20Newman.pdf

News, B. B. C. (11 December, 2003). *Call centers 'bad for India.'* Retrieved from www.bbcnews.co.uk

Ngan, R., & Ma, S. (2008). The relationship of mobile telephony to job mobility in China's Pearl River Delta. *Know Techn Pol*, *21*, 55–63. doi:10.1007/s12130-008-9046-8

Nordenstreng, K., & Hannikainen, L. (1984). *The Mass Media Declaration of UNESCO*. Norwood, NJ: Ablex Publishing Corporation.

Noronha, E., & d'Cruz, P. (2007). Reconciling dichotomous demands: Telemarketing agents in Bangalore and Mumbai, India. *Qualitative Report*, *12*(2), 255–280.

Norrie, J. (2007, April 14). A nation's youth all thumbs with PCs. *The Age.* Retrived July 20, 2008, from http://www.theage.com.au/articles/ 2007/04/13/1175971346787.html

Norris, P. (2001). *Digital divide: civic engagement, information poverty and the Internet in democratic societies*. New York: Cambridge University Press.

November 21, 2009, from https://www.cia.gov/library/public ations/the-world-factbook/geos/bc.html

November 4, 2007, from http://news.zdnet.com/2100-1035-5905677.html

NTT. (2000, August 7). *"i-mode" subscriptions top 10 million.* Retrieved August 1, 2008, from http://www.nttdocomo.com/ pr/2000/001031.html

NTT. (2006). Keyword search service expands mobile convenience. In *Mobility*. Tokyo: Nippon Telegraph & Telephone.

O'connell, P. L. (2005, October 5). Korea's high-tech utopia, where everything is observed. *New York Times*. Retrieved November 15, 2009, from http://www.nytimes.com/2005/10/05/t echnology/techspecial/05oconnell.html

O'Keefe, G., & Sulanowski, B. (1995). More than just talk: Uses and gratifications and the Telephone. *Journalism & Mass Communication Quarterly*, *75*, 922–933.

O'Regan, N. (2008). Heed The Music Lessons. *The Sunday Business Post* [online], Sunday April 6th. Retrieved from http://archives.tcm.ie/businesspost/2008/04/06/story31725.asp

O'Reilly, T. (2005). *What Is Web 2.0: Design Patterns and Business Models for the Next Generation of Software*. Retrieved 10 September 2009 from

Oba, G., & Chan-Olmsted, S. M. (2005). The development of cable television in East Asian countries. *Gazette: The International Journal for Communication Studies*, *67*(3), 211–237. doi:10.1177/0016549205052225

O'Donnell, C. (2009, March 15). The everyday lives of video game developers: Experimentally understanding underlying systems/structures. *Text.Serial.Journal*. Retrieved December 3, 2009, from http://journal.transformativeworks.org /index.php/twc/article/viewArticle/73

OECD. (1997). *Information technology outlook 1997*. Paris: OECD.

OECD. (2002). *Information technology outlook 2002*. Paris: OECD.

OECD. (2009). Broadband penetration rates [from http://www.oecd.org/sti/ict/broadband]. *Top (Madrid)*, *5*, Retrieved March 6, 2010.

OECD. (2008). *Households with broadband access, 2004-2008. Percentage of all households*. Retrieved March 6, 2010, from http://www.oecd.org/sti/ict/broadband

Ohiwa, H. (2004). Keyboard training in an information-oriented society. *IFIP International Federation for Information Processing*, *14*, 147–152. doi:10.1007/1-4020-8136-7_15

Ohmae, K. (1998). Akio Morita. *Time*, *152*(23), 100.

Ohta, S. (1999). The shift to Internet EDI and Web EDI. *JEDIC Newsletter, 39*. Retrieved August 1, 2008, from http://www.ecom.jp/e_jedic/activity/ news44-35/n39/koen1_e.htm

Okabe, D., & Ito, M. (2005). *Keitai* in public transportation. In Ito, M., Okabe, D., & Matsuda, M. (Eds.), *Personal, portable, pedestrian* (pp. 205–217). Cambridge, MA: MIT Press.

Okabe, D., & Ito, M. (2006). Keitai in Public Transportation. In Ito, M., Okabe, D., & Matsuda, M. (Eds.), *Personal, portable, pedestrian: Mobile phones in Japanese life* (pp. 205–218). Cambridge, MA: The MIT Press.

Okada, T. (2005). Youth culture and the shaping of Japanese mobile media: Personalization and the *keitai* Internet as multimedia. In Ito, M., Okabe, D., & Matsuda, M. (Eds.), *Personal, portable, pedestrian* (pp. 41–60). Cambridge, MA: MIT Press.

OLPC. (2009). *One Laptop Per Child.* Retrieved June 15, 2009, from http://laptop.org

Olson, G. M., & Olson, J. S. (2000). Distance matters. *Human-Computer Interaction*, *15*(2), 139–178. doi:10.1207/S15327051HCI1523_4

Onishi, N. (2008, January 20). Thumbs race as Japan's best sellers go cellular. *New York Times*. Retrieved March 10, 2010, from http://www.nytimes.com/2008/0 1/20/world/asia/20japan.html

Open Net Initiative Bulletin. Pulling the plug; A technical review of the Internet shutdown in Burma. (n.d.). Retrieved April, 12, 2008, from http://opennet.net/sites/opennet.net/files /ONI_Bulletin_Burma_2007.pdf

OSISA. (2009). *Electricity for Africa?* Retrieved June 15, 2009, from http://www.osisa.org/node/4164

Osorio, A. (2008). Xbox Song Downloads Topping 3.8 Million Monthly. *Digital Music News* [online], Tuesday 21st October 2008. Retrieved from http://www.digitalmusicnews.com/stories/102108msft

Osorio, A. (2008a). MySpace Educating Masses on Chinese Democracy. *Digital Music News* [online], Thursday 20th November. Retrieved from http://www.digitalmusicnews.com/stories/112008myspace/?searchterm=Chinese%20Democracy%20on%20MySpace

Otmazgin, N. (2008). Contesting soft power: Japanese popular culture in East and Souteast Asia. *International relationships of the Asia-Pacific, 8*, 73-101.

Oyang, Y. S., & Lee, C. H. (2004) *The Lineage II* locks out the overseas IP, the players criticize the NCSoft Taiwan as cheater. *ET Today,* Retrieved May 20, 2006, from http://www.ettoday.com/2004/03/16/10846-1600618.htm (In Chinese)

Pakulski, J. (1997). Cultural citizenship. *Citizenship Studies, 1*(1), 73–86. doi:10.1080/13621029708420648

Pal, M., & Buzzanel, P. (2008). The Indian Call Center Experience: A Case Study in Changing Discourses of Identity, Identification, and Career in a Global Context. *Journal of Business Communication, 45*(1), 31–60. doi:10.1177/0021943607309348

Palen, L., Salzman, M., & Youngs, E. (2001). Discovery and integration of mobile communications in everyday life. *Personal and Ubiquitous Computing, 5*, 109–122. doi:10.1007/s007790170014

Papacharissi, Z. (2002). The virtual sphere: The internet as a public sphere. *new media & society, 4*(1), 9-27.

Papastergiadis, N., McQuire, S., & Martin, M. (2009). Large Screens, Cultural Creativity and the Global City. In Lim, W. S. (Ed.), *Collection of Essays on Asian Design Cultures* (pp. 42–53). Singapore: Select Books.

Park, S. H. (1995). Negotiation between Dream Work and Samsung Failed. *Chosun Ilbo., 24*(March), 9.

Park, S. H. (2007) Samsung's reinvestment in the film business it ended in 9 years ago, *Chosun Ilbo*, 18 September.

Parker, R. S., Schaefer, A. D., & Hermans, C. M. (2006). An investigation into teens' attitudes towards fast-food brands in general: A cross-cultural analysis. *Journal of Foodservice Business Research, 9*(4), 25–39. doi:10.1300/J369v09n04_03

Parks, L. (2004). Kinetic screens: Epistemologies of movement at the interface. In Couldry, N., & McCarthy, A. (Eds.), *Mediaspace: Place, scale and culture in a media age*. London: Routledge.

Parks, L. (2004). Flexible microcasting: Gender, generation, and television-Internet convergence. In Spigel, L., & Olson, J. (Eds.), *Television after TV* (pp. 133–156). Durham, NC: Duke University Press.

Parsons, P. R. (2003). Horizontal integration in the cable television industry: History and context. *Journal of Media Economics, 16*(1), 23–40. doi:10.1207/S15327736ME1601_3

Patel, R. (2006). Working the Night Shift: Gender and the Global Economy. *ACME: An International E-Journal for Critical Geographies, 5*(1), 9–27.

Peberdy, S. (2000). Mobile entrepreneurship: Informal sector cross-border trade and street trade in South Africa. *Development Southern Africa, 17*, 201–219. doi:10.1080/713661400

Pedersen, P. E., & Ling, R. (2003). *Modifying adoption research for mobile Internet service adoption: Cross-disciplinary interactions.* Paper presented at the 36th Hawaii International Conference on System Sciences (HICSS 2003), Hawaii, January.

Penfold, C. (2007). Trade union responses to white-collar off-shoring. *Economic and Labour Relations Review*, *18*(1), 53–77.

Penfold, C. (2008). Off-shoring and decent work: Worlds apart? *International Journal of Comparative Labour Law and Industrial Relations*, *24*(4), 573–594.

Penfold, C. (July, 2009). Off-shored services workers: labor law and practice in India. *Economic and Labor Relations Review*.

Perez, C. (1983). Structural Change and the Assimilation of New Technologies in the Economic and Social System. *Futures*, *15*(4), 357–375. doi:10.1016/0016-3287(83)90050-2

Peters, J. D. (2001). Witnessing. *Media Culture & Society*, *23*, 707–723. doi:10.1177/016344301023006002

Peters, J. D. (1999). *Speaking Into the Air: A History of the Idea of Communication*. Chicago: University of Chicago Press.

Petrovic, P. (2005). Between visibility and invisibility: Baudrillard, Jean-Luc Marion, and Lance Olsen's gild imagined by chance. *Extrapolation*, *46*(2), 249–258.

Phillipson, R. (1992). *Linguistic imperialism*. Oxford: Oxford University Press.

Phone problems: Cellular phone etiquette becomes a major issue. (2000, July 5). Retrieved October 15, 2006, from http://web-japan.org/trends 00/honbun/tj000704.html

Picard, R. (2005). Unique Characteristics and Business Dynamics of Media Products. *Journal of Media Business Studies*, *2*(2), 61–69.

Picard, R. (2002). Delusions of Grandeur: The Real Problems of Concentration in Media. In Demers, D. (Ed.), *Global Media News Reader* (pp. 33–48). Spokane, WA: Marquette Books.

Picard, R. (1998). Media Concentration, Economics, and Regulation. In Graber, D., McQuail, D., & Norris, P. (Eds.), *The News of Politics* (pp. 193–217). Washington, D.C: Congressional Quarterly Press.

Pieterse, J. (2004). *Globalization and culture: Global mélange*. Lanham, MD: Rowman and Littlefield Publishers.

Pippin, T. (1999). *Apocalyptic bodies: The Biblical End of the World in Text and Image*. New York: Routledge.

Pisani, F. (2006, Winter). Journalism and Web 2.0: Tomorrow's potential readers are using the web in ways we can hardly imagine, and we want to remain significant for them. We need to understand how. *Nieman Reports*, *60*(4), 42–45.

Plauché, M., & Nallasamy, U. (2007). Speech interfaces for equitable access to information technology. *Information Technologies and International Development*, *4*(1), 69–86. doi:10.1162/itid.2007.4.1.69

Pollack, A. (1984, June 7). The keyboard stymies Japan. *New York Times*, D17.

Pool, I. de Sola. (1983). *Technologies of Freedom: On Free Speech in an Electronic Age*. Cambridge, MA: Belknap Press/Harvard University Press.

Pope, C. (2009). Has Music Had It's Day? *The Irish Times*, Monday April 27th, 15.

Poremba, C. (2003). *Player as author: Digital games and agency*. Unpublished master's thesis. Simon Fraser University, Vancouver, Canada.

Postigo, H. (2003). From pong to planet quake: Post-industrial transitions from leisure to work. *Information Communication and Society*, *6*(4), 593–607. doi:10.1080/1369118032000163277

Postigo, H. (2009). Of mods and modders: Chasing down the value of fan-based digital game modifications. *Games and Culture*, *2*(4), 300–313. doi:10.1177/1555412007307955

Pradhan, J. P., & Abraham, V. (2005). Social and cultural impact of outsourcing: Emerging issues from Indian call centres. *Harvard Asia Quarterly*, *9*(3), 22–30.

Prahalad, C. K. (2004). *The Fortune at the Bottom of the Pyramid: Eradicating Poverty through Profits*. Upper Saddle River, NJ: Wharton School Publishing.

Pratt, A. C. (2005). Cultural Industries and Public Policy. *International Journal of Cultural Policy*, *11*(1), 31–44. doi:10.1080/10286630500067739

Preston, P. (2001). *Reshaping Communication: Technology, Info & Social Change*. Thousand Oaks, CA: Sage.

Preston, P. (2008). *Making the News*. London: Routledge.

Preston, P., & Flynn, R. (2000). Rethinking Universal Service: Citizenship Consumption Norms and the Telephone. *The Information Society*, *16*(2), 91–98. doi:10.1080/01972240050032843

Preston, P., Kerr, A., & Cawley, A. (2009). Digital media sector innovation in the knowledge economy: rethinking knowledge inputs and policies. *Information Communication and Society*, *12*(4), 994–1014. doi:10.1080/13691180802578150

Preston, P., & Grisold, A. (1995). Unpacking the Concept of Competition in Media Policy Making: The Case of Ireland & Austria. In Corcoran, F., & Preston, P. (Eds.), *Democracy and Communication in the New Europe: Change & Continuity in East and West* (pp. 67–96). Creskill, NJ: Hampton Press.

PricewatehouseCoopers. (2007). *Global entertainment and media outlook: 2007-2011*. New York: PricewaterhouseCoopers LLP.

Purcell, K. (1997). Towards a communication dialectic: Embedded technology and the enhancement of place. *Sociological Inquiry*, *67*(1), 101–112. doi:10.1111/j.1475-682X.1997.tb00432.x

Qiu, J. L., & Chan, J. M. (2009). Toward the study of new media events. *The Chinese Journal of Communication and Society*, *9*, 19–38.

Qiu, J. (2008). Wireless Working-Class ICTs and the Chinese Informational City. *Journal of Urban Technology*, *3*, 57–77. doi:10.1080/10630730802677954

Radbourne, J. (1997). Creative nation – A policy for leaders or followers? An evaluation of Australia's 1994 cultural policy statement. *Journal of Arts Management. Law and Society*, *26*(4), 271–284.

Raessens, J. (2005). Computer games as participatory media culture. In Raessens, J., & Goldstein, J. (Eds.), *Handbook of computer game studies* (pp. 373–388). Cambridge, MA: The MIT Press.

Ramesh, B. P. (2004). *Labour in Business Process Outsourcing: A Case Study of Call-Centre Agents*. Noida, India: V. V. Giri National Labour Institute.

Randall, V. (2008). Analytical approaches to the study of politics in the developing world. In Peter, B., & Vicky, R. (Eds.), *Politics in the developing world*. Oxford, UK: Oxford University Press.

Rangers, S. M. (1999, March). Madagascar: A waning tradition. *UNESCO Courier, 52,* 35. Retrieved December 23, 2008, from http://www.unesco.org/courier /1999_03/pdf/courier.pdf

Rao, S., & Wasserman, H. (2007). Global media ethics revisited: A postcolonial critique. *Global Media and Communication*, *3*(1), 29–50. doi:10.1177/1742766507074358

Raymond, M. A., Shao, A. T., & Taylor, C. (1999). Shifting advertising appeals in Taiwan. *Journal of Advertising Research*, *39*(6), 61–70.

Reardon, M. (2005, October 21). The lowdown on mobile TV. *ZDNet News*. Retrieved

Remo, C., & Sheffield, B. (2008, July 18). Redefining Game Narrative: Ubisoft's Patrick Redding On Far Cry 2. *Gamasutra*. Retrieved December 4, 2009, from http://www.gamasutra.com/view/ feature/3727/redefining_game_narrative_.php

Ren, M. (2003). Qingshaonian wenhua yihua xianxiang de shehuixue fenxi [Sociological analysis of the cultural alienation of youth and juveniles]. *Contemporary Young Research*, *1*, 21–24.

Ren, X. (2006). Chengshi qingnian xiaofei wenhua jiedu [Interpretation of urban youth consuming culture]. *Journal of Beijing Youth University for Political Science*, *1*, 23–26.

Reuters. (2009). Makers of online games in China look overseas. *New York Times*, October 29, 2009. Retrieved March 26, 2010 from http://www.nytimes.com/2009/10/27/business/global/27iht-game.html?_r=1

Rice, C. (2005). Experience and criticality: Returning to Federation Square. *The Journal of Architecture*, *10*(3), 323–333. doi:10.1080/13602360500162253

Rice, R. E., & Katz, J. E. (2003). Comparing internet and mobile phone usage: Digital divides of usage, adoption and dropouts. *Telecommunications Policy*, *27*, 597–623. doi:10.1016/S0308-5961(03)00068-5

Rifkin, J. (2005). When markets give way to networks… everything is a service. In Hartley, J. (Ed.), *Creative industries* (pp. 361–373). Oxford: Basil Blackwell.

Ritzer, G. (2000). *The McDonaldization of Society*. Thousand Oaks, CA: Pine Forge Press.

Robertson, R. (1995). Glocalization: Time-space and homogeneity-heterogeneity. In Featherstone, M., Lash, S., & Robertson, R. (Eds.), *Global modernities* (pp. 25–44). London: Sage.

Robins, J. A. (1993). Organization as strategy: Restructuring production in the film industry. *Strategic Management Journal, 14* (Special Issue: Corporate Restructuring), 103-118.

Robison, R., & Goodman, D. S. G. (Eds.). (1996). *The new rich in Asia: Mobile phones, McDonalds and middle-class revolution*. London: Routledge.

Rogers, R. (2006). From cultural exchange to transculturation: A review and reconceptualization of cultural appropriation. *Communication Theory*, *16*(4), 474–503. doi:10.1111/j.1468-2885.2006.00277.x

Rogers, E. M. (2003). *Diffusion of innovations* (5th ed.). New York: Free Press.

Rogers, E. M., & Kincaid, D. L. (1981). *Communication networks: Toward a new paradigm for research*. New York: Free Press.

Rogers, E. M. (1995). *Diffusion of innovations* (5th ed.). New York: Free Press.

Rosaldo, R. (1999). Cultural citizenship, inequality, and multiculturalism. In Torres, R. D., Miron, L. F., & Inda, J. X. (Eds.), *Race, identity and citizenship: A reader* (pp. 79–100). Cambridge, MA: Blackwell.

Rose, F. (2009). Why I published Those Cartoon. *Design*, *99*, 34–35.

Roseman, D. (2005). The WTO and telecommunications services in China: Three years on. *Info*, *7*(2), 25–48. doi:10.1108/14636690510587207

Rosenthal, R. (Ed.). (1968). *McLuhan: Pro & con*. New York: Funk & Wagnalls.

Ross, S. M. (2008). *Beyond the box: Television and the Internet*. Malden, MA: Blackwell.

Ross, C., Orra, E., Sisica, M., Arseneaulta, J., Simmeringa, M., & Orra, R. (2009)... *Computers in Human Behavior*, *25*(2), 578–586. doi:10.1016/j.chb.2008.12.024

Roth, S. D. (1982). Ray casting for modeling solids. *Computer Graphics and Image Processing*, *18*(2), 109–144. doi:10.1016/0146-664X(82)90169-1

Rubin, A. M. (2002). The uses-and-gratifications perspective of media effects. In Bryant, J., & Zillmann, D. (Eds.), *Media effects: Advances in theory and research* (2nd ed., pp. 525–548). Mahwah, NJ: Lawrence Erlbaum Associates.

Ruggiero, T. E. (2000). Uses and gratifications theory in the 21st century. *Mass Communication & Society*, *3*, 3–37. doi:10.1207/S15327825MCS0301_02

Ruggiero, T. (2000). Uses and gratifications theory in the 21st century. *Mass Communication & Society*, *3*(1), 3–37. doi:10.1207/S15327825MCS0301_02

Rundle, G. (2002, August 27). Building the past. *The Bulletin with Newsweek, 120*(6339), 13.

Russell, S. (30-01-2006). Tone-deaf to customer complaints, Dell opens yet another call center in India. *ZDNet*. Retrieved 2008-06-05.

Rutter, J., & Bryce, J. (2005). *Consumer Engagement with Counterfeit Goods*. Report for the Organised Crime Task Force, Northern Ireland Office. Retrieved from http://www.allianceagainstiptheft.co.uk/downloads/pdf/Fake-Nation.pdf

Ryan, M.L. (1994). Immersion vs. Interactivity: Virtual reality and literary theory. *Postmodern Culture, 5*(1). Retrieved March 2, 2006 from Project MUSE database.

Samsung Group. (2007). Retrieved from http://www.mind-advertising.com/kr/samsung_kr.htm

Samsung. (n.d.). *Samsung Introduction.* Retrieved from http://www.samsung.co.kr/about/ir_annual.jsp

Samuel, J., Shah, N., & Hadingham, W. (2005). Mobile communications in South Africa, Tanzania and Egypt: Results from Community and Business Surveys. *Africa: The Impact of Mobile Phones. The Vodafone Policy Paper Series, 2*(March), 44–52.

Sandoval, G. (2009). Sony joins YouTube and Universal on Vevo video site. *cnet* [online]. Retrieved from http://news.cnet.com/8301-1023_3-10257359-93.html

Sartre, J. (1956). *Being and Nothingness. (Trans. H. Barnes)*. New York: Washington Square Press.

Sasikumar, S. K., & Varma, U. K. (2004). *Information and Communication Technology and Decent Work: Study of India's Experience*. Noida, India: V. V. Giri National Labour Institute.

Scannell, P. (1996a). For a phenomenology of radio and television. *The Journal of Communication, 45*(3), 4–19. doi:10.1111/j.1460-2466.1995.tb00741.x

Scannell, P. (1996b). *Radio, television & modern life; A phenomenological approach*. Oxford, UK: Blackwell.

Schaufeli, W. B., & Bakker, A. B. (2004). Job demands, job resources, and their relationship with burnout and engagement: a multi-sample study. *Journal of Organizational Behavior, 25*(3), 293–315. doi:10.1002/job.248

Schieck, A. F. g. (2006, February). Towards an integrated architectural media space. *First Monday, 4*. Retrieved October 20, 2009, from http://www.uic.edu/htbin/cgiwrap/bin/o js/index.php/fm/rt/printerFriendly/1550/1465

Schiller, D. (2007). *How to think about Information.* Urbana, IL: University of Illinois Press.

Schiller, H. (1989). *Culture Inc.* New York: Oxford University Press.

Schiller, H. (1984). *Information and the Crisis Economy.* Norwood, NJ: Ablex Publishing Corporation.

Schiller, H. (1998). Striving for communication dominance: A half-century review. In Thussu, D. K. (Ed.), *Electronic empires: Global media and local resistance.* London: Arnord.

Schroder, K., Drotner, K., Kline, S., & Murray, C. (2003). *Researching Audiences*. London, UK: Hodder Arnold.

Schwartz-Cowan, R. (1985). How the refrigerator got its hum. In Wajcman, J., & MacKenzie, D. (Eds.), *The Social Shaping of Technology*. Buckingham, UK: Open University Press.

Scott, A. J. (2004). Hollywood and the world: The geography of motion-picture distribution and marketing. *Review of International Political Economy, 11*(1), 33–61. doi:10.1080/0969229042000179758

Sekula, A. (1984). *Photography against the grain: Essays and photo works 1973-1983.* Halifax: The Press of Nova Scotia College of Art and Design.

Sellers, M. (2006). Designing the experience of interactive play. In Vorderer, P., & Bryant, J. (Eds.), *Playing video games: Motives, responses, and consequences* (pp. 9–22). Mahwah, NJ: Lawrence Erlbaum Associates, Publishers.

Sen, F., & Shiel, M. (2006). From business process outsourcing (BPO) to knowledge process outsourcing (KPO): some issues. *Human Systems Management, 25*, 145–155.

Seshagiri, S., Sagar, A., & Joshi, D. (2007). Connecting the 'Bottom of the Pyramid' – An Exploratory Case Study of India's Rural Communication Environment. *WWW 2007, May 8-12, 2007*, Alberta, Canada.

Shanzhaiji. (2009). *Daihao MI Obama* [Coded MI Obama]. Retrieved January 16, 2009, from http://www.shanzhaiji.cn/info /20090116/7035.html

Sheffield, B. (2007, March 5). All For Games: An Interview with Warren Spector. *Gamasutra*. Retrieved December 3, 2009, from http://www.gamasutra.com/featur es/20070305/sheffield_01.shtml

Sheth, J. (2009). The call center couple: India's new middle class. *Journal of Indian Business Research*, *1*(1), 10–13. doi:10.1108/17554190910963163

Shim, D. (2006). Hybridity and the rise of Korean popular culture in Asia. *Media Culture & Society*, *28*(1), 25–44. doi:10.1177/0163443706059278

Shim, D. (2005). Globalization and cinema regionalization in East Asia. *Korea Journal*, *45*(4), 233–260.

Shim, H., & Ho, T. (2009). Translation of 'America' during the early Cold War period: A Comparative study on the history of popular music in South Korea and Taiwan. *Inter-Asia Cultural Studies*, *10*(1), 83–102. doi:10.1080/14649370802605274

Shim, D. B. (2002). South Korean Media Industry in the 1990s and the Economic Crisis. *Prometheus*, *20*(4), 337–350. doi:10.1080/08109020210000233336

Shim, D. (2008). The growth of Korean cultural industries and the Korean wave. In Chua, B. H., & Iwabuchi, K. I. (Eds.), *East Asian pop culture: Analyzing the Korean wave*. London: Eurospan. [distributor]

Shim, D. (2006). Hybridity and the rise of Korean popular culture in Asia. *Journal of media culture and society, 28*(1), 25-44.

Shin, D. (2006). Prospect of Mobile TV: Another Bubble or Killer Application. *Telematics and Informatics*, *23*, 253–270. doi:10.1016/j.tele.2005.08.001

Shirky, C. (2008). *Here comes everybody*. Keynote presented at the Aspen Ideas Festival, June 30-July 8th, Aspen Coldrado. Retrieved 20 December 2008 from http://fora.tv/2008/07/06/Clay_Shirky_o n_Social_Networks_like_Facebook_and_MySpace#chapter_01

Silverstone, R. (1994). *Television and everyday life*. London: Routledge.

Silverstone, R. (1999). What's New about New Media?: Introduction. *New Media & Society*, *1*(10), 10–12. doi:10.1177/1461444899001001002

Singh, J. (2000). Performance productivity and quality of frontline employees in service organizations. *Journal of Marketing*, *64*(2), 15–34. doi:10.1509/jmkg.64.2.15.17998

Singh, J., Goolsby, J. R., & Rhoads, G. K. (1994). Behavioral and psychological consequences of boundary spanning burnout for customer service representatives. *JMR, Journal of Marketing Research*, *31*(4), 558–569. doi:10.2307/3151883

Singhal, A., & Dearing, J. W. (2006). *Communication of innovations: A journey with Ev Rogers*. Thousand Oaks, CA: Sage Publications.

Siong, Z. M. B., Mellor, D., Moore, K. A., & Firth, L. (2006). Predicting intention to quit in the call center industry: does the retail model fit? *Journal of Managerial Psychology*, *21*(3), 231–243. doi:10.1108/02683940610659579

Sjogaard, G., Lundberg, U., & Kadefors, R. (2000). The role of muscle activity and mental load in the development of pain and degenerative processes at the muscle cell level during computer work. *European Journal of Applied Physiology*, *83*, 99–105. doi:10.1007/s004210000285

Slaatta, T. (2006, February). Urban screens: Towards the convergence of architecture and audiovisual media. *First Monday, 4*. Retrieved October 20, 2009, from http://www.uic.edu/htbin/cgiwrap/bin/ ojs/index.php/fm/rt/printerFriendly/1549/1464

Slack, J. D., & Wise, J. M. (2006). Cultural studies and communication technologies. In Lievrouw, L., & Livingstone, S. (Eds.), *The handbook of new media*. Thousand Oaks, CA: Sage.

Sloterdijk, P. (1987). *Critique of cynical reason*. Minneapolis: University of Minnesota Press.

Snider, M. (2009, August 18). Back to the future for id Software, 'Wolfenstein'. *USA Today*. Retrieved December 3, 2009, from http://www.usatoday.com/life/lifestyle/2009-08-17-wolfenstein_N.htm

Snijders, T. (2001). The Statistical Evaluation of Social Network Dynamics. *Sociological Methodology*, *31*, 361–395. doi:10.1111/0081-1750.00099

Sobchack, V. C. (2004). *Carnal thoughts: Embodiment and moving image culture*. Berkeley, CA: University of California Press.

Sones, B. E. (1999, June 30). *Interview with John Carmack*. Retrieved December 3, 2009, from http://web.archive.org/web/20030515172424/ http://www.cdmag.com/articles/021/004/jc_interview.html

Sonntag, S. K. (2008). Appropriating Identity or Cultivating Capital? Global English in Offshoring Service Industries. *Anthropology of Work Review*, *26*(1), 13–20. doi:10.1525/awr.2005.26.1.13

Sonntag, S. K. (2009). Linguistic globalization and the call center industry: Imperialism, hegemony or cosmopolitanism? *Language Policy*, *8*, 5–25. doi:10.1007/s10993-008-9112-9

Sontag, S. (1977). *On photography*. New York: Farrar, Struas & Giroux.

Sontag, S. (2003). *Regarding the pain of others*. London: Penguin.

Sony Corporation. (2007). *About Sony*. Retrieved from http://www.sony.net/SonyInfo/CorporateInfo/

Sony. (2007). *Annual Report 2007*. Tokyo: Sony.

Sony. (2009). *Annual Report 2009*. Tokyo: Sony.

Sotamaa, O. (2007). Let Me Take You To the Movies. *Convergence*, *13*(4), 383–401.

Soukup, C. (2006). Hitching a ride on a star: Celebrity, fandom, and identification on the World Wide Web. *The Southern Communication Journal*, *71*(4), 319–337.

Southern Weekly. (2008). To defend the spirit of *shanzhai* mobile phones. Retrieved August 4, 2008, from http://news.qq.com/a/20080804/000440.htm

Spivak, G. (2009). The new subaltern: A silent interview. In During, S. (Ed.), *The cultural reader* (3rd ed.). New York: Routledge.

Square, F. (2009). *Fed square annual report – June 2009*. Melbourne, Australia: Fed Square Pty Ltd.

Srinivasan, J. (2007). The role of trustworthiness in information service usage: The case of Parry information kiosks in Tamil Nadu, India. In *Proceedings of the International Conference on Information and Communication Technologies for Development (ICTD)*, (pp. 345-352). New York: ACM Press.

Srivastava, L. (2005). Bobile phone and the evolution of social behaviour. *Behaviour & Information Technology*, *24*, 111–129. doi:10.1080/01449290512331321910

Stafford, T. F., & Stafford, M. R. (2001). Identifying motivations for the use of commercial web sites. *Information Resources Management Journal*, *14*, 22–30.

Starner, T. (2004). Keyboards redux: Fast mobile text entry. *IEEE Pervasive Computing / IEEE Computer Society [and] IEEE Communications Society*, *3*(3), 97–101. doi:10.1109/MPRV.2004.1321035

Stein, L. (2008). Speech without rights: The status of public space on the Internet. *Communication Review*, *11*(1), 1–23. doi:10.1080/10714420801888385

Stein, S. (2008). Six Reasons Why First-Person Shooters. *Esquire*. Retrieved December 3, 2009, from http://www.esquire.com/the-side/opinion/six-reasons-why-shooters-suck

Steinberg, B. (2008, August 11). Big 3 TV nets follow consumers out of the living room; ABC, NBC and CBS set up screens at grocery stores, gas stations, malls. *Advertising Age*, *79*(41), 4.

Steuer, J. (1992). Defining virtual reality: dimensions determining telepresence. *The Journal of Communication*, *42*(4), 73–93. doi:10.1111/j.1460-2466.1992.tb00812.x

Steuer, J. (1992). Defining virtual reality: Dimensions determining telepresence. *The Journal of Communication*, *42*(4), 73–93. doi:10.1111/j.1460-2466.1992.tb00812.x

Stevenson, D. (2000). *Arts and organisation: making Australian cultural policy*. St Lucia, Australia: University of Queensland Press.

Stevenson, N. (2003). *Cultural citizenship: cosmopolitan questions*. Maidenhead, UK: Open University Press.

Stillman, S. (2008, Jan 22). The whole world was watching. *Truthdig.com*. Retrieved from http://www.truthout.org/docs_ 2006/012308I.shtml

Storey, J. (2003). *Inventing popular culture: From folklore to globalization*. Malden, MA: Blackwell Publishing.

Storper, M. (1989). The transition to flexible specialisation in the US film industry: External economies, the division of labour, and the crossing of industrial divides. *Cambridge Journal of Economics*, *13*(2), 273–305.

Storsul, T., & Stuedahl, D. (Eds.). (2007). *Ambivalence Towards Convergence: Digitalization and Media Change*. Nordicom.

Straw, W. (2000). Proliferating screens. *Screen*, *41*(1), 115–119. doi:10.1093/screen/41.1.115

Striphas, T., & McLeod, K. (2006). Strategic improprieties: Cultural studies, the everyday, and the politics of intellectual properties. *Cultural Studies*, *20*(2/3), 119–144. doi:10.1080/09502380500495635

Strover, S. (2005). United States: Cable television. March 1, 2007, from: http://www.museum.tv/home.php

Sturken, M., & Cartwright, L. (2001). *Practices of looking: An introduction to visual culture*. Oxford: Oxford University Press.

Suh, Y. K. (2005). *Satellite-DMB: Successful strategy for industrialization of new technology*. Paper presented at Korea IT Leaders Forum, Seoul, Korea.

Tabuchi, H. (2007, November 5). Japanese bypassing PCs in favor of an array of electronic gadgets. *Seattle Times*. Retrieved March 27, 2010, from http://seattletimes.nwsource.com/html /businesstechnology/2003994240_bt-japanpcs05.html

Tabuchi, H. (2009a, July 19). Why Japan's cellphones haven't gone global. *New York Times*, B1.

Tabuchi, H. (2009b, November 29). In Japan, an odd perch for Google: Looking up at the leader. *New York Times*, B1.

Tai, Z. (2006). *The Internet in China: Cyberspace and Civil Society*. New York: Routledge.

Takeuchi, K. (2005). The challenge of JanJan in participatory news. *OhmyNews*. Retrieved April 15, 2009, from http://english.ohmynews.com/articleview/arti cle_view.asp?menu=c10400&no=234393&rel_no=1

Tamber, C. (1995). The financial interest and syndication rules - take two. *CommLaw Conspectus*, *3*, 107–120.

Tar, K. (2001). *Toleration, diversity, and global justice*. University Park, PA: Pennsylvania State.

Taylor, J. (2000). Problems of photojournalism: Realism, the nature of news and the humanitarian narrative. *Journalism Studies*, *1*(1), 129–143. doi:10.1080/146167000361212

Taylor, P., & Bain, P. (2000). Entrapped by the "electronic panopticon"? Worker resistance in the call centre. *New Technology, Work and Employment*, *15*(1), 2–18. doi:10.1111/1468-005X.00061

Taylor, P., & Bain, P. (2005). India calling to the far away towns: the call centre labour process and globalization. *Work, Employment and Society*, *19*(2), 261–282. doi:10.1177/0950017005053170

Taylor, T. L. (2005). *Play between worlds: Exploring online game culture*. Cambridge, MA: MIT Press.

Taylor, P. & Bain, P. (1999). An assembly line in the head: work and employee relations in the call centre. *Industrial Relations Journal*.

Taylor, P., Scholarios, D., Noronha, E., & d'Cruz, P. (2007). *Union formation in Indian call centres/BPO--the attitudes and experiences of UNITES members*, University of Strathclyde Business School and Indian Institute of Management, Ahmedabad. Retrieved from http://www.union-network.org

Techno-Z, F. H. (1997). *The Content Challenge: Electronic Publishing and the New Content Industries - Research report commissioned and published by EC*. DG XIII/E.

Teng, R. (2005). Digital multimedia broadcasting in Korea. *In-Stat*. Retrieved July 21, 2005, from http://www.instat.com/promos/05/IN0502469WHT.pdf

Tester, K. (1994). *Media, culture and morality*. London: Routledge.

Thayer, A., & Kolko, B. E. (2004). Localization of digital games: The process of blending for the global game market. *Technical Communication, 51*(4), 477–488.

The Apple Daily (January 11, 2006). Editorial: Yao mad (in Chinese). *Apple Daily*. Retrieved April 17, 2010, from http://tw.nextmedia.com/applenews/art icle/art_id/2329059/IssueID/20060111

The Call Center Answer Team reaches out to the industry for to crack a tough nut. (2003, July 30). Q&A: How Many Calls Should I Monitor. Retrieved from www.callcentermagazine.com. Retrieved 2008-06-05.

The Economist. (2000, 23 September). Solving the paradox: Productivity growth and information technology. *Economist Newspaper, US Edition,* pp. 15.

The Epoch Times (July 31, 2005). Seven cable channels failed in the review (in Chinese). *The Epoch Times*. Retrieved April 17, 2010, from http://www.epochtimes.com/b5/5/8/1/n1003832.htm

The Liberty Times. (January 13, 2006). Editorial: Damn on the pro-China group (in Chinese). *Liberty Times*. Retrieved April 17, 2010, from http://www.libertytimes.com.tw/ 2006/new/jan/13/today-f1.htm

The Localization Industry Standards Association. (2006). *Frequently asked questions about LISA and the localization industry.* Switzerland: The Localization Industry Standards Association. Retrieved, July 7, 2006 http://www.lisa.org/info/faqs.html

The Pew Research Center for the People and the Press. (2007). *The Pew Global Attitudes Project*. Retrieved October 4, 2007, from http://www.pewglobal.org

The United Daily (August 22, 2005). Taike, Hanru (in Chinese). *United Daily,* p. A2.

The United Daily (January 2, 2002). Editorial: Liberalization, best for state TV (in Chinese). *United Daily,* p. A2.

The United Daily (March 20, 2005). Editorial: Korea wins over Taiwan? (in Chinese). *United Daily,* p. A2.

The Weakling Kicks Back. (1999, July 3)... *Economist, 352*, 56.

Theorell, T., Hasselhorn, H. M., Vingård, E., & Andersson, B. (2000). Interleukin 6 and cortisol in acute musculoskeletal disorders: results from a case-referent study in Sweden. *Stress Medicine, 16*, 27–35. doi:10.1002/(SICI)1099-1700(200001)16:1<27::AID-SMI829>3.0.CO;2-#

Thomas, A. (2007). *Youth online: Identity and literacy in the digital age*. New York: Peter Lang.

Thomas, A. (2004). *The Dynamics of Games* (4th ed.). London: Informa Media.

Thompson, J. B. (1995). *The media and modernity: A social theory of the media*. Cambridge, UK: Polity Press.

Thussu, D. (2006). *International Communication: change and continuity* (2nd ed.). London: Arnold.

Thussu, D. (2006). Mapping global media flow and contra-flow. In Thussu, D. (Ed.), *Media on the move: Global flow and contra-flow* (pp. 11–32). London: Routledge.

Toffler, A. (1980). *The third wave*. New York: Morrow.

Tomlinson, J. (1991). *Cultural Imperialism: A Critical Introduction*. London: Printer Publishers.

Tomlinson, J. (1999). *Globalization and culture*. Chicago: The University of Chicago Press.

Tomlinson, J. (1997). Cultural globalization and cultural imperialism. In Mohammadi, A. (Ed.), *International communication and globalization* (pp. 170–190). London: Sage.

Tomlinson, J. (2002). Interests and identities in cosmopolitan politics. In Vertovec, S., & Cohen, R. (Eds.), *Conceiving cosmopolitanism: Theory, context, and practice* (pp. 240–253). Oxford, UK: Oxford University Press.

Townsend, A. (2007). Seoul: birth of a broadband metropolis. *Environment and Planning. B, Planning & Design*, *34*, 396–413. doi:10.1068/b32036t

Trepte, S., Ranné, N., & Becker, N. (2003). Personal Digital Assistants—Patterns of user Gratifications. *Communications*, *28*, 457–473. doi:10.1515/comm.2003.029

Tse, D. (1996). Understanding Chinese people as consumers: Past findings and future propositions. In Bond, M. (Ed.), *The Handbook of Chinese Psychology* (pp. 352–363). Hong Kong: Oxford University Press.

Tsui, L. (2001). Big mama is watching you: Internet control and the Chinese government. *The economist*, (January): 10–14.

Tsui, L. (2003). The Panopticon as the Antithesis of a Space of Freedom. *China Information*, *17*(2), 65–82. doi:10.1177/0920203X0301700203

Turner, B. S. (2002). Outline of a general theory of cultural citizenship. In Stevenson, N. (Ed.), *Culture and citizenship* (pp. 11–32). London: Sage.

Twomey, C. (2007). Emaciation or emasculation: Photographic images, white masculinity and captivity by the Japanese in World War Two. *Journal of Men's Studies*, *15*(3), 295–310. doi:10.3149/jms.1503.295

UK Department of Trade and Industry. (1998). *Regulating Communications: Approaching Convergence in the Information Age* [Green Paper on "Convergence"]. London: Department of Trade and Industry.

UNAIDS. (2009). *09 AIDS epidemic update*. Retrieved November 15, 2010, from http://data.unaids.org/pub/Report/20 09/JC1700_Epi_Update_2009_en.pdf

UNESCO. (2009). *UNESCO Institute for Statistics*. Retrieved June 15, 2009, from http://www.uls.unesco.org

United Nations. (2008). *Creative Economy Report 2008*. Retrieved April, 1, 2008 from: http://www.unctad.org/creative-economy

Ureta, S. (2008). Mobilising poverty?: Mobile phone use and everyday spatial mobility among low-income families in Santiago, Chile. *The Information Society*, *24*, 83–92. doi:10.1080/01972240701883930

USA, Central Intelligent Agency (CIA). (2009). *The world factbook*. Botswana: Retrieved.

Ushahidi.com. (2009). *Crowdsourcing Crisis Information (FOSS)*. Retrieved June 15, 2009, from http://www.ushahidi.com/

van Der Wurff, R. (2008). The impact of the internet on media content. In Küng, L., Picard, R. G., & Towse, R. (Eds.), *The internet and the mass media* (pp. 65–85). Oxford, UK: Sage Publications.

Van Dijck, J. (2009). Users like you? Theorizing agency in user-generated content. *Media Culture & Society*, *31*(1), 41–60. doi:10.1177/0163443708098245

van Elteren, M. (2006). Rethinking Americanization abroad: Toward a critical alternative to prevailing paradigms. *Journal of American Culture*, *29*(3), 345–367. doi:10.1111/j.1542-734X.2006.00376.x

Victoria, A. (2003). *Creative Capacity +*. Retrieved June 20, 2008, from http://www.arts.vic.gov.au/content/Public/ About_Us/Who_We_Are/Policy_and_Planning/Creative_Capacity_.aspx

Virilio, P. (2000). *The information bomb*. London: Verso.

Wajcman, J., & MacKenzie, D. (Eds.). (1985). *The Social Shaping of Technology*. Buckingham, UK: Open University Press.

Waldfogel, J. (2009). Lost on the web: Does web distribution stimulate or depress television viewing? *Information Economics and Policy*, *21*(2), 158–168. doi:10.1016/j.infoecopol.2008.11.002

Wallace, C. M., Eagleson, G., & Waldersee, R. (2000). The sacrificial HR strategy in call centres. *International Journal of Service Industry Management*, *11*(2), 174–184. doi:10.1108/09564230010323741

Wallerstein, I. (1974). The Rise and Future Demise of the World Capitalist System: Concepts for Comparative Analysis. *Comparative Studies in Society and History*, *16*(4), 387–415. doi:10.1017/S0010417500007520

Wang, S. (1998). Where are the audience of domestic dramas? [in Chinese]. *Advertising Magazine*, *85*, 127–130.

Wang, Y. (2008). Shanzhaiji yu guochanshouji de fazhan [*Shanzhai* mobile phones and the development of mobile phones made in China] [Humanities and Social Sciences]. *Journal of Hubei University of Economics*, *12*, 71–72.

Wang, B. (January 12, 2006). Yao: policy distorted (in Chinese). *Liberty Times.* Retrieved April 17, 2010, from http://www.libertytimes.com.tw/ 2006/new/jan/12/today-p12.htm

Wang, C. (August 20, 2005). Not boosting culture industry, Taiwan lags behind (in Chinese). *United Daily,* p. A15.

Wang, S. (June 24, 2002). Behind the Korean wave (in Chinese). *United Daily,* p. A15.

Wareham, J., Levy, A., & Shi, W. (2004). Wireless diffusion and mobile computing: Implications for the digital divide. *Telecommunications Policy*, *28*(5-6), 439–457. doi:10.1016/j.telpol.2003.11.005

Wasson, H. (2007). The networked screen: moving images, materiality, and the aesthetics of size. In Marchessault, J., & Lord, S. (Eds.), *Fluid screens, expanded cinema* (pp. 74–95). Toronto, Canada: University of Toronto Press.

Waters, W. (2001). *Globalization,* (2nd ed., p. 6). London: Routledge.

Webster, J. G. (2005). Beneath the veneer of fragmentation: Television audience polarization in a multichannel world. *The Journal of Communication*, *55*(2), 366–382. doi:10.1111/j.1460-2466.2005.tb02677.x

Wee, C. H., Tan, S. J., & Cheek, K. H. (1995). Non-price determinants of intention to purchase counterfeit goods: An exploratory study. *International Marketing Review*, *12*(6), 19–46. doi:10.1108/02651339510102949

Weil, S. (2002). *Making museums matter*. Washington, DC: Smithsonian Institution Press.

Westgaard, R. H. (1999). Effects of physical and mental stressors on muscle pain. *Scandinavian Journal of Work, Environment & Health*, *25*(4), 19–24.

White paper 2005 & 2006. Retrieved December 10, 2007, from http://www.dynamicitkorea.com/ media/index.jsp?nodeId=1914

Who's on the Line? Women in Call Centres Project. (n.d.). Atlantic Centre of Excellence for Women's Health. *Health Canada*. Available at www.acewh.dal.ca, Retrieved 2008-06-05

Wilk, S. L., & Moynihan, L. M. (2005, Sep). Sep Display rule "regulators": the relationship between supervisors and worker emotional exhaustion. *The Journal of Applied Psychology*, *90*(5), 917–927. doi:10.1037/0021-9010.90.5.917

Williams, R. (1975). *Television: Technology and cultural form*. New York: Shocken. doi:10.4324/9780203450277

Williams, R. (1974). *Television: Technology and cultural form*. London: Fontana. doi:10.4324/9780203426647

Williams, D. (2006). A brief social history of game play. In Vorderer, P., & Bryant, J. (Eds.), *Playing video games: Motives, responses, and consequences* (pp. 197–212). Mahwah, NJ: Lawrence Erlbaum Associates, Publishers.

Williams, M. (2006, July 5). More mobile users than wired in Japan. *InfoWorld.* Retrieved May 2, 2009, from http://www.infoworld.com/d/networking/ more-mobile-internet-users-wired-in-japan-259

Winckler, E. (1994). Cultural policy on postwar Taiwan. In Harrell, S., & Huang, C. (Eds.), *Cultural change in postwar Taiwan* (pp. 111–134). Oxford, UK: Westview Press.

Winifred, P. (2004). *Who's On the Line? Indian Call Center Agents Pose as Americans for U.S.-Outsourced Firms.* Paper presented at the annual meeting of the American Sociological Association, Hilton San Francisco, Available at www.allacademic.com retrieved dated 2010-05-07.

Winseck, D. (2002). Netscapes of power: Convergence, consolidation and power in the. Canadian mediascape. *Media Culture & Society, 24*(6), 795–819. doi:10.1177/016344370202400604

Winseck, D. (2002). Lost in Cyberspace: Convergence, consolidation and power in the Canadian mediascape. *Media, culture and society.*

Winston, B. (1998). *Media technology and society. A history: From the telegraph to the internet*. New York: Routledge.

Winston, B. (1995). How are media born and developed? In Downing, J., Mohammadi, A., & Sreberny-Mohammadi, A. (Eds.), *Questioning the media. A critical introduction* (pp. 54–74). London: Sage.

Winter, J., & Wedemeyer, D. (2009). The roots of the right to communicate and emerging participatory policy. In Dakroury, A., Eid, M., & Kamalipour, Y. R. (Eds.), *The right to communicate: Historical hopes, global debates, and future premises* (pp. 43–70). Dubuque, IA: Kendall/Hunt Publishing.

Witcomb, A. (2007). The materiality of virtual technologies: A new approach to thinking about the impact of multimedia in museums. In Cameron, F., & Kenderdine, S. (Eds.), *Theorizing digital cultural heritage: A critical discourse*. Cambridge, MA: MIT Press.

Wolf, M. J. (2000). *The Entertainment Economy: how mega-media forces are transforming our lives*. London: Penguin.

Wolfe, C. R. (2001). Learning and teaching on the World Wide Web. In Wolfe, C. R. (Ed.), *Learning and teaching on the World Wide Web* (pp. 11–22). London: Academic Press.

World Bank. (2006). *Facts and Figures*. Retrieved November 21, 2009, from World Development Indicators 2006.

World Internet Project. (2004, January 14). Retrieved January, 9, 2007, from http://www.anderson.ucla.edu/x965.xml

World's largest companies. (2004). *Fortune, 150*(2), 163-180.

World's largest companies. (2005). *Fortune, 152*(2), 119-139.

Worth, T. (2009). Best bets, global: Strait to stardom. *Billboard.com*, January 24. Retrieved from NexusNexus on October 31, 2009

Wright, T. A., & Hobfoll, S. E. (2004). Commitment, psychological well-being and job performance: an examination of conservation of resources (COR) theory and job burnout. *Journal of Business and Management, 9*(4), 389–406.

Xi, G. (2008). Woguo shouji yonghushu 6.08yi [The number of the mobile phone users has reached 608 million]. *People.com.* Retrieved February 43, 2009, from http://www.chinaz.com/News /IT/09243XT2008.html

Xin, H. (2010). Domestic online game industrial chain getting mature. *China Economic Net*, March 5. Retrieved April 3, 2010 from http://en.ce.cn/Insight/201003/05/t20100305_21064568.shtml

Yamada, (2006). Casualties: Narrative and images of the war on Iraq. *International Journal of Health Services, 36*(2), 410–415. doi:10.2190/6PXW-LQ3B-DWN6-XD97

Yang, F. C. (2008). Rap(p)ing Korean wave: National identity in question. In Chua, B. H., & Iwabuchi, K. I. (Eds.), *East Asian pop culture: Analysing the Korean wave* (pp. 191–216). Hong Kong: Hong Kong University.

Yankelovich, W. W., Roberts, P., Wessler, M., Kaplan, J., & Provino, J. (2004). Meeting Central: Making distributed meetings more effective. In *Proceedings of the Conference on Computer-supported Cooperative Work (pp.* 419-428). New York: ACM Press.

Ye, D. (2006). Shouji waiguansheji de yulehua qushi [Entertainment tendency of mobile phone contour design]. *Packing Engineering, 27*(3), 178–180.

Ye, Q. (2005). Shehuixueshiyexia dangdai chengshi qingnian de xiaofeiguan [Contemporary urban youth's consumption value under the sociology perspective]. *Journal of Hefei University of Technology, 19*(5), 144–148.

Yeah, W. (2002). *The management strategies of importing foreign programs: The example of vedioland* (in Chinese). Unpublished master dissertation, Minchuang University.

Yim, H. (2002). Cultural Identity and Cultural Policy in South Korea. *International Journal of Cultural Policy*, *8*(1), 37–48. doi:10.1080/10286630290032422

Yin, S. (2009). Shanzhai shi dianfu ["*Shanzhai*" subversion], *21Century. Business Review (Federal Reserve Bank of Philadelphia)*, *04*, 20.

Yoon, J. H. (2007). KDDI's Unified Portal Site – au one: A brief analysis on KDDI's unified portal site for mobile phones and PCs. *Research On Asia (ROA) Group White Paper*. Retrieved December 12, 2007, from http://www.infoedge.com/product_ type.asp?product=RO-0019

Yu, H. (2007). Blogging Everyday Life in Chinese Internet Culture. *Asian Studies Review*, *31*(4), 423–433. doi:10.1080/10357820701710724

Yu, J. (2005). Shouji duanxin shishang dui daxuesheng jingying wenhua de chongji [Shocks to college students elite culture from mobile phone short messages]. *Journal of Yuxi Teachers'. College*, *21*(10), 88–90.

Yu, S., & Zhang, S. (2005). Chuangzao renxinghua de shouji shijie [Creating human-based mobile phone world]. *Chinese Journal of Ergonomics*, *11*(4).

Yu, Y. (2007). Qingnian wenhua shiye kan qingnian shishang [Youth fashion from youth culture view]. *Journal of Beijing Youth University for Political Science*, *16*(2), 12–15.

Zelizer, B. (1998). *Remembering to forget: Holocaust memory through the camera's eye*. Chicago: University of Chicago Press.

Zelizer, B. (Ed.). (2001). *Visual Culture and the Holocaust*. New Jersey: Rutgers University Press.

Zhang, L., & Wu, W. (2008). Dangdai chengshi qingnian tezheng yuqi yingxiang yinsu de fenxi [Characteristics of contemporary urban youth and the analysis of influencing factors]. *Journal of Shanxi College for Youth Administrators*, *21*(4), 9–11.

Zhang, Z. (2004). "Yule" shouji fazhan de xinqushi ["Entertainment" as the new trend of mobile phone development]. *Computer Knowledge and Technology*, *6*, 26–28.

Zhang, L. L. (2006). Behind the "Great Firewall": Decoding China's internet media policies from the inside. *Convergence: The International Journal of Research into New Media Technologies*, *12*(3), 271–291. doi:10.1177/1354856506067201

Zhao, B., Yin, Q., & Han, K. (2006). Daxuesheng shouji shiyong qingkuang diaocha [Survey of the usage of mobile phones among university students]. *Market Research*, *10*, 15–19.

Zhao, X. (2007). Qianxi dangdai qingnian shenghuo fangshi de tezheng ji fazhan qushi [The analysis of characteristics and developing trend of young people's life style]. *Youth Study*, *2*, 18–20.

Zhao, Y. Z. (2004). The state, the market, and media control in China. In Thomas, P. N., & Nain, Z. (Eds.), *Who owns the media?: Global trends and local resistances* (pp. 179–212). London, New York: Zed Books.

Zhong, D. (2009). *Youyizhong moshi jiao shanzhai* [There is one model called *Shanzhai*]. Wuhan, China: Wuhan University Press.

Zhong, Y. (2007). Houxiandai xiaofei: qingshaonian yule yawenhua jiqi fasheng [Postmodern consumption: Youths' entertaining subculture and their occurrence]. *Current Youth Study*, *1*, 31–34.

Zhou, L., & Hui, M. (2003). Symbolic value of foreign products in the People's Republic of China. *Journal of International Marketing*, *11*(2), 36–58. doi:10.1509/jimk.11.2.36.20163

About the Contributors

Dal Yong Jin finished his Ph.D. degree from the Institute of Communications Research at the University of Illinois at Urbana Champaign and taught at Simon Fraser University in Canada for several years. His major research and teaching interests are on media convergence and de-convergence, cultural policies and industries, new media and game studies, globalization and media, and the political economy of media and culture. He is the author of two books entitled Hands On/Hands Off: The Korean State and the Market Liberalization of the Communication Industry (Hampton Press) and Korea's Online Gaming Empire (MIT Press) and his recent work has appeared in several scholarly journals, including *Media, Culture and Society, Games and Culture, Telecommunications Policy, Television and New Media, Information Communication and Society* and *Javnost-the Public.*

* * *

Mark Bailey's background is in human computer interaction and the psychological side of human performance. Since joining IBM in 2007 his research has focused on designing for groups that are usually marginalized, such as people with disabilities, older adults, low-literacy and illiterate users, and people in developing countries. Mark's current work includes creating applications for low-end phones – devices that are now in the hands of billions of people. These users are among the "next billion" users of IT, whose numbers are growing faster than any other user group and who are driving tremendous innovation in the IT space. Mark earned his Master's degree in Computer Science from the University of Oregon.

Pi-Chun Chang is an Assistant Professor in Department of East Asian Culture and Development at National Taiwan Normal University (NTNU) in Taiwan. She got her Ph.D. from Department of Communication, State University of New York at Buffalo (SUNY at Buffalo) and was a lecturer in Asian Studies at SUNY at Buffalo. Her research interests are East Asian cultural studies, cultural policy studies, studies on Chinese discourses and communication technology and social change. She has published "Does China Have Alternative Modernity? An Examination of Chinese Modernization Discourse" Communication for Development and Social Change; "Globalized Chinese Cinema and Localized Western Theory" China Media Research and a monograph An Examination on Chinese Alternative Discourses: Perspectives from Nationalism, Postmodernism and Neo-Confucianism.

Peichi Chung teaches in the Communications and New Media Programme at the National University of Singapore. Her research interests include Asian cinema, interactive entertainment media and cultural studies. She has published journal articles and book chapters on the online game industries in South

Copyright © 2011, IGI Global. Copying or distributing in print or electronic forms without written permission of IGI Global is prohibited.

Korea, China and Singapore. Her current research project examines Asian regionalization in the online game industries of East Asia and Southeast Asia.

Jim Christensen is a Senior Technical Staff member at IBM's T.J. Watson Research Center in Hawthorne New York. He is the primary developer of the Picture Talk Web and voice clients. He has held both management and technical positions in IBM, and worked on a broad range of projects including applications for voice, image and text, on tools for program development and debugging, and at all levels of systems infrastructure. Jim graduated from the University of Illinois in 1977 with a Masters degree in Computer Science.

Aliaa Dakroury is Assistant Professor at the Department of Communication, at the University of Ottawa. She is the author of Communication and Human Rights (2009), co-editor of The Right to Communicate: Historical Hopes, Global Debates, and Future Premises (2009), editor of the Fall 2008 of the Global Media Journal - American Edition (on the Right to Communicate), and co-editor of Introduction to Communication and Media Studies (2008). Dr. Dakroury is the author of numerous peer-reviewed journal articles and book chapters. Her work appears in The Canadian Journal of Communication; Journal of Intergroup Relations; Culture, Language, and Representation Journal; Reconstruction; American Journal of Islamic Social Sciences; and The Journal of International Communication. She is the winner of the Canadian Communication Association's 2005 Van Horne Award, and a member of many human rights organizations, and serves on various journal editorial boards. She was recently nominated as honorary expert in the Islamic Resource Bank a joint project of the American Association of Muslim Social Scientists and the International Institute of Islamic Thought.

Catalina Danis is a Research Staff Member in the Social Computing Group at IBM's T. J. Watson Research Center in Hawthorne, NY. Her research focus is on understanding how people work, individually and as members of groups and organizations. She then works with designers and application developers to incorporate her research findings into new software systems and follows up with studies of the system's impact on users. Since joining IBM in 1987, Catalina has worked in the areas of automatic speech recognition, collaboration technologies and high performance computing. She publishes in the CHI, CSCW and DIS communities. Catalina holds a Ph.D. in Cognition and Communication from the University of Chicago.

Jason B. Ellis is a researcher in the Social Computing Group at the IBM T.J. Watson Research Center in New York. His research focuses on the design, implementation, and analysis of social computing applications that facilitate collaboration among diverse user populations. Examples include online gaming communities, inter-generational communication, and the grassroots teams in open source. His current work explores using social computing to empower the "next billion" users of IT -- those in developing countries who do not yet benefit from computing technology. He has published in key conferences such as ACM CHI, CSCW, DIS and learning sciences conferences CSCL and ICLS. Jason earned his Ph.D. in Computer Science at Georgia Tech.

Thomas Erickson is an interaction designer and researcher at IBM Research in New York, to which he telecommutes from his home in Minneapolis. His primary interest is in studying and designing systems that enable distributed groups to interact productively over networks; his approach involves using minimalist visualizations called social proxies to provide cues about activities of participants. He is the primary designer of Picture Talk's user interface. His approach to systems design is shaped by methods developed in HCI, theories and representational techniques from architecture and urban design, and theoretical and analytical approaches from rhetoric and sociology. Over the last two decades Erickson has published about fifty peer reviewed articles and chapters, and been involved in the design of over a dozen systems ranging from research prototypes to commercial products. He is co-editor of HCI Remixed, a book of essays on works that have influenced the HCI community that was published by MIT Press in 2008.

Robert (Rob) Farrell is a Research Staff Member in the Social Computing Group at IBM T. J. Watson Research Center in Hawthorne, NY. His research focuses on information organization, human learning, and collaboration. He is the primary developer of the Picture Talk smartphone cliient. He current work explores the use of social navigation in voice interfaces as part of the group's focus on mobile technology for the "next billion" IT users. Since joining IBM in 1988, Rob has been a group manager, principal investigator, and technical lead. He holds Master of Science and Master of Philosophy degrees in Computer Science from Yale University. Prior to IBM, he worked as a Member of Technical Staff at Bell Communications Research in Morristown, NY. He is the author of over fifty publications in the fields of human-computer interaction, information science, and artificial intelligence.

Anthony Y.H. Fung is a professor in the School of Journalism and Communication at the Chinese University of Hong Kong. He received his Ph.D. from the School of Journalism and Mass Communication at the University of Minnesota in 1998. His research interests focus on cultural industries and cultural studies, popular music, gender and youth identity, cultural studies, political economy of culture and new media technologies. Recently, he has been conducting research on Asian game industries and China's media globalization. His new books are Global Capital, Local Culture: Transnational Media Corporations in China (2008), and Policies for the Sustainable Development of the Hong Kong Film Industry (2009) (co-authored with Chan and Ng).

Sam Hinton is a senior lecturer in Media Arts and Production at the University of Canberra, Australia. His research focuses on games and the games industry, the history and social uses of media technologies and electronic visualization and the arts. He is also a keen amateur programmer and was involved in the development of Urban Terror, a mod for Quake 3 from 1999 to 2002 (a mod which is still around and is now freely available).

Larissa Hjorth is an artist, digital ethnographer and senior lecturer in the Games Programs at RMIT University. Since 2000, Hjorth has been researching and publishing on gendered customising of mobile communication, gaming and virtual communities in the Asia–Pacific — these studies are outlined in her book, Mobile Media in the Asia-Pacific (London, Routledge). Hjorth has published widely on the topic in national and International journals in journals such as Games and Culture journal, Convergence journal, Journal of Intercultural Studies, Continuum, ACCESS, Fibreculture and Southern Review and in 2009 co-edited two Routledge anthologies, Gaming Cultures and Place in the Asia–Pacific region (with Dean Chan) and Mobile technologies: from Telecommunication to Media (with Gerard Goggin). In 2009 Hjorth began her ARC discovery fellowship with Michael Arnold exploring the role of the local and online with communities in the region. This three-year cross-cultural case study will locations such as Tokyo, Seoul, Shanghai, Singapore, Manila, and Melbourne. In 2010 Hjorth releases Games & Gaming textbook with berg.

Yasmin Ibrahim is a Reader in International Business and Communications at Queen Mary, University of London. Her ongoing research on new media technologies explores the cultural dimensions and social implication of the diffusion of ICTs in different contexts. Beyond new media and digital technologies she writes on political communication and political mobilisation from cultural perspectives. Her other research interests include globalization, visual culture and memory studies.

Sun Jung is a Research Fellow in the School of Communication and the Arts at Victoria University, Australia. Her publications include Globalized Masculinities in South Korean Popular Culture (HKUP 2010) and essays such as "The Shared Imagination of Bishonen: Reading DBSK, Youtube.com and Transcultural New Media Consumption," in Intersections (2009), "Chogukjeok Pan-East Asian Soft Masculinity," in Complicated Currents: Media Flows and Soft Power in East Asia (2009), and "Bae Yong-Joon, Hybrid Masculinity and the Counter-Coeval Desire of Japanese Female Fans," in Particip@tions (2006).

Inkyu Kang is an Assistant Professor at Penn State University, The Behrend College. He holds a PhD in Communication Arts (Media and Cultural Studies) from the University of Wisconsin-Madison in 2009. His research interests include visual communication, cultural studies, and new media technologies. Kang is also a technology and culture columnist for OhmyNews, the world's first participatory journalism site. He has contributed to the Seoul-based news medium since 2002. As a journalist, he writes widely about digital media and popular culture, receiving the "Journalist of the Year" award in 2008. His columns, articles and photos were published as an anthology in the same year. He lives in Erie, PA.

Wendy A. Kellogg manages the Social Computing Group at IBM's T. J. Watson Research Center. Topics addressed by the group have included social translucence, computer-mediated communication, and recently social computing for green IT and the next billion users. Kellogg holds a Ph.D. in Cognitive Psychology and publishes in HCI and CSCW. She serves on the editorial board of ACM's Transactions on Computer-Human Interaction (ToCHI) and has chaired a variety of key conferences and technical programs and venues (CHI, DIS, CSCW). Wendy served on the National Academy of Science's Computer Science and Telecommunications Board (CSTB). She was named ACM Fellow in 2002 for contributions to social computing and human-computer interaction, and elected to the CHI Academy and IBM Academy in 2008.

Do Kyun Kim is an Assistant Professor in the Department of Communication at the University of Louisiana – Lafayette. He is a diffusion expert who has studied and practiced knowledge and skills of diffusion of information and innovations for social change. His application of diffusion knowledge has not limited to a certain area of academic and social boundaries as his academic history including degrees in economics, political science and communication demonstrates. Accordingly, his research and teaching interests include designing and implementing communicative diffusion strategies to purposively change individual attitudes and behaviors and analyzing dynamics between the individual changes and social transformation. His research articles have been published in many academic journals, and he recently published a book, Hallyu: Korean media and pop culture waves in Asia and beyond, dealing with the diffusion of Korean media products and culture in the new Millennium.

Andrew Kirk is a lecturer in the Department of Speech Communication at Southern Illinois University in Carbondale, Illinois, USA. This chapter is an amended version of a an essay presented through the Media Studies Interest Group at the Central States Communication Association conference proceedings in 2009. It received the Samuel L. Becker Award—the Top Student Paper prize—at that event. Kirk would like to thank Drs. Jonathan M. Gray, Nilanjana R. Bardhan, and Craig Gingrich-Philbrook for their guidance and encouragement while writing this essay and for their feedback in the months afterwards. He would also like to thank his partner, John David Lewis, for his continued support and patience, and his parents for not giving up on him despite his stellar gift for testing and re-testing parental patience. Kirk's primary research and pedagogical interests are in intercultural communication, popular culture, and gender and sexuality.

Francis L. F. Lee is associate professor at the School of Journalism and Communication at the Chinese University of Hong Kong. His research covers a wide range of issues in journalism studies, political communication, and analyses of public opinion and public discourses. His work has appeared in such journals as Harvard International Journal of Press/Politics, International Journal of Public Opinion Research, Communication Research, Journalism and Mass Communication Quarterly, Discourse & Society, among others. His work related to the politics of popular culture, meanwhile, has appeared in Sociology of Sports Journal, Popular Communication, and Media, Culture & Society.He is currently an associate editor of the Chinese Journal of Communication (published by Routledge). He is also the author of Media, Social Mobilization, and Mass Protests in Post-colonial Hong Kong: The Power of a Critical Event (with Joseph M. Chan as a co-author, to be published in November 2010 by Routledge).

Seung-Hyun Lee is an Assistant Professor of Media Studies at the University of North Carolina, Greensboro. She received her Ph.D. in Mass Communications from the University of Wisconsin-Madison in 2008. Her main research areas include the impacts, roles and users of digital and interactive media, such as the Internet, digital television and mobile multimedia, in the social, national, and international context; online journalism and citizen journalism for local communities; diffusion and use of new media; multimedia reporting and production with civic engagement. She is expanding her research area to mobile journalism and global media use, particularly international mobile multimedia use in developed and developing countries.

Hongye Li received a MA degree in Communication and New Media from the Department of Communication, University of Macau, in 2009. She is currently working as an Account Executive in AcrossChina Communication Group, a public relation agency in Beijing.

Luzhou Li is a Ph.D. student in the Institute of Communications Research at the University of Illinois at Urbana-Champaign. She received her Master of Philosophy (M.Phil.) in Communication from the School of Journalism and Communication at the Chinese University of Hong Kong, and her Bachelor's degree in journalism from the School of Journalism at Fudan University, Shanghai, China. Her research interests focus on critical and cultural studies of television and new media in developmentally transitional society. Her works have been presented at international conferences including the

NCA (National Communication Association) and the IAMCR (International Association of Media and Communication Research).

Ying-Chia H. Lin is an assistant professor in the Department of Communication Arts at Fu-Jen Catholic University, Taiwan. She received her PhD degree from the University of Washington, Seattle in Communication in 2006. She is interested in cultural analysis of media in general, but particularly in digital games, and transnational circulation of digitalized media content. Her recent research focuses on gamer culture and media representation of digital games.

Lihyun Lin is an Associate Professor of Journalism at National Taiwan University and a visiting scholar at the University of Westminster, London. Born in Taipei, she obtained her MA from National Chengchi University's Graduate Institute of Journalism, and her PhD from the Communication and Media Research Institute (CAMRI) of the University of Westminster, London. Her main research interests include the media of East Asia, media and political transition, intellectual history of media studies, and the origins and future of public service broadcasting. Her research projects have been sponsored by the National Science Council of the Taiwan government. Among her publications, the book History of Communication Research in Taiwan (2004) won a prize awarded by the Tseng Hsu-Pai Journalism Foundation. Her research papers, entitled Studies of Communication History in Taiwan" (1999) and "Authoritarian State and Television: A Comparative Study Between Taiwan and South Korea"(2005), have won best paper awards from the Chinese Communication Society in Taiwan.

Hudson Moura is a Canadian-Brazilian film scholar. He acquired his PhD in Literature and Film from the University of Montreal and completed postdoctoral research in Intercultural Cinema at Simon Fraser University. His research focuses primarily on transnational cinemas and their attempt to depict/represent exile and diaspora as well as on the effects of globalization on film production. Over the past few years he also has been interested in the proliferation of screen culture a phenomenon resulting from the potential of the digitization of cinema and access to an increased number of screen formats. Dr. Moura has taught at several universities such as University of British Columbia, Simon Fraser University, and University of Montreal in Canada; Federal University of São Carlos in Brazil; and University of Algarve in Portugal. He has presented papers at several film and media conferences and regularly publishes his work that he has written in English, French, and Portuguese in peer-reviewed journals and books in Canada, the United States, Brazil, France, and Portugal.

Paschal Preston is a full professor and current Head of School in the School of Communications at Dublin City University. His main research and teaching interests include: the institutional and political-economy aspects of the media; innovation in media sector as multi-dimensional process; the interfaces between technological and socio-economic and cultural change; regulation of communication services; changing role of mediated communication in international relations; mapping new ICT, media/cultural industries and the information/knowledge economy more broadly.

Jim Rogers is a post-doctoral research fellow based at the School of Communications, Dublin City University (DCU). His current research programme, under the 'Knowledge Society' strand of the Irish Social Sciences Platform (ISSP) focuses on the theme of innovation in the digital economy. He is also a part-time lecturer at the School of Communication, DCU. Core research and teaching interests include political economy of media/cultural industries (with particular interests in the areas of music and broadcasting); approaches to techno-social change; and approaches to international communication.

Maheswar Satpathy (FIAHP) completed his master's degree in Applied (Clinical) Psychology from University of Delhi with the distinction of being the Gold Medalist in 2007. At present he is working on a project of community mental health at Indian Institute of Technology Kanpur as a UGC Research Fellow. He has been trained in neuropsychology at All India Institute of Medical Sciences (AIIMS) and has completed a Certificate Course in HIV/AIDS Management (Le Countre de Seda, France). He has total of 4 years of diverse and rich work experience with National Council for Educational Research and Training (NCERT), New Delhi, Amity University, ICFAI University and Lovely Professional University. Recently he has been selected for the prestigious Australian Leadership Awards (ALA) Scholarship from SAARC region to join research work at National centre in HIV Social Research, University of New South Wales (UNSW), Australia by Australian Government. He has also presented several scientific papers at International & National level Forums and has published many research papers in refereed Journals and Books as chapters. His research interest concerns the areas of Social Policy, Community Mental Health, Sexuality and LGBT studies, Male Psychology and emerging cultures.

Sergio Sparviero is an Irish Social Sciences Platform (ISSP) post-doctoral research fellow at Dublin City University. He holds a bachelor degree in economics from Université libre de Bruxelles (ULB), a master in development studies from University College Dublin (UCD) and a PhD in communication from Dublin City University (DCU). He is a scholar of economics and political economy of communication and in the last few years, he lectured on media economics and on the history of innovations in information industries at Dublin City University. His recent publications include journal articles focussed on the importance of Baumol cost disease for understanding the economic value of creativity and, more generally, economic trends in media industries ("Creative Inputs as the Cause of Baumol's Cost Disease: The Example of Media Services" in 2009, and "Creativity and the positive reading of Baumol cost disease" in 2010).

Zixue Tai, PhD, University of Minnesota-Twin Cities, is an Assistant Professor in the School of Journalism and Telecommunications at the University of Kentucky. He teaches undergraduate and graduate courses in media effects, global media systems, advanced multimedia, and video game studies. His research interests address a multitude of issues in the new media landscape of China. He is the author of The Internet in China: Cyberspace and Civil Society (Routledge, 2006), and has also published in journals such as International Communication Gazette, Journalism & Mass Communication Quarterly, New Media & Society, Journal of Communication, Journal of Retailing and Consumer Services, and Psychology & Marketing.

D. Ndirangu Wachanga PhD (Interdisciplinary PhD in information science, University of North Texas) is assistant professor of journalism and information science at the University of Wisconsin-Whitewater. He has written many passionate pieces, particularly about Kenyan and other African media, and political and Diaspora issues; yet, he is able to set aside that narrow focus and work with the same enthusiasm on theoretical issues in media, Diaspora studies, and information science. His book, Cultural Identity and New Communication Technologies: Political, Ethnic and Ideological Implications, is scheduled for release early 2011.Dr. Wachanga has written for the Kenyan media, wrote for the North Texas Daily, and produced programs for the North Texas radio and TV stations. He served as a bilingual (English and Swahili) editor with the East African Educational Publishers, the largest publishing company in East and Central Africa. His research has focused on new media and identity, Diaspora, technology and information seeking behavior, media ethics, African media systems and representation. His research findings have been published in peer reviewed journals as well as presented at international conferences.

Mei Wu, PhD (Concordia University) Associate Professor and Acting Head, Department of Communication, Faculty of Social Sciences & Humanities, University of Macau. A founding member of the MA programme in Communication and New Media at the University of Macau, she specializes in new media and communication studies in Chinese societies, particularly the Internet and mobile telephony in China. Her publications include: "Dancing thumbs: mobile telephony in contemporary China," in China's Information and Communications Technology Revolution (Routledge, 2009), "Magic Wand: Mobile phones and Fujian Entrepreneurs in China," International Journal of Sociotechnology and Knowledge Development (1[4], 2009), "Measuring Political Debate on the Chinese Internet Forum," Javnost-The Public (15, 2008). Her new book "Pushing Hands on the Chinese Internet," which documents for the first time the concept of viral marketing operations in China, will come out in the summer of 2010.

Audrey Yue is Senior Lecturer in Cinema and Cultural Studies at the University of Melbourne, Australia. Her publications include Ann Hui's Song of the Exile (2010, in press), AsiaPacifiQueer: Rethinking Gender and Sexuality (2008) and Mobile Cultures: New Media in Queer Asia (2003).

Haifang Zeng is a doctoral student in the School of Film & TV Arts and Technology at Shanghai University, China. She studied in the University of Kentucky as a visiting student with the sponsorship of China Scholar Council during the school year of 2009-2010. Her research interests focus on international and cross cultural communication, especially the comparative study of newspaper and new media industries in the U.S. and China. Her articles have appeared in several academic journals in China.

Index

Copyright © 2011, IGI Global. Copying or distributing in print or electronic forms without written permission of IGI Global is prohibited.

L

M

N

O

P

V

W

X

Y

Z

Breinigsville, PA USA
29 December 2010
252342BV00004B/1-62/P